Recent Advances in Freshwater Crustacean Biodiversity and Conservation

Crustacean Science and Research
Series Editors

Stefan Koenemann
University of Veterinary Medicine, Hannover, Germany

Ingo S. Wehrtmann
University of Costa Rica, San Jose

The *Crustacean Science and Research* series publishes internationally significant contributions to the biology of Crustacea. The thematic focus of individual volumes includes particular aspects from various fields of research, such as molecular biology, comparative morphology, developmental biology, systematics, phylogenetics, natural history, evolution, palaeontology, zoogeography conservati on biology, (eco-) physiology, ecology, extreme environments, behavioural biology, and fisheries and aquaculture.

For more information about this series, please visit: https://www.crcpress.com/Crustacean-Science-and-Research/book-series/CRCCRUSTAISS

Recent Advances in Freshwater Crustacean Biodiversity and Conservation

Edited by
Tadashi Kawai and D. Christopher Rogers

CRC Press
Taylor & Francis Group
Boca Raton London New York

CRC Press is an imprint of the
Taylor & Francis Group, an **informa** business

First edition published 2021
by CRC Press
6000 Broken Sound Parkway NW, Suite 300, Boca Raton, FL 33487-2742

and by CRC Press
2 Park Square, Milton Park, Abingdon, Oxon, OX14 4RN

© 2021 Taylor & Francis Group, LLC

CRC Press is an imprint of Taylor & Francis Group, LLC

Library of Congress Cataloging-in-Publication Data

A catalog record for this title has been requested

ISBN: 978-0-367-44350-4 (hbk)
ISBN: 978-0-367-68904-9 (pbk)
ISBN: 978-1-003-13956-0 (ebk)
Typeset in Times
by Deanta Global Publishing Services, Chennai, India

Contents

Editor Bio

Tadashi Kawai is chief researcher at the Hokkaido Research Organization, Central Fisheries Research Institution (Japan). His main research program is "Environmental monitoring for the conservation of native ecosystems". He received his Ph.D. from the Kyushu University in Fukuoka, Fukuoka Prefecture, Japan. Dr. Tadashi covers biology and systematic of freshwater decapod crustacean (particularly freshwater crayfish) and conservation of wetlands on a global scale. He has over 100 peer-reviewed scientific publications in crustacean biology and 10 monographs on conservation biology of freshwater decapod crustacean and a popular book (*Freshwater Crayfish: a Global Overview*, published by CRC Press, Taylor & Francis Group) for conservationists. Dr. Tadashi is on the editorial board of *Crustacean Research* and President of International Association of Astacology (2018–). He provides his service for global conservation of freshwater crustaceans.

D. Christopher Rogers is a research zoologist at the University of Kansas with the Kansas Biological Survey and is affiliated with the Biodiversity Institute, with various research projects all over the world. He received his Ph.D. from the University of New England in Armidale, NSW, Australia. Dr. Rogers specializes in freshwater and terrestrial crustaceans (particularly Branchiopoda and Malacostraca) and the invertebrate fauna of seasonally astatic wetlands on a global scale. He has more than 170 refereed publications in crustacean taxonomy and invertebrate ecology, as well as published popular and scientific field guides and identification manuals to freshwater invertebrates. Christopher is an Associate Editor of the *Journal of Crustacean Biology* and a founding member of the Southwest Association of Freshwater Invertebrate Taxonomists. He has been involved in aquatic invertebrate conservation efforts all over the world.

Contributors

Ada Acevedo-Alonso
Department of Biology, Northern
Michigan University
Michigan, U.S.A.

Sérgio L. S. Bueno
Instituto de Biociências, Universidade
de São Paulo
São Paulo, Brazil

Martha R. Campos
Universidad Nacional de Colombia,
Instituto de Ciencias
Naturales-ICN
Bogotá, Colombia

Jason Coughran
Faculty of Mathematics, Physical
Sciences and Life Sciences,
Sheridan Institute of Higher
Education
Perth, Australia

Keith A. Crandall
Computational Biology Institute,
George Washington University
Virginia, U.S.A.

Neil Cumberlidge
Department of Biology, Northern
Michigan University
Michigan, U.S.A.

Mikhail E. Daneliya
Department of Biosciences, University
of Helsinki
Helsinki, Finland

Henri J. Dumont
Institute of Hydrobiology, Jinan
University
Jinan, People's Republic of China

Henrik Glenner
Biological Institute, University of Bergen
Bergen, Norway

Michał Grabowski
Department of Invertebrate Zoology and
Hydrobiology, University of Łódź
Łódź, Poland

Frederic Grandjean
Laboratoire Ecologie and Biologie des
Interactions (EBI), Université de
Poitiers
Poitiers, France

Brooke Grubb
Auburn University
Alabama, U.S.A.

Jens T. Høeg
Section for Marine Biology, University
of Copenhagen
Copenhagen, Denmark

David M. Hudson
The Maritime Aquarium at Norwalk
Connecticut, U.S.A.

Lauren E. Hughes
Natural History Museum
London, U.K.

Kamil Hupało
Aquatische Ökosystemforschung,
Universität Duisburg-Essen
Essen, Germany

Ioannis Karaouzas
Hellenic Centre for Marine Research,
Institute of Marine Biological
Resources and Inland Waters
Anavyssos, Greece

Philippe Keith
Laboratoire Biologie des Organismes et
    Écosystèmes Aquatiques (BOREA),
    Muséum national d'Histoire naturelle
Paris, France

Werner Klotz
Rum, Austria

Vjacheslav S. Labay
Sakhalin branch of the Federal State
    Budget Scientific Institution,
    Russian Federal Research Institute
    of Fisheries and Oceanography
    ("SakhNIRO")
Yuzhno-Sakhalinsk, Russia

Carlos A. Lasso
Instituto de Investigación de Recursos
    Biológicos Alexander von Humboldt
Bogotá, Colombia

Sue Lindsay
Faculty of Science and Engineering,
    Macquarie University, Australia

Tomasz Mamos
Department of Invertebrate Zoology
    and Hydrobiology, University of
    Lodz
Łódź, Poland

Gérard Marquet
Laboratoire Biologie des Organismes et
    Écosystèmes Aquatiques (BOREA),
    Muséum national d'Histoire
    naturelle
Paris, France

Valentin de Mazancourt
Center for Integrative Biodiversity
    Discovery, Museum für Naturkunde,
    Leibniz Institute for Evolutions and
    Biodiversity Science
Berlin, Germany

Benjamin Mos
National Marine Science Centre, School of
    Environment, Science and Engineering,
    Southern Cross University
Coffs Harbour, Australia

Sameer M. Padhye
Systematics, Ecology and Conservation
    Lab, Zoo Outreach Organization
Tamil Nadu, India

Gillian Phillips
The Maritime Aquarium at Norwalk
Connecticut, U.S.A.

David J. Rees
Biological Institute, University of
    Bergen
Bergen, Norway

Tomasz Rewicz
Department of Invertebrate Zoology
    and Hydrobiology, University of
    Lodz
Łódź, Poland

Christoph D. Schubart
Zoology and Evolutionary Biology,
    University of Regensburg
Regensburg, Germany

Roberto M. Shimizu
Instituto de Biociências, Universidade
    de São Paulo
São Paulo, Brazil

David B. Stern
Computational Biology Institute,
    George Washington University
Virginia, U.S.A.

Fabio Stoch
Evolutionary Biology & Ecology,
    Université Libre de Bruxelles
Brussels, Belgium

Carroll M. Teresa
Department of Natural Science,
    University of South Carolina
    Beaufort
South Carolina, U.S.A.

Milena R. Wolf
Instituto de Biociências, Universidade
    Estadual Paulista (UNESP)
São Paulo, Brazil

Anna Wysocka
Faculty of Biology, University of
    Gdansk
Gdansk, Poland

Karl J. Wittmann
Abteilung für Umwelthygiene,
    Medizinische Universität Wien
Vienna, Austria

Denis S. Zavarzin
Sakhalin branch of the Federal State
    Budget Scientific Institution,
    Russian Federal Research Institute
    of Fisheries and Oceanography
    ("SakhNIRO")
Yuzhno-Sakhalinsk, Russia

CHAPTER 1

Conservation Biology of Freshwater Crustaceans
Introduction

D. Christopher Rogers and Tadashi Kawai

CONTENTS

1.1 INTRODUCTION

Crustaceans originated in marine systems and have since invaded freshwater habitats multiple times. Freshwater crustaceans are extremely diverse and display a plethora of adaptations to these systems. These animals are important and characteristic components of most inland aquatic habitats, yet surprisingly little is known about their biogeography and ecology. However, freshwaters are impacted by human activities, such as habitat conversion and alien introductions. Freshwater systems are easily affected, causing extinctions or mass mortalities. As a result, large numbers of freshwater crustacean species have been listed as threatened or endangered, although most taxa are data deficient.

Much on the conservation biology of freshwater crustaceans has been published, contributing toward needed conservation activities. However, so far, most of effort has been focused only on the larger decapods: shrimp, prawns, crabs, and crayfish. The focus of this symposium is the conservation of all freshwater crustaceans, but we especially wanted to include the typically underrepresented groups, such as branchiopods, copepods, ostracods, barnacles, amphipods, isopods, and mysids.

Toward this end, we cochaired a symposium titled "Conservation Biology of Freshwater Crustaceans", at the Ninth International Crustacean Congress, which was held at the Smithsonian Institution in Washington, DC, during 22–25 May 2018. We specifically solicited contributions from diverse backgrounds, including taxonomy,

1

systematics, morphology, physiology, behavior, ecology, genetic and genomic, bio-mechanics, neurology, and developmental biology. We also encouraged integrative approaches, relationships between biodiversity and environment, and effect from climate change or alien species as important subjects for this effort. But specifi-cally, we wanted strong introductions into those crustacean taxa that have had the least conservation attention. Thus, this book focuses on minor crustacean groups and regionally endemic groups, all from brackish water and freshwater. The meat of that conference, as well as some separate, invited contributions, is presented here in the following chapters.

Unfortunately, we were not able to find authors to cover all groups and all regions. Several important groups such as ostracods, bathynellaceans, and anispi-daceans were not addressed here as were the conservation needs of many regions, such as Asia, Africa, and much of the Americas.

We expect this book to be used by undergraduate and graduate students, as well as researchers in universities, agencies, and NGOs. Scientific professional research-ers, science educators, conservationists, and government conservation policy makers will utilize this resource. We hope that college, university, and city libraries will also use this book as an important reference on invertebrate and crustacean conservation. Furthermore, this book will have uses regarding aquaculture, fisheries, and invasive species as some of the taxa discussed are economically important.

This book should be used as a companion volume to *A Global Overview of the Conservation of Freshwater Decapod Crustaceans*, edited by T. Kawai and N. Cumberlidge (2016). Our book complements Kawai and Cumberlidge (2016), in that it addresses crustacean groups not previously treated and provides additional infor-mation beyond that presented in the earlier work.

The majority of the chapters are concerned with distributions and diversity. Several chapters discuss the origins of regional faunas and the threats to their con-tinued existence, while some demonstrate diversity previously overlooked. These chapters exhibit many conceptual approaches and perspectives, but also cogently demonstrate that our knowledge of freshwater crustaceans is still very much in its infancy. There is still much more to do. We have only scratched the surface of fresh-water crustacean biodiversity, barely understanding the relationships, the evolution, and the evolutionary responses of these organisms. We cannot conserve, we cannot protect, what we do not know exists.

REFERENCE

Kawai, T. and Cumberlidge, N. 2016. *A Global Overview of the Conservation of Freshwater Decapod Crustaceans*. Springer International Publishing AG, Switzerland.

CHAPTER **2**

Phylogenetic Analyses Suggest a Single Origin of Freshwater Barnacles

Henrik Glenner, Jens T. Høeg, David J. Rees, and Christoph D. Schubart

CONTENTS

2.1 INTRODUCTION

Cirripedes are known from a vast variety of habitats, ranging from the deep sea to the splash zone in the rocky intertidal, and many of them live attached to other organisms, with lifestyles ranging from suspension feeders attached to the skin of marine mammals to obligate parasites of other crustaceans (Fig. 2.1). Most other crustacean taxa with a similarly wide diversity of lifestyles and habitats include speciose lineages with successful freshwater radiations, e.g., decapods, peracarids, and copepods. It is therefore remarkable that only 4 of about 1,500 described cirripede species have been reported from truly freshwater habitats. The acorn barnacle *Amphibalanus improvisus* can be found in the Baltic Sea at very low salinities but cannot tolerate true freshwater. The four entirely freshwater cirripedes all belong to the parasitic Rhizocephala, within the Sacculinidae of the suborder Kentrogonida (Annandale, 1911, Feuerborn, 1931, Boschma, 1966, 1967, Andersen et al., 1990). Thus, expansion into freshwater habitats

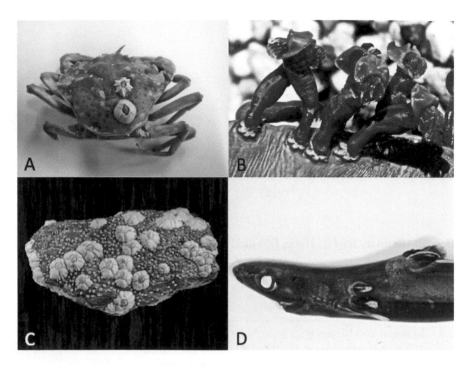

Figure 2.1   Habitat diversity of barnacles. (A) *Balanus crenatus* and *Astrominius modes-*
tus on the dorsal carapace of a European shore crab, *Carcinus maenas*. (B)
Xenobalanus globicipitis on the skin of a dolphin. (C) Newly settled juveniles and
adult specimens of the high intertidal barnacle, *Semibalanus balanoides*, on a
stone. (D) The parasitic shark barnacle, *Anelasma squalicola*, on a velvet belly
lantern shark, *Etmopterus spinax*.

is an extremely rare evolutionary event within Cirripedia, indicating that the adaptation to this environment has been a tremendous challenge for the taxon. The most par-
simonious scenario would suggest that a common rhizocephalan ancestor underwent
a complete marine-to-freshwater transition and then, following an adaptive radiation,
speciated into the four extant freshwater rhizocephalan species. The present taxonomy,
however, does not indicate this evolutionary event, and the four species are classified
into two different rhizocephalan genera, viz., *Sesarmaxenos* (2) and *Ptychascus* (2).
In order to test whether this taxonomy reflects the phylogeny, we collected specimens
from the two genera to perform a thorough phylogenetic analysis based on molecular
data and a morphological analysis of key larval characters.

2.2 MATERIAL AND METHODS

2.2.1 Morphological Analyses: Material Examined

Ptychascus barnwelli (Andersen et al., 1990) on the fiddler crab *Minuca mordax*
(Smith, 1870), Caribbean coast of Costa Rica, Central America, are deposited as

material at the Zoological Museum at the University of Copenhagen (ZMUC). Larvae of *Ptychascus glaber* and *Ptychascus barnwelli* are also deposited with types at ZMUC on SEM stubs.

Ptychascus glaber (Boschma, 1933) on the crabs *Aratus pisonii* (H. Milne-Edwards, 1837) and *Armases benedicti* (Rathbun, 1897) are from Brazil, Natural History Museum London.

Sesarmaxenos gedehensis (Feuerborn, 1932) on the crabs *Bresedium philippinense* (Rathbun, 1914), *Sesarmops impressus* (Milne-Edwards, 1837), and *Sesarmops* sp., all from Kawasan Falls, Cebu, Philippines, are collected by Alexey V. Rybakov in 2004. Cypris larvae of *S. gedehensis* were extracted from a parasite in the collections of ZMUC and mounted on stubs for SEM. These stubs are also deposited in the ZMUC as documentation of the present study.

We had no access to original material of *Sesarmaxenos monticola* (Annandale, 1911), but it is very likely synonymous with *S. gedehensis* (see results and discussion for detailed information).

Sacculina insueta (Boschma, 1966) on the crab *Ptychognathus riedelii* (Milne-Edwards, 1868), Kawasan Falls, Cebu, Philippines, were collected by A.V. Rybakov in 2004.

The parasite larvae of *Ptychascus* examined here were those extracted from the mantle cavity (externa) of the parasites and mounted for SEM by Andersen et al. (1990). In addition, we examined cypris larvae of *S. gedehensis*, extracted from a parasite in the Natural History Museum Copenhagen, and nauplius larvae extracted from one of the parasites of *S. insueta*. As comparative material, we examined cypris larvae of other sacculinids from our personal collections. Preparation for SEM analysis was as described in Andersen et al. (1990). For cyprids of *S. gedehensis*, the very detailed drawings in Feuerborn (1933) also provided morphological information.

2.2.2 Material Included in the Molecular Analyses

Collection and host information:

- Freshwater/terrestrial environments
  - *Sesarmaxenos gedehensis* on crabs of *Sesarmops* collected from Kawasan Falls, Cebu, Philippines (Rybakov 2004).
  - *S. gedehensis* on the crab *Sesarmops* sp. collected from Kawasan Falls, Cebu, Philippines (Rybakov 2004).
  - *S. gedehensis,* on the crab *Bresedium philippinense* collected from Kawasan Falls, Cebu, Philippines (Rybakov 2004).
  - *Ptychascus barnwelli* on the fiddler crab *Minuca mordax*, Costa Rica, collected by Christoph Schubart in 2014.
- Marine environments
  - *Sacculina bicuspidata* on the crab *Omalacantha bicornuta* (Latreille, 1825) (Majoidea, "speck-claw decorator crab"), Bocas del Toro, Panama, Caribbean. Given to us by Dr. Mark E. Torchin, Smithsonian Tropical Research Institute; sample dates lost.

- *Sacculina compressa* Boschma, 1931 on the crab *Ozius tuberculosus* (H. Milne-Edwards, 1834) (Oziidae), Panglao Island, Philippines (Rybakov, 2004).
- *Sacculina insueta* on the crab *Ptychognathus riedelii* (A. Milne-Edwards, 1868) (Varunidae), Kawasan Falls, Cebu, Philippines (Rybakov, 2004).
- From GenBank, we obtained DNA sequences of our studied genes for 6 thoracican barnacles and 21 rhizocephalans as outgroup species (Table 2.2).

2.2.3 Molecular Methods

Total genomic DNA was extracted from ~1 mm^3 of tissue from the mantle of individual externae using the Qiagen DNeasy Blood & Tissue Kit following the Qiagen DNeasy Protocol for Animal Tissues, 7/2006.

DNA fragments from two nuclear genes (28S and 18S) and one mitochondrial ribosomal gene (16S) were amplified and sequenced using the primers indicated in Table 2.1. Almost complete coverage (~1800 base pair [bp] unaligned sequence) of the nuclear 18S ribosomal RNA gene was achieved together with a ~2 kilobase (KB) fragment of nuclear 28S. For the mitochondrial 16S rDNA, two primer pairs were utilized, yielding approximately 500 bp (H621 and L12247L), and 360 bp (16Sar-L and 16Sbr-H) (see Table 2.1).

All PCR reactions were carried out using a Bio-Rad C1000 Thermal Cycler in 25 μl volumes containing 1 μl of DNA extract, 2.5 μl 10X PCR buffer, 1.2 μl of dNTP mixture (2.5 μM each), 1 μl of each 10 μM primer, and 0.75 U of Takara polymerase. Conditions for all amplifications were as follows: initial denaturation at 94°C for 5 min, then 35 cycles of 30 s denaturation at 94°C, 1 min primer annealing at 52°C, and 1 min extension at 72°C, with a final 7-min 72°C extension. All PCR products were visualized on 1% agarose gels and stored at 4°C prior to purification and

Table 2.1 **Primer Details for Gene Amplification (18S, 28S, and 16S)**

Gene	Primer	Sequence (5′ to 3′	Reference
18S	329 (F) a- (R) 345+ (F) UnivF15 (R)	TAATGATCCTTCCGCAGGTT CAGCMGCCGCGGTAATWC GCATCGTTTAHGGTT CTGCCAGTAGTCATATGC	Spears et al. (1992) Spears et al. (1992) Spears et al. (1992) Frischer et al. (2002)
28S	1274 (F) 28S1a (F) FF (R) 28S4 (R) rD6.2b (R) rD5b (R)	GACCCGTCTTGAAACACGGA CCCSCGTAAYTTAAGCATAT GGTGAGTTGTTACACACTCCTTAG CCTTGGTCCGTGTTTCAAGAC AATAKKAACCRGATTCCCTTTCGC CCACAGCGCCAGTTCTGCTTAC	Nunn et al. (1996) [=D3A] Friedrich & Tautz (1997) [=D1F] Modified from Hillis & Dixon (1991) Crandall et al. (2000) [=rD4b]] Whiting (2002) Mallatt & Sullivan (1998)
16S	16Sar-L (F) 16Sbr-H (R) H621 (F) L12247L (R)	CGCCTGTTTATCAAAAACAT CCGGTCTGAACTCAGATCACGT CYGTGCAAAGGTAGCATA TTAATYCAACATCGAGGTCRC	Palumbi et al. (1991) Palumbi et al. (1991) Tsuchida et al. (2006) Tsuchida et al. (2006)

All primers presented as 5¢-3¢. Primer directions are indicated in parentheses by F (forward) and R (reverse).

Table 2.2 **GenBank Accession Numbers for Ingroup and Outgroup Taxa Involved in Phylogenetic Analyses. Taxonomy in this table follows Høeg et al. 2019.**

Taxon	16S	18S	28S
Heterosaccus dollfusi	FJ481949	EU082413	EU082333
Heterosaccus lunatus	FJ481947	EU082414	EU082334
Ibla quadrivalvis	AY520755	AY520655	AY520621
Lepas anatifera	GU993670	FJ906773	GU993603
Loxothylacus panopaei	FJ481956	AY265364	/
Loxothylacus texanus	/	L26517	/
Peltogaster paguri	/	EU082415	EU082335
Poecilasma kaempferi	/	EU082410	EU082329
Polyascus gregaria	JN616263	AY265363	GU190705
Polyascus plana	FJ481954	AY265368	GU190698
Polyascus polygenea	/	AY265362	GU190704
Pottsia serenei	/	DQ826567	GU190702
Ptychascus barnwelli			
Sacculina carcini	FJ481957	AY265366	AY520622
Parasacculina yatsui	/	AY265361	GU190706
Parasacculina leptodiae	FJ481952	AY265365	/
Parasacculina oblonga	FJ481953	AY265367	GU190699
Parasacculina shiinoi	KF539761	KF539758	KF539759
Parasacculina sinensis	/	AY265360	GU190707
Sacculina upogebiae	KF539762	KF539758	KF539760
Sesarmaxenos gedehensis	KF561270	KF561254	KF561261
Sesarmaxenos gedehensis	KF561271	KF561255	KF561262
Sesarmaxenos gedehensis	KF561272	KF561256	KF561263
Sesarmaxenos gedehensis	KF561273	KF561257	KF561264
Sacculina insueta	KF561274	KF561258	KF561265
Sacculina insueta	KF561275	KF561259	KF561266
Parasacculina bicuspidata	/	KF561260	KF561267
Parasacculina compressa	KF561276	/	KF561268
Thompsonia littoralis	/	DQ826573	/

sequencing. PCR products were cleaned by the addition of 0.1 μl (1 U) exonuclease I, 1 μl (1 U) of shrimp alkaline phosphatase, and 0.9 μl of ddH$_2$O to 8 μl of PCR product. This was carried out by incubation at 37°C for 30 min and deactivation of the enzymes at 85° for 15 min. Sequence reactions were performed using the BigDye v.3.1 Cycle Sequencing kit (Applied Biosystems, Inc., Norwalk, CT, USA) with the same primers used for initial PCR amplification. Both strands of all PCR products were sequenced using an ABI 3730 capillary sequencer.

All sample PCR products were sequenced in both directions in order to improve accuracy and aligned using ClustalW (Thompson et al., 1994) implemented in eBioX version 1.5.1 (www.ebioinformatics.org) with characters equally weighted and using default parameters. Additional data for outgroup and ingroup taxa were taken from GenBank (accession numbers for all taxa included in alignments and phylogenetic analyses can be found in Table 2.2). Following minor improvements by eye performed in Mesquite version 2.6 (Maddison & Maddison, 2009), duplicates were made of alignments for each gene prior to further analyses. One duplicate of the data set was analyzed without further manipulation, and the second was run through Gblocks (Castresana, 2000; implemented in Seaview: Gouy et al., 2010) to remove hypervariable and potentially problematic regions in the alignment. Gblocks criteria used for this were for a less stringent selection, allowing for gaps within blocks and smaller final blocks. The two treatments (Gblocks and non-Gblocks) for each gene alignment were then concatenated in Seaview (Gouy et al., 2010) before final phylogenetic analyses.

Following alignment with ClustalW, lengths of individual gene data sets were 2,268 bp for 18S, 1,958 bp for 28S, and 558 bp for 16S; the concatenated non-Gblocks data set was 4,784 bp). Following removal of the more variable and problematic regions of the alignment with Gblocks (as well as regions for which data was available for a limited number of taxa), the data sets were 1,668, 694, and 327 bp for 18S, 28S, and 16S, respectively. The concatenated Gblocks data set was 2,689 bp. Sequence data for all samples have been submitted to GenBank (see Table 2.2 for accession numbers).

Best-fit nucleotide substitution models were inferred for each individual gene and the concatenated data set using JModeltest v.2.1.4 (Darriba et al., 2012).

The Akaike information criterion (AIC) results from JModeltest indicated the same best-fit models for both Gblocks and non-Gblocks data sets: TIM3+I+G for 18S, TIM1+I+G for 16S, and GTR+I+G for 28S and the concatenated data sets. For Bayesian analyses where a specific model could not be implemented (i.e., TIM1/TIM3), it was substituted by the closest available over-paramaterized model. Following these results, the final analyses proceeded with the concatenated data set. Phylogenetic analyses were performed on the concatenated data set using Bayesian methods coupled with Markov chain Monte Carlo (MCMC) inference, as implemented in MrBayes v3.2 (Ronquist et al., 2012). For these analyses, two independent runs were performed, each consisting of four chains and proceeding for 10 million generations sampling every 1,000 generations. Results were visualized in Tracer v. 1.5.0 (Drummond & Rambaut, 2007), and proper mixing of the MCMC was assessed by calculating the effective sampling size (ESS) for each parameter. For each data set, the maximum clade credibility tree (MCC; the tree with the largest product of posterior clade probabilities) was selected from the posterior tree distribution (after removal of 25% burn-in).

2.3 RESULTS AND DISCUSSION

In view of the variety of habitats occupied by species of Cirripedia, it is remarkable that only four species have been reported from true freshwater conditions, and it is of particular interest that all these belong to the rhizocephalan family Sacculinidae (see Annandale, 1911, Feuerborn, 1931, 1933, Boschma, 1967). Unfortunately, all four species are extremely rare, which has greatly impeded any close investigation. High prevalence of sacculinids is reported from a few brachyuran crabs (Veillet, 1945, Reisser & Forward, 1991), but in most crab species known to host a rhizocephalan barnacle, parasitized individuals only constitute a tiny fraction of the overall crab population. In order to find infected individuals, therefore, it is often necessary to inspect a very large number of crabs. If crab species are rare and/or found in difficult to access habitats, which is the case for the freshwater rhizocephalans, finding parasites obviously becomes even more challenging. Thus, it is due to the rarity of the material and not the lack of interest that this topic has not been addressed before. After years of collection efforts, we have now accumulated sufficient material of freshwater-inhabiting and marine rhizocephalans to finally address the problem by performing a morphological and molecular phylogenetic study. Of the four semiterrestrial/freshwater species of sacculinids in question, only *Sesarmaxenos monticola* is not represented in the species assembly. However, a careful reading of Annandale (1911) and Feuerborn (1931, 1933) revealed that there are no published morphological differences between *S. monticola* and *S. gedehensis*. Based on consultation with our deceased colleague Dr. A.V. Rybakov, we therefore believe that the two species are conspecific. If true, this would entail that our study indeed comprises all rhizocephalan species known from freshwater. In the molecular examination, we have extracted DNA from *Ptychascus barnwelli* and *Sesarmaxenos gedehensis*, while the morphological analysis included cypris larvae from three semiterrestrial/freshwater species, viz., *S. gedehensis*, *P. barnwelli*, and *P. glaber*.

2.3.1 Phylogenetic Analysis

To investigate the phylogenetic position of the freshwater rhizocephalans within Cirripedia, we conducted a molecular phylogenetic analysis of a carefully selected collection of rhizocephalan and thoracican outgroup species using nuclear (18S and 28S) and mitochondrial (16S) ribosomal genes as phylogenetic markers (Table 2.1). DNA sequences from six rhizocephalan species are new for this study, while DNA sequences from additional 22 species have been obtained from GenBank (Table 2.2). We performed eight phylogenetic analyses, with and without hypervariable regions of nucleotides (see method section), and including/excluding DNA sequences from members of the rhizocephalan suborder Akentrogonida, which are notoriously problematic to align due to a large number of deletions and insertions in ribosomal genes (author's experience). All three data sets were analyzed with Bayesian inferences of phylogeny and maximum likelihood methods. The clade support values and the tree topology of the phylogenetic trees changed to some extent across the different data sets and analytical methods, but the overall phylogeny is in agreement with earlier studies of parasitic barnacles (Spears et al., 1994, Glenner & Hebsgaard, 2006, Glenner et al., 2010, Pérez-Losada et al., 2012). Figure 2.2 shows the most commonly occurring tree topology with support values from the Bayesian inference of phylogeny, with the most variable DNA sequence regions excluded.

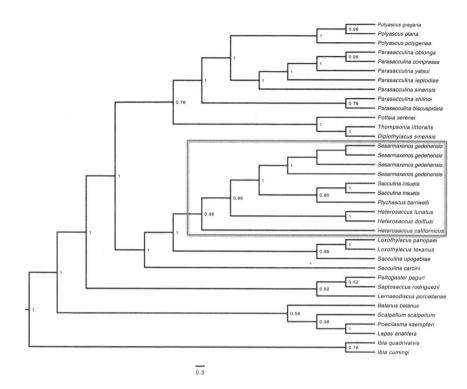

Figure 2.2   Phylogenetic position of the freshwater barnacles. Phylogeny based on Bayesian inference of 32 barnacle species. Six thoracican barnacles are serving as outgroups to 26 rhizocephalan species (including three freshwater-dwelling rhizocephalan species *Sesarmaxenos gedehensis*, *Sacculina insueta*, and *Ptychascus barnwelli*). These three barnacle species form a strongly supported monophyletic clade within a paraphyletic *Heterosaccus* genus. Together, *Heterosaccus* and the freshwater barnacles form a well-supported monophyletic clade, indicated with a blue box. Numbers at the branches of the tree indicates posterior probabilities.

2.3.2 A Single Freshwater Invasion

The freshwater rhizocephalans (*Ptychascus barnwelli*, four specimens of *Sesarmaxenos gedehensis* and *Sacculina insueta*) were recovered as a monophyletic group in seven of eight analyses within a paraphyletic genus *Heterosaccus*. Two species of *Loxothylacus*, *L. texanus* and *L. panopaei*, formed together with *Sacculina upogebiae* a monophyletic sister clade to *Heterosaccus* and the clade of freshwater species. A single species, *Sacculina carcini*, branches off basally in the clade forming a sister taxon to the remaining taxa (Fig. 2.2). Seven of the eight analyses placed *S. insueta* and the two specimens of *Ptychascus barnwelli* in a sister clade to four *Sesarmaxenos gedehensis* specimens. The position of *S. insueta* as sister group to *Ptychascus* is interesting, but hardly surprising. *Sacculina insueta* was not originally described as a freshwater rhizocephalan (Boschma, 1966), but it parasitizes crabs of the genus *Ptychognathus*. Members of this genus are found in both brackish water, estuarine regions, and in true freshwater lakes and ponds (Ng et al., 2008), wherein they are believed to have been able to complete their entire life cycle. The life cycle of *Ptychognathus riedelii*, the host species of the two specimens of *S. insueta*, in our study is not known, but it is likely that the larvae of the crab can develop in brackish water and/or even freshwater environments, an argument that can also be adopted for the parasite larvae. Thus, our analyses provide strong support for a monophyletic clade of parasites that share an extended tolerance for living in brackish water/freshwater environments, probably inherited from a common brackish-water-tolerant *Heterosaccus*-like ancestor. At least for *Sesarmaxenos*, it is certain that the entire parasite life cycle, including the free-swimming larval phase, takes place in a true freshwater habitat. In the following discussion, this monophyletic group of brackish, semiterrestrial, and freshwater rhizocephalans, for convenience, will be termed "freshwater rhizocephalans".

The phylogenetic position of a species of the genus *Sacculina* deeply within *Heterosaccus* and *Loxothylacus* as sister taxon to *Pthycascus* once again demonstrates the messy taxonomic state of *Sacculina* and the need for a profound revision of this genus, as well as the entire family Sacculinidae (see Glenner & Hebsgaard, 2006).

A paraphyletic status of *Heterosaccus* has already been suggested by Pérez-Losada et al. (2008). But in contrast to the latter study, where *Heterosaccus* was paraphyletic relative to the genus *Loxothylacus*, our analyses excluded the species of *Loxothylacus* from *Heterosaccus*. Together, *H. lunatus* and *H. dolfusii* form a sister clade to the monophyletic freshwater rhizocephalans, with *H. californicus* as a basally departing outgroup (Fig. 2.2). All known species of *Heterosaccus* are marine and parasitize exclusively marine spider (Majoidea) or swimming (Portunuidea) crabs (Glenner et al., 2008). The hosts of the freshwater rhizocephalans, i.e., sesarmids, varunids, ocypodids, majids, and portunuids, are distantly related and must have evolved their semiterrestrial/freshwater lifestyle independently (e.g., Schubart et al., 2000, for Sesarmidae). The coevolution between hosts and freshwater parasites is complicated and has most likely involved several host shifts. This makes the co-phylogeny problematic to reconstruct and scenarios for how members of *Sesarmidae* and *Ptychognathus* became freshwater-dwelling species and coevolved with their rhizocephalan parasites highly speculative. Nevertheless, our phylogenetic analyses suggest that freshwater rhizocephalans are monophyletic and appear to have originated within *Heterosaccus*. This implies that apomorphies for *Heterosaccus* might also be found in the freshwater species. An examination of externae (adult parasites – Fig. 2.3A) from representative species of the outgroup genus *Loxothylacus*, the species of *Heterosaccus*, and the three freshwater rhizocephalans did not reveal any obvious morphological characters as candidates for synapomorphies for the clade. This is no surprise, considering the extremely reduced adult morphology of all sacculinid rhizocephalans that leave only a few traits available for taxonomy (Figure 2.3a) and due to the fact that adaptations to life in a hypoosmotic environment most likely require physiological rather than morphological adaptations.

In contrast, a likely adaptation for the new freshwater lifestyle is found in the abbreviated development of *P. barnwelli*, *P. glaber*, and *S. gedehensis* (also *S.*

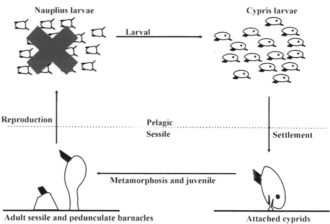

Figure 2.3 The externa and life cycle of freshwater barnacles.(A) The ventral side showing two *Sacculina insueta* externas under the abdomen of a freshwater/brackish
water crab *Ptychognathus* sp.(B) All adult barnacles are permanently attached
as adults. They normally release nauplius larvae to disperse, which through six
nauplius stages eventually develop into the final larva stage, the cypris larvae, the
settling stage of cirripedians. In the two freshwater-dwelling rhizocephalan genera, *Ptychascus* and *Sesarmaxenos*, all nauplius larvae stages are missing (indicated with a red cross) and the eggs hatch directly as cypris larvae. In *Sacculina
insueta* (a), nauplius larvae are still present in the life cycle.

monticola, possibly synonymous with *S. gedehensis* – see discussion above). While *H.
californicus* and all other known members of the marine genus *Heterosaccus* release
nauplius larvae, species of *Pthycascus* and *Sesarmaxenos* hatch and are released as
cypris larvae (Fig. 2.3B). This entails that in *Ptychascus* and *Sesarmaxenos*, the four
to six nauplius stages forming part of the ontogeny in almost all barnacle species

have been eliminated. A similarly abbreviated life cycle is rare in cirripedes and occurs only in members of the rhizocephalan order Akentrogonida (Høeg & Lützen, 1993), in most acrothoracican barnacles and in thoracican barnacles from the deep sea. Interestingly, the mantle cavity of the two specimens of *S. insueta* contains what appear to be functional nauplius larvae in stages I and II, revealing that dispersing via nauplius larvae still plays a role in the life cycle of *S. insueta*. In rhizocephalans, it is quite common that the molt from stage I to II occurs in the mantle cavity, whence both stages are expelled simultaneously. The stage II nauplii of *S. insueta* had well-developed appendages with natatory setae and well-developed frontolateral horns, indicating that these larvae are capable of free swimming. Consequently, the reduction of the larval nauplius stages must have evolved independently twice in *Sesarmaxenos* and *Pthycascus*. Rhizocephalan nauplii are always lecithotrophic and thus only serve for dispersal. The absence of nauplii in *Sesarmaxenos* and *Ptychascus* most likely reflects an adaptation to the microcosm of a freshwater pond, where dispersal is irrelevant or even disadvantageous. The absence of nauplii entails that the larvae can settle and metamorphose immediately after their release (Andersen et al., 1990). The presence of nauplius larvae in the life cycle of *S. insueta* might reflect that this species is exposed to larger and more variable environments, such as river deltas where some dispersal capacity may confer an advantage for the species. Thus, *S. insueta* in many respects resembles *S. gregaria* on the mitten crab *Eriocheir* sp., which also have nauplii that develop and disperse in brackish water/ saltwater where they are released when the host migrates to river mouths.

Features in cypris morphology examined by SEM have previously proved to be of high value in rhizocephalan phylogeny (Glenner et al., 2010), and our examination of *Sesarmaxenos*, *Pthycascus*, and *Heterosaccus* cyprids revealed small but significant features in support of our molecularly derived tree (Fig. 2.4). In kentrogonid rhizocephalans, the male cyprids have a large-sized aesthetasc on the third antennular segment, whereas both male and female cyprids carry a smaller aesthetasc on the small, fourth segment (Glenner et al., 1989; *S. carcini* in Fig. 2.4). In species of *Heterosaccus*, the male-specific aesthetasc is present but very small, and the fourth segmental aesthetasc is rather small in cyprids of both sexes. This pattern is also found in *Ptychascus* and *Sesarmaxenos*. The male-specific aesthetasc on the third segment is rather small in *P. barnwelli*, although larger than in *Heterosaccus*, and it is absent altogether in both *P. glaber* and *S. gedehensis*. We emphasize that male cyprids can easily be identified from single larvae, even in the absence of an aesthetasc, by the presence of a spinous process at the tip of the attachment disk (Høeg, 1987, Glenner et al., 1989). In contrast, cyprids *Loxothylacus* and *S. carcini* have the normal pattern with a large-sized male-specific aesthetasc on the third segment (see also Andersen et al., 1990) (Fig. 2.4D–F). Another feature is a flap-shaped and villus-covered extension from the posterior (proximal) end of the attachment disk in male cyprids. This flap occurs in *Heterosaccus* (see Walker & Lester, 2000, Glenner et al., 2010), *Ptychascus*, and *Sesarmaxenos*, but is not present in *Loxothylacus*, *S. carcini*, or in any other rhizocephalan cyprids examined by us (Glenner et al., 1989, Pasternak et al., 2004). We therefore suggest that both the presence of such a flap and the reduction or loss of the third segmental aesthetasc are putative apomorphies

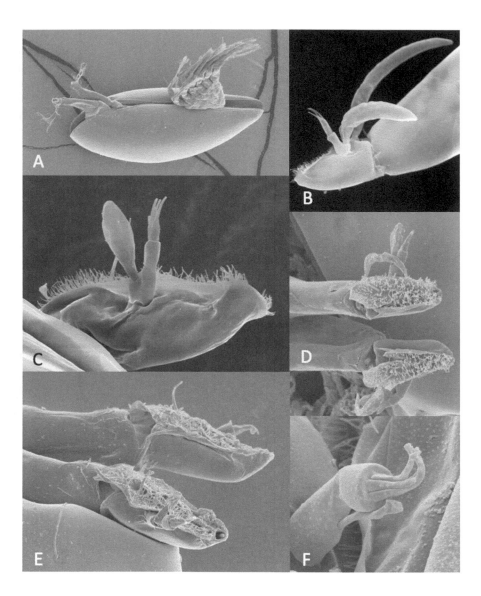

Figure 2.4   Cypris anatomy of freshwater rhizocephalans. (A) The general anatomy of a typical barnacle larvae. *Balanus rosea* in ventral view with the extruding antennules
to the left and the six pairs of thoracic appendages to the right. (B) The second,
third, and fourth antennule segments of a male *Sacculina carcini* in ventral view.
(C) The third and fourth antennular segments of a male *Heterosaccus californicus* cypris larva (ventrolateral view). (D) The ventral attachment part of the third
segment of a male *Ptychascus barnwelli* larva. (E) The ventral view of the attachment disks of the two pairs of third antennules of a male *Ptychascus glaber* cypris
larva. (F) A close-up of the fourth antennular segment of a male *S. gedehensis*
cypris larva.

for a clade comprising species of *Heterosaccus*, *Ptychascus*, *Sesarmaxenos*, and *Sacculina insueta*. Unfortunately, we had no cyprids of *S. insueta* for our analysis, which could further support this hypothesis. According to our phylogeny (Fig. 2.2), a completely cladistic classification would subsume the freshwater species (*Ptychascus*, *Sesarmaxenos*, and *S. insueta*) in *Heterosaccus* or, alternatively, erect a new genus for the freshwater species and a new and monotypic genus for *H. californicus*. Nevertheless, we shall delay any formal taxonomic steps for a comprehensive analysis of a wide range of scacculinid and other rhizocephalan species. Until then, it should be reminded that generic names in Rhizocephala confer little or no information on either phylogenetic relationship or biology.

2.4 SUMMARY

Our results suggest a single origin of freshwater rhizocephalans. The absence of nauplius larvae is probably an adaptation to a habitat, where dispersal capacity is redundant or even disadvantageous. We found no obvious morphological adaptations to the freshwater/terrestrial environment in the adult parasites. The cypris larvae of *Sesarmaxenos* and *Pthycascus* have subtle but consistent morphological modifications of the antennules, which are shared with species of *Heterosaccus*. Species of *Pthycascus*, *Sesarmaxenos*, and *Sacculina insueta* form a well-supported monophyletic group within a paraphyletic *Heterosaccus*.

ACKNOWLEDGMENTS

We are grateful to our close friend and colleague Dr. Alexey V. Rybakov for having collected material for us in the Philippines and performed much of the SEM analysis on cyprids reported here. His untimely death by an accident in 2013 just after full recovery from a serious disease has sadly prevented his coauthorship of this paper. JTH was financially supported by the Danish Agency for Independent Research and the Carlsberg Foundation.

REFERENCES

Andersen, M., Bohn, M., Hoeg, J. & Jensen, P. 1990. Cyprid ultrastructure and adult morphology In *Ptychascus barnwelli*, new species, and *Ptychascus glaber* (Cirripedia, Rhizocephala), parasites on semiterrestrial crabs. *Journal of Crustacean Biology* 10:20–8.

Annandale, N. 1911. Note on a rhizocephalous crustacean from fresh water and on some specimens of the order from Indian seas. *Records of the Indian Museum* 6:1–4.

Boschma, H. 1966. A rhizocephalan parasite of a crab of the genus *Ptychognathus* from Japan. *Proceedings of the Koninklijke Nederlandse Akademie van Wetenschappen Series C-Biological and Medical Sciences* 69:1–7.

Boschma, H. 1967. On two specimens of rhizocephalan parasite *Ptychascus glaber* Boschma from Island Trinidad. *Proceedings of the Koninklijke Nederlandse Akademie Van Wetenschappen Series C-Biological and Medical Sciences* 70:321–23.

Castresana, J. 2000. Selection of conserved blocks from multiple alignments for their use in phylogenetic analysis. *Molecular Biology & Evolution* 17:540–52.

Crandall, K.A., Harris D.J. & Fezner Jr., J.W. 2000. The monophyletic origin of freshwater crayfish estimated from nuclear and mitochondrial DNA sequences. *Proceedings of the Royal Society of London B* 267:1679–86.

Darriba, D., Taboada G.L., Doallo, R. & Posada, D. 2012. jModelTest 2: More models, new heuristics and parallel computing. *Nature Methods* 9(8):772.

Drummond, A.J. & Rambaut, A. 2007. BEAST: Bayesian evolutionary analysis by sampling trees. *BMC Evolutionary Biology* 7:214.

Feuerborn, H. 1931. Ein Rhizocephale und zwei Polychaeten aus dem Süßwasser von Java und Sumatra. *Verhandlungen der Internationalen Vereinigung von Theoretischen und Angewandten Limnologie* 5:618–60.

Feuerborn, H. 1933. Das Cyprisstadium des Süßwasserrhizocephalen *Sesarmaxenos*. *Verhandlungen der Deutschen Zoologischen Gesellschaft* 35:127–38.

Friedrich, M. & Tautz D. 1997. Evolution and phylogeny of the Diptera: A molecular phylogenetic analysis using 28S rDNA sequences. *Systematic Biology* 46:674–98.

Frischer, M.E., Hansen, A.S., Wyllie, J.A., Wimbush, J., Murray, J. & Nierzwicki-Bauer, S.A. 2002. Specific amplification of the 18S rRNA gene as a method to detect zebra mussel (*Dreissena polymorpha*) larvae in plankton samples. *Hydrobiologia* 487:33–44.

Glenner, H. & Hebsgaard, M. 2006. Phylogeny and evolution of life history strategies of the parasitic barnacles (Crustacea, Cirripedia, Rhizcephala). *Molecular Phylogenetics and Evolution* 41:528–38.

Glenner, H., Hoeg, J., Klysner, A. & Larsen, B. 1989. Cypris ultrastructure, metamorphosis and sex in seven families of parasitic barnacles (Crustacea, Cirripedia, Rhizocephala). *Acta Zoologica* 70:229–42.

Glenner, H., Høeg, J.T., Stenderup, J. & Rybakov, A.V. 2010. The monophyletic origin of a remarkable sexual system in akentrogonid rhizocephalan parasites: A molecular and larval structural study. *Experimental Parasitology* 125:3–12.

Glenner, H., Thomsen, P.F., Rybakov, A.V., Galil, B.S. & Hoeg, J.T. 2008. The phylogeny of rhizocephalan parasites of the genus *Heterosaccus* using molecular and larval data (Cirripedia: Rhizocephala; Sacculinidae). *Israel Journal of Ecology & Evolution* 54:223–38.

Gouy, M., Guindon, S. & Gascuel, O. 2010. SeaView version 4: A multiplatform graphical user interface for sequence alignment and phylogenetic tree building. *Molecular Biology and Evolution* 27:221–24.

Hillis, R.E. & Dixon, M.T. 1991. Ribosomal DNA: Molecular evolution and phylogenetic inference. *Quarterly Review of Biology* 66:411–53.

Høeg, J.T. 1987. The relation between cypris ultrastructure and metamorphosis in male and female *Sacculina carcini* (Crustacea, Cirripedia). *Zoomorphology* 107:299–311.

Høeg, J. & Lützen, J. 1993. Comparative morphology and phylogeny of the family Thompsoniidae (Cirripedia, Rhizocephala, Akentrogonida), with descriptions of three new genera and seven new species. *Zoologica Scripta* 22:363–86.

Høeg, J.T., Noever, C., Rees, D.A., Crandall, K., Glenner, H. 2020. A new molecular phylogeny-based taxonomy of parasitic barnacles (Crustacea: Cirripedia: Rhizocephala), *Zoological Journal of the Linnean Society* 190(2):632–653. https://doi.org/10.1093/zoolinnean/zlz140.

Maddison, W.P. & Maddison, D.R. 2009. Mesquite: A modular system for evolutionary analysis. Version 2.6 http://mesquiteproject.org.

Mallatt, J. & Sullivan J. 1998. 28S and 18S rDNA sequences support the monophyly of lampreys and hagfishes. *Molecular Biology and Evolution* 15:1706–18.

Ng, P.K., Guinot, D. & Davie, P.J. 2008. Systema Brachyurorum: Part 1. An annotated checklist of extant brachyuran crabs of the world. *The Raffles Bulletin of Zoology* 17:1–286.

Nunn, G., Theisen, B., Christensen, B. & Arctander, P. 1996. Simplicity-correlated size of the nuclear 28S ribosomal RNA D3 expansion segment in the crustacean order Isopoda. *Journal of Molecular Evolution* 42:211–23.

Palumbi, S.R., Martin, A., Romano, S., McMillan, W.O., Stice, L. & Grabowski, G. 1991. *The Simple Fools Guide to PCR. A Collection of PCR Protocols, Version 2.* Honolulu, University of Hawaii.

Pasternak, Z., Garm, A. & Høeg, J.T. 2004. The morphology of the chemosensory aesthetasc-like setae used during settlement of cypris larvae in the parasitic barnacle *Sacculina carcini* (Cirripedia: Rhizocephala). *Marine Biology* 146:1005–13.

Pérez-Losada, M., Harp, M., Høeg, J.T., Achituv, Y., Jones, D., Watanabe, H. & Crandall, K.A. 2008. The tempo and mode of barnacle evolution. *Molecular Phylogenetics and Evolution* 46:328–46.

Pérez-Losada, M., Hoeg, J.T. & Crandall, K.A. 2012. Deep phylogeny and character evolution in Thecostraca (Crustacea: Maxillopoda). *Integrative and Comparative Biology* 52:430–42.

Reisser, C. & Forward, R. 1991. Effect of salinity on osmoregulation and survival of a rhizocephalan parasite, *Loxothylacus panopaei*, and its crab host, *Rhithropanopeus harrisii*. *Estuaries* 14:102–06.

Ronquist, F., Teslenko, M., van der Mark, P., Ayres, D.L., Darling, A., Höhna, S., Larget, S., Liu, L., Suchard, M.A. & Huelsenbeck, J.P. 2012. MrBayes 3.2: Efficient Bayesian phylogenetic inference and model choice across a large model space. *Systematic Biology* 61:539–42.

Schubart, C.D., Cuesta, J.A., Diesel R & Felder, D.L. 2000. Molecular phylogeny, taxonomy, and evolution of nonmarine lineages within the American grapsoid crabs (Crustacea: Brachyura). *Molecular Phylogenetics and Evolution* 15:179–90.

Spears, T., Abele, L.G. & Applegate, M.A. 1994. Phylogenetic study of cirripedes and selected relatives (Thecostraca) based on 18S rDNA sequence analysis. *Journal of Crustacean Biology* 14:641–56.

Spears, T., Abele, L.G. & Kim, W. 1992. The unity of the Brachyura: A phylogenetic study based on rRNA and rDNA sequences. *Systematic Biology* 41:446–51.

Thompson, J.D., Higgins, D.G. & Gibson, T.J. 1994. ClustalW: Improving the sensitivity of progressive multiple sequence alignment through sequence weighting, position specific gap penalties and weight matrix choice. *Nucleic Acids Research* 22:4673–80.

Tsuchida, K., Lützen, J. & Nishida, M. 2006. Sympatric three-species infection by *Sacculina* parasites (Cirripedia: Rhizocephala: Sacculinidae) of an intertidal grapsoid crab. *Journal of Crustacean Biology* 26:474–79.

Veillet, A. 1945. Recherches sur le parasitisme des crabes et les galathées par les rhizocéphales et des epicarides. *Bulletin de l'Institut Océanographique de Monaco* 22:193–341.

Walker, G. & Lester, R. 2000. The cypris larvae of the parasitic barnacle *Heterosaccus lunatus* (Crustacea, Cirripedia, Rhizocephala): Some laboratory observations. *Journal of Experimental Marine Biology and Ecology* 254:249–57.

Whiting, M.F. 2002. Mecoptera is paraphyletic: Multiple genes and phylogeny of Mecoptera and Siphonaptera. *Zoologica Scripta* 31:93–104.

CHAPTER 3

Malacostraca (Arthropoda: Crustacea) of Fresh and Brackish Waters of Sakhalin Island: The Interaction of Faunas of Different Origins

Vjacheslav S. Labay

CONTENTS

3.1 INTRODUCTION

Questions on the zoogeographic zoning of Sakhalin Island on various freshwater fauna have long attracted researchers attention (Taranet 1938, Berg 1949, Kruglov and Starobogatov 1993, Safronov and Nikiforov 1995, Starobogatov 1996, Chereshnev 1998, Nikiforov 2001, Prozorova 2001). Most recently, Bogatov et al. (2007) analyzed various groups of plants and invertebrates; some freshwater taxa were considered, i.e., mollusks, plecopterans, and chironomids. The Schmidt Line is the most important biogeographical boundary dividing the northeastern and southwestern parts of the island (Fig. 3.1) (Berg 1949, Bogatov et al. 2007). An analysis of terrestrial and freshwater fauna and flora distribution ranges has demonstrated another 14 units. The secondary most important biogeographical barriers after the

Schmidt line

Figure 3.1   A schematic map of Sakhalin Island.

Schmidt Line are those running along the southern border of the North Sakhalin Lowland (demarcating the Tym-Poronaysky district) and the barrier present at the Isthmus (Figs. 3.1 and 3.2).

Here, I focus on zoogeographic analysis of the malacostracan fauna in the fresh and brackish waters of Sakhalin Island. Zoogeographic analysis of the

Figure 3.2   A schematic map of lagoon reservoirs of Sakhalin Island: 1 – Baikal Bay, 2 – Pomr' Bay, 3 – Kuegda Bay, 4 – Neurtu Bay, 5 – Tropto Bay, 6 – Kolendu Bay, 7 – Khanguza Bay, 8 – Ketu Bay, 9 – Urkt Bay, 10 – Ekhabi Bay, 11 – Odoptu Bay, 12 – Piltun Bay, 13 – Chaivo Bay, 14 – Nyiskyi Bay, 15 Nabil Bay, 16 – Lunskyi Bay, 17 – Nevskoe Lake, 18 – Izmenchivoye Lake, 19 – Tunaicha Lake, 20 – Busse Lake, 21 – Vavajskaya lake system, 22 – Chibisanskaya lake system, 23 – Ptych'e Lake, 24 – Aynskoe Lake.

malacostracans provides a number of advantages over freshwater ichthyofauna, mollusks, and amphibiotic insects, because it is possible to describe the zoogeographical regionalization of Sakhalin Island not only as a set of climatically and geographically isolated distribution ranges, but also as a result of a long historical process. For example:

1. absence of out-water developmental stages (for example, in amphibiotic insects) for crossing of watersheds;
2. lack of a pelagic or parasitic larvae, as in mollusks (only prawns have pelagic larval phases among fresh and brackish water Malacostraca in Sakhalin Island);
3. lack of pronounced feeding and spawning migrations, as in many fish (the crab *Eriocheir japonica* and the amphipod *Sternomoera rhyaca* characterized by catadromous migrations, are a few exceptions among Malacostraca), and;
4. the presence of pronounced physiologically determined salinity barriers that prevents the resettlement of freshwater and brackish water crustaceans through coastal marine waters (Khlebovich 1974, 2015).

These features more easily allow the definition of zoogeographical zonation of Sakhalin Island, based not only on climatic and geographical characteristics, but also on historical and geological backgrounds. I provide a zoogeographical description of the fresh and brackish waters malacostracan fauna of Sakhalin Island based on analysis of distribution ranges and origin of individual groups.

3.2 MATERIALS AND METHODS

Materials for this work were the collections of the hydrobiological expeditions of the Sakhalin State Pedagogical Institute (1994–1996), Sakhalin State University (2015–2017), and Sakhalin Scientific Research Institute of Fisheries & Oceanography (1996–2014, 2017–2018), and samples collected by the author and colleagues during 1991–2018 from various reservoirs of Sakhalin Island. In addition, literature and archival data on the benthos were also used (Chernjavskiy 1882, 1883, Brazhnikov 1907, Derzhavin 1927, 1930a, b, Birstein and Vinogradov 1934, Ushakov 1934, 1948, Ueno 1935a, b, 1940, Gurjanova 1936, 1951, 1962, Schellenberg 1936, 1937a, b, Myadi 1938, Birstein 1939, 1940, 1951, 1955, Vinogradov 1950, Lomakina 1955, 1958, Bulycheva 1957, Kluchareva et al. 1964, Brodsky 1974, Kussakin 1974, 1979, Tzvetkova 1975, Levanidov 1980, Safronova and Safronov 1980, Volova and Koz'menko 1984, Garkalina 1982, Kafanov 1984, Zvyagintsev 1985, Tabunkov et al. 1988, Starobogatov 1995, 1996, Starobogatov and Vasilenko 1995, Labay 1996a, b, 1997, 1999, 2002a, b, 2003a, b, 2004a, b, 2005, 2007, 2008, 2009, 2010, 2011a, b, c, 2012, 2014, 2015a, b, 2016, Ivankov et al. 1999, Ivleva et al. 1999, Labay et al. 2000, 2004, 2010, 2013, 2014, 2015, 2016a, b, Labay and Pecheneva 2001, Kafanov and Pecheneva 2002, Pecheneva et al. 2002, Samatov et al. 2002, Kafanov et al. 2003, Zhivoglyadova and Labay 2003, Labay and Rogotnev 2005, Sidorov 2005, Pecheneva and Labay 2006, Labay and Barabanschikov 2009, Marin 2013, Kawai et al. 2013, 2016, Labay and Labay 2014).

For zoogeographical names compilation for species distribution ranges, the approach of Kafanov and Kudryashov (2000) was applied.

The Sörensen coefficient was used to compare lists of hydrobionts ($I_{x,y}$, %) (Geographia i monitoring bioraznoobrazia, 2002):

$$I_{x,y} = \frac{2c}{a+b} * 100$$

c – the number of common species in the compared districts x and y;

a and b – the number of species in districts x and y, respectively.

The initial matrices clustering was performed according to the method of unweighted pair group average (Duran and Odell 1974).

In relation to salinity (classification by Khlebovich [1974, 2015]), all species are divided into the marine-euryhaline species, living at salinities up to 22–26 psu (practical salinity units) and briefly entering brackish waters (the boundary is called the β-chorohaline zone); the brackish water species in salinity ranges from 22–26 psu to 5–7 psu (the lower bound corresponds to the α-horohaline zone); the oligohaline species ranges from 5–7 psu to 0.5 psu; and the freshwater species living at salinities is less than 0.5 psu.

3.3 BRIEF PHYSICAL-GEOGRAPHIC CHARACTERISTICS OF SAKHALIN ISLAND

Sakhalin Island is elongated in the meridional direction from 45°54′N to 54°25′E latitude, its length is 948 km, and the maximum width is 160 km (at the latitude of the town of Lesogorsk) and the minimum width is 26 km (Isthmus Poyasok).

The island is separated from the continent by the Tatar Strait in the Sea of Japan and the Strait of Nevelskoy, the narrowest width of which is 7.5 km at the Liman of Amur River and the Sakhalin Bay, and also from the Hokkaido Island (Japan) by the Strait of Laperuz (with the shortest distance of 41 km). Sakhalin Island is surrounded by the Sea of Japan (Tatar Strait) in the west and the Sea of Okhotsk on the remaining coasts. The water exchange between the Sea of Okhotsk and the Sea of Japan is through the Strait of Laperuz. Water exchange between the seas through the Nevelsky Strait does not have a significant role because of its shallow depth (1–27 m), narrow width, and the desalinated waters of the Amur River estuary entering from the north (Fig. 3.1).

The west coast is slightly indented, with no significant bays and coves, whereas in the northwest there are two large lagoons in Baikal Bay and Pomr Bay. In the north, the island bears a narrow isthmus into the Schmidt Peninsula. The northeast coast to the south of 51°N is replete with closed meridional bays of the lagoon type. In the east, the Terpeniya Cape separates the large Terpeniya Bay. In the northern part, Terpeniya Bay adjoins the oligohaline lagoon Nevskoe Lake. The main part of the island is connected by a narrow isthmus with its southern "tail-shaped" part at

48° N. The Cape of Krilion and the Cape of Aniva delimit Aniva Bay in the south. The lagoon Lake Busse adjoins the northeast corner of Aniva Bay.

Several small islands are in close proximity to Sakhalin. Moneron Island is located 50 km southwest of Gornozavodsk in the Tatar Strait, and Tyuleniy Island is located 15 km south of the tip of Terpeniya Cape; freshwaters are only present on Moneron.

The southwestern coast is warmed by the southern Tsushima Current, which varies seasonally. The warm water influx from the south increases in summer, and the current reaches the top of the Tatar Strait. The southern inflow weakens in the autumn–winter period, and the current from the north intensifies due to prevailing northwesterly winds. The Tsushima Current forms a year-round cyclonic circulation around Moneron Island. This enables a number of subtropical and temperate species to survive on the island and in its coastal waters. The Soya Current (a branch of the Tsushima Current) flows from the Sea of Japan to the Sea of Okhotsk through the Strait of Laperuz along the coast of Hokkaido and has a warming effect on the southern tip of Cape Krilion. The waters of the Amur River go through the Amur estuary to Sakhalin Bay and warm the northwestern coast of the island. The cold East Sakhalin Current runs along the eastern coast from north to south (Atlas Sakhaliskoy oblasti 1967, Sakhalin Region 1994).

3.4 SPECIES COMPOSITION AND BIOGEOGRAPHIC CHARACTERISTICS OF MALACOSTRACA

The Malacostraca fauna of Sakhalin Island includes 72 species in general, of which 56 are marine euryhaline and brackish, 30 are recorded from freshwater and oligohaline water, and 14 are common for both biotopes (Table 3.1). Amphipoda comprises the largest group (43 species). The Decapoda and Isopoda are relatively minor components: 11 and 9 species, respectively. Island distribution ranges of Sakhalin crustaceans are shown in Figs. 3.3–3.22.

Certain difficulties arise in the malacostracan zoogeographical description of Sakhalin Island: (1) the geographical biotope separation; (2) the difference in the historical origin of individual distribution ranges, and; (3) the concomitant genesis of certain species. Each is considered separately.

3.4.1 The Geographical Biotope Separation

The island relief is heterogeneous. The northern part is flat (excluding the Schmidt Peninsula), with a large number of marshes, overgrown lakes, and rivers with low flows. Brackish water basins are represented by large formations as the Amur River estuary, the Nevelsky Strait, and the northern lagoons. Most rivers have no direct connection with the sea and flow into one or another brackish water lagoons. The middle and southern parts of the island are represented by low mountains with a highly developed river network. Most rivers here are foothill and mountain types,

Table 3.1 **Species Composition of Malacostraca from Freshwater and Brackish Water of Sakhalin Island**

No.	Taxon	Relation to salinity*	Notes
	Phylum Arthropoda		
	Subphylum Crustacea		
	Class Malacostraca		
	Order Mysida		
	Mysidae Haworth, 1825		
1	*Archaeomysis grebnitzkii* Czerniavsky, 1882	m-eu	
2	*Neomysis awatschensis* (Brandt 1851)	fr, ol, br	
3	*Neomysis czerniavskii* Derzhavin, 1913	m-eu	
4	*Neomysis mirabilis* (Czerniavsky 1882)	m-eu	
5	*Neomysis rayii* (Murdoch 1885)	m-eu	
	Order Cumacea		
	Diastylidae Bate, 1856		
6	*Diastylis lazarevi* (Lomakina 1955)	m-eu, br	
7	*Diastylopsis dawsoni* Smith, 1880 f. *calmani* Derzhavin, 1926	m-eu	
	Lampropidae Sars, 1878		
8	*Lamprops korroensis* Derzhavin, 1923	fr, ol	
9	*Lamprops sarsi* Derzhavin, 1926	m-eu	
	Order Amphipoda		
	Suborder Caprellidea		
	Caprellidae Leach, 1814		
10	*Caprella cristibrachium* Mayer, 1903	m-eu	
	Suborder Gammaridea		
	Anisogammaridae Bousfield, 1977		
11	*Anisogammarus* cf. *pugettensis* (Dana 1853)	m-eu	
12	*Eogammarus kygi* (Derzhavin 1923)	fr, ol	
13	*Eogammarus barbatus* (Tzvetkova 1965)	m-eu, br, fr, ol	Characteristic of significant upward migrations
14	*Eogammarus tiuschovi* (Derzhavin 1927)	br	
15	*Jesogammarus* (*Annanogammarus*) *annandalei* (Tattersall 1922)	fr	
16	*Locustogammarus hirsutimanus* (Kurenkov & Mednikov 1959)	fr, ol	
17	*Locustogammarus intermedius* (Labay 1996b)	fr	

(Continued)

Table 3.1 (Continued) **Species Composition of Malacostraca from Freshwater and Brackish Water of Sakhalin Island**

No.	Taxon	Relation to salinity*	Notes
18	*Locustogammarus locustoides* (Brandt 1851)	m-eu	
19	*Spasskogammarus spasskii* (Bulycheva 1952)		
	Atylidae Lilljeborg, 1865		
20	*Atylus collingi* (Gurjanova 1938)	m-eu	
	Calliopiidae G.O. Sars, 1893		
21	*Calliopius laeviusculus* (Kroyer 1838)	m-eu	
	Corophiidae Leach, 1814		
22	*Crassicorophium bonellii* (Milne Edwards 1830)	m-eu	
23	*Monocorophium steinegeri* (Gurjanova 1951)	m-eu	
	Dogielinotidae Gurjanova, 1953		
24	*Dogielinotus moskvitini* (Derzhavin 1930b)	br	
25	*Haustorioides gurjanovae* Bousfield & Tzvetkova, 1982	m-eu, br	
26	*Haustorioides magnus* Bousfield & Tzvetkova, 1982	m-eu, br	
	Gammaridae Leach, 1814		
27	*Gammarus lacustris* G.O. Sars, 1864	fr	
28	*Gammarus setosus* Dementieva, 1931	m-eu	
29	*Gammarus wilkitzkii* Birula, 1897	m-eu	
	Haustoriidae Stebbing, 1906		
30	*Eohaustorius washingtonianus* (Thorsteinson, 1941)	m-eu	
31	*Eohaustorius eous* (Gurjanova, 1951)	m-eu	
	Ischyroceridae Stebbing, 1899		
32	*Ischyrocerus commensalis* Chevreux, 1900	m-eu	
	Kamakidae Myers & Lowry, 2003		
33	*Kamaka derzhavini* Gurjanova, 1951	m-eu, br	
34	*Kamaka kuthae* Derzhavin, 1923	fr, ol, br	
	Melitidae Bousfield, 1973		
35	*Melita nitidaformis* (Labay 2003a)	ol	
36	*Melita shimizui sakhalinensis* (Labay 2016a)	m-eu, br	
	Photidae Boeck, 1871		
37	*Photis nataliae* Bulycheva, 1952	m-eu	
38	*Photis spasskii* Gurjanova, 1951	m-eu	

(Continued)

Table 3.1 (Continued) **Species Composition of Malacostraca from Freshwater and Brackish Water of Sakhalin Island**

No.	Taxon	Relation to salinity*	Notes
	Phoxocephalidae Sars, 1891		
39	*Grandifoxus longirostris* (Gurjanova 1938)	m-eu	
	Pontogeneiidae Stebbing, 1906		
40	*Paramoera anivae* Labay, 2012	m-eu	Interstitial, noted at the source of streams on the littoral
41	*Pontogeneia rostrata* Gurjanova, 1938	m-eu	
42	*Sternomoera moneronensis* Labay, 1997	fr	
43	*Sternomoera rhyaca* (Kuribayashi et al. 1996)	fr, ol, m-eu	Passing catadromous species
	Pontoporeiidae Dana, 1852		
44	*Monoporeia affinis* (Lindström 1855)	m-eu	
45	*Pontoporeia femorata* Krøyer, 1842	m-eu	
	Pseudocrangonyctidae Holsinger, 1989		
46	*Pseudocrangonyx bochaensis* (Derzhavin 1927)	fr	Underground
47	*Pseudocrangonyx relicta* Labay, 1999	fr	Underground
48	*Pseudocrangonyx susunaensis* Labay, 1999	fr	Underground
49	*Pseudocrangonyx birsteini* Labay, 1999	fr	Underground
	Talitridae Rafinesque, 1815		
50	*Platorchestia joi* Stock & Biernbaum, 1994	m-eu, br, ol	Supralittoral
51	*Platorchestia pachypus* (Derzhavin 1937)	m-eu, br, ol	Supralittoral
52	*Traskorchestia ochotensis* (Brandt 1851)	m-eu, br, ol	Supralittoral
	Order Isopoda		
	Suborder Asellota		
	Asellidae Rafinesque, 1815		
53	*Asellus* (*Asellus*) *levanidovorum* Henry & Magniez, 1995	fr	
	Suborder Cymothoida		
	Cymothoidae Leach, 1818		
54	*Ichthyoxenus amurensis* (Gerstfeldt 1858)	fr, ol	Fish ectoparasite
	Suborder Oniscidea		
	Ligiidae Leach, 1814		

(*Continued*)

Table 3.1 (Continued) **Species Composition of Malacostraca from Freshwater and Brackish Water of Sakhalin Island**

No.	Taxon	Relation to salinity*	Notes
55	*Ligia cinerascens* Budde-Lund, 1885	m-eu, br, ol, fr	Supralittoral
	Suborder Sphaeromatidea		
	Sphaeromatidae Latreille, 1825		
56	*Gnorimosphaeroma ovatum* (Gurjanova 1933)	m-eu, br, ol	
57	*Gnorimosphaeroma kurilensis* Kussakin, 1974	ol, fr	
58	*Gnorimosphaeroma noblei* Menzies, 1954	m-eu	
	Suborder Valvifera		
	Chaetiliidae Dana, 1849		
59	*Saduria entomon* (Linnaeus 1758)	m-eu	Spawns on littoral at freshwater outlets
	Idoteidae Samouelle, 1819		
60	*Idotea gurjanovae* Kussakin, 1974	m-eu	
61	*Idotea ochotensis* Brandt, 1851	m-eu	
	Order Decapoda		
	Palaemonidae Rafinesque, 1815		
62	*Palaemon modestus* (Heller 1862)	fr, ol	
63	*Palaemon paucidens* (de Haan 1841)	fr, ol, br	
64	*Palaemon sinensis* (Solland 1911)	fr, ol	
	Crangonidae Haworth 1825		
65	*Crangon amurensis* Bražnikov, 1907	m-eu, br, ol	
66	*Crangon dalli* (Rathbun 1902)	m-eu	
	Cambaridae Hobbs, 1942		
67	*Cambaroides schrenckii* (Kessler 1874)	fr	
	Camptandriidae Stimpson, 1858		
68	*Deiratonotus cristatum* (de Man 1895)	m-eu, br	
	Cheiragonidae Ortmann, 1893		
69	*Telmessus cheiragonus* (Tilesius 1812)	m-eu	
	Upogebiidae Borradaile, 1903		
70	*Upogebia major* (De Haan 1841)	m-eu, br	
	Varunidae H. Milne Edwards, 1853		
71	*Eriocheir japonica* (de Haan 1850)	m-eu, br, ol, fr	Passing catadromous species
72	*Hemigrapsus takanoi* (Asakura & Watanabe 2005)	m-eu	

* m-eu: marine euryhaline; br: brackish water; ol: oligosaline; fr: freshwater.

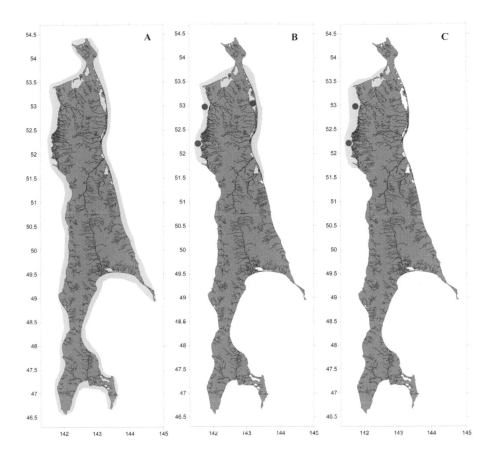

Figure 3.3   The island distribution range of freshwater and brackish water Mysida: A –
Archaeomysis grebnitzkii (Chernjiavskiy 1882), *Neomysis awatschensis* (Brandt
1851), and *Neomysis mirabilis* (Chernjiavskiy 1882; B – *Neomysis czerniavskii*
Derzhavin 1913); C – *Neomysis rayii* (Murdoch 1885). Here and below, the shad-
ing shows the distribution of the species in the sea coast, river estuaries, and
coastal lagoons; circles show collecting sites for rare species in inland reservoirs.

flowing directly into the sea. Brackish water basins are mainly relict lakes. The
Malacostraca fauna is defined according to biotopic differences. For example, the
subterranean *Pseudocrangonyx* are common in the mountainous regions of the
entire Far East (Birshtein 1955, Sidorov 2008), but are absent in the northern lowland part of the island.

3.4.2 The Difference in the Historical Origin of Regions of the Sakhalin Island

There were several geological epochs when the modern territory of the island or
its separate parts belonged to one or another formation. The most ancient period of
relief formation is the Permo-Carbon glacial epoch (about 300 million years ago)

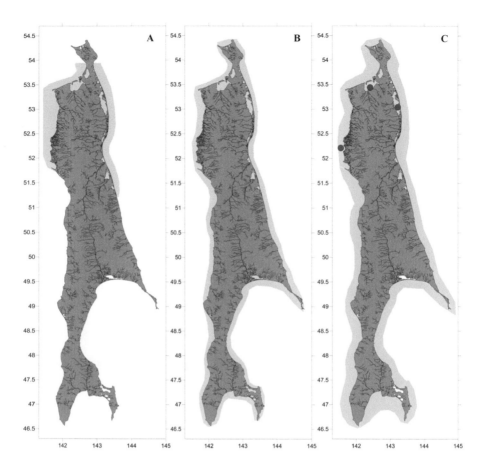

Figure 3.4 The island distribution range of freshwater and brackish water Cumacea:
A – *Diastylis lazarevi* (Lomakina 1955) and *Diastylopsis dawsoni* f. *calmani*
(Derzhavin 1926); B – *Lamprops korroensis* (Derzhavin 1923); C – *Lamprops
sarsi* (Derzhavin 1926).

with a warmer climate, similar to modern conditions (Matjushkov et al. 2014). This
period was an epoch of development and dispersal of ancient crustacean groups (for
example, crangonictid amphipods).

Broad continental contact with the mainland occurred in the late Pliocene–
Pleistocene, when continental regional uplift caused a regression of the upper
Pliocene Sea and the drainage of vast expanses of lowland shelf territories (Geologia
1970). Repeated contacts of southern Sakhalin with northern Japan took place during the Pliocene–Pleistocene, with the territorial disunity of the middle and southern
parts of Sakhalin, due to marine transgressions. A river basin with a top in the area
of modern Tatar Strait, which existed in the Pliocene–Pleistocene regression, was of
great importance for the formation of Malacostraca fauna (Lindberg 1972, Pacific
Ocean 1982).

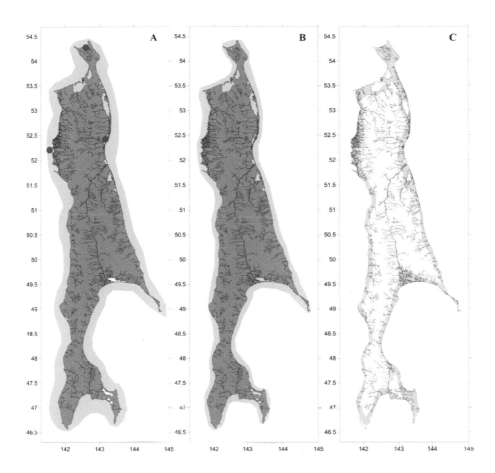

Figure 3.5 The island distribution range of freshwater and brackish water Amphipoda: A – *Caprella cristibrachium* (Mayer 1903); B – *Anisogammarus* cf. *pugettensis* (Dana 1853); C – *Eogammarus kygi* (Derzhavin 1923) and *Eogammarus barbatus* (Tzvetkova 1965).

The Sea of Japan and the Sea of Okhotsk were characterized by brackish lagoon conditions during the Pleistocene interglacial Nom phase (Lindberg 1972).

The strongest regression of the Sea of Okhotsk and the Sea of Japan led to the connection of northern Sakhalin with the mainland in the Pleistocene (75–12,000 years ago) (Khudjakov 1972, Korotkiy et al. 1997, Bezverkhniy et al. 2002). The lower segment of the paleo-Amur Valley was located between the Schmidt Peninsula in the north and the northeast mountains in the south at the modern alluvial plains of northern Sakhalin. The overlap of the Amur River bed by sandy sediments and its turning north into the Sakhalin Bay had occurred at the border of the Pleistocene–Holocene.

Numerous contacts between brackish water and freshwater with pure marine environment occur even at present, contributing to the penetration of marine-euryhaline species into new biotopes. This process is most active in the Amur estuary

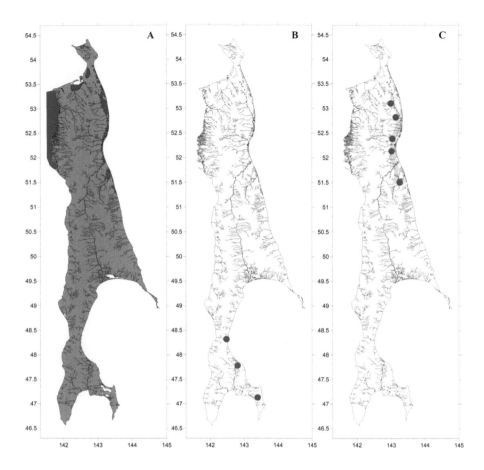

Figure 3.6   The island distribution range of freshwater and brackish water Amphipoda: A –
              Eogammarus tiuschovi (Derzhavin 1927); B – *Jesogammarus (Annanogammarus)*
              annandalei (Tattersall 1922); C – *Locustogammarus hirsutimanus* (Kurenkov &
              Mednikov 1959).

and Nevelsky Strait, where significant seasonal variations in salinity are observed
(Sailing 2003).

The stages of Sakhalin Island relief formation and freshwater and brackish water
contact with adjacent regions largely determine the differences in species complex
genesis. The island provided an area for expansion, origin, and contact of various
groups over historical times. Each group has its own habitat area. In total, there are
eight such groups:

1. The oldest Pangea group with a single genus *Pseudocrangonyx*. Species of
   Pseudocrangonyx (*P. bochaensis*, *P. relicta*, *P. susunaensis*, and *P. birsteini*)
   belong to the oldest crangonyctoids, and are the oldest freshwater malacostracan
   representatives on the island. This group developed and flourished on Pangea
   (Barnard and Barnard 1983, Holsinger 1989). The territories belonging now to

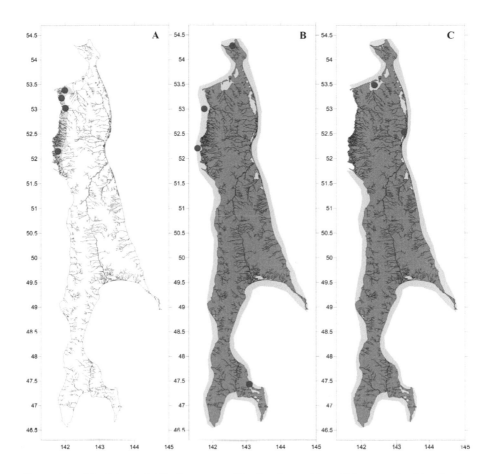

Figure 3.7   The island distribution range of freshwater and brackish water Amphipoda: A –
Locustogammarus intermedius (Labay 1996b); B – *Locustogammarus locustoides* (Brandt 1851); C – *Spasskogammarus spasskii* (Bulycheva 1952).

Berringia, including Alaska, were separated from the European portion by the
Turgai Strait as well as the shallow West Siberian Sea, remnants of the Tetis paleo-
sea, which existed until the end of the Eocene. The rest of North America was
separated by the Arctic Ocean (Marincovich et al. 1990). Under these conditions,
the East Asian crangonyctoids gave rise to the Pseudocrangonyctidae. The expansion of *Pseudocrangonyx* to Sakhalin can be confidently attributed to the Miocene–
Pliocene time, when there was a land bridge from the continent at modern Susunai
and Tonino–Aniva mountains. The later marine transgressions separated individual
groups, giving rise to modern *Pseudocrangonyx* species. Undescribed species may
yet occur in semi-subterranean waters of the Tonino–Aniva Ridge and Schmidt
Peninsula, although none were found during a survey of several watercourses in the
Schmidt Peninsula by the author in 2006.
2. A separate group that occupies the entire inner part of the island is made up of
freshwater crustaceans of continental origin, which penetrated into the island in the

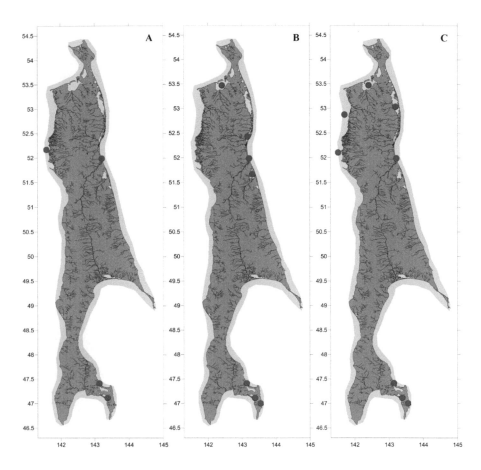

Figure 3.8 The island distribution range of freshwater and brackish water Amphipoda: A – *Atylus collingi* (Gurjanova 1938); B – *Calliopius laeviusculus* (Kroyer 1838); C – *Crassicorophium bonellii* (Milne Edwards 1830).

late Pliocene–Pleistocene era. *Gammarus lacustris* invaded from the Sarmatian–Balkan speciation center during the Oligocene–Pliocene era (Tzvetkova 1975, Dedju 1980). The species reached the eastern borders of Asia, moving along the edge of the glaciers during the Pliocene–Early Pleistocene, through slow-flowing watercourses and oligo-β-mesotrophic lakes. Its invasion into northern Eurasia took place in the interglacial periods along the existing river systems, oriented, as now, from south to north. The current distribution of the species on Sakhalin is an indirect confirmation of the above: from the Tym–Poronay River basin, further south along the eastern shore and along the basins of relatively large rivers such as Nayba, Susuya, Lyutoga. At the same time, it is absent from the Schmidt Peninsula eastern side, southwestern Sakhalin, along the eastern and western shores of Aniva Bay, and the mid-east coast from Terpeniya Cape to Lunsky Bay (Fig. 3.10B). It is characteristic that *G. lacustris* occupies only high sources and small tributaries in mountain rivers and is absent in the lower reaches (author's data).

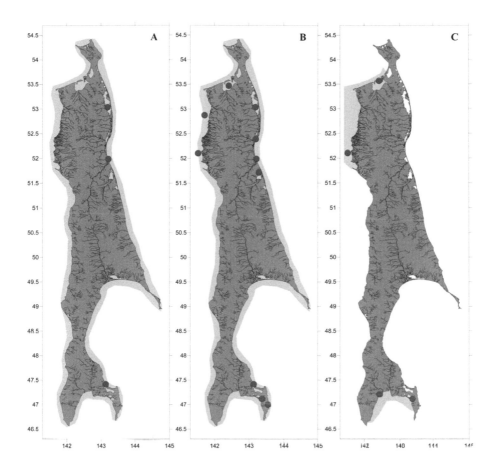

Figure 3.9 The island distribution range of freshwater and brackish water Amphipoda: A – *Monocorophium steinegeri* (Gurjanova 1951); B – *Dogielinotus moskvitini* (Derzhavin, 1930b); C – *Haustorioides gurjanovae* (Bousfield & Tzvetkova 1982).

Sternomoera rhyaca occupies the biotopes of *G. lacustris* on the Aniva Bay and southwestern Sakhalin coasts (Fig. 3.15A), and this species is the main competitor of *G. lacustris* in small streams.

Asellus levanidovorum belongs to a species group of the Bering center of speciation. *Asellus* was formed here in the Miocene–Early Pliocene (about 5 million years ago) (Levanidov 1980). This genus was widely distributed in northern Eurasia and North America in the Pliocene. Subsequent global Pleistocene glaciations, which covered most of northern Eurasia and North America (Imbrie and Imbrie 1979), destroyed most of the genus distribution range. This led to the isolation of the southwestern and southeastern Eurasian groups, where the specialized *A. aquaticus* and *A. hilgendorfii* developed. The latter prospered in the middle Far East, since the Pleistocene. *A. levanidovorum* was derived from *A. hilgendorfii* in the marginal territories, including Sakhalin. It occurs commonly in stagnant and weakly flowing acidified waters (Fig. 3.18A) (Labay 1999, 2005: as *A. hilgendorfi*; Sidorov 2005).

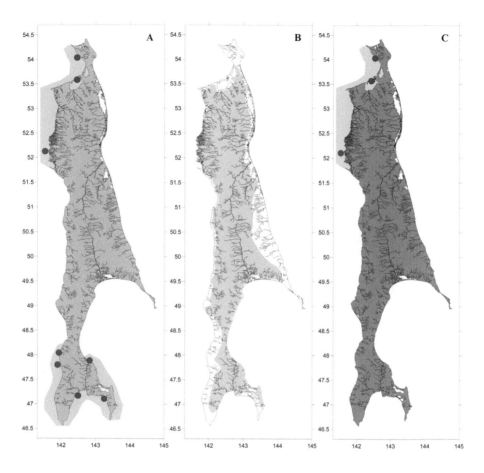

Figure 3.10   The island distribution range of freshwater and brackish water Amphipoda: A –
              Haustorioides magnus (Bousfield & Tzvetkova 1982); B – *Gammarus lacustris*
              G.O. Sars, 1864; C – *Gammarus setosus* (Dementieva 1931) and *Gammarus*
              wilkitzkii (Birula 1897).

The warmwater components of the Indo-West Pacific origin are freshwater
prawns *Palaemon paucidens* and *P. sinensis* (Holthuis 1950), which penetrated
into southern Sakhalin freshwaters possibly during the Pliocene–Pleistocene, when
repeated contacts with northern Japan took place. The invasion of these species
to the north was limited by watersheds. *P. paucidens*, as a relatively euryhaline
species, was able to penetrate into middle Sakhalin river systems only at the end
of the Pleistocene–early Holocene. This happened after the formation of modern
Poronaysky and Tymovsky river basins, most likely during the regression of the
late Wurm (about 15,000 years ago), when the sea level was 100–120 m lower than
the present, and extensive brackish water contacts appeared in the southeastern
Sakhalin upper shelf. This species probably penetrated northward to Aynskoe Lake
along the west coast and further at the same time, when the Sea of Japan salinity
was lower than at present (Pletnev et al. 1988). The late species penetration time on
Sakhalin Island is inferred by weak morphological differentiation between different

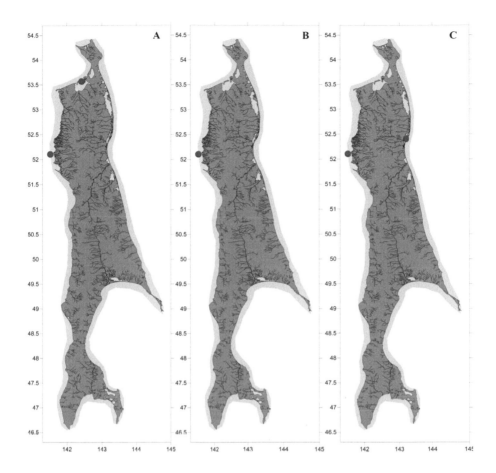

Figure 3.11   The island distribution range of freshwater and brackish water Amphipoda: A
– *Eohaustorius washingtonianus* (Thorsteinson 1941); B – *Eohaustorius eous*
(Gurjanova 1951); C – *Ischyrocerus commensalis* (Chevreux 1900).

habitats on Sakhalin, the Primorsky Region, and Japan (Labay and Barabanschikov
2009).

3. Relics of the vanished Pliocene–Pleistocene river basin in the modern Sea of
Japan include *Sternomoera moneronensis* from Moneron Island mountain streams
and *S. rhyaca* from freshwater of Sakhalin. Other species of this genus inhabit
the mountain streams of central and northern Japan (Barnard and Karaman 1991,
Kuribayashi et al. 1994, 1996). All representatives of *Sternomoera* are confined to
mountain and foothill streams, which are ancient and practically unchanged in the
historical time biotope (Barnard and Barnard 1983, Kuribayashi et al. 1994, 1996).
The modern geographical distribution suggests that the formation and expansion
of *Sternomoera* into mountain streams occurred during the Pliocene–Pleistocene
regression in the brackish-lake-river basin that existed at the future Sea of Japan.
The catadromous migration behavior in *S. rhyaca* indicates this (Kuribayashi et al.
2006). Later marine transgressions led to the geographical isolation of distribution

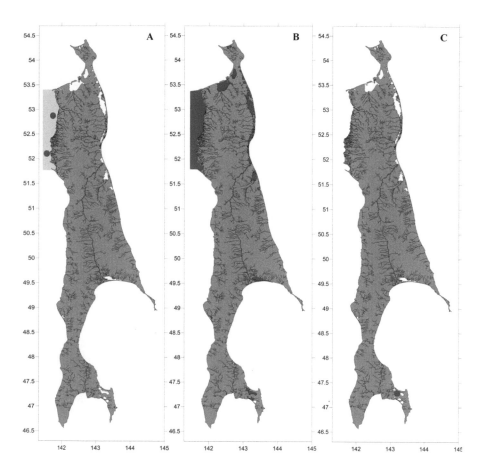

Figure 3.12   The island distribution range of freshwater and brackish water Amphipoda: A
– *Kamaka derzhavini* (Gurjanova 1951); B – *Kamaka kuthae* (Derzhavin 1923);
C – *Melita nitidaformis* (Labay 2003a).

range parts and the selection of the current species. This assumption is supported
by the few species on Japanese islands and Moneron Island, which had extensive
contact during the Pliocene–Pleistocene (Geologa 1970). Perhaps, the same transgressions have destroyed other species from this basin. Interestingly, the modern
distribution of *Sternomoera* contradicts the Melioransky–Lindberg hypothesis
(Melioranskiy 1936, Lindberg 1955) about the connection of paleo-Amur and the
paleo-Sea of Japan at the beginning of the Quaternary period at modern Lake Kizi.
However, it confirms the modern geological concept, since in that epoch there
existed an elevation that divided the paleo-Amur and the coast, and the freshwater-
continental sediments are not found here (Khudjakov et al. 1972).

4. Northern paleo-Amur group

  Penetration of the paleo-Amur species *Cambaroides schrenckii* and *Ichtyoxenus
amurensis* to north Sakhalin probably occurred in the Pleistocene–Holocene during a period of the late Wurm regression. The crayfish *C. schrenckii* is widely

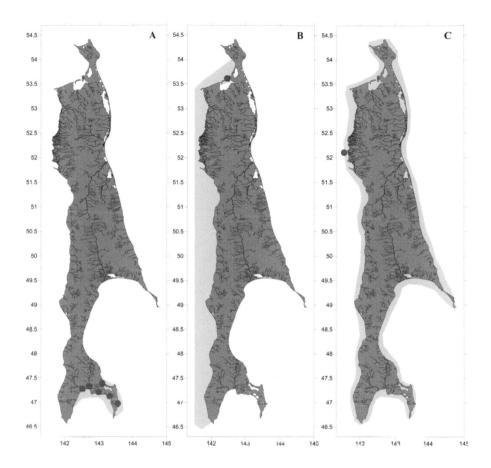

Figure 3.13   The island distribution range of freshwater and brackish water Amphipoda: A
– *Melita shimizui sakhalinensis* (Labay 2016); B – *Photis nataliae* (Bulycheva
1952); C – *Photis spasskii* (Gurjanova 1951).

distributed from the north to the south to Aleksandrovsk-Sakhalinsky and to
Piltun Bay (Fig. 3.21A). It is limited in its distribution only to the lower segment
of the Amur River basin and the northern part of Sakhalin, where it evolved in the
Pleistocene. Sakhalin Island was connected to the continent at this time in the area
of the modern Nevelskoy Strait, and modern northern Sakhalin was in the lower
segment and delta of the Amur River. The parasitic isopod *I. amurensis* is associated with three species of fish: *Leuciscus waleckii*, *Coregonus ussuriensis*, and
Cyprinus carpio haematopterus (e.g., Kussakin 1979). These fish species are relics
of the paleo-Amur (Lindberg 1955, 1972, Safronov and Nikiforov 1995), which
also indicates the affiliation of *I. amurensis* to a palaeo-Amur relic.
    5. The Northern Arctic brackish group is one of the youngest elements among the
       Sakhalin fauna, i.e., *Saduria entomon*, *Monoporeia affinis*, and *Gammarus wilkitz-
       kii*. These species have settled in the northern brackish waters up to the northern
       desalinated part of the Tatar Strait along the west coast and to Cape Terpeniya along
       the eastern coast (Figs. 3.10C, 3.15B, and 3.19C). These species are the marginal

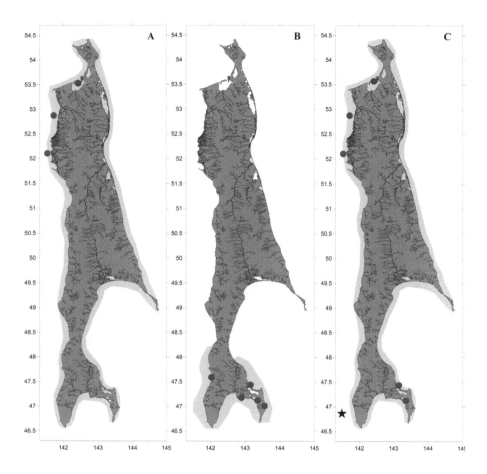

Figure 3.14   The island distribution range of freshwater and brackish water Amphipoda: A –
Grandifoxus longirostris (Gurjanova 1938); B – *Paramoera anivae* (Labay 2012);
C – *Pontogeneia rostrata* (Gurjanova 1938) and (black star symbol) *Sternomoera
moneronensis* (Labay 1997).

eastern group of the Siberian high-Arctic fauna, which was formed from the Kara
Sea fauna, a peripheral part of the Arctic basin during the ice ages under the influence of alternating desalination and salinization phases. Arctic relicts had invaded
into the North Pacific in the first half of the late Pleistocene during interglacial periods. The Sea of Okhotsk regression in the late Pleistocene in the late Wurm allowed
this group to invade through temporally formed coastal shallow water along
lagoon's shores of the Sea of Okhotsk. The wide connections of the Sea of Okhotsk
with the Pacific Ocean at the Kuril Straits prevented global desalination. Thus, all
these species are marine euryhaline. The land bridge at modern Nevelskoy Strait
limited Arctic invaders further south along the Sakhalin west coast (Bezverkhniy
et al. 2002). The sharp drop in depth of the Sea of Okhotsk southern trough, which
at that time adjoined directly to the coast in southern present-day Cape Terpeniya,
impeded this group along the eastern shore of the island. The postglacial transgression formed the final geographic pattern of distribution of the Far Eastern group of

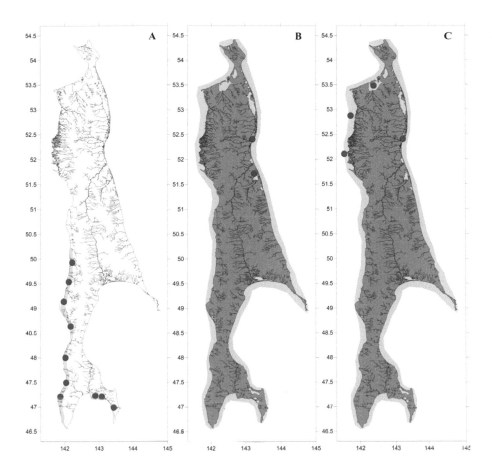

Figure 3.15   The island distribution range of freshwater and brackish water Amphipoda: A –
Sternomoera rhyaca (Kuribayashi et al. 1996); B – *Monoporeia affinis* (Lindström
1855); C – *Pontoporeia femorata* Krøyer, 1842.

Arctic relics. The group's invasion to the south was limited by the general warming
of the waters, which is presumed in that these species migrate at the shelf waters to
greater depths than in the Arctic region (Gurjanova 1936, 1951, author unpublished
data). Transition of certain species (*S. entomon, M. affinis*) to brackish waters in
the Far East is a recurring phenomenon, since they were originally freshwater and
brackish water inhabitants in Europe, and *S. entomon* reproduces exclusively in
desalinated waters (Garkalina 1982).

6. The relict brackish Middle Pleistocenous group (the Sea of Japan and the Sea of
Okhotsk) comprise the largest brackish malacostracan group. Many have preserved
ancient, primitive features (Zarenkov 1965, Tzvetkova 1975) and inhabit brackish
biotopes. Even species that have adapted to freshwater can live in oligohaline and
brackish water and are found only in water bodies directly or genetically associated with the sea: these are *Lamprops korroensis, Neomysis awatschensis, Kamaka
kuthae, Eogammarus kygi, Locustogammarus intermedius, Jesogammarus*

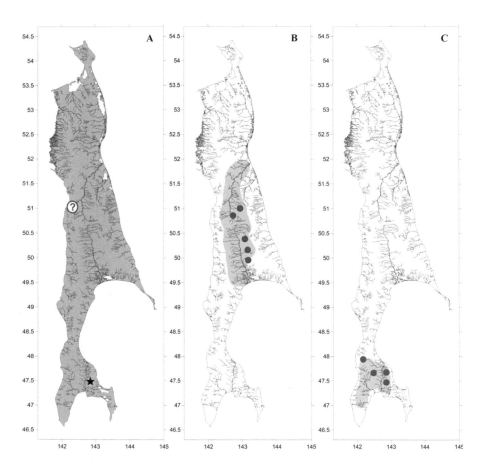

Figure 3.16 The island distribution range of subterranean freshwater Amphipoda: A – *Pseudocrangonyx bochaensis* (Derzhavin 1927) and (black star symbol) *Pseudocrangonyx birsteini* (Labay 1999); B – *Pseudocrangonyx relicta* (Labay 1999); C – *Pseudocrangonyx susunaensis* (Labay 1999).

annandalei, and *Gnorimosphaeroma kurilensis*. Other species are confined mainly to brackish water, although they are capable of tolerating wide salinity variations: i.e., *Diastylopsis dawsoni* f. *calmani*, *Diastylis lazarevi*, *Neomysis mirabilis*, *Gnorimosphaeroma ovatum*, *G. noblei*, *Eogammarus tiuschovi*, *E. barbatus*, *Locustogammarus locustoides*, *Dogielinotus moskvitini*, *Crangon amurensis*, and *Deiratonotus cristatum*. The Pleistocene brackish water species fell into the paleo-seas from the coastal shallow waters of the Pacific Ocean, although their early history is different. Mysids, the isopod *G. noblei*, the amphipods of the Anisogammaridae, *D. moskvitini*, and the shrimp *C. amurensis* evolved in the boreal paleo-Pacific, as supposed on their modern distribution. In contrast, the crabs *Eriocheir japonica* and *D. cristatum* have a tropical–subtropical origin from the Indo-West Pacific (Holthuis 1950, Starobogatov 1972). The northern boundary of the *E. japonica* distribution range formed in the late Wurm–early Holocene. The geographic boundaries that prevented the invasion of Arctic species to the south

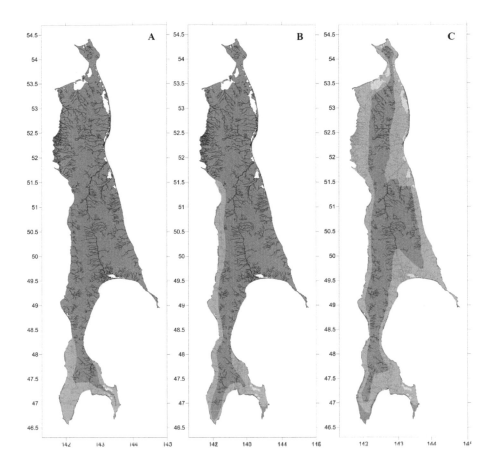

Figure 3.17   The island distribution range of supralittoral Amphipoda: A – *Platorchestia joi*
Stock & Biernbaum, 1994; B – *Platorchestia pachypus* (Derzhavin 1937); C –
Traskorchestia ochotensis (Brandt 1851).

(see above) also limited the *E. japonica* invasion into the north. This is confirmed
by the modern range of the species stretching north to the Amur estuary and to
Cape Terpeniya. The Pleistocene brackish water species entered newly formed seas,
replete with island archipelagos, and were subsequently separated by the advanc-
ing marine transgression. Here they formed a number of new local endemics in
geographical isolation: e.g., *G. kurilensis*, *L. intermedius*, and *J. annandalei*. Thus,
Sakhalin Island is the contact region of two groups of the Pleistocene brackish
water Malacostraca: (1) southern, mostly subtropical and temperate origins, most
common in the southern part of the island, and (2) boreal species, occupying the
entire coast.
7. Modern invaders from marine to fresh and brackish waters. The penetration of
   warmwater tropical–subtropical low-boreal species to the north can be accurately
   linked to one of the warm Holocene phases. There are two opposing views about the
   period of the climatic optimum on Sakhalin Island. Khotynskiy (1977) suggested
   the thermal maximum with the early Holocene (8,000–9,000 years ago) based on

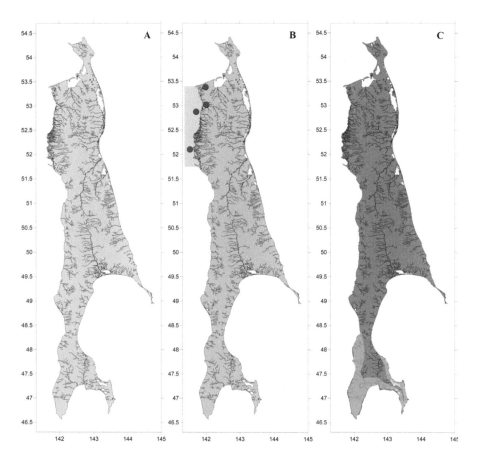

Figure 3.18   The island distribution range of freshwater and brackish water Isopoda: A –
Asellus (*Asellus*) *levanidovorum* Henry & Magniez, 1995; B – *Ichthyoxenus amurensis* (Gerstfeldt 1858); C – *Ligia cinerascens* Budde-Lund, 1885.

the spore-pollen spectra of the section on Cape Wandi (northwest Sakhalin). This time period appears to be significantly older than the Atlantic temperature optimum. Mikishin and Gvozdeva (1996) identified the warm phases of the Atlantic and subboreal periods of the Holocene (7,800 to 2,200 years ago) based on spore-pollen spectra from southeastern Sakhalin. Both viewpoints appear opposed, but they do not take local conditions into account. The warm Tsushima Current penetrated into the Sea of Japan 12,000–13,000 years ago and the Laperuz Strait formed 11.000–12,000 years ago, both based on the sea level change curve over the past 35,000 years (Geologicheskoe razvitie Yaponskich ostrovov 1968, Milliman and Emery 1968, Mikishin and Gvozdeva 1996). The depth needed for the penetration of the Soya Current into the Laperouse Strait occurred about 8,000 thousand years ago. The Soya Current conveys warm waters mostly up to the Tatar Strait of the Sea of Japan. The annual Soya Current flow through the Laperuz Strait is from 7.32 Sv (19,237 km³/year) to 10.8 Sv (28,382 km³/year) according to modern estimates (Saveliev et al. 2002, Chastikov et al. 2003). We calculated the amount

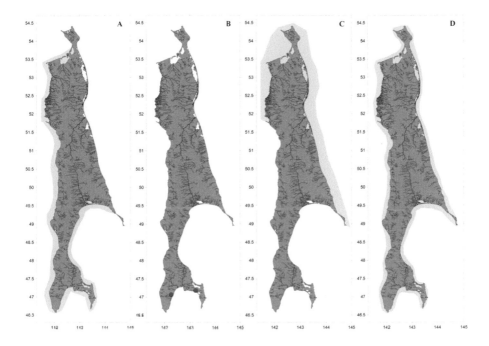

Figure 3.19   The island distribution range of freshwater and brackish water Isopoda: A –
Gnorimosphaeroma ovatum (Gurjanova 1933) & *Gnorimosphaeroma noblei*
(Menzies 1954); B – *Gnorimosphaeroma kurilensis* Kussakin, 1974; C – *Sadurla
entomon* (Linnaeus 1758); D – *Idotea gurjanovae* Kussakin, 1974 and *Idotea
ochotensis* Brandt, 1851.

of heat transfer to be 5.65×10^{16} to 8.33×10^{16} kcal, from the annual difference
of averaged water temperature of 2.86°C throughout the layers between the Soya
Current and the Laperouse Strait (Pishchalnik and Bobkov 2000). The mean annual
temperature difference of 3.25°C between the Soya Current and the northern part of
the Tatar Strait is even more impressive, and this could provide heat transfer from
south to north in the amount of 6.42×10^{16} to 9.47×10^{16} kcal. The transfer of the
thermophilic fauna to the northern part of the Sea of Japan probably took place
in parallel with the heat transfer. The later transgression isolated some refugia for
thermophilic fauna, the most significant of which is the Amur estuary, including the
Nevelsky Strait.

The previous group is close to the group of marine inhabitants who are able
to tolerate the significant desalination: i.e., *Neomysis cherniawskii*, *N. rayi*,
Idotea gurjanovae, *I. ochotensis*, and *Upogebia major*. These species are common in the Amur estuary and in the Nevelskoy Strait (Figs. 3.3B, 3.3C, 3.19D,
and 3.22A), which is an area of modern expansion of marine crustaceans into
brackish waters.

8. Ground supralittoral species form another group of recent freshwater introduction.
Traskorchestia ochotensis is common not only on the coast, but also occurs more
than 20 km inland (author's data). The species follows river beds to their sources
at an altitude of 300 m above sea level. *Platorchestia joi* is found along the shores

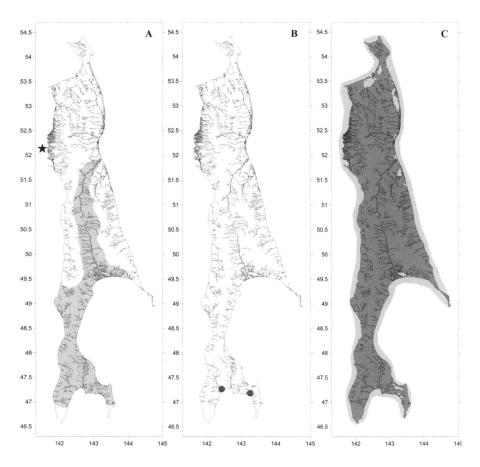

Figure 3.20   The island distribution range of freshwater and brackish water Decapoda: A –
Palaemon paucidens (de Haan 1841) and (black star symbol) *Palaemon modes-*
tus (Heller 1862); B – *Palaemonetes sinensis* (Solland 1911); C – *Crangon*
amurensis Bražnikov, 1907 and *Crangon dalli* Rathbun, 1902.

of lagoons and lakes associated with the sea. *Ligia cinerascens* is common on the
rocky supralittoral of the south and was discovered by the author in rocky streams
near the coast.

The amphipod *Melita nitidaformis* is endemic to the lagoon lake Tunaicha (Fig.
3.12C) in oligosaline waters. Its evolution is directly related to the development of
the lake. The species originated from the coastal-marine *Melita* after separation of
the lake from the sea by a sand spit and the formation of the Krasnoarmeiskaya channel, about 5,000 years ago (Mikishin et al. 1995).

There is some similarity between the distribution ranges of typically freshwater
taxa with initially brackish and/or marine taxa. For example, the range of paleo-
Amur species along latitudinal boundaries is similar to the distribution range of the
Arctic fauna. The distribution range of the subtropical low-boreal brackish water

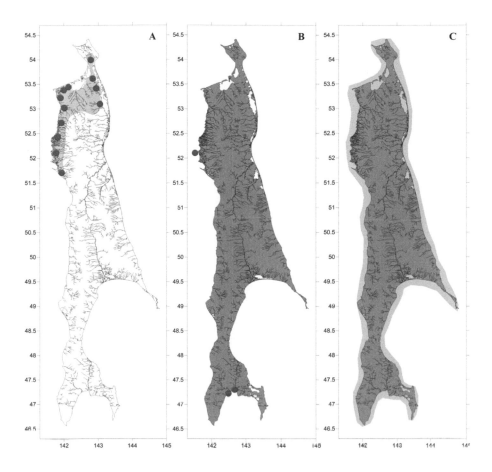

Figure 3.21   The island distribution range of freshwater and brackish water Decapoda: A –
Cambaroides schrenckii (Kessler 1874); B – *Deiratonotus cristatum* (de Man
1895); C – *Telmessus cheiragonus* (Tilesius 1812).

invaders largely overlaps with the Hokkaido fauna distribution range. However, they
cannot be combined.

Zoogeographic typification in marine areas is characterized by initially brackish water and marine species (Kafanov and Kudryashov 2000): the northern part
of the island belongs to the Lamut province of the high-boreal Aleutian subregion
of the Pacific boreal region, and the southern part belongs to the Ainu subregion of
the same (Kussakin 1979). The border between these two subregions lies in western Sakhalin through the Nevelskoy Strait and the Amur estuary, and in eastern
Sakhalin along the Terpeniya Peninsula. This boundary is quite conditional, since
the brackish water lagoons of Sakhalin are warm water relative to surrounding seas
and are refugia for warmwater species (Kafanov et al. 2003). The Amur estuary and
the Nevelskoy Strait are the warmest areas among Sakhalin coastal waters (TeraScan
Satellite Station of the Sakhalin Institute of Fisheries and Oceanography, kindly

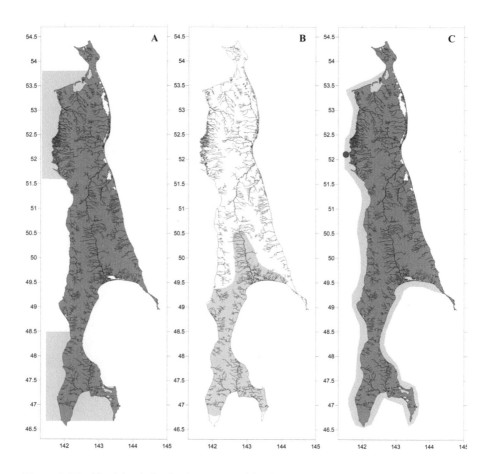

Figure 3.22   The island distribution range of freshwater and brackish water Decapoda: A
– *Upogebia major* (De Haan 1841); B – *Eriocheir japonica* (de Haan 1850); C –
Hemigrapsus takanoi (Asakura and Watanabe 2005).

provided by Zh. R. Tshay) and also provides similar refugia: i.e., *D. cristatum*, *E. japonica*, and *U. major*.

Zoogeographical zonation is independent for each marine brackish water and freshwater oligohaline fauna (Mironov 2013, Zhirkov 2017). Nine main distribution patterns are distinguished when comparing Sakhalin brackish water species. The largest number of species occur in all brackish water lagoons and estuaries. This distribution range comprises the majority of the species on the island: i.e., mysids *Archaeomysis grebnitzkii*, *N. awatschensis*, *N. mirabilis*; cumacea *L. korroensis*, *L. sarsi*; amphipods *Caprella cristibrachium*, *Anisogammarus* cf. *pugettensis*, *E. barbatus*, *E. tiuschovi*, *L. locustoides*, *Spasskogammarus spasskii*, *Atylus collingi*, *Calliopius laeviusculus*, *Crassicorophium bonellii*, *Monocorophium steinegeri*, *D. moskvitini*, *Eohaustorius washingtonianus*, *E. eous*, *Ischyrocerus commensalis*, *K. kuthae*, *Photis spasskii*, *Grandifoxus longirostris*, *Pontogeneia rostrata*, *M. affinis*,

Pontoporeia femorata, *T. ochotensis*, isopods *I. ochotensis*, *I. gurjanovae*; sand shrimp *Crangon dalli*, *C. amurensis*; and the crab *Telmessus cheiragonus* (Figs. 3.3A, 3.4B and C, 3.5A–C, 3.6A, 3.7B and C, 3.8, 3.9A and B, 3.11, 3.12B, 3.13C, 3.14A and C, 3.15B and C, 3.17C, 3.18D, 3.20C, and 3.21C).

The second pattern type is characterized by species exclusively in northern Sakhalin (along the west coast to the south, including the Nevelskoy Strait and the Amur estuary, and south to Terpeniya Cape along the east coast). This distribution range has the mysid *Neomysis czerniavskii*, the cumaceans *D. lazarevi*, *D. dawsoni*, and the isopod *S. entomon* (Figs. 3.3B, 3.4A, and 3.19C). The third distribution range type seems to be a special case of the second type, covering the coast of northeastern Sakhalin, with the amphipod *L. hirsutimanus* (Fig. 3.6C).

A separate type of distribution range exists for brackish water forms, which unites the Amur estuary and the Nevelsky Strait (the fourth type), i.e., *N. rayi*, *Kamaka derzhavini*, and *Palaemon modestus* (Figs. 3.3C, 3.12A, and 3.20A).

The fifth distribution range type is similar to the previous one and unites species ranges along the northwestern coast, i.e., *Gammarus setosus* and *Gammarus wilkitzkii* (Fig. 3.10C).

The sixth distribution range type unites southern Sakhalin with the Amur estuary and Sakhalin Bay along the western coast to Cape Terpeniya. The isopods *G. ovatum*, *G. noblei*, and the crab *Hemigrapsus takanoi* have this distribution (Figs. 3.19A and 3.22C).

The seventh type of distribution range is similar to the previous one from Cape Tyk to Cape Terpeniya. This distribution range includes the supralittoral amphipod *Platorchestia pachypus*, the freshwater prawn *P. paucidens*, and the crab *Eriocheir japonica* (Figs. 3.17B, 3.20A, and 3.22B).

Warmwater species occupy the brackish waters of southern Sakhalin to the Isthmus (the eighth distribution range type), i.e., amphipods *Melita shimizui sakhalinensis*, *Paramoera anivae*, *P. joi*, and isopod *L. cinerascens*.

The ninth distribution range type is a group of warmwater species with a broken distribution range that covers the brackish waters of southern Sakhalin and the Amur estuary refugium as well as the Nevelskoy Strait, i.e., amphipods *Haustorioides gurjanovae*, *H. magnus*, and decapods *D. cristatum* and *U. major*.

The distribution of brackish water Malacostraca is determined by many factors (see above), resulting in two mutually opposite processes. First, there is a decrease in species diversity from south to north. This phenomenon is due to the concentration in the south of Japanese–Hokkaido species and warmwater brackish water species. Second, there is a decrease in diversity from the northwestern to the south and east. This is due to the concentration of warmwater brackish species in the Amur estuary and the Nevelsky Strait coexisting with the rich cold water fauna of brackish water Malacostraca.

Five faunal districts are recognized by overlaying the above distribution ranges: 1 – the southern district occupies the southern part of the island to the Isthmus Poyasok; 2 – the southern intermediate district, its northern boundary near the top of Terpeniya Bay along the eastern shore (including the mouth of the Poronay River) and near the city of Aleksandrovsk-Sakhalinsky along the western shore; 3 – the northeastern district, including the eastern part of the Schmidt Peninsula and the mouth of the Tym River; 4 – the northwest district (the boundaries coincide with those of the island range of the sakhalin crayfish – see above); and 5 – the Amur estuary and the Nevelskoy Strait (Fig. 3.23).

The southern district and the united district of the Amur estuary with the Nevelskoy Strait are most unique. The southern district has the highest degree of

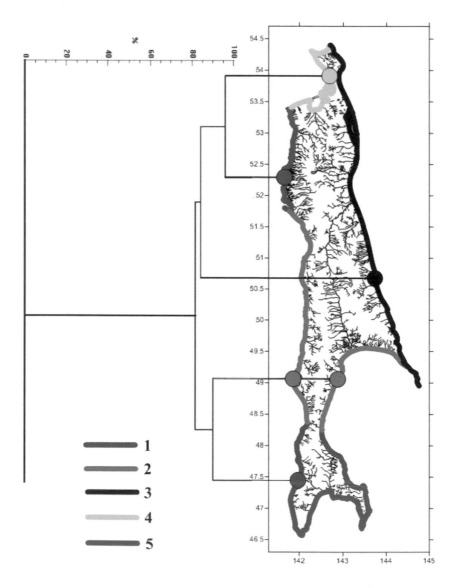

Figure 3.23 The dendrogram of the faunistic similarity of Malacostraca from brackish waters of Sakhalin Island (Sorensen coefficient). Faunistic districts: 1 – Southern, 2 – Southern Intermediate., 3 – Northeastern, 4 – Northwestern, 5 – Amur estuary and the Nevelskoy Strait.

endemism, at 9% (4 species of 46). The southern intermediate district (38 species) is derived from the southern district. This district is characterized by the absence of warmwater species: *H. gurjanovae, H. magnus, M. shimizui sakhalinensis, P. anivae, Pl. joi, L. cinerascens, D. cristatum,* and *U. major.*

The Amur estuary and the Nevelskoy Strait are characterized by a unique combination of 47 warmwater and Arctic-boreal species. The prawn *P. modestus* is noted only here. Several warmwater species cooccur in this area with the southern and southern intermediate faunistic districts, i.e., *G. ovatum, G. noblei, H. magnus, H. gurjanovae, U. major, E. japonica,* and *D. cristatum.* Some Arctic-boreal species are also found only in the Amur estuary and in the Nevelsky Strait, i.e., *N. rayi* and *K. derzhavini.* Other cold water species are common to the northeast district, i.e., *D. dawsoni* and *S. entomon.* There are species lacking here and are specific only for the northwestern district (45 species), which is similar to the Nevelsky Strait and the Amur estuary. The prawn *P. modestus,* the crab *D. cristatum,* and the amphipod *K. derzhavini* are all absent from this area.

The northeastern district has only 36 species, and lacks all warmwater species. The brackish water faunal districts comprise four clusters with at least 90% similarity (Fig. 3.23, Table 3.2). The first cluster unites the northwestern and the Amur estuary with the Nevelskoy Strait (96% similarity), with a mix of cold water Arctic highly boreal and warmwater subtropical–temperate fauna. The second cluster includes the southern and the southern intermediate (90% similarity).

This fits with the previous scheme of Guryanova (1964) and Kussakin (1979) for marine isopods. The association of the southern and the southern intermediate zoogeographical districts belonging to the Ainu/North-Japanese subregion of the Pacific boreal region is also supported here. The union of the three northern zoogeographical districts correlates with the Beringia subregion of the same region.

Two less dependent centers are observed in brackish water malacostracan distribution (Fig. 3.24): the south Sakhalin with a predominantly warmwater brackish water fauna and the Amur estuary with the Nevelskoy Strait with a mixed specific thermophilic-cryophilic brackish water fauna. The southern intermediate district is similar to the southern district. The northwestern and the northeastern districts cold

Table 3.2 Matrix of Faunal Similarity (%) of Faunistic Districts of Brackish Malacostraca of Sakhalin Island According to Sørensen Ratio (Notation as in Fig. 3.23), Below the Diagonal Is the Number of Common Species between Compared Areas

Faunistic districts	1	2	3	4	5
1	–	90	76	84	82
2	38	–	84	84	80
3	31	31	–	86	82
4	38	35	35	–	96
5	38	34	34	44	–

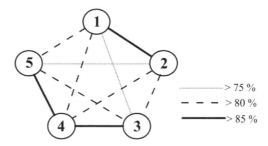

Figure 3.24 A graph of similarities of the faunistic districts of brackish water Malacostraca of Sakhalin Island (notation as in Fig. 3.23).

water fauna are similar to the Amur estuary and the Nevelskoy Strait. A decrease in diversity is observed when moving away from those centers. The boundary between the zoogeographic parts (the Beringian and the North-Japanese subregions of the Pacific boreal region) passes along Cape Terpeniya on the eastern shore and at the Nevelsky Strait on the western shore of Sakhalin Island. This boundary was established by late Pleistocene events (regression of the sea and the total cooling).

Similarly, the comparison of fresh and oligohaline Malacostraca demonstrates ten distribution range types. Part of the species are distributed throughout the island. The isopod *A. levanidovorum* and the amphipod *G. lacustris* are among the freshwater species. A similar type of distribution ranges, covering the entire coast, is observed among brackish water species living in water bodies associated with the sea. This type of area is characterized by *N. awatschensis, L. korroensis, E. kygi, E. barbatus, T. ochotensis, K. kuthae,* and *C. amurensis.*

The parasitic isopod *I. amurensis,* amphipod *L. intermedius,* and crayfish *C. schrenckii* live in freshwaters of northwestern Sakhalin (the second type of distribution range). The boundaries coincide with the distribution of *C. schrenckii.* Probably, *P. modestus* can be attributed to this distribution range type.

The third distribution range type covers the northeastern. Such distribution is noted for the side swimmer *L. hirsutimanus.*

The fourth distribution range type includes the entire south to the Amur estuary along the west coast and to Cape Terpeniya along the east coast. The isopod *G. ovatum* has this type of distribution.

The freshwater prawn *P. paucidens* lives in watercourses, freshwater and brackish lakes of southern Sakhalin, as well as in the Poronay and Tym Rivers (the fifth distribution range type). Although only one species has such a distribution among the freshwater Malacostraca, there are several species with a similar range in other aquatic invertebrates such as the bivalve *Kurilinaia.*

The sixth distribution range type is similar to the two previous ones and occupies the southern part to the Isthmus Poyasok along the east coast and to Cape Tyk and to the city of Aleksandrovsk-Sakhalinskiy along the west coast. This distribution range type is characterized by *S. rhyaca* and *P. pachypus.*

The seventh distribution range type has the crab *E. japonica* in southern Sakhalin and in the Poronay River.

The eighth distribution range type includes southern Sakhalin to the Isthmus Poyasok. It is characterized by *L. cinerascens* and *Pl. joi*. The ninth distribution range type includes the distribution of the isopod *G. kurilensis*, amphipod *J. annandalei*, and prawn *Palaemonetes sinensis*, which are quite close to the previous distribution ranges. All are found exclusively in the Aniva Bay basin. Several malacostracans are endemic to particular water bodies or aquatic systems, e.g., *Sternomoera moneronensis*, *M. nitidaformis*, *P. birsteini*, *P. susunaensis*, and *P. relicta*. The first four species are endemic to southern Sakhalin, and the latter species characterizes the Tym–Poronay water system (the tenth type of distribution range).

The freshwater and brackish taxa decrease in diversity from south to north, due to the number of Japanese–Hokkaido species in the south. Inversely, a diversity decrease from the south and to the east is also observed. This is explained by the presence of paleo-Amur freshwater species in the water bodies of northwestern Sakhalin. The central areas are the least diverse, especially in the Tym–Poronay River basin (here it is conditionally appropriate to combine these river systems into one, since in the area of Palevsky heights, there are numerous direct contacts between these rivers). However, these areas are characterized by endemic *Pseudocrangonyx* species.

Six faunal districts were defined by overlaying the above distribution ranges: 1 – the southern district occupies the southern part of the island to the Isthmus Poyasok; 2 – the southwestern intermediate faunistic district, the northern boundary of which passes to the north of the city of Aleksandrovsk-Sakhalinsky along the western shore; 3 – the southeast intermediate district, the northern boundary of which passes near the top of the Terpeniya Bay (including the mouth of the Poronay River) along the eastern shore; the boundary between the southwestern intermediate and the southeast intermediate districts is the Kamyshoviy ridge watershed; 4 – the Tym–Poronay district, the boundaries, which coincide with the basins of the rivers of the same name, including Nevskoe Lake; 5 – the northwestern region, the boundaries of the district coincide with the borders of the island distribution range of *C. schrenckii* (see above); 6 – the northeastern district, including the eastern part of the Schmidt Peninsula and the mouth of the Tym River (Fig. 3.25).

The southern and the northwestern districts are unique in terms of faunal community and are characterized by the greatest degree of endemism (39%, 9 of 23 species). The isopod *G. kurilensis*, also found in the south of the Kuril Islands, can be considered a conditional endemic for the district. The northwestern district (15 species) is distinguished by paleo-Amur species: *I. amurensis* and *C. schrenckii*. *Locustagammarus intermedius*, also found in the lower reaches of the Somon River (Chikhachev Bay), can be classified as a conditional endemic of the district.

The southeast (12 species) and the Tym–Poronay (13 species) districts are the most impoverished. The southeastern district is a depleted derivative of the southern faunistic district. The presence of warmwater species as represented by the prawn *P. paucidens* (not noted at these latitudes in other districts), the crab *E. japonica* (feeding migrant up to 100 km from the mouth of the Poronay River), and an endemic species of the genus *Pseudocrangonyx* is a sufficient basis for isolation of the combined river basin of Tym–Poronay to a separate district.

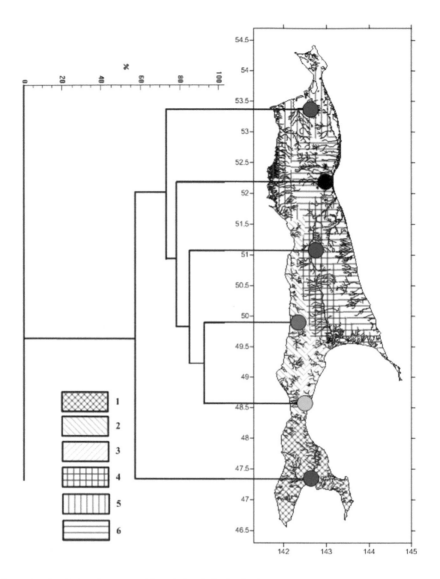

Figure 3.25 A dendrogram of faunistic similarity for Malacostraca from freshwater and oli-
gosaline water of Sakhalin Island (Sorensen coefficient). Faunistic districts: 1
– Southern, 2 – Southwest Intermediate, 3 – Southeast Intermediate, 4 – Tym–
Poronay, 5 – Northwestern, 6 – Northeastern.

The southwestern faunistic district (14 species) is also a depleted derivative from
the southern district, but, unlike the southeastern district, several warmwater species
are noted in it, i.e., *S. rhyaca* and *P. pachypus*.

The northeastern region (10 species) is distinguished by the absence of both
warmwater southern species of the Hokkaido fauna and paleo-Amur relicts. It is also

distinguished by the presence of the amphipod *L. hirsutimanus*, which is specific to this district.

Although the traditional biogeographic zoning scheme of the Sakhalin freshwater fish fauna (Safronov and Nikiforov 1995, Nikiforov and Safronov 1996) is closest to our assessment, ours differs, as it does not unify the Western and Eastern districts.

The scenario presented by Bogatov et al. (2007) is not confirmed for Malacostraca. This scheme is too fractional for the southern part of Sakhalin Island, and this fragmentation was not supported in the present analysis.

The faunistic districts form three clusters at a similarity level >75% (Fig. 3.25; Table 3.3). The first cluster includes the northwestern and northeastern districts. The second cluster includes the only southern district. The other districts are combined into a single cluster, which can be characterized as intermediate districts. I refer the first cluster to the Orelian province of the Amur subregion of the Sino-Indian region, and the second cluster (southern Sakhalin) belongs to the Aniva province of the Japanese subdomain of the same region (initial selection made by Kruglov and Starobogatov 1993 and Starobogatov 1996). This dissection by province coincides with the opinion of many authors, although the province names are not always the same (Chereshnev 1998, Prozorova 2001). The border between the zoogeographic units is blurred. Between them, there are transition zones, which include the combined basin of Tym and Poronay Rivers.

In the similarity matrix, several districts are clearly identified (Tables 3.3, and 3.4). The southern district is associated with the southwestern intermediate and southeastern intermediate faunistic districts. The northwestern district is connected with the northeastern intermediate and Tym–Poronay intermediate districts.

We also observed two slightly dependent diversity centers: the southern Sakhalin with predominantly the Hokkaido–Sakhalin freshwater fauna, and the northwestern Sakhalin with the specific paleo-Amur fauna (Fig. 3.26). The southwestern, southeastern, and Tym–Poronay districts are weakly connected with the selected centers, demonstrating their transitional significance. A decrease in diversity with distance from those centers was observed. In the zoogeographic section boundaries,

Table 3.3 **Matrix of Faunal Similarity (%) of Faunistic Districts of Fresh and Oligosaline Waters Malacostraca of Sakhalin Island According to Sørensen Ratio (Notation as in Fig. 3.25), Below the Diagonal Is the Number of Common Species between the Compared Areas**

Faunistic districts	1	2	3	4	5	6
1	–	76	69	61	54	55
2	14	–	92	81	71	75
3	12	12	–	88	77	82
4	11	11	11	–	67	78
5	10	10	10	9	–	75
6	9	9	9	9	9	–

Table 3.4 **Matrix of Differences (S_d – S^*) of Faunistic Districts of Freshwater and Oligohaline Malacostraca of Sakhalin Island (Notation as in Fig. 3.25)**

Faunistic districts	1	2	3	4	5	6
1		–	21	19	-4	-8	-9
2			–	2	-2	–	-4
3				–	-1	-1	-3
4					–	18	1
5						–	1
6							–

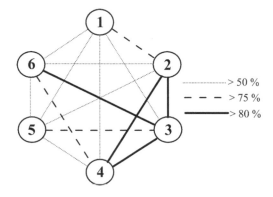

Figure 3.26 A graph of similarities between faunistic districts of freshwater and oligosaline water Malacostraca from Sakhalin Island (notation as in Fig. 3.25).

the Orelian and Aniva provinces run from Cape Terpeniya, along the united Tym–Poronay River basin and the Nevelsky Strait. This border was established by late Pleistocene events (regression of the sea and the general cooling). The Schmidt Line, which is significant for defining the terrestrial fauna and flora (Bogatov et al. 2007), is not clearly applied to fresh and oligohaline malacostracan fauna.

3.5 CONCLUSION

The freshwater and brackish water Malacostraca faunas of Sakhalin Island contains 72 species of different origins. This was not only due to climatic differences in the northern and southern parts of the island, but also due to a complex history of relief formation. This brought a great variety of geographical Malacostraca distribution ranges, resulting in nine main types of the brackish water Malacostraca and ten types of freshwater Malacostraca. Similarly, five faunistic districts for the brackish water fauna and six districts for the freshwater fauna are described by matching of distribution ranges. In both cases, two centers of species diversity are distinguished: the northern and the southern. The association of the southern and the southern intermediate zoogeographical brackish districts belong to the Ainu/North-Japanese subregion of the Pacific boreal region. The union of the three northern zoogeographical districts correlates to the Bering subregion of the same region. The northern Sakhalin freshwater fauna belongs to the Orelian province of the Amur subregion of the Sino-Indian region, the southern Sakhalin freshwater fauna belongs to the Aniva province of the Japanese subdomain of the same region. The southwestern, the southeastern, and Tym–Poronay districts are weakly connected with the selected centers, which identify their transitional significance.

ACKNOWLEDGMENT

The author sincerely thanks Zh. R. Tshay of SakhNIRO who kindly assisted with the research and the manuscript.

REFERENCES

Atlas Sakhaliskoy oblasti (Atlas of the Sakhalin Region). 1967. Moscow, Glavnoe upravlenie geodezii i kartographii pri Sovete ministrov SSSR. (in Russian).

Barnard, J.L. and Barnard, C.M. 1983. *Freshwater Amphipoda of the World*. Vol. 1, 2. Mt. Vernon, VA, Hayfield Associates.

Barnard, J.L. and Karaman, G.S. 1991. The families and genera of marine gammaridean Amphipoda (except marine gammaroids). Part 1. *Records of the Australian Museum Supplement* 13:1 417.

Berg, L.S. 1949. *Ryby presnykh vod SSSR i sopredel'nykh stran. Chast 2.* (Freshwater fishes of the USSR and of neighboring countries. Part 2). Academy of Science USSR: Series: Oprediliteli po faune SSSR (Keys to the fauna of the USSR). Moscow–Leningrad, Zoological Institute, Academy of Science USSR. 29:467–926. (in Russian).

Bezverkhniy, V.L., Pletnev S.P. and Nabiullin, A.A. 2002. *Outline of Geological Structure and Development of the Kuril Island System and Adjacent Regions.* Flora and Fauna of Kuril Island (Materials of International Kuril Island Project). Vladivostok, Dalnauka:9–22. (in Russian).

Birstein, J.A. 1939. Materialy po geographicheskomu rasprostraneniyu vodnykh zhivotnykh v SSSR. 5. O nekotorykh osobennostykh geographicheskogo rasprostraneniya presnovodnykh Malacostraca Dalnego Vostoka. (Materials on the geographical distribution of aquatic animals in the USSR. 5. On some features of the geographical distribution of freshwater Malacostraca of Far East). *Zoological Journal* 18:54–69. (in Russian).

Birstein, J.A. 1940. *Vysshie raki (Malacostraca). Zhizn presnykh vod v SSSR* (Freshwater life in the USSR). 1. Moscow–Leningrad, Academy of Science USSR: 403–430. (in Russian).

Birstein, J.A. 1951. *Presnovodnye osliki (Asellota). Phauna SSSR. Rakoobraznye.* (Freshwater lice (Asellota). Fauna of the USSR. Crustacea. 7. Moscow–Leningrad, Academy of Science USSR. (in Russian).

Birstein, J.A. 1955. Genus *Pseudocrangonyx* Akatsuka et Komai (Crustacea, Amphipoda) in USSR. In *Bulletin of the Moscow Society of Naturalists (New Series.).* Moskow, Department of Biology, Moscow State University. 60:77–84. (in Russian).

Birstein, J.A. and Vinogradov, L.G. 1934. Presnovodnye Decapoda SSSR I ikh geographiches-koe rasprostranenie (predvaritel'noe soobshchenie) (Die Süsswasserdecapoden der USSR und ihre geographische Verbreitung (Vorläufige Mitteilung)). Zoologicheskiy Zhurnal (*Zoological Journal, Moscow*) 13:39–70.

Bogatov, V.V., Yu Storozhenko, S. Yu. Barkalov, V. et al. 2007. Biogeography of Sakhalin Island by the example of the distribution of terrestrial and freshwater biota. In *Theoretical and Practical Problems of Studying of Invertebrates Associations: In Memory*, ed, Y.I. Starobogatov. Moscow: KMK Scientific Press Ltd. 193–224. (in Russsian).

Brazhnikov, V. 1907. *Materialy po faune russkikh vostochnykh morey, sobrannye shkhunoy "Storozh" v 1899–1902 gg.* (Materials on the fauna of the Russian eastern seas, col-lected by the schooner "Storozh" 1899–1902). Zapiski Imperatorskoy Akademii Nauk. VII seria po phiziko-matematicheskomu otdeleniyu. (De L'Academie Imperiale des Sciences de St.-Petersbourg. 8 sereie. Class Physico-mathermatique). 20 (6):1–115. (in Russian).

Brodsky, S.Y. 1974. Crayfishes (Crustacea, Astacidae) of the Soviet Union. Communication III. Distribution of crayfishes of *Astacus, Cambaroides* and *Austropotamobius* genera. *Vestnik Zoologii* 6:43–49. (in Russian).

Bulycheva, A.I. 1957. *Morskie blokhi SSSR i sopredelnykh vod (Amphipoda, Talitroidea).* (Sea fleas of the USSR and adjacent waters (Amphipoda, Talitroidea)). Opredeliteli po faune SSSR (Keys to the fauna of the USSR). Moscow–Leningrad: Academy of Science USSR. (in Russian).

Chastikov, V., Kantakov, G. Shevchenko, G. and Sedaeva, O. 2003. Long-term direct mea-surements of currents on the southwestern shelf of Sakhalin island. In Proceedings of the 18th International Symposium on Okhotsk Sea and Sea Ice. Mombetsu, Hokkaido, Japan:265–70.

Chereshnev, I.A. 1998. *Biogeography of Freshwater Fish Fauna in the Russian Far East.* Vladivostok, Dal'nauka. (in Russian).

Chernjavskiy, V. 1882. *Monografia mizid preimushchestvenno Rossiyskoy imperii.* (A mono-graph of mysids, mainly of the Russian Empire) 1. In Proceedings of the St. Petersburg Society of Naturalists, St. Petersburg. (in Russian).

Chernjavskiy, V. 1883. *Monografia mizid preimushchestvenno Rossiyskoy imperii.* (A mono-graph of mysids, mainly of the Russian Empire) 3. In Proceedings of the St. Petersburg Society of Naturalists, St. Petersburg. (in Russian).

Dedju, I.I. 1980. *Amphipody presnych i solonovatych vod yugo-zapada SSSR* (Fresh and brackish amphipods of the south-west of the USSR). Kishinev: Stiynitsa. (in Russian).

Derzhavin, A.N. 1927. Novye formy presnovodnych gammarid Ussuriyskogo kraya (New forms of freshwater gammarids of the Ussuri region). *Russkiy Gidrobiologicheskiy Zhurnal (Russian Hydrobiological Journal)* 6(8–10):156–79. (in Russian).

Derzhavin, A.N. 1930a. Articheskie elementy v faune Peracarida Japonskogo morja (Arctic elements in the Peracarida fauna of the Sea of Japan). *Hydrobiological Journal* 8:326–9. (in Russian).

Derzhavin, A.N. 1930b. Presnovodnye Malacostraca Dalnego Vostoka SSSR (Freshwater Malacostraca from Far East of the USSR). *Hydrobiological Journal* 9:1–8. (in Russian).

Duran, B.C. and Odell, P.L. 1974. *Cluster Analysis: A Survey.* London, Springer.

Garkalina, N.N. 1982. Osobennosti raspredeleniya i biologia morskogo tarakana *Mesidotea entomon* v Amurskom limane (Distributional features and biology of the *Mesidotea entomon* marine cockroach in the Amur estuary). In Biologia sel'fovykh zon Mirovogo okeana: Tezisy dokladov Vtoroy vsesousnoy konferentzii po morskoy biologii (Vladivostok, September 1982). (Biology of the shelf zones of the World Ocean: Abstracts of the Second All-Union Conference on Marine Biology (Vladivostok, September 1982)). Vladivostok. 1:13–14. (in Russian).

Geographia i monitoring bioraznoobrazia (Geography and Monitoring of Biodiversity). 2002. Moscow, Izdatel'stvo nauchnogo i metodicheskogo centra (Publisher of Scientific and Methodological Center). (in Russian).

Geographia Sakhalinskoy oblasti (Geography of Sakhalin Region) / editor Litenko N.L.. 1992. Yuzhno-Sakhalinsk. (in Russian).

Geologia SSSR. 1970. Tom 33. *Ostrov Sakhalin. Geologicheskoe opisanie.* (Geology of the USSR. Vol. 33. Sakhalin Island. Geological description). Moscow, Nedra. (in Russian).

Geologicheskoe razvitie Yaponskich ostrovov (Geological development of Japanese islands). 1968. Moscow: Mir. (in Russian).

Gurjanova, E.F. 1936. *Ravnonogie raki dalnevostochnych morey. Fauna SSSR. Rakoobraznye.* (Isopods of Far Eastern seas. The fauna of USSR. Crustacea). 7. St. Petersburg-Lenigrard, Academy of Science USSR. (in Russian).

Gurjanova, E.F. 1951. *Bokoplavy morei SSSR i sopredel'nykh vod (Amphipoda–Gammaridea)* (Side-swimmers of the seas of the USSR and adjacent waters (Amphipoda–Gammaridea)). Academy of Science USSR. Opredeliteli po Faune SSSR (Keys to the fauna of the USSR). 41. St. Petersburg-Leningrard, Academy of Science USSR. (in Russian).

Gurjanova, E.F. 1962. *Bokoplavy severnoy chasti Tikhogo okeana (Amphipoda— Gammaridea). Chast 1.* (Side-swimmers (Amphipoda–Gammaridea) of the northern part of Pacific. Part 1). Opredeliteli po Faune SSSR (Keys to the fauna of the USSR). 74. St. Petersburg-Lenigrard, Academy of Science USSR. (in Russian).

Gurjanova, E.F. 1964. *Zoogeographicheskoe rayonirovanie Mirovogo okeana. Donnaya fauna materikovoy otmeli* (Zoogeographical zoning of the World Ocean. Bottom fauna of the continental shoals). Phiziko-geographicheskiy atlas mira (Physico geographical atlas of the world). Moscow: Academy of Science and GUGK GGK USSR. Map 63B. (in Russian).

Holsinger, J.R. 1989. Allocrangonyctidae and Pseudocrangonyctidae, two new families of Holarctic subterranean amphipod crustaceans (Gammaridea), with comments on their phylogenetic and zoogeographic relationships. *Proceeding of the Biological Society of Washington* 102:947–59.

Holthuis, L.B. 1950. The Decapoda of the Siboga Expedition. Pt. X. The Palaeamonidae collected by the Siboga and Snellius expeditions, with remarks on other species. Part I. Subfamily Palaemoninae. *Siboga Expeditie, Leiden* 39(A9):1–268.

Imbrie, J. and Imbrie, K.P. 1979. *Ice Ages: Solving the Mystery.* London: Macmillan.

Ivankov, V.N., Andreeva, V.V.Tyapkina, M.V.Rukhlov F.N. and Fadeeva, N.P. 1999. *Biology and Feeding Base of Juvenile Pacific Salmons during the Early Period of Ocean Life.* Vladivostok, Far Eastern State University Press. (in Russian).

Ivleva, I.V., Labay, V.S., Raschepkina, E.V., Shtyrtz, L.A. and Shulga, O.P. 1999. Macrozoobenthos communities of the Sladkoe Lake. *Transactions of Sakhalin Scientific Research Institute of Fisheries and Oceanography* 2:95–99. (in Russian).

Kafanov, A.I. 1984. Bentos lagun severo-vostochnogo Sakhalina (Бентос из лагун Северо-Востока Сахалина). Itogi issledivaniy po voprosam ratzionalnogo ispolzovaniya i okhrany biologitcheskikh resursov Sakhalina i Kurilskikh ostrovov: Tezisy docladov 2-y nauchno-prakticheskoy konferetzii (Results of studies on the rational use and protection of the biological resources of Sakhalin and Kuril Islands: Abstracts of the 2nd Scientific Practical Conference). Yuzhno-Sakhalinsk, Geographicheskoe obshchestvo SSSR:147–50. (in Russian).

Kafanov, A.I. and Kudryashov, V.A. 2000. *Marine Biogeography: A Text-Book.* Moscow: Nauka. (in Russian).

Kafanov, A.I. and Pecheneva, N.V. 2002. Sostav i proiskhozhdenie bioty lagun severo-vostochnogo Sakhalina (Composition and origin of the biota lagoons of northeastern Sakhalin). *Izvestia TINRO* 130:297–328. (in Russian).

Kafanov, A.I., Labay, V.S. and Pecheneva, N.A. 2003. *Biota and Bottom Communities of the Northeast Sakhalin.* Yuzhno-Sakhalinsk: Sakhalin Institute of Fishery and Oceanology. (in Russian).

Karaman, G.S. 1991. New survey of described and cited freshwater *Gammarus* species (Fam. Gammaridae) from Soviet Union with redescription of two taxa. *Poljopriverda i Sumarstvo* 27:37–73.

Kawai, T., Labay, V.S. and Filipova, L. 2013. Taxonomic re-examination of *Cambaroides* (Decapoda: Cambaridae) with a redescription of *C. schrenckii* from Sakhalin Island Russia and phylogenetic discussion of the Asian cambarids based on morphological characteristics. *Journal of Crustacean Biology* 33:702–17.

Kawai, T., Min, G.-S., Baravanshchikov, E., Labay, V.S. and Ko, H.S. 2016. Asia. In: *Freshwater Crayfish: A Global Overview*, ed. T. Kawai, Z. Faulkes, and G. Scholtz. Boca Raton, FL, CRC Press, Taylor & Francis Group:313–68.

Khlebovich, V.V. 1974. *Kriticheskaya solenost' bioloicheskikh protsessov* (Critical Salinity of Biological Processes). Leningrad, Nauka. (in Russian).

Khlebovich, V.V. 2015. Applied aspects of the concept of critical salinity. *Biology Bulletin Reviews* 5:562–67.

Khotinskiy, N.A. 1977. *Golocen Severnoy Evrasii (Holocene of Northern Eurasia).* Moscow: Nauka. (in Russian).

Khudjakov, G.I., Denisov, E.P. Korotkiy, A.M. et al. 1972. *Yug Dal'nego Vostoka* (South of Far East). Moscow, Nauka. (in Russian).

Kluchareva, O.A., Koreneva, T.A., Sokol'skaya, N.L. and Starobogatov, Y.I. 1964. Donnye bespozvonochnye ozer Yuzhnogo Sakhalina (Bottom invertebrates of lakes in South Sakhalin). In *Ozera yuzhnogo Sakhalina i ich ichtyofauna* (Lakes of southern Sakhalin and their ichthyofauna), ed. O.A. Kluchareva. Moscow, Moscow State University:169–89. (in Russian).

Korotkiy, A.M., Grebennikova, T.A., Pushkar', V.S., Razzhigaeva, N.G., Volkov, V.G., Ganzey, L.A., Mokhova, L.M., Bazarova V.B. and Makarova. T.R. 1997. Klimaticheskie smeny na territorii yuga Dalnego Vostoka v pozdnem Pleistocene–Golocene (Climate changes in the territory of the south of the Far East in the Late Pleistocene–Holocene). *Vestnik of the Far East Branch of the Russian Academy of Sciences* 3:121–43. (in Russian).

Kruglov, N.D. and Starobogatov, Ya., I. 1993. Guide to recent molluscs of northern Eurasia. 3. Annotated and illustrated catalogue of species of the family Lymnaeidae (Gastropoda Pulmonata Lymnaeiformes) of Palaearctic and adjacent river drainage areas. Part 1. *Ruthenica* 3:65–92.

Kuribayashi, K., Ishimaru, S. and Mawatari, S.F. 1994. Redescription of *Sternomoera yezoensis* (Ueno, 1933) (Amphipoda: Eusiridae) with reference to sexual dimorphism on pleopod 2. *Crustacean Research* 23:79–88.

Kuribayashi, K., Katakura, H., Kyono, M., Dick, M.H. and Mawatari, S.F. 2006. Round-trip catadromous migration in a Japanese amphipod, *Sternomoera rhyaca* (Gammaridea: Eusiridae). *Zoological Science* 23:763–74.

Kuribayashi, K., Mawatari, S.F. and Ishimaru, S. 1996. Taxonomic study on the genus *Sternomoera* (Crustacea: Amphipoda), with redefinition of *S. japonica* (Tattersall, 1922) and description of a new species from Japan. *Journal of Natural History* 30:1215–37.

Kussakin, O.G. 1974. Fauna i ecologia ravnonogikh rakoobraznykh (Crustacea, Isopoda) litorali Kurilskikh ostrovov (Fauna and ecology of isopod crustaceans (Crustacea, Isopoda) in the littoral of the Kuril Islands). In *Rastitelnyi i zhivotnyi mir litorali Kurilskikh ostrovov* (Flora and fauna of the Kuril Islands littoral). Novosibirsk, Nauka:27–75. (in Russian).

Kussakin, O.G. 1979. *Morskie i solonovatovodnye ravnonogie rakoobraznye (Isopoda) kholodnykh i umernnykh vod Severnogo polushariya. Podotryad Flabellifera.* (Marine and brackish water isopod crustaceans (Isopoda) of cold and temperate waters of the Northern Hemisphere. Suborder Flabellifera). Opredeliteli po Faune SSSR (Keys to the fauna of the USSR). St. Petersburg-Leningrad, Nauka:122. (in Russian).

Labay, V.S. 1996a. K faune vysshikh rakov presnykh poverkhnostnykh vod severo-zapadnogo Sakhalina (To the fauna of Malacostraca of fresh surface water of northwestern Sakhalin). *Transactions of Sakhalin Scientific Research Institute of Fisheries and Oceanography*. 1:65–76. (in Russian).

Labay, V.S. 1996b. Soobshchestva makrobentosa "goljanovykh" ozer severo-zapadnogo Sakhalina (Communities of macrobenthos of "minnow" lakes of northwest Sakhalin). Tezisy nauchno-prakticheskoy konferencii molodykh issledovateley "Nauka segodnya: Problem i perspektivy". Estestvoznanie. (Theses of the scientific-practical conference of young researchers "Science today: Problems and prospects". Natural science). Yuzhno-Sakhalinsk, Sakhalin State University:11–12. (in Russian).

Labay, V.S. 1997. *Sternomoera moneronensis* sp. n. (Amphipoda, Eusiridae) from freshwater of Moneron Island. *Russian Journal of Zoology* 76:754–58. (in Russian).

Labay, V.S. 1999. The Atlas-key of high Crustacea (Crustacea Malacostraca) of Sakhalin fresh and brackish waters. *Transactions of Sakhalin Scientific Research Institute of Fisheries and Oceanography* 2:59–73. (in Russian).

Labay, V.S. 2002a. Some patterns of *Kamaka kuthae* Derzhavin, 1923 population (Amphipoda Corophiidae) from the Piltun lagoon. *Transactions of Sakhalin Scientific Research Institute of Fisheries and Oceanography* 4:277–83. (in Russian).

Labay, V.S. 2002b. Three species of the genus *Psuedocrangonyx* Akatsuka et Komai, 1922 (Crustacea: Amphipoda) from subterranean fresh waters of the island of Sakhalin. *Arthropoda Selecta* 10:289–96.

Labay, V.S. 2003a. A new species of *Melita* Leach (Amphipoda: Melitidae) from oligosaline waters of Russian Far East. *Zootaxa* 356:1–8.

Labay, V.S. 2003b. *Sternomoera yezoensis* Ueno, 1933 (Crustacea, Amphipoda, Eusiridae) a new species for Russia from fresh waters of the southern Sakhalin Island. *Transactions of Sakhalin Scientific Research Institute of Fisheries and Oceanography* 5:99–105. (in Russian).

Labay, V.S. 2004a. Macrozoobenthos of the Nevelskoy Strait. *Transactions of Sakhalin Scientific Research Institute of Fisheries and Oceanography* 6:305–30. (in Russian).

Labay, V.S. 2004b. *Paracleistostoma cristatum* De Man, 1895 (Crustacea: Decapoda), a crab species new for the fauna of Russia from the estuarine waters of the South Sakhalin. *Russian Journal of Marine Biology* 30:56–60.

Labay, V.S. 2005. Fauna of the Malacostraca (Crustacea) from the fresh and brackish water of Sakhalin Island. In *Flora and fauna of Sakhalin Island* (Materials of International Sakhalin Island Project). Part 2. Vladivostok, Dalnauka:64–87. (in Russian).

Labay, V.S. 2007. Benthos distribution in the lower rithral of the Poronai River under the impact of some abiotic environmental factors. *Transactions of Sakhalin Scientific Research Institute of Fisheries and Oceanography* 9:184–206. (in Russian).

Labay, V.S. 2008. Macrobenthos composition and structure of Vavajskaja system lakes (southern Sakhalin). *Vladimir Ya. Levanidov's Biennial Memorial Meetings* 4:224–38. (in Russian).

Labay, V.S. 2009. Response of macrozoobenthos of lagoon lake Izmenchivoye (Sakhalin Island) to the discontinuance of water exchange with the sea. *Russian Journal of Marine Biology* 35:279–87.

Labay, V.S. 2010. Distribution of Macrozoobenthos in the Metarithral of the Salmon River of the Sakhalin Island. *Hydrobiological Journal* 46:12–26.

Labay, V. S. 2011a. Composition and distribution of macrozoobenthos in Lake Nevskoye (Sakhalin Island). *Transactions of Sakhalin Scientific Research Institute of Fisheries and Oceanography* 12:152–66. (in Russian).

Labay, V.S. 2011b. 5.4. Conservation biology of freshwater decapods in Sakhalin and Kuril Islands. In *Kani Zrigani* (Shrimp, crab, and crayfish), eds. T. Kawai, and K. Nakata., Ebi Tokyo, Seibutsukenkyusha:419–434. (in Japanese).

Labay, V. S. 2011c. Zoogeographical outline of Malacostraca (Crustacea) fauna from the fresh and brackish water of Sakhalin Island. *Transactions of Sakhalin Scientific Research Institute of Fisheries and Oceanography* 12:131–51. (in Russian).

Labay, V.S. 2012. *Paramoera anivae*, a new species of Eusiridae Stebbing, 1888 (Crustacea: Amphipoda: Gammaridea) from the Okhotsk Sea. *Zootaxa* 3475:69–85.

Labay, V.S. 2014. Seasonal dynamics of macrozoobenthos of Ljutoga floodplain lake (southern Sakhalin). *Vladimir Ya. Levanidov's Biennial Memorial Meetings.* 6:360–8. (in Russian).

Labay, V.S. 2015a. Macrozoobenthos of small lowland lakes of Sakhalin Island. *Izvestia TINRO* 183:145–55. (in Russian).

Labay, V.S. 2015b. Species composition of macrozoobenthos in lagoons of Sakhalin Island. *Izvestia TINRO* 183:125–44. (in Russian).

Labay, V.S. 2016. Review of amphipods of the *Melita* group (Amphipoda: Melitidae) from the coastal waters of Sakhalin Island (Far East of Russia). III. Genera *Abludomelita* Karaman, 1981 and *Melita* Leach, 1814. *Zootaxa* 4156:1–73.

Labay, V.S. and Barabanschikov, E.A. 2009. A morphological variety of freshwater shrimps (Crustacea: Decapoda: Palaemonidae) of Sakhalin Island and adjacent territories. In The Crustacean Society Summer Meeting in Tokyo, Japan and the 47th Annual Meeting of the Carcinological Society of Japan. Program and Abstracts. September 20–24, 2009. Shinagawa, Tokyo, University of Marine Science and Technology:53.

Labay V.S. and Labay, S.V. 2014. Daily vertical migrations of Malacostraca (Crustacea) in lagoon lake Ptych'e (southern Sakhalin). *Vladimir Ya. Levanidov's Biennial Memorial Meetings* 6:369–79. (in Russian).

Labay, V.S. and Pecheneva, N.V. 2001. Sravnitrlnaya kharakteristika raspredelenia, sostava i struktury presnovodnogo zoobentosa lagun Piltun i Nyiskiy zaliv (severo-vostochnyi Sakhalin) (Comparative characteristics of the distribution, composition and structure of freshwater zoobenthos in lagoons Piltun and the Nijsky Bay (northeast Sakhalin)). *Vladimir Ya. Levanidov's Biennial Memorial Meetings* 1:55–64. (in Russian).

Labay, V.S. and Rogotnev, M.G. 2005. Composition, structure and seasonal dynamics of macrobenthos in the Tunaycha Lake (South Sakhalin). *Vladimir Ya. Levanidov's Biennial Memorial Meetings* 3:62–94. (in Russian).

Labay, V.S., Atamanova, I.A., Zavarzin, D.S., Motylkova, I.V., Moukhametova, O.N. and Nikitin, V.D. 2014. *Reservoirs of Sakhalin Island: From Lagoons to Lakes.* Yuzhno-Sakhalinsk, Sakhalin Regional Museum. (in Russian).

Labay, V.S., Dairova, D.S., Kurilova, N.V. and Shpil'ko, T.S. 2013. Macrobenthos of Baikal Bay (Sakhalin Island). *Transactions of the Sakhalin Research Institute of Fisheries and Oceanography* 14:211–36. (in Russian).

Labay, V.S., Kurilova, N.V. and Shpilko, T.S. 2016a. Seasonal variability of macrozoobenthos in a lagoon having a periodic connection with the sea (Ptich'e Lake, southern Sakhalin). *Biological Bulletin (in Russian)* 43:145–59.

Labay, V.S., Latkovskaya, E.M., Pecheneva, N.V. and Krasavcev, V.B. 2000. Features of the structural organization of macrozoobenthos in the lagoon with a pronounced gradient of abiotic factors. In *Fundamental Problems of Water and Water Resources at the Turn of the Third Millennium.* Materials of the International Scientific Conference September 3–7, 2000. Tomsk:539544. (in Russian).

Labay, V.S., Rogotnev, M.G. and Shpilko, T.S. 2004. Vertical distribution and seasonal dynamics of macrobenthos in the Tunaicha Lake ground, South Sakhalin. *Research of Water Biological Resources of Kamchatka and of the Northwest Part of Pacific Ocean: Selected Papers.* 7:111–21. (in Russian).

Labay, V.S., Zavarzin, D.S., Motylkova, I.V., et al. 2016b. *Water Biota of Tunaicha Lake (Southern Sakhalin) and Conditions of It Dwelling.* Yuzhno-Sakhalinsk, Sakhalin Scientific Research Institute of Fisheries and Oceanography. (in Russian).

Labay, V.S., Zavarzin, D.S., Moukhametova O.N., et al. 2010. *Plankton and Benthos of Vavajskaya Lakes System (Southern Sakhalin) and Conditions of Their Dwelling.* Yuzhno-Sakhalinsk, Sakhalin Scientific Research Instute of Fisheries & Oceanography. (in Russian).

Labay, V.S., Zhivogljadova, L.A., Polteva, A.V., et al. 2015. *Watercourses of Sakhalin Island: Life in the Running Water.* Yuzhno-Sakhalinsk, Sakhalin Regional Museum. (in Russian).

Levanidov, V.Y. 1980. Novyc vidy i rasprostranenie vodjanykh oslikov *Asellus* s. str. (Isopoda, Asellidae) na severo-vostoke Azii (New species and distribution of water licc *Asellus* s. str. (Isopoda, Asellidae) in northeast Asia). In *Fauna presnykh vod Dalnego Vostoka* (Fauna of fresh waters of the Far East). Vladivostok, Far Eastern Science Centre, Academy of Science USSR:13–23. (in Russian).

Lindberg, G.U. 1955. *Chetvertichnyi period v svete biogeographicheskikh dannykh* (Quaternary period in the light of biogeographic data). Moscow–Leningrad, Nauka. (in Russian).

Lindberg, G.U. 1972. *Krupnye kolebaniya urovnja okeana v chetvertichnyi period* (Large fluctuations in sea level in the Quaternary). St. Petersburg-Leningrad, Nauka. (in Russian).

Lomakina, N.B. 1955. Kumovye raki (Cumacea) dalnevostochnykh morey (Cumaceans (Cumacea) of the Far Eastern Seas). *Trudy Zoologicheskogo instituta RAN (Proceedings of the Zoological Institute of the USSR Academy of Sciences)* 13:166–218. (in Russian).

Lomakina, N.B. 1958. *Kumovye raki (Cumacea) morey SSSR* (Cumaceans (Cumacea) of the seas of the USSR). Moscow–Leningrad, Nauka. (in Russian).

Marin, I.N. 2013. *Atlas of Decapod Crustaceans of Russia.* Moscow, KMK Scientific Press. (in Russian).

Marincovich, L., Brouwers, E.M., Hopkins, D.M. and McKenna, M.C. 1990. Late Mezozoic and Cenozoic paleogeographic and paleoclimatic history of the Arctic Ocean Basin, based on shallow-water marine faunas and terrestrial vertebrates. The geology of North America, V. L. The Arctic Ocean region. The Geological Society of America 23:403–26.

Matjushkov, G.V., Solovjev, A.V. and Melnikov, O.A. 2014. *Sakhalin Island Geological Past.* Yuzhno-Sakhalinsk, Sakhalin Regional Museum. (in Russian).

Melioranskiy, V.A. 1936. Materialy k morphologii severnogo Sikhote-Alinja (Materials on the morphology of the northern Sikhote-Alin). *Izvestia Gosudarstvennogo geographicheskogo obshchestva (News of the State Geographical Society)* 68:928–35. (in Russian).

Mikishin, Y.A. and Gvozdeva, I.G. 1996. *The Natural Evolution in the South-Eastern Part of Sakhalin Island in Holocene.* Vladivostok, Far Eastern State University. (in Russian).

Mikishin, Y.A., Rybakov, V.F. and Brovko, P.F. 1995. Southern Sakhalin. Tunaicha Lake. In *The History of Lakes of North Asia* (Series: History of lakes). St. Petersburg, Nauka:112–20. (in Russian).

Milliman, J.D. and Emery, K.O. 1968. Sea level during the past 35000 years. *Science* 162:1121–23.

Mironov, A.N. 2013. Biotic complexes of the Arctic ocean. *Invertebrate Zoology* 10:3–48.

Miyadi, D. 1938. Ecological studies on marine relics and landlocked animals in inland waters of Nippon. *The Philippine Journal of Science* 65:238–48.

Nikiforov, S.N. 2001. *The Ichthyofauna of Fresh Waters of Sakhalin Island and Its Formation.* Avtoreferat dissertation, kandidata biologicheskikh nauk. Vladivostok, Institute of Marine biology, Far Eastern Branch, Academy of Science of Russia. (in Russian).

Nikiforov, S.N. and Safronov, S.N. 1996. Vozmozhnye geneticheskie svjazi fauny ryb Amura i presnych vodoemov Sakhalina (Possible genetic links of fish fauna from the Amur and from fresh water bodies of Sakhalin). Materialy nauchno-prakticheskoy konferencii molodykh issledovateley (Materials of the scientific-practical conference of young researchers). Yuzhno-Sakhalinsk, Yuzhno-Sakhalinsk State Pedagogical Institute:39–44. (in Russian).

Pacific Ocean, ed. L. Galerkin, M.S. Barash. 1982. Moscow, Mysl':1–320. (in Russian).

Pecheneva, N.V. and Labay, V.S. 2006. Macrozoobenthos of the lagoon-type Lake Izmenchivoye (southeastern Sakhalin). *Transactions of Sakhalin Scientific Research Institute of Fisheries and Oceanography* 8:67–88. (in Russian).

Pecheneva, N.V., Labay, V.S. and Kafanov, A.I. 2002. Bottom communities of Nyivo Lagoon (northeastern Sakhalin). *Russian Journal of Marine Biology* 28:225–34.

Pishchalnik, V.M. and Bobkov, A.O. 2000. *Oceanographical ATLAS of the Sakhalin Shelf. Part 1.* Yuzhno-Sakhalinsk, Sakhalin State University. (in Russian).

Pletnev, S.P., Grebennikova, T.A. and Kisilev, I.V. 1988. Paleo-salinity of the Sea of Japan in Late Wurm. In *Quantitative Parameters of the Natural Environment in the Pleistocene.* Vladivostok, Pacific Institute of Geology, Far Eastern Branch, Academy of Science USSR:26–40. (in Russian).

Prozorova, L.A. 2001. Features of the distribution of freshwater mollusk fauna in the Far East of Russia and its biogeographical zoning. *Vladimir Ya. Levanidov's Biennial Memorial Meetings.* 1:112–25. (in Russian).

Safronov, S.N. and Nikiforov, S.N. 1995. Species composition and distribution of ichthyofauna of fresh and brackish waters of Sakhalin (Report). In Materials of the XXX Scientific and Methodical Conference of Teachers of Yuzhno-Sakhalinsk State Pedagogical Institute, April 1995. Part 2. Yuzhno-Sakhalinsk, Yuzhno-Sakhalinsk State Pedagogical Institute:112–24. (in Russian).

Safronova, R.K. and Safronov, S.N. 1980. Zoobenthos and feeding of Amur carp from lakes of the Okhotsk group of southern Sakhalin. In *Distribution and rational using of aquatic zoological resources of the Sakhalin Island and the Kuril Islands.* Vladivostok, Far Eastern Science Center, Academy of Science USSR:22–31. (in Russian).

Sailing directions of the Tatar Strait, Amur Estuary and the Laperuz Strait, ed. Yu.V. Starkov. 2003. St. Petersburg, Main Department of Navigation and Oceanography of Defense Department of RF. (in Russian).

Sakhalin Region. 1994. Geographical overview. Yuzhno-Sakhalinsk, Sakhalin Book Publishing House. (in Russian).

Samatov, A.D., Labay, V.S., Motylkova, I.V., et al. 2002. Short characteristic of water biota of Tunaicha Lake (Southern Sakhalin) in summer period. *Transactions of Sakhalin Scientific Research Institute of Fisheries and Oceanography* 4:258–69. (in Russian).

Saveliev, A.V., Danchenkov, M.A. and Hong, G-H., 2002. Volume transport through the La-Perouse (Soya) strait between the East Sea (Sea of Japan) and the Sea of Okhotsk. *Ocean and Polar Research* 24:147–52.

Schellenberg, A. 1936. Die Amphipodengattungen um *Crangonyx*, ihre Verbreitung und ihre Arten. *Mitteilungen aus dem Museum Naturkunde in Berlin. Zoologisches Museum und Institut* 22:31–43.

Schellenberg, A. 1937a. Kritische Bemerkungen zur Systematik der Süßwassergammariden. *Zoologische Jahrbücher (Systematik)* 69:469–516.

Schellenberg, A. 1937b. Schlüssel und Diagnosen der dem Süsswasser *Gammarus* nahestehenden Einheiten ausschließlich der Arten des Baicalsees und Australiens. *Zoologischer Anzeiger* 117:267–80.

Sidorov, D.A. 2005. Freshwater lice fauna (Crustacea, Isopoda, Asellidae) of Far East and adjacent lands. *Vladimir Ya. Levanidov's Biennial Memorial Meetings.* 3:255–74. (in Russian).

Sidorov, D.A. 2008. *Freshwater Hypogean Higher Crustaceans (Crustacea: Malacostraca) of Far East of Russia.* Avtoreferat dissertation. kandidata biologicheskikh nauk. Vladivostok, Institute of Biology and Soil Science, Far Eastern Branch, Russian Academy of Science. (in Russian).

Starobogatov, Y.I. 1972. Crabs of the littoral of the Gulf of Tonkin. In *The Fauna of the Gulf of Tonkin and the Conditions of Its Existence.* Issledovania fauny morey (Studies of the Fauna of the Seas). Leningrad St. Petersburg, Nauka, 10.333–358. (in Russian).

Starobogatov, Y.I. 1995. Amphipoda, Isopoda. In *Key to Freshwater Invertebrates of Russia and Adjacent Lands.* 2. Crustacea. Leningurard-St. Petersburg, Nauka:167–73, 184–206. (in Russian).

Starobogatov, Y.I. 1996. Taxonomy and geographical distribution of crayfishes of Asia and East Europe (Crustacea Decapoda Astacoidei). *Arthropoda Selecta* 4:3–25.

Starobogatov, Y.I. and Vasilenko, S.V. 1995. Decapoda. In *Key to Freshwater Invertebrates of Russia and Adjacent Lands.* 2. Crustacea. St. Petersburg, Nauka:174–83. (in Russian).

Tabunkov, V.D., Averintzev, V.G., Sirenko, B.I. and Sheremetevskiy, A.I. 1988. Composition and structure of the bottom population of the Nabil and Piltun lagoons (Nort-Eastern Sakhalin). In *Biota and communities of the Far Eastern seas: Lagoons and bays of Kamchatka and Sakhalin.* Vladivostok, Far Eastern Branch, Academy of Science USSSR:7–30. (in Russian).

Taranetz, A.Y. 1938. K zoogeographii Amurskoy perekhodnoy oblasti na osnove izuchenia presnovodnoy ichtyofauny (To the zoogeography of the Amur transition region based on the study of freshwater ichthyofauna). *Bulletin of the Far Eastern Branch of Academy of Science USSR* 28:99–115. (in Russian).

Tzvetkova, N.L. 1975. *Pribrezhnye gammaridy severnykh i dal'nevostochnykh morey SSSR i sopredel'nykh vod. Rody Gammarus, Marinogammarus, Anisogammarus, Mesogammarus* (Amphipoda, Gammaridae) (Coastal gammarids of the northern and far eastern seas of the USSR and adjacent waters. Genera *Gammarus*, *Marinogammarus*, *Anisogammarus*, *Mesogammarus* (Amphipoda, Gammaridae)). Leningrad, Nauka. (in Russian).

Ueno, M. 1935a. Crustacea collected in the lakes of southern Sakhalin. *Annotationes Zoologicae Japanenses* 15:88–94.

Ueno, M. 1935b. Limnological reconnaissance of southern Sakhalin. II. Zooplancton. *Bulletin of the Japanese Society of Scientific Fisheries* 4:190–94.

Ueno, M. 1940. Some freshwater amphipods from Manchoukuo, Corea and Japan. *Bulletin of the Biogeografical Society of Japan* 10:63–85.

Ushakov, P.V. 1934. K faune opresnennykh vod Amurskogo limana i Sakhalinskogo zaliva Yaponskogo morja (To the fauna of the desalinated waters of the Amur Estuary and the Sakhalin Bay of the Sea of Okhotsk). *Bulletin of Pacific Committee*, Academy of Science USSR 3:39–40. (in Russian).

Ushakov, P.V. 1948. The fauna of invertebrate of the Amur estuary and the neighboring desalinated areas of the Sakhalin Bay. In *Memory of Academician*, ed. S.A. Zernov. Moscow and Leningrad, Academy of Science USSR:175–91. (in Russian).

Vinogradov, L.G. 1950. Opredelitel' krevetok, rakov i krabov Dal'nego Vostoka (Key to shrimp, crayfish and crabs of the Far East). *Izvestia TINRO* 33:179–350. (in Russian).

Volova, G.N. and Koz'menko, V.B. 1984. Benthos of Chayvo lagoon (Sakhalin Island). In *Fauna and Ecology of Marine Organisms*. Vladivostok, Far Eastern State University:125–36. (Manuscript deposited in VINITI. No. 3651–84). (in Russian).

Zarenkov, N.A. 1965. Revision of the genera *Crangon* Fabricius and *Sclerocrangon* G. O. Sars (Decapoda–Crustacea). *Russian Journal of Zoology* 44:1761–75. (in Russian).

Zhirkov, I.A. 2017. *Bio-geographia obshchaya i chastnaja: sushi, morja i kontinental'nych vodoemov* (General and particular bio-geography: Land, sea and continental water bodies). Moscow, KMK Scientific Press. (in Russian).

Zhivoglyadova, L.A. and Labay, V.S. 2003. Some biological and taxonomic peculiarities of the Sakhalin river crawfish *Cambaroides sachalinensis* of northern Sakhalin lakes. *Hydrobiological Journal* 39:58–68. (Originally published in 2002. *Gidrobiologicheskiy Zhurnal* 38:35–44).

Zvyagintsev, A.Y. 1985. Obrastanie sudov pribrezhnogo i portovogo plavania v rayone ostrova Sakhalin. In *Bentos i usloviya ego sushchestvovania na shaelfovykh zonakh Sakhalina* (Fouling of coastal and port vessels in the area of Sakhalin Island). Vladivostok, Far Eastern Scientific Center of the USSR Academy of Sciences:102–117. (in Russian).

CHAPTER 4

Notes on Australian Marsh-hoppers (Protorchestiidae: Amphipoda: Crustacea)

Lauren E. Hughes

CONTENTS

4.1 INTRODUCTION

The problem of confirming the morphology of the largest growth form, or even species designation, of continuously developing males has been highlighted specifically in talitrids since Müller (1864). Until the present study, male specimens attributed to *Cochinorchestia* Lowry and Peart, 2010 were of a comparatively small body size (*C. lindsayae* – 6.5 mm; *C. metcalfeae* – 7 mm; *C. morrumbene* – 9 mm; *C. poka* – 8 mm; *C. tulear* – 8 mm). In other Australian and New Zealand talitrid genera, adult males are known to reach over 20 mm (*Bellorchestia*, see Hughes and Ahyong 2017), with the largest known talitrid males around 40 mm (*Orchestia aucklandiae* Spence-Bate 1862; see Hurley 1957).

Lowry and Peart (2010) established *Cochinorchestia* as a monotypic genus of marsh-hopper based on the poorly described *Parorchestia notabilis* Barnard (1935) listing the sexual dimorphism state of male specimens as unknown. Later, Lowry and Springthorpe (2015) included a further six species in *Cochinorchestia* and updated the generic diagnosis with the male pereopods considered to be nonsexually dimorphic. *Microrchestia* was established by Bousfield, 1984 to accommodate the marsh-hopper *Parorchestia macrochela* Bousfield, 1971 from the Bismark Archipelago, highlighting a number of undescribed species across various global locations (Lowry and Peart 2010).

In this study, illustrations are provided to support new distribution records of marsh-hoppers in Australia. Based on the larger sized specimens at hand, the study further documents variation with growth stage in sexually dimorphic features and robust setal counts. These findings prompt a re-evaluation of marsh-hopper genera, as the smaller body size individuals of *Microrchestia* Bousfield, 1984 are seen to sequentially develop into a larger growth form attributable to *Cochinorchestia*. Therefore, the diagnosis of *Microrchestia* is updated here as the senior synonym of *Cochinorchestia*.

The salt marsh and mangrove habitats are a significant feature of the world's coastlines (Bunting et al. 2018), and further, the diversity of animals inhabiting these habitats is most likely undersampled, given the challenges of access to these environments. The material documented across three species, *Microrchestia metcalfeae* (Lowry and Springthorpe 2015) nov. comb., *M. poka* (Lowry and Springthorpe 2015) nov. comb., and *M. watsonae* (Lowry and Peart 2010) is a small but significant contribution to the life history of these marsh-hoppers. The relationship of Cochinorchestia & Microrchestia was presented as both sister taxa and a complex in Lowry & Springthorpe (2019), their work being based on a literature review of the geographical distribution of species in these genera and prior knowledge of the present study on morphology. The data presented here provides the empirical data to support with *Cochinorchestia* now placed as a junior synonym of *Microrchestia*. This synonymy resolves the configuration of these genera as a complex, and instead presents a greater metamorphosis in ontogeny than seen elsewhere in the Amphipod with perhaps the exception of the Leucothoidae.

4.2 MATERIAL AND METHODS

Material examined for this study included unsorted Amphipoda from coastal and wetland regions in the Northern Territory and Queensland from existing museum collections. Specimens are lodged in the Queensland Museum, Australia (QM). Material was dissected in 80% ethanol. Slides were made using Aquatex™ mounting agent. Standard abbreviation on the plates are as follows: A, antenna; G, gnathopod; Md, mandible; P, pereopod; T, telson; U, uropod; Ur, urosome; L, left and R, right.

4.3 SYSTEMATICS

Protorchestiidae Myers & Lowry, 2020

Microrchestia Bousfield, 1984
Microrchestia Bousfield, 1984: 202.
Cochinorchestia Lowry and Peart, 2010: 21. – Lowry and Springthorpe 2015: 160–161; 178, 179 (key).
Type species. *Parorchestia macrochela* Bousfield, 1971, original designation.
Generic diagnosis (modified after Lowry and Peart 2010 and Lowry and Springthorpe 2015). Head: eyes medium (greater than 1/5 to 1/3 head length). Antenna 1: long, subequal to or reaching slightly beyond antenna 2 peduncle. Antenna 2:

peduncular articles slender; peduncular article 3 without ventral process; flagellum apical article minute, virgula divina present or absent. Mandible left: lacinia mobilis 4-cuspidate. Maxilliped: palp article 2 distomedial lobe well developed, article 4 small, well defined. Gnathopod 1: subchelate; posterior margin of merus, carpus, and propodus, each with lobe covered in palmate setae. Gnathopod 2: subchelate in male, mitten-shaped in female; merus and carpus free; dactylus not modified distally, blunt. Pereopods 3–7: simplidactylate. *Pereopod 4* dactylus slender, similar to pereopod 3. Pereopods 5–7: without setae along posterior margin of the dactylus. Pereopod 7: sexually dimorphic, enlarged; merus, carpus, and propodus elongate in males, without rows of short setae along posterior margin of dactyli. Pleonites 1–3: without dorsal spines. Pleopods 1–3: well developed. Epimera 1–3: ventral margin without slits. Uropod 1: rami without apical pear-shaped setae; outer ramus slender, with or without marginal robust setae. Uropod 3: well developed; ramus shorter than peduncle. *Telson* entire, with 2–6 apical robust setae per lobe.

Species composition. *Microrchestia bousfieldi* (Lowry and Peart 2010); *M. lindsayae* (Lowry and Springthorpe 2015) nov. comb.; *M. macrochela* (Bousfield 1971) type species; *M. metcalfeae* (Lowry and Springthorpe 2015) nov. comb.; *M. morini* (Peethambaran Asari 1998) nov. comb.; *M. morrumbene* (Lowry and Springthorpe 2015) nov. comb.; *M. notabilis* (Barnard 1935) nov. comb.; *M. ntensis* (Lowry and Springthorpe 2015); *M. poku* (Lowry and Springthorpe 2015) nov. comb.; *M. similis* (Bousfield 1971); *M. tulear* (Lowry and Springthorpe 2015) nov. comb. and *M. watsonae* (Lowry and Peart 2010).

Discussion. The holotype of *C. poka* is an 8-mm male, which is now considered an immature male, as specimens reported here include males of *C. poka* up to 18 mm in body length. These larger male individuals have sexually dimorphic pereopods, which was not previously known (Fig. 4.12).

Material of *Microrchestia metcalfeae, M. poka,* and *M. watsonae,* studied here at 12, 18, and 19 mm, respectively, have sexually dimorphic pereopods (Figs. 4.4, 4.5, 4.8, and 4.12). In these large males, pereopod 7 becomes much longer than pereopod 6. Based on these three species, it is possible that male specimens of *M. morrumbene* nov. comb. and *M. tulear* nov. comb. also attain a larger body size, which will also show dimorphism.

The presence or absence of marginal robust setae along the outer ramus of the uropods is a generic level character elsewhere in the Talitridae, yet results from this study show that robust setae are absent in juvenile males while present in the larger specimens (Fig. 4.6). This is similar to findings for telsonic setae in *Bellorchestia* (Lowry and Serejo 2008), which are known to increase in number with individual size. Setal counts were previously used as a species-level character in *Bellorchestia* (Hughes and Lindsay Chapter 5).

Microrchestia watsonae Lowry and Peart, 2010

Microrchestia watsonae Lowry and Peart, 2010: 30–37, Figs. 4.7–4.12. Lowry and Springthorpe, 2015: 154 (regional key) (Figs. 4.1–4.7 and 4.14).

Type locality. Ferriers Creek, Lizard Island, Great Barrier Reef, Far North Queensland, Australia.

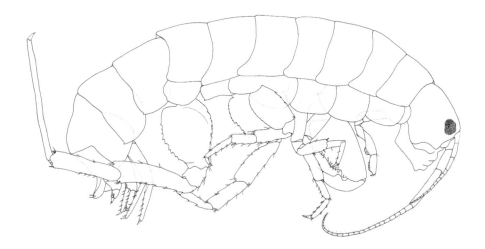

Figure 4.1   *Microrchestia watsonae* habitus male, 19 mm, QM W29430, Port Stewart, Stewart River, Far North Queensland, Australia.

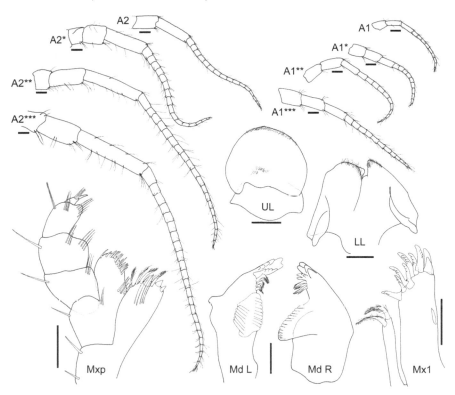

Figure 4.2   *Microrchestia watsonae* male 10 mm, QM W29427; male* 13 mm, QM W29428; male** 14 mm, QM W29429; male***, 19 mm, QM W29430; Port Stewart, Stewart River, Far North Queensland, Australia. Scales 0.2 mm.

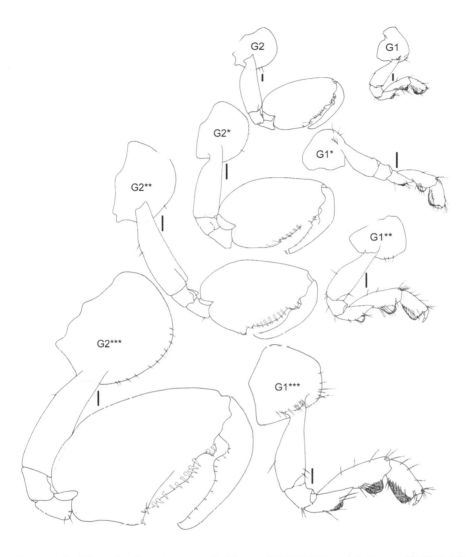

Figure 4.3 *Microrchestia watsonae* male 10 mm, QM W29427; male* 13 mm, QM W29428; male** 14 mm, QM W29429; male***, 19 mm, QM W29430; Port Stewart, Stewart River, Far North Queensland, Australia. Scale 0.2 mm.

Material examined. Male, 19 mm, dissected, 2 slides (remaining parts in micro-vial), QM W29430; male, 14 mm, dissected, 4 slides, QM W29429; B male, 13 mm, dissected, 3 slides, QM W29428; male, 10 mm, dissected, 3 slides, QM W29427; female, 10 mm, dissected, 3 slides, QM W29431; 3 juvenile males, QM W27016, Port Stewart, Stewart River, Far North Queensland, tidally inundated mangroves, in and under logs, 6 November 1982, coll. P. Davie.

Description. Based on male, 19 mm, QM W29430.

Head: eye medium size (greater than 1/5 to 1/3 head length). Antenna 1: long, reaching beyond peduncular article 5 of antenna 2. Antenna 2: slightly less than

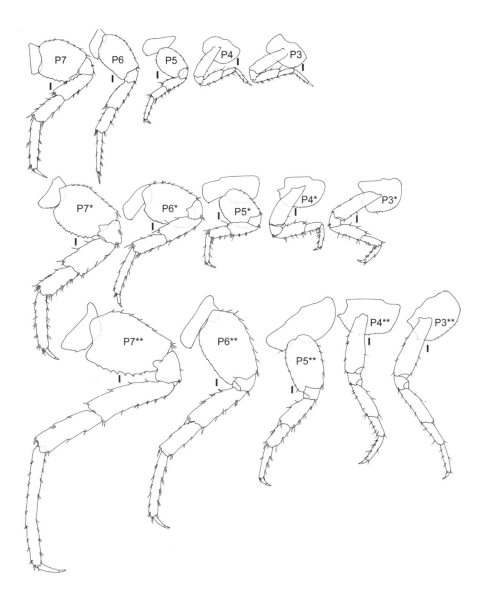

Figure 4.4 *Microrchestia watsonae* male 10 mm, QM W29427; male* 13 mm, QM W29428; male** 14 mm, QM W29429, Port Stewart, Stewart River, Far North Queensland, Australia. Scale 0.2 mm.

half body length (0.45 × length); peduncular articles slender; article 5 longer than article 4; flagellar articles final article large, cone-shaped forming a *virgula divina*. Labrum: with apical setal patch; epistome without robust setae. Labium: distolateral setal tuft present, distomedial setal tuft present; without inner plates. Mandible: left lacinia mobilis 4-cuspidate. Maxilla 1: with 1-articulate palp small. Maxilliped: palp article 2 distomedial lobe well developed; article 4 small well defined. Pereon.

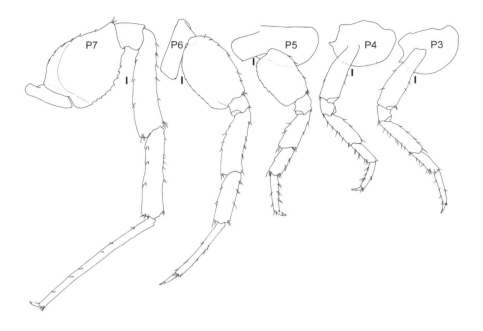

Figure 4.5 *Microrchestia watsonae* male, 19 mm, QM W29430; Port Stewart, Stewart River, Far North Queensland, Australia. Scale 0.2 mm.

Gnathopod 1: sexually dimorphic; subchelate; coxa smaller than coxa 2; merus posterior margin with palmate lobe; carpus longer than propodus, 1.7 as long as propodus, posterior margin with palmate lobe; propodus subrectangular, twice as long as broad, anterior margin with 4 groups of robust setae, posterior margin with palmate lobe, palm transverse; dactylus simplidactylate, shorter than palm, without anterodistal denticular patch. Gnathopod 2: sexually dimorphic; subchelate; basis slender; ischium anterior margin with lateral and medial rounded lobe, lobes dissimilar in size; carpus triangular, reduced, enclosed by merus and propodus; propodus 1.9 × as long as wide, palm subacute, 55% along posterior margin, lined with robust setae, posteroproximal corner with groove, without cuticular patch at corner of palm, posterodistal corner with tooth; dactylus curved, subequal in length to palm, with anteroproximal bump, apically blunt. Pereopods 2–4: coxae wider than deep. Pereopods 3–7: simplidactylate, dactylus without anterodistal denticular patch. Pereopod 3: carpus length twice width. Pereopod 4: significantly shorter than pereopod 3; carpus significantly shorter than carpus of pereopod 3, length 1.8 × width; dactylus slender, without anterodistal denticular patch. Pereopod 5: merus length twice width; carpus length twice width; propodus distinctly longer than carpus, length 4.1 × width; dactylus slender, length 3 × width. Pereopod 6: much shorter than pereopod 7; coxa posterior lobe rectilinear, posteroventral corner subquadrate, posterior margin perpendicular to ventral margin; merus length 2.8 × width; carpus length 2.5 × width; propodus distinctly longer than carpus, length 10.5 × width. Pereopod 7: sexually dimorphic, enlarged; basis posterior margin

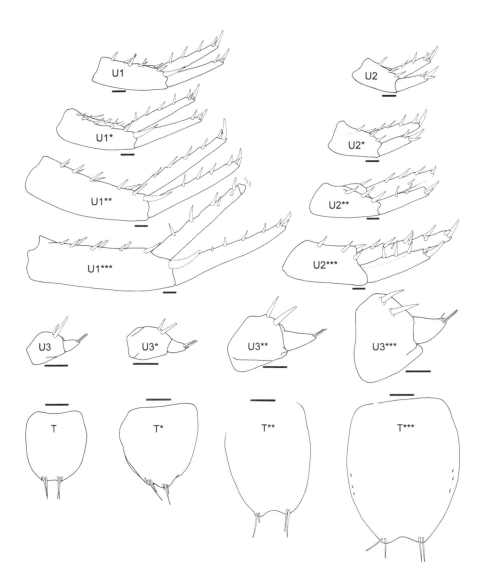

Figure 4.6 *Microrchestia watsonae* male 10 mm, QM W29427; male* 13 mm, QM W29428;
male** 14 mm, QM W29429; male***, 19 mm, QM W29427; Port Stewart, Stewart
River, Far North Queensland, Australia. Scale 0.2 mm.

convex, strongly serrate, lateral sulcus pronounced proximally; merus length 4 ×
width; carpus length 3.8 × width; propodus greatly elongate, distinctly longer than
carpus, length 24 × width.

Pleon. Pleonites 1–3: without dorsal carina. Pleopods 1–3: well developed, rami
segmented. *Epimeron 1* posterior margin convex, posteroventral corner with small
acute tooth. Epimera 2–3: posterior margin straight, posteroventral corner with
small acute tooth. Uropod 1: not sexually dimorphic; peduncle with 9 robust setae;

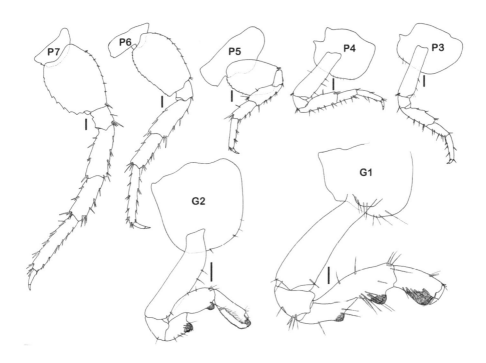

Figure 4.7 *Microrchestia watsonae* female, 10 mm, QM W29430, Port Stewart, Stewart River, Far North Queensland, Australia. Scale 0.2 mm.

inner ramus subequal in length to outer ramus, with 6 marginal robust setae in 2 rows; outer ramus with 4 marginal robust setae in 1 row. Uropod 2: peduncle with 4 robust setae; inner ramus longer than outer ramus, with 7 marginal robust setae in 2 rows; outer ramus with 3 marginal robust setae in 1 row. Urosomite 3: without dorsal carina. Uropod 3: well developed; peduncle dorsally concave, margin accommodating telson, depth greater than length, 1.4 × length, with 2 robust setae; ramus linear, not fused to peduncle, shorter than peduncle, 1.1 × as long as broad, without marginal setae. Telson: subovate, longer than broad, length extending beyond uropod 3 peduncle, lateral margins straight, entire, apically notched, dorsal midline absent, with 2 apical robust setae per lobe.

Female (sexually dimorphic characters). Gnathopod 1: subchelate; merus, carpus and propodus posterior margin with palmate lobe; propodus palm near transverse; dactylus not reaching end of palm. Gnathopod 2: mitten-shaped; basis not expanded anteromedially; ischium without lobe; carpus well developed (not enclosed by merus and propodus), length 2.2 × width; propodus 3 × width, palm obtuse; dactylus shorter than palm, posterior margin smooth. Pereopod 7: not sexually dimorphic; basis lateral sulcus absent; merus length twice width; carpus length 2.1 × width; propodus slightly longer than carpus, length 4.5 × width.

Allometric growth variation. Sexually dimorphic male pereopod 7 begins developing in males ~13 mm with the merus, carpus, and propodus becoming elongate

(Figs. 4.4 and 4.5). The number of marginal robust setae on the outer ramus of uropods 1 and 2 increases from 0 to 4 and 3, respectively, in males 10–19 mm (Fig. 4.6). The number of teeth on the lacinia mobilis is four in both the smallest and largest males.

Remarks. *Microrchestia watsonae* can be distinguished from its congeners by the male gnathopod 2 dactylus posterior margin having a proximal lobe. The pereopod 7 propodus is particularly elongate in comparison to other species, the length being 24 times the width in males of 19 mm.

Habitat. Mangroves.

Distribution. Australia, Far North Queensland: Ferriers Creek, Lizard Island (Lowry and Peart 2010); Port Stewart (current study).

Microrchestia metcalfeae (Lowry and Springthorpe 2015) nov. comb. (Figs. 4.8 and 4.14).

Cochinorchestia metcalfeae (Lowry and Springthorpe 2015: 164–169) (Figs. 4.7–4.10).

Type locality. Opposite Channel Island, Wickham Point, Darwin Harbour, Northern Territory, Australia.

Material examined. Northern Territory: 1 male, 12 mm, dissected, parts in microvial, QM W26984, East Alligator River mouth, Kakadu National Park (12.07°S 132.32°E), estuarine littoral mangroves, dense *Rhizopora* sp. about 3 m in height, tall dead emergent *Rhizopora* sp. trunks to about 10 m, substrate very moist, 16 June 1982, coll. P. Davie (T5Q3, K2374); 1 male juvenile specimen, QM W26989, Point Farewell, East Alligator River, Kakadu National Park (12.05°S 132.34°E), estuarine littoral mangroves, low *Rhizopora* sp., soft mud, much organic detritus below surface, about 25% coverage of permanent shallow pools, salinity in pools (just after high tide) ranged from 38–44 ppt., 10 June 1981, coll. P. Davie (T1Q3, K2586); 1 juvenile male, QM W26975, Alligator River, Kakadu National Park, Northern

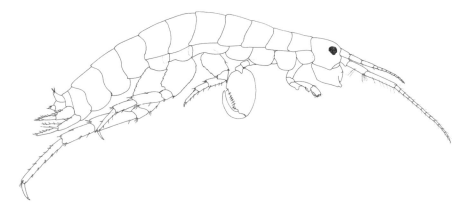

Figure 4.8   *Microrchestia metcalfeae* Lowry and Springthorpe, 2015, habitus male, 12 mm, QM W26984, East Alligator River mouth, Kakadu National Park, Northern Territory.

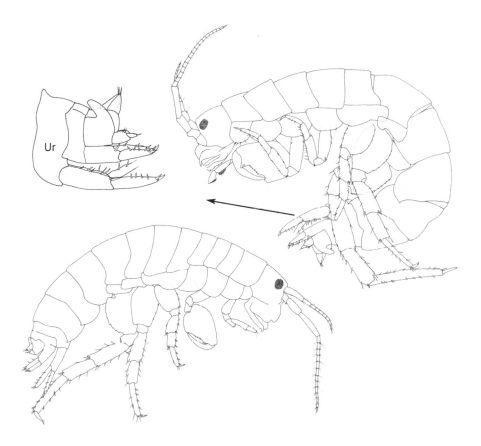

Figure 4.9   *Microrchestia poka* (Lowry and Springthorpe 2015) top habitus and Ur, male, 18
mm, QM W5197, Serpentine Creek; bottom habitus male, 11 mm, QM W9471,
Susan River, South East Queensland, Australia. Scale 0.2 mm.

Territory, 17 June 1982 (T5Q2, K2396); 3 specimens (1 male, 1 gravid female, and 1
juvenile), QM W26983, about 2 km upstream, toward mouth near Flying Fox Island,
Tributary of East Alligator River (12.12°S 132.41°E), estuarine littoral mangroves, in
cracks in mud, dry surface but moist underneath, 21 June 1982, coll. P. Davie (St27).

Allometric growth variation. Pereopod 7 in a 12-mm male is sexually dimorphic
with the merus, carpus, and propodus enlarged and elongated (Fig. 4.8).

Remarks. *Microrchestia metcalfeae* was originally described from specimens up
to 10 mm in body length. The material identified in this study includes males up to
12 mm. The larger male specimen confirms the development of a sexual dimorphic
pereopod 7. Gnathopod 2 of the large male specimens drawn here is less convex and
straighter than the 7 mm holotype. Records reported here only slightly extend the
known distribution of *M. metcalfeae* along the coast of tropical northern Australia
from Darwin Harbour west to Kakadu National Park.

Distribution. Australia, Northern Territory: Darwin Harbour (Lowry and
Springthorpe, 2015), Kakadu National Park (current study).

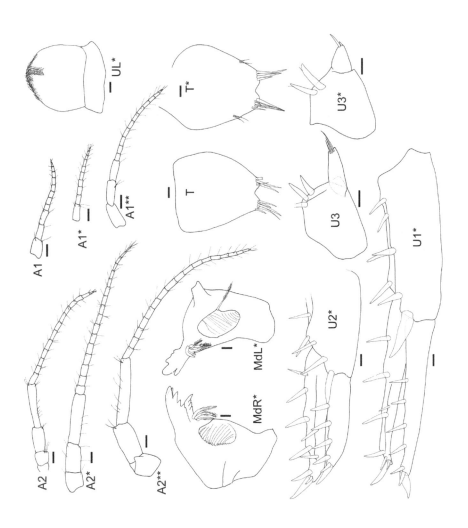

Figure 4.10   *Microrchestia poka* (Lowry and Springthorpe 2015) juvenile male, 8.7 mm, QM W29433, Serpentine Creek; male*, 11 mm, QM W9471, Susan River, male**, 18 mm, QM W5197, Serpentine Creek, South East Queensland, Australia. Scales A1, A2, 0.2 mm, all other parts 0.1 mm.

Microrchestia poka (Lowry and Springthorpe 2015) nov. comb. (Figs. 4.9–4.14).
  Orchestia sp. (Ledoyer, 1979: 173).
  Cochinorchestia poka. (Lowry and Springthorpe 2015: 173–175) (Fig. 4.14).
  Material examined. Queensland: male, 11 mm, dissected, 4 slides, QM W9471, Susan River, South East Queensland (no GPS), in wet mud adjacent to rotten mangrove *Avicennia* sp. limb, 1970, coll. unknown (site 9); male, 18 mm, dissected, parts in microvial, QM W5197, Serpentine Creek, South East Queensland (no GPS), 10 August 1972, coll. Campbell et al. (transect 3/L, site W); 1 juvenile male, 8.7 mm, dissected 2 slides, QM W29433, Serpentine Creek, South East Queensland (no GPS), log litter, 2 August 1972, coll. Campbell et al. (transect 1, site F); 1 juvenile male, 9 mm, QM W5125, Serpentine Creek, South East Queensland (no GPS), log litter, 2 August 1972, coll. Campbell et al. (transect 1, site F); 2 males (10 and 11.2 mm),

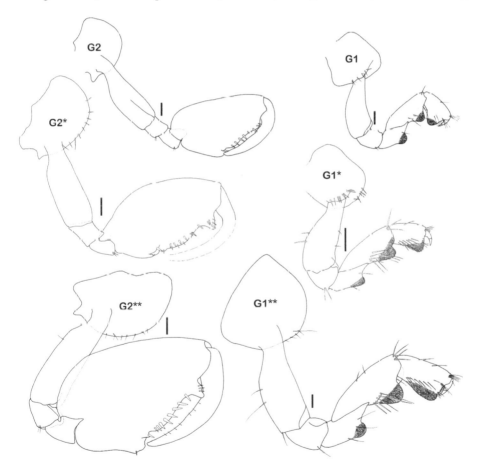

Figure 4.11 *Microrchestia poka* (Lowry and Springthorpe 2015) juvenile male, 8.7 mm, QM W29433, Serpentine Creek; male*, 11 mm, QM W9471, Susan River, male**, 18 mm, QM W5197, Serpentine Creek, South East Queensland, Australia. Scale 0.2 mm.

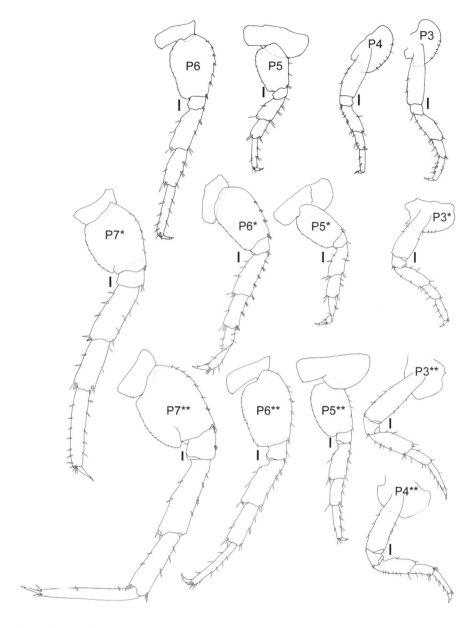

Figure 4.12 *Microrchestia poka* (Lowry and Springthorpe 2015) juvenile male, 8.7 mm, QM W29433, Serpentine Creek; male*, 11 mm, QM W9471, Susan River, male**, 18 mm, QM W5197, Serpentine Creek, South East Queensland, Australia. Scale 0.2 mm.

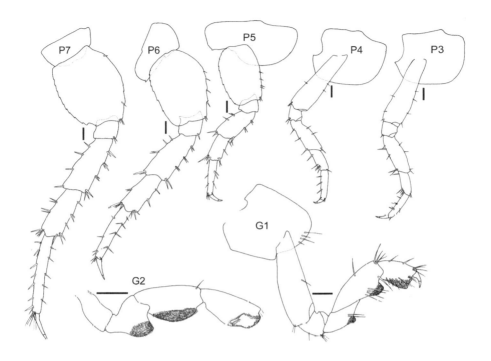

Figure 4.13 *Microrchestia poka* (Lowry and Springthorpe 2015) female, 11 mm, QM W29432, Serpentine Creek, South East Queensland, Australia. Scale 0.2 mm.

QM W5128, Serpentine Creek, South East Queensland (no GPS), 2 August 1972, coll. Campbell et al. (transect 1, site F); 1 female, 11 mm, dissected, 4 slides, QM W29432, Serpentine Creek, South East Queensland, log sample, 1 August 1972, coll. Campbell et al. (transect 1, site C); 5 specimens (4 juveniles and 1 juvenile male, 5, 5.5, 6, and 9 mm), QM W5116, Serpentine Creek, South East Queensland, log sample, 1 August 1972, coll. Campbell et al. (transect 1, site C); 2 males (15 and 16 mm), QM W5253, Jacksons Creek, Cribb Island, South East Queensland (no GPS), 12 October 1972, coll. Campbell et al. (transect 2, site D); 1 juvenile male, QM W16884, Andoon Creek, near Weipa, Queensland (12°34'S 141°52'E), 3 November 1990, coll. P. Davie and J. Short.

Allometric growth variation. Urosomite 3 wing-like dorsal expansions first appear in males of 10 mm in body length. The expansions begin to recurve, projecting anteriorly, in males 11 mm and are wholly recurved in males 18 mm. The sexually dimorphic pereopod 7 in males begins elongation and expansion in males of ~11 mm.

Remarks. Lowry and Springthorpe (2015) erected the name *Cochinorchestia poka* for material reported by Ledoyer (1979) as *Orchestia* sp. The description of *Cochinorchestia poka* was based on a literature assessment with the original illustrations of Ledoyer (1979) reproduced. Comparison of males from 8 mm (holotype) to 18 mm shows that pereopod 7 is sexually dimorphic. The holotype male of *M. poka*

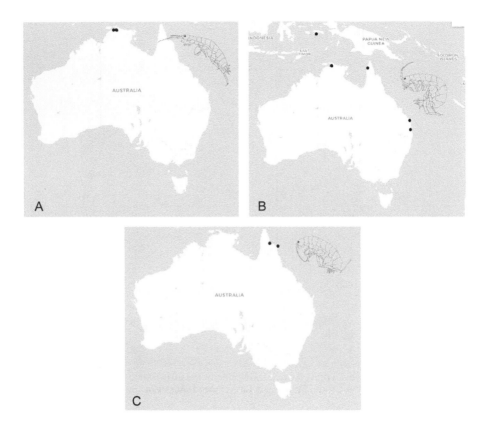

Figure 4.14   Distribution map of (A) *Microrchestia metcalfeae* Lowry and Springthorpe, 2015, (B) *Microrchestia poka* Lowry and Springthorpe, 2015, and (C) *Microrchestia watsonae* Lowry and Peart, 2010.

is 8 mm in length and is now considered an immature specimen. The size range of material examined here shows that the unique wing-like dorsal expansions on urosomite 3 continue to develop with the allometric growth being recurved in the largest known male (18 mm). This unusual character first illustrated by Ledoyer (1979) is not seen elsewhere within the Talitrids.

The present study extends the distribution of *M. poka* from Ambon, Indonesia, to Australia. *Microrchestia poka* nov. comb. is recorded from mangrove habitats at both the type locality and four locations in southern Queensland. The material from Australia provides a considerable range extension with collections from the eastern and western coastline of Queensland. Few collections are known from coastal region between these locations in tropical northern Australia due to the inaccessibility of the extensive, dense mangrove and swampy wetlands. These dominant mangrove habitats suggest that *M. poka* could be present along this coastline.

Habitat. Supralittoral, mangroves.

Distribution. Indonesia: Ambon (Ledoyer 1979). Australia: Queensland: Susan River, Serpentine Creek, Cribb Island, Weipa (current study).

ACKNOWLEDGMENTS

I am grateful to Darryl Potter (QM) for loan and curation of material. This research is supported by the Australia Biological Resources Study (ABRS) grant RF 215–51.

REFERENCES

Barnard, K.H. 1935. Report on some Amphipoda, Isopoda, and Tanaidacea in the collections of the Indian Museum. *Records of the Indian Museum* 37:279–319.

Bousfield, E.L. 1971. Amphipoda of the Bismarck Archipelago and adjacent Indo-Pacific islands (Crustacea). *Steenstrupia* 1:255–93.

Bousfield, E.L. 1984. Recent advances in the systematics and biogeography of landhoppers (Amphipoda: Talitridae) of the Indo-Pacific Region. *Bishop Museum Special Publication* 72:171–210.

Bunting, P., Rosenqvist, A., Lucas, R., Rebelo, L.M., Hilarides, L., Thomas, N., Hardy, A., Itoh, T., Shimada, M. and Finlayson, C. 2018. The global mangrove watch—A new 2010 global baseline of mangrove extent. *Remote Sensing* 10:1669.

Hurley, D.E. 1957. Terrestrial and littoral amphipods of the genus *Orchestia*. Family Talitridae. *Transactions of the Royal Society of New Zealand* 85:149–99.

Ledoyer, M. 1979. Expédition Rumphius II (1975) Crustacés parasites, commensaux, etc. (Th. Monod and R. Serene, ed.) VI. Crustacés Amphipodes Grammariens. *Bulletin du Muséum National d'Histoire Naturelle, Paris, Series 4, Section A* 1:137–81.

Lowry, J.K. and Peart, R. 2010. The genus *Microrchestia* (Amphipoda: Talitridae) in eastern Australia. *Zootaxa* 2349:21–38.

Lowry, J.K. and Springthorpe, R.T. 2015. Coastal Talitridae (Amphipoda: Talitroidea) from north-western Australia to Darwin with a revision of the genus *Cochinorchestia* Lowry and Peart, 2010. *Zootaxa* 3985:151–202.

Lowry, J.K., and Springthorpe, R.T. 2019. Talitrid amphipods from India, East Africa and the Red Sea (Amphipoda, Senticaudata, Talitroidea, Talitridae). *Zootaxa*, *4638*(3), 351–378.

Müller, F. 1864. *Für Darwin*. Wilhelm Engelmann, Leipzig. [in German]

Myers, A.A. and Lowry, J.K. 2020. A phylogeny and classification of the Talitroidea (Amphipoda, Senticaudata) based on interpretation of morphological synapomorphies and homoplasies. *Zootaxa*, *4778*(2), 281–310.

Peethambaran Asari, K. 1998. *Brackish Water Amphipods of the Parangipettai Coast*. Lyla, P.S., Velvizhi, S. and Ajmal Khan, S. (eds.). Centre of Advanced Study in Marine Biology, Annamalai University, Parangipettai, 1–80.

CHAPTER 5

New and Known Species of *Bellorchestia* Serejo and Lowry, 2008 from Australia (Talitridae: Amphipoda: Crustacea)

Lauren E. Hughes and Sue Lindsay

CONTENTS

5.1 INTRODUCTION

Talitrid amphipods are a dominant component of the invertebrate assemblage on sandy beaches in Tasmania, southern Australia, and New Zealand (Hurley 1956, Richardson et al. 1991, Richardson et al. 1997a, b, Serejo and Lowry 2008). Living in burrows within the supralittoral zone, these talitrids are dependent on marine subsidies as a food source and are in turn a prey item for various invertebrate predators, reptiles, and shore birds (Richardson and Araujo 2015).

The genus *Bellorchestia* was established for eight species by Serejo and Lowry (2008), with Lowry (2012) describing one additional species. Recent redescription of the type material of *B. pravidactyla* (Haswell 1880) by Hughes and Ahyong (2017), following discovery of Haswell's syntypes, which had been presumed lost, showed that the generic type, *B. richardsoni* Serejo and Lowry, 2008, along with *B. mariae* Lowry, 2012, were junior synonyms of *B. pravidactyla*. It is evident that a second little-known species, *Talorchestia marmorata* Haswell, 1880, should be transferred out of *Bellorchestia* following the assessment of recently found type material; this taxonomic move is dealt with elsewhere (Hughes and Lowry, in prep.). This chapter focuses on four new species of *Bellorchestia* described from Tasmania: *Bellorchestia lutruwita* sp. nov., *B. needwonnee* sp. nov., *B. palawakani* sp. nov., and *B. reliquia* sp. nov. Additional distribution records are also provided for *B. pravidactyla*. There are currently five species of *Bellorchestia* in Australia and a key to the Australian species is provided.

5.2 MATERIAL AND METHODS

The majority of material used for this study came from talitrid research projects led by Alastair Richardson, and significantly from two grants from the National Estate Grants Program where Colin Shepherd is acknowledged here as the main collector in the field. Specimens are lodged in the Australian Museum, Sydney (AM), Natural History Museum London (B.M. [N.H.]), and The Tasmanian Museum and Art Gallery (TMAG). Material was dissected in 80% ethanol. Slides were made using Aquatex™ mounting agent. Specimens were prepared for scanning electron microscopy as follows: specimens were sonicated in a 10% solution of the surfactant TWEEN 80 to remove detritus before being transferred back to 80% ethanol; preserving solution was sequentially advanced in 5% increments from 80% to 100% ethanol; the material was critical point dried, mounted individually on pins or stubs and gold sputter coated. Images were captured using a Zeiss EVO LS15 scanning electron microscope (SEM) with Robinson backscatter detector and a JEOL JSM 6480LA SEM. The collector abbreviation ARTP stands for Alastair Richardson Talitrid Projects. Standard abbreviation on the plates are as follows: A – antenna; C – coxa; Dv – divina virgula; Ep – epimeron; G – gnathopod; H – head; Md – mandible; Mx – maxilla; Mxp – maxilliped; P – pereopod; T – telson; U – uropod; Ur – urosome; l – left and r – right. NB. Myers & Lowry (2020) place *Bellorchestia* in both the new family Talitroidae and also Talitridae, the former placement appears to be in error based on other citations within the manuscript where the genus name *Bellorchestia* appears within the Talitrinae subfamily.

5.3 SYSTEMATICS

Talitridae Rafinesque, 1815

    Bellorchestia Serejo and Lowry, 2008
    Talorchestia Dana, 1853: 851. – Stebbing, 1906: 543 (in part). – Stephensen, 1948: 7 (in part). – Hurley, 1956: 359 (in part).
    Bellorchestia Serejo and Lowry, 2008: 169. – Lowry, 2012: 2. –Myers and Lowry, 2020: 303–304 (lapsus p. 300).
    Type species. *Bellorchestia pravidactyla*, as senior synonym of *B. richardsoni*, original designation by Serejo and Lowry, 2008 (see also Hughes and Ahyong 2017).
    Generic diagnosis (modified after Serejo and Lowry 2008 and Lowry 2012).
    Male. Eyes medium (1/5 to 1/3 head length). Antenna 1: short, not longer than article 4 of antenna 2 peduncle. Antenna 2: peduncular articles well developed, not

geniculate, incrassate; peduncular article 3 without ventral process. Labrum: with robust setae. Mandible: left lacinia mobilis 4-6-dentate. Maxilliped: palp article 2 distomedial lobe well developed; article 4 reduced, button-shaped. Gnathopod 1: sub to parachelate; carpus posterodistal corner without protuberance covered in palmate setae; propodus palm posterodistal corner defined by large protuberance covered in palmate setae. Gnathopod 2: subchelate; basis narrow; merus- and carpus-free; dactylus not attenuated distally. Pereopods 3–7: cuspidactylate. Pereopod 4: carpus significantly shorter than carpus of pereopod 3. Pereopods 5–7: without setae along posterior margin of the dactylus. Pereopods 6–7: weakly sexually dimorphic, without row of short setae along posterior margin of dactylus. Pereopod 6: longer than Pereopod 7. Pereopod 7: basis lateral sulcus present, slightly pronounced. Gills lobate and/or convoluted; gills 3–5 smaller than gills 2 and 6. Pleonites 1–3: without dorsal spines. Pleopods 1–3: well developed. Epimera 1–3: without vertical slits. Uropod 1: rami without apical spear-shaped setae; outer ramus slender; with marginal robust setae. Uropod 3: well developed; ramus subequal to or longer than peduncle. Telson: apically incised, with 7–10+ robust setae per lobe. Female gnathopod 2: basis linear. Oostegites 2–5: setae with simple straight tips.

Species composition: *B. lutruwita* sp. nov.; *B. needwonnee* sp. nov.; *B. palawakani* sp. nov.; *B. pravidactyla* (Haswell 1880); *B. quoyana* (Milne Edward 1840); *B. reliquia* sp. nov.; and *B. spadix* (Hurley 1956).

Remarks. The above diagnosis is based on Lowry (2012) and expanded to include characters originally defined in the written diagnosis and Table 2 of Serejo and Lowry (2008). All five Australian species of *Bellorchestia* occur in Tasmania with only *B. pravidactyla* and *B. palawakani* sp. nov. being recorded on the Australian mainland (Hughes and Ahyong 2017).

A number of poorly described Australian and New Zealand species are excluded from *Bellorchestia* based on recent research (*Talorchestia chathamensis* Hurley 1956; *T. kirki* Hurley 1956; *T. marmorata* Haswell 1880; *T. tumida* Thomson 1885); these will be dealt with more thoroughly in separate forthcoming works. Notably, *Talorchestia marmorata* Haswell, 1880 requires placement in a different genus following an examination of wet specimens that has confirmed the presence of palmate setal lobes on gnathopods 1 and 2, which excludes *T. marmorata* from *Bellorchestia*. Further, *T. chathamensis* is more suitably placed alongside its sister taxon *T. telluris* Bate, 1862, an oversight in earlier works on *Bellorchestia*. Similarly, new characters are recognized in the closely aligned *T. kirki* and *T. tumida* Thomson, 1885, prompting their placement elsewhere.

The number of robust setae on the telson and uropods is used as a species-level character within the Talitridae (Myers and Lowry, 2020). Within *Bellorchestia*, these setal counts are not a robust species-level character since considerable overlap is observed between species and there is variation with growth stage. Recognition of this intraspecific variation was made possible, thanks to the large amount of material examined from multiple locations and through time.

The presence of robust setae on the labrum unites *Bellorchestia* and *Hermesorchestia*, whereas the presence of palmate lobes on the propodus of the male gnathopod 1 separates *Bellorchestia* from *Hermesorchestia*, in which lobes are present on both the carpus and propodus. Further, male sexual dimorphism is seen in the weakly expanded merus and carpus on both pereopods 6 and 7 in *Bellorchestia*, while in *Hermesorchestia* dimorphism in the male is extremely pronounced, but with grossly expanded articles of pereopod 7 only. Study of male *Bellorchestia pravidactyla* from extensive collections across the species confirms the slight sexually dimorphic development of pereopod 7 and is not an artifact of unobserved adult male form. Specimens of *B. pravidactyla* up to 24.5 mm lack a grossly expanded pereopod 7 basis or carpus, whereas male *H. alastairi* Hughes and Lowry, 2017 of 11 mm have already developed these characters. Therefore, it must be assumed that male dimorphism in *Bellorchestia* does not progress to the degree seen for the closely related *Hermesorchestia*. With the small number of species in each genus, it is tempting to unite these taxa; however, such variation in sexually dimorphic characters is strong evidence for different reproductive behaviors.

Within the genus *Bellorchestia* sensu stricto, only four species (*B. lutruwita*, *B. pravidactyla*, *B. quoyana*, and *B. spadix*) are known to have slight sexually dimorphic variation in the merus and carpus of pereopod 7, with a thickening from a laminar article to a three-dimensional cylinder shape (Hurley 1956, Fig. 5.11; Hughes and Ahyong 2017, Figs. 5.4 and 5.9). This is in contrast to other sexually dimorphic talitrid genera where sexual variation is much more pronounced with articles showing either an obvious laminar broadening (*Orchestia cavimana* Heller 1865) or developed as a novel structure (*Hermesorchestia alastairi* Hughes and Lowry, 2017, *T. chathamensis*). The absence of records of variation in the other *Bellorchestia* species may be a result of under sampling, given the challenges of recognizing penultimate males.

Bellorchestia lutruwita sp. nov. (Figs. 5.1–5.10)

Type material. Holotype male, 23.3 mm, dissected, parts SEM stubs, whole animal pin mount, AM P.97655, Cox Bight Beach, just west of Point Eric, Tasmania (43°29′33″S, 146°14′26″E), cast kelp at EHWN, 4 February 1990, coll. A. Richardson and S.M. Eberhard (TA143). Paratypes: female, 13 mm, dissected, parts SEM stubs, whole animal pin mount, AM P.97656; male, 18 mm, dissected, 1 slide, AM P.99317; female, 20.8 mm, dissected, AM P.99318; 20 specimens, AM P.96907, same location as holotype. Paratypes 20 specimens, AM P.96909, Cox Bight Beach, just west of Point Eric, Tasmania (43°29′33″S, 146°14′26″E), cast kelp at extreme high water spring, 4 February 1990, coll. A. Richardson and S.M. Eberhard (TA144). Paratypes: male, dissected, 18.5 mm, AM P.99320; male, dissected, 21 mm, AM P.99321; male, dissected, 13.5 mm, AM P.99319; 20 specimens, AM P.96910, Cox Bight Beach, just west of Point Eric, Tasmania (43°29′33″S, 146°14′26″E), sand burrows at extreme high water spring, 4 February 1990, coll. A. Richardson and S.M. Eberhard (TA145). Paratypes 20 specimens, AM P.96913, Cox Bight Beach, just west of Point Eric, Tasmania (43°29′33″S, 146°14′26″E), sand burrows in sandfall at base of peat cliff just above extreme high water spring, 0 m, 4 February 1990, coll. A. Richardson and S.M. Eberhard (TA146).

Additional material examined. Tasmania: 2 specimens, AM P.99959, Cox Bight (43°31′S, 146°14′E), 27 June 1997, coll. ARTP; many specimens, AM P.99960,

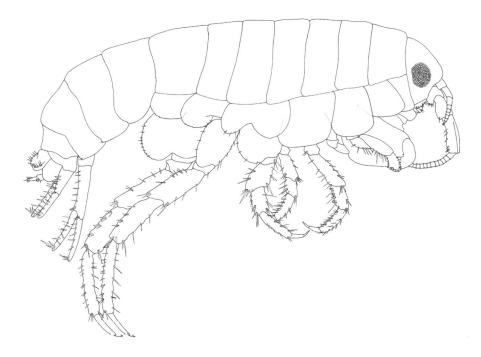

Figure 5.1   *Bellorchestia lutruwita* sp. nov., male, 18 mm, AM P.99317, Cox Bight Beach, just west of Point Eric, Tasmania.

South Cape Rivulet (43°34′11″S, 146°46′12″E), 23 February 1997, coll. ARTP; many specimens, AM P.99961, Finns Beach (43°32′41″S, 146°53′28″E), 20 January 1995, coll. ARTP; many specimens, AM P.96950, Bramble Cove, Bathurst Channel, 43°19′21″S, 146°00′12″E, under seaweed at low tide on beach, 17 February 1987, coll. J.H. Waterhouse (TA248); 1 male, AM P.99322, Lighthouse Bay, Bruny Island, 43°29′16″S, 147°08′59″E, burrows in clay and wet sand, 1 m above beach flat, sedges, *Correa* sp., 16 November 1993, coll. A. Richardson, R. Swain, C. Shepherd, and M. Nelson (TA340); 1 male, AM P.99323, South Cape Bay (43°36′04″S 146°47′04″E), on vegetation at night (feeding), on ridge immediately above extreme high water spring, 21 August 1988, coll. A. Richardson and J. Tupascz (TA197); many specimens, AM P.96986, Ocean Beach, Strahan, 42°08′47″S, 145°15′39″E, toward high water, under fresh cast strapweed cf *Ecklonia* sp., 30 January 1991, coll. M. Moore and K. Heiden (TA286).

Type locality. Cox Bight Beach, just west of Point Eric, Tasmania, Australia, 43°29′33″S, 146°14′26″E.

Etymology. Named from Lutruwita, the indigenous word for Tasmania.

Description. Based on type material. Head: eye medium size (greater than 1/5 to 1/3 head length). Antenna 1: short, reaching end of peduncular article 4 of antenna 2. Antenna 2: slightly less than half body length (0.45 × length); peduncular articles slender, with many large robust setae; article 5 longer than article 4; flagellar articles final article large, cone-shaped forming a *virgula divina*. Labrum: with apical setal

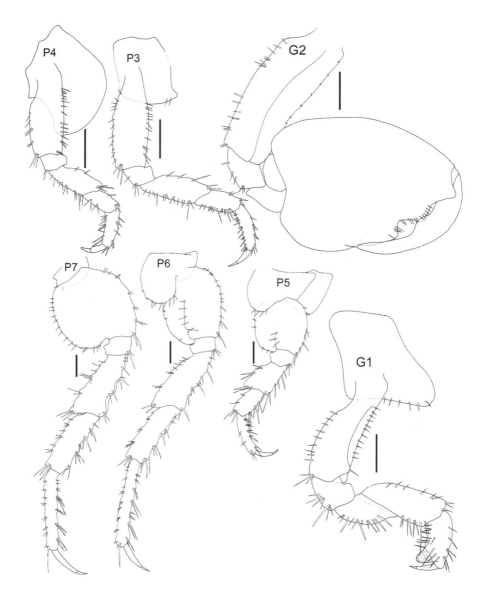

Figure 5.2 *Bellorchestia lutruwita* sp. nov., male, 18 mm, AM P.99317, b male, 18.5 mm, AM P.99320, c male, 21 mm, AM P.99321, Cox Bight Beach, just west of Point Eric, Tasmania. Scales 2 mm.

patch; epistome with many robust setae (damaged in SEM image Fig. 5.13, AM P.97655). Mandible: left lacinia mobilis 5-dentate. Labium: distolateral setal tuft present, distomedial setal tuft present; without inner plates. Maxilla 1: with 1-articulate small palp. Maxilliped: palp article 2 distomedial lobe well developed; article 4 fused with article 3.

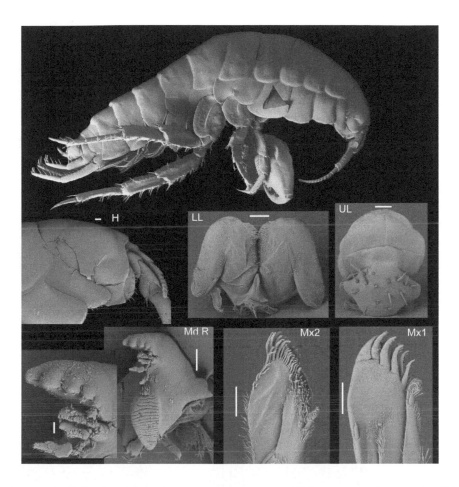

Figure 5.3   *Bellorchestia lutruwita* sp. nov., male, 18 mm, AM P.99317, b male, 18.5 mm, AM
P.99320, c male, 21 mm, AM P.99321, Cox Bight Beach, just west of Point Eric,
Tasmania. Scales 2 mm.

Pereon. Gnathopod 1: sexually dimorphic; subchelate; coxa subequal to coxa 2; carpus longer than propodus, 1.8 as long as propodus; propodus subrectangular, twice as long as broad, anterior margin with 4–5 groups of robust setae, posterior margin with palmate lobe, palm transverse; simplidactylate, dactylus overreaching palm, without anterodistal denticular patch. Gnathopod 2: sexually dimorphic; subchelate; basis slender; ischium anterior margin with weak lateral and medial rounded lobe; carpus triangular, reduced, enclosed by merus and propodus; propodus 1.7 × as long as wide, palm weakly subacute, extending to 45% along posterior margin, lined with robust setae, with well-developed midpalmar sinus; dactylus curved, subequal in length to palm, midposterior margin with bump, apically acute, closing into dactylar socket. Pereopods 3–7: asymmetrically tricuspidate laterally, dactylus without anterodistal denticular patch. Pereopods 3–4: coxae deeper than wide. Pereopod 3: carpus length twice width. Pereopod 4: significantly shorter than pereopod 3; carpus

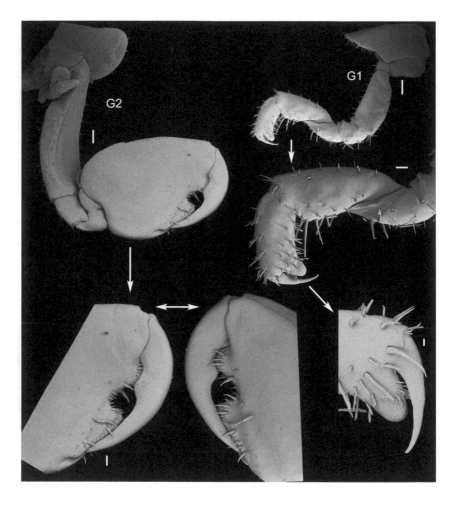

Figure 5.4   *Bellorchestia lutruwita* sp. nov., male, 18 mm, AM P.99317, b male, 18.5 mm, AM
P.99320, c male, 21 mm, AM P.99321, Cox Bight Beach, just west of Point Eric,
Tasmania. Scales 2 mm.

significantly shorter than carpus of pereopod 3, length 1.5 × width; dactylus posterior margin thickened with 1 proximal and 1 distal projections. Pereopod 5: merus
length 1.1 × width; carpus length 1.2 × width; propodus distinctly longer than carpus, length 2.2 × width; dactylus long, slender, length 3 × width. Pereopods 6–7:
not sexually dimorphic. Pereopod 6: longer than pereopod 7; coxa posterior lobe
posteroventral corner rounded, posterior margin perpendicular to ventral margin;
merus not expanded, length 1.5 × width; carpus not expanded, length 2.1 × width;
propodus length 3.5 × width, subequal to carpus length. Pereopod 7: basis expanded,
produced posteriorly, posterior margin convex, lined with small robust setae; merus
length 1.5 × width; carpus length twice width; propodus length 3.1 × width, subequal to carpus length.

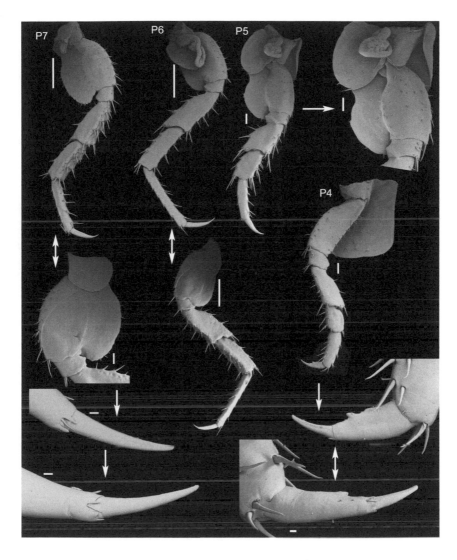

Figure 5.5   *Bellorchestia lutruwita* sp. nov., female, 20.8 mm, AM P.99318, Cox Bight Beach, just west of Point Eric, Tasmania. Scales 2 mm.

Pleon. Pleopods 1–3: well developed, rami segmented. Epimeron 1 –3: postero-ventral corner weakly convex, minutely serrate, lined with robust setae. Uropod 1: peduncle with 15+ robust setae, without apical spear-shaped setae; rami positioned directly on top of each other; inner ramus subequal in length to outer ramus, with 25 marginal robust setae in 3 rows; outer ramus with 12+ marginal robust setae 2 rows, both rows on outer margin. Uropod 2: peduncle with 7+ robust setae; inner ramus subequal in length to outer ramus, with 11 marginal robust setae in 2 rows; outer ramus with 4 marginal robust setae in 1 row. Urosomite 3: subrectangular, not enclosing around telson, broader than deep, 1.2 × as broad as deep. Uropod 3:

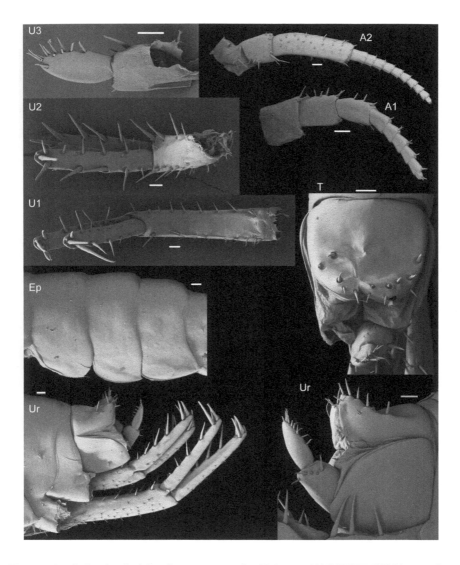

Figure 5.6　*Bellorchestia lutruwita* sp. nov., male, 23.3 mm, AM P.97655, SEM image, Cox Bight Beach, just west of Point Eric, Tasmania. Scales H, Ul, LL, Md, Mx1, Mx2, Mxp, 100 μm; Md incisors, 20 μm.

vestigial; peduncle not dorsally concave, length 1.1 × depth, with 4 robust setae; ramus linear, partially fused to peduncle (when viewed dorsally), subequal to peduncle, 2.5 × as long as broad, with 11–12 marginal setae in two lateral rows and cluster of apical setae. Telson: subovate, broader than long, lateral, and apical margins convex, incised apically, dorsal midline weakly, with around 10–15 dorsal, marginal, and apical robust setae per lobe; length subequal to uropod 3 peduncle.

　　Female (sexually dimorphic characters). Gnathopod 1: simple; propodus posterior margin without palmate lobe; propodus palm near transverse. Gnathopod 2:

Figure 5.7 *Bellorchestia lutruwita* sp. nov., male, 23.3 mm, AM P.97655, SEM image, Cox Bight Beach, just west of Point Eric, Tasmania. Scales Gn1, Gn2, A2, 200 µm; A1, Gn1 palm, Gn2 palm, 100 µm; Gn1 in part, 20 µm.

mitten-shaped; basis expanded anteriomedially; ischium without lobe; carpus well developed (not enclosed by merus and propodus), length 2.5 × width; propodus 2.1 × width, palm obtuse; dactylus shorter than palm.

Variation. There is little morphological variation with growth stage in males.

Remarks. The male is damaged with the upper epistome missing in SEM image (Fig. 5.13, AM P.97655); however, robust setae were confirmed as present on the upper lip, on all other wet specimens.

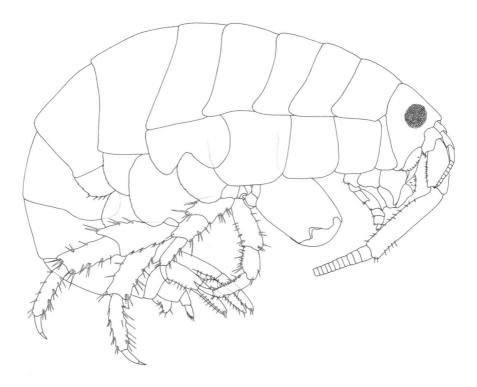

Figure 5.8 *Bellorchestia lutruwita* sp. nov., male, 23.3 mm, AM P.97655, SEM image, Cox
Bight Beach, just west of Point Eric, Tasmania. Scales P4 dactylus, P5 dactylus,
P7 dactylus, P7 basis, 200 µm; P3 dactylus, 100 µm; P3, P4, P5, P6, P7, 1 mm.

Bellorchestia lutruwita sp. nov. is closely related to *B. pravidactyla* in the
development of the male gnathopod 2 palm sculpturing (proximally concave and
distal margin with a low ridge or teeth), pereopod 4 dactylus and large number of
robust setae on the telson. *Bellorchestia lutruwita* sp. nov can be distinguished
from *B. pravidactyla* by the hump on the posterior margin of the male gnathopod 2 dactylus, which is recurved in shape, while the later species has no hump
and the dactylus posterior margin is more concave than recurved with the apical
margin falcate.

Bellorchestia lutruwita sp. nov. differs from *B. needwonnee* sp. nov. by the angle
of the gnathopod 2 palm being more transverse. Overall these species are very similar in the palm sinus and dactylar bump of gnathopod 2, only in close inspection of
specimens is it clear that the palm angles are very different. Pereopod 6 propodus
being shorter than carpus is another strong character separating these species but
easily overlooked. *Bellorchestia lutruwita* sp. nov. is known to cooccur at sites with
all other Australian *Bellorchestia* species.

Ecology. Burrowing in sand.

Distribution. Tasmania: Cox Bight; Finns Beach; Bruny Island, Bathurst Channel;
South Cape Bay; Strahan (current study).

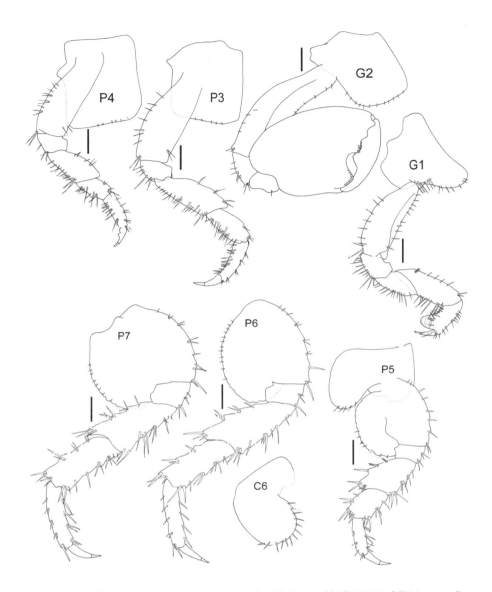

Figure 5.9   *Bellorchestia lutruwita* sp. nov., male, 23.3 mm, AM P.97655, SEM image, Cox Bight Beach, just west of Point Eric, Tasmania. Scales Ep, 200 μm; Ur in part, Ur dorsal view, Ur1, Ur2, Ur3, 100 μm.

Bellorchestia needwonnee sp. nov. (Figs. 5.11–5.17)

Type material. Holotype male, 14.3 mm, dissected, parts SEM stubs, whole animal pin mount, AM P.96911, Cox Bight Beach, just west of Point Eric, Tasmania (43°29′33″S, 146°14′26″E), sand burrows at extreme high water spring, 4 February 1990, coll. A. Richardson and S.M. Eberhard (TA145). Paratype female, 15.4 mm, dissect, parts SEM stubs, whole animal pin mount, AM P.97657, Cox Bight Beach,

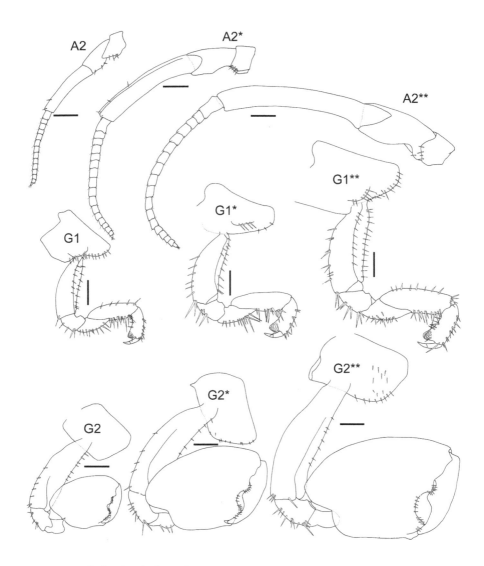

Figure 5.10 *Bellorchestia lutruwita* sp. nov., paratype female, 13 mm, AM P.97656, SEM image, Cox Bight Beach, just west of Point Eric, Tasmania. Scales Gn1, Gn2, Ep, 200 μm; Gn2 in part, Oost, Ur, Ur dorsal view, 100 μm; Gn1 palm, Gn2 palm, 20 μm.

just west of Point Eric, Tasmania (43°29′33″S, 146°14′26″E), sand burrows at extreme high water spring, 4 February 1990, coll. A. Richardson and S.M. Eberhard (TA145).

Additional material examined. Tasmania: Male, 13 mm, dissected, 1 slide, AM P.99324, Lighthouse Bay, Bruny Island (43°29′16″S, 147°08′59″E), burrows in clay and wet sand, 1 m above beach flat, sedges, *Correa* sp., 16 November 1993, coll. A. Richardson, R. Swain, C. Shepherd, and M. Nelson (TA342); female, 14.3 mm, dissected, AM P.99325, Lighthouse Bay, Bruny Island (43°29′16″S, 147°08′59″E),

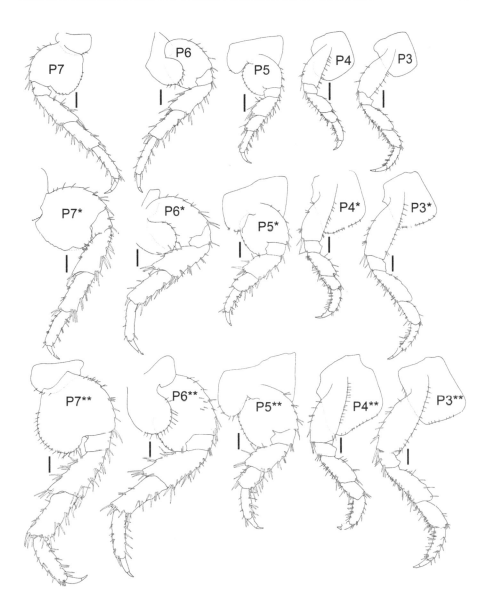

Figure 5.11   *Bellorchestia needwonnee* sp. nov., male, 13 mm, AM P.99324, Lighthouse Bay, Bruny Island, Tasmania.

burrows in clay and wet sand, 1 m above beach flat, sedges, *Correa* sp., 16 November 1993, coll. A. Richardson, R. Swain, C. Shepherd, and M. Nelson (TA342); 2 females, AM P.97025, Lighthouse Bay, Bruny Island (43°29′16″S, 147°08′59″E), burrows in clay and wet sand, 1 m above beach flat, sedges, *Correa* sp., 16 November 1993, coll. A. Richardson, R. Swain, C. Shepherd, and M. Nelson (TA342); 2 specimens (1 male, 1 female), AM P.97309, Lighthouse Bay, Bruny Island (43°29′16″S, 147°08′59″E),

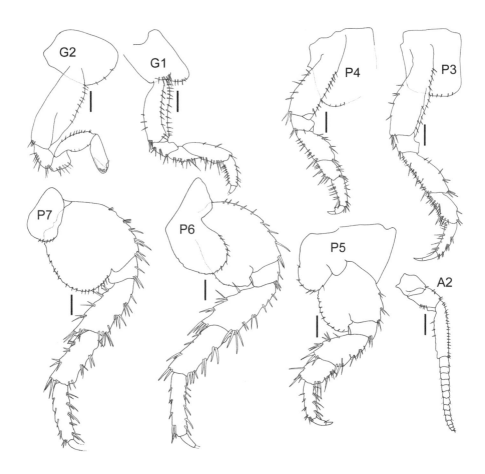

Figure 5.12   *Bellorchestia needwonnee* sp. nov., male, 13 mm, AM P.99324, Lighthouse Bay, Bruny Island, Tasmania. Scales 2 mm.

burrows in clay and wet sand, 1 m above beach flat, sedges, *Correa* sp., 16 November 1993, coll. A. Richardson, R. Swain, C. Shepherd, and M. Nelson (TA341); 1 male, AM P.99326, Ocean Beach, Strahan (42°08′47″S, 145°15′39″E), toward high water, under fresh cast strapweed, macroalga cf *Ecklonia* sp., 30 January 1991, coll. M. Moore and K. Heiden (TA286); 1 specimen, AM P.99709, Arthur Beach (41°04′48″S, 144°40′11″E), 20 May 1997, coll. ARTP; few specimens, AM P.99712, Broken Arm Beach, King Island (40°04′S, 144°02′E), 23 March 1997, coll. ARTP; 3 specimens, AM P.99719, Cloudy Bay, Bruny Island (43°27′22″S, 147°15′12″E), 24 March 1995, coll. ARTP; 3 specimens, AM P.99720, Cloudy Bay, Bruny Island (43°27′35″S, 147°13′12″E), 17 May 1997, coll. ARTP; many specimens, AM P.99721, Colliers Beach, King Island (40°06′S, 143°59′E), 23 March 1997, coll. ARTP; 1 specimen, AM P.99722, Cox Bight (43°31′S, 146°14′E), 27 June 1997, coll. ARTP; 4 specimens, AM P.99723, Deephole Bay (43°27′23″S, 146°58′09″E), 20 January 1995, coll. ARTP; 1 specimen, AM P.99724, Fortescue Bay (43°08′14″S, 147°57′15″E), 26 April 1995,

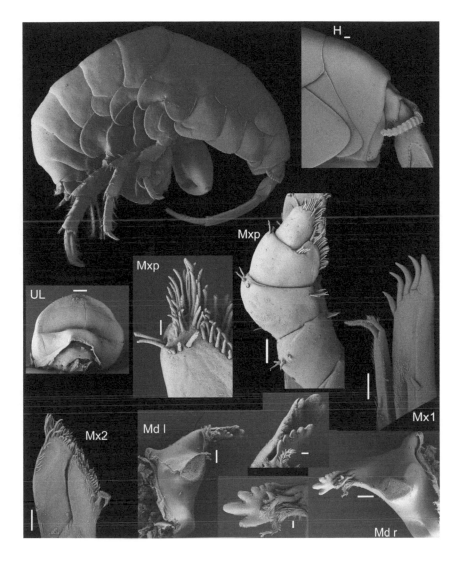

Figure 5.13   *Bellorchestia needwonnee* sp. nov., male, 14.3 mm, AM P.96911, SEM image,
Cox Bight Beach, just west of Point Eric, Tasmania. Scales H, Ul, LL, Md, Mx1,
Mx2, Mxp, 100 μm; Md incisor, 20 μm.

coll. ARTP; 3 specimens, AM P.99725, Little Jetty Beach, Bruny Island (43°27′33″S,
147°09′21″E), 6 March 1995, coll. ARTP; 2 specimens, AM P.99726, Nine Mile
Beach, King Island (39°44′S, 144°06′E), 24 March 1997, coll. ARTP; 2 specimens,
AM P.99727, Nye Bay (43°03′S, 145°40′E), 11 April 1997, coll. ARTP; 1 specimen,
AM P.99728, Ocean Beach, Strahan (42°08′47″S, 145°15′39″E), 19 June 1997, coll.
ARTP; 1 specimen, AM P.99729, Perkins Island (40°46′12″S, 145°03′00″E), 1 June
1997, coll. ARTP; 4 specimens, AM P.99730, Sandy Cape Beach (41°22′S, 144°46′E),
12 April 1997, coll. ARTP; many specimens, AM P.99731, St Albans Bay (40°57′S,

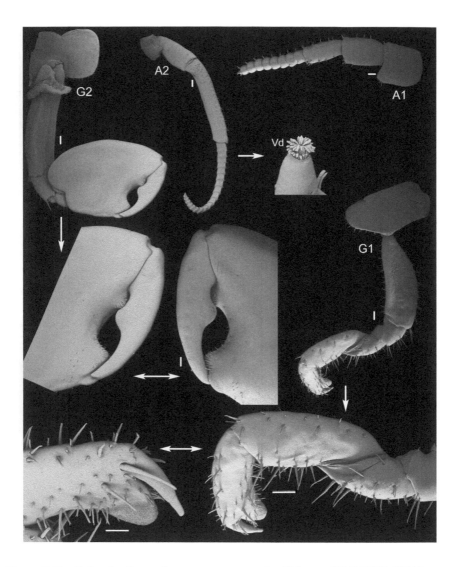

Figure 5.14   *Bellorchestia needwonnee* sp. nov., male, 14.3 mm, AM P.96911, SEM image,
Cox Bight Beach, just west of Point Eric, Tasmania. Scales Gn1, Gn2, 200 μm;
Gn in part, 100 μm; Gn1 palm, 20 μm.

147°18′E), 2 May 1997, coll. ARTP; 1 specimen, AM P.99732, Spain Bay (43°22′S, 145°58′E), 27 June 1997, coll. ARTP; 1 specimen, AM P.99733, White Beach, King Island (39°36′29″S, 144°00′45″E), 24 March 1997, coll. ARTP; 4 specimens, AM P.99734, Yellow Rock Beach, King Island (39°40′11″S, 143°54′36″E), 24 March 1997, coll. ARTP.

Type locality. Cox Bight Beach, just west of Point Eric, Tasmania, 43°29′33″S, 146°14′26″E.

Figure 5.15 *Bellorchestia needwonnee* sp. nov., male, 14.3 mm, AM P.96911, SEM image, Cox Bight Beach, just west of Point Eric, Tasmania. Scales P7 basis, P5 basis, P4, 200 μm; P5, P7 dactylus, 200 μm; P4 dactylus 100 μm; P6, P7, 1 mm.

Etymology. Named for the indigenous mob "Needwonnee", whose territory includes the type locality, Cox Bight Beach.

Description. Based on type material. Head: eye medium size (greater than 1/5 to 1/3 head length). Antenna 1: short, not reaching midpoint of peduncular article 5 of antenna 2. Antenna 2: slightly less than half body length (0.45 × length); peduncular articles slender, with many large robust setae; article 5 longer than article 4; flagellar articles final article large, cone-shaped forming a *virgula divina*. Labrum: with apical setal patch; epistome with many robust setae. Mandible: left lacinia mobilis

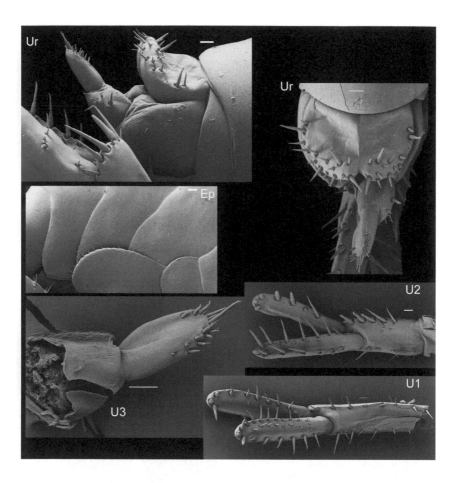

Figure 5.16   *Bellorchestia needwonnee* sp. nov., male, 14.3 mm, AM P.96911, SEM image,
Cox Bight Beach, just west of Point Eric, Tasmania. Scales Ep, Ur, 200 μm; A1,
A2, Ur in part, Ur dorsal view, Ur1, Ur2, 100 μm.

4-dentate (based on additional material examined). Labium: distolateral setal tuft
present, distomedial setal tuft present; without inner plates. Maxilla 1: with 1-articu-
late small palp. Maxilliped: palp article 2 distomedial lobe well developed; article 4
fused with article 3 (based on additional material examined).

Pereon. Gnathopod 1: sexually dimorphic; subchelate; coxa subequal to coxa
2; carpus longer than propodus, 1.6 as long as propodus; propodus subrectangular,
twice as long as broad, anterior margin with 4–5 groups of robust setae, posterior margin with palmate lobe, palm transverse to slightly oblique; simplidactylate,
dactylus overreaching palm, without anterodistal denticular patch. Gnathopod 2:
sexually dimorphic; subchelate; basis slender; ischium anterior margin with weak
lateral and medial rounded lobe; carpus triangular, reduced, enclosed by merus
and propodus; propodus 1.2 × as long as wide, palm subacute, 55% along posterior
margin, with midpalmar sinus, lined with robust setae; dactylus curved, subequal

Figure 5.17 *Bellorchestia needwonnee* sp. nov., female, 15.4 mm, AM P.97657, SEM image, Cox Bight Beach, just west of Point Eric, Tasmania. Scales Gn1, Gn2, 200 µm; Gn1 in part, Gn2 in part, T, 100 µm; Gn1 palm, 30 µm; Gn1 palm, Gn2 in part, 20 µm.

in length to palm, midposterior margin with bump, apically acute, closing into dactylar socket. Pereopods 3–7: asymmetrically tricuspidate laterally, dactylus without anterodistal denticular patch. Pereopods 3–4: coxae deeper than wide. Pereopod 3: carpus length twice width. Pereopod 4: significantly shorter than pereopod 3; carpus significantly shorter than carpus of pereopod 3, length 1.5 × width; dactylus posterior margin thickened with 1 proximal and 1 distal projections. Pereopod 5: merus length 1.1 × width; carpus length 1.5 × width; propodus distinctly longer than carpus, length 3 × width; dactylus long, slender, length 5 × width. Pereopods 6–7: not sexually dimorphic. Pereopod 6: longer than pereopod 7; coxa posterior lobe posteroventral corner rounded, posterior margin perpendicular to ventral margin; merus not expanded, length twice width; carpus not expanded, length 2.8 × width; propodus length 6.5 × width, longer than carpus. Pereopod 7: basis expanded, produced posteriorly, posterior margin convex, lined with small robust setae; merus length twice width; carpus length 2.8 × width; propodus length 5.1 × width, distinctly longer than carpus.

Pleon. Pleopods 1–3: well developed, rami segmented. Epimeron 1: posterior margin convex, smooth; posterodistal corner with small tooth. Epimeron 2: posterior margin straight, smooth; posterodistal corner with small tooth. Epimeron 3: posteroventral corner produced, smooth; posterodistal corner with small tooth. Uropod 1: peduncle with 15+ robust setae, without apical spear-shaped setae; rami positioned directly on top of each other; inner ramus subequal in length to outer ramus, with 4 marginal robust setae in 2 rows; outer ramus with 1 marginal robust setae 1 row. Uropod 2: peduncle with 11 robust setae; inner ramus subequal in length to outer ramus, with 4 marginal robust setae in 1 row; outer ramus with 9 marginal robust setae in 2 rows. Urosomite 3: subrectangular, not enclosing around telson, broader than deep, 1.2 × as broad as deep. Uropod 3: vestigial; peduncle not dorsally concave, length 1.1 × depth, with 5 robust setae; ramus linear, not fused to peduncle, subequal to peduncle, 2.5 × as long as broad, with 7–9 marginal setae and cluster of apical setae. Telson: subovate, broader than long, lateral margins straight to convex, apical margins convex, incised apically, dorsal midline absent or weakly present, with around 5–8 dorsal, marginal, and apical robust setae per lobe; length subequal to uropod 3 peduncle.

Female (sexually dimorphic characters). Gnathopod 1: simple; propodus posterior margin without palmate lobe; propodus palm near transverse. Gnathopod 2: mitten-shaped; basis expanded anteromedially; ischium without lobe; carpus well developed (not enclosed by merus and propodus), length 3.1 × width; propodus 2.4 × width, palm obtuse; dactylus shorter than palm.

Variation. The male gnathopod 2 palmar sinus and dactylar hump becomes more pronounced with increasing body size.

Remarks. With few specimens at hand, it is tempting to place *B. needwonnee* sp. nov. as an aberrant form of *B. lutruwita* sp. nov.; however, the points of variation between specimens are species-level characters elsewhere in the genus *Bellorchestia*. While only a few specimens of *B. needwonnee* sp. nov. were available, the significant amount of material of *B. lutruwita* sp. nov. for study showed that these characters did not vary within or between populations.

Bellorchestia needwonnee sp. nov. differs from *B. lutruwita* sp. nov. and *B. pravidactyla* in the more shallow proximal sinus on the male gnathopod 2 propodus and the larger ridges/subquadrate teeth defining the palm. Both *B. needwonnee* sp. nov. and *B. lutruwita* sp. nov. have a proximal hump on the male dactylus posterior margin, which is absent in *B. pravidactyla*. *Bellorchestia needwonnee* sp. nov. is known to cooccur at sites with all other known Australian *Bellorchestia* species.

Ecology. Burrowing in clay and wet sand.

Distribution. Tasmania: Cox Bight Beach; Strahan; Bruny Island, Arthur Beach, King Island, Perkins Island, Sandy Cape Beach, St Albans Bay, Spain Bay (current study).

Bellorchestia palawakani sp. nov. (Figs. 5.18–5.25)
Type material. Holotype male, 13 mm, dissected, parts SEM stub mounts, whole animal SEM pin mount, AM P.97658, Macintyres Beach, 10 km south of Falmouth, Tasmania (41°34′23″S, 148°18′31″E), aggregating on cast kelp at low tide, at night, 27 December 1994, coll. A. Richardson (TA352). Paratypes: female, 15.7 mm, dissected, parts SEM stub mounts, whole animal SEM pin mount, AM P.97659; 11 specimens (8 males, 3 non-gravid females), AM P.97014, same location as holotype. Paratypes: male, 17 mm, dissected, 1 slide, AM P.99327; 20 specimens, AM P.97034, Macintyres Beach, 10 km south of Falmouth, Tasmania (41°34′23″S, 148°18′31″E), on dry sand at high water, at night, 27 December 1994, coll. A. Richardson (TA353).

Additional examined. Tasmania: 4 specimens (1 male, 3 non-gravid females), AM P.96975, Mariposa Beach, south of Falmouth (41°31′52″S, 148°16′49″E), in

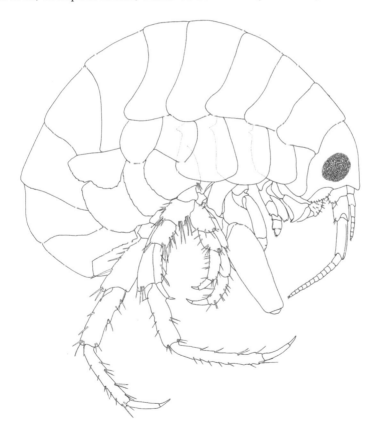

Figure 5.18　*Bellorchestia palawakani* sp. nov., male, 17 mm, AM P.99327, Macintyres Beach, 10 km south of Falmouth, Tasmania.

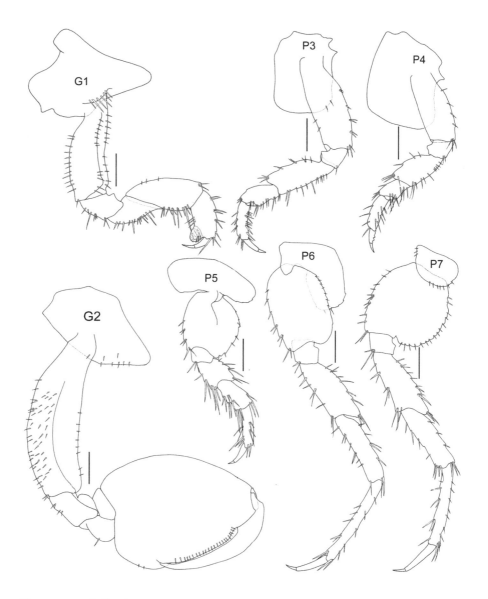

Figure 5.19   *Bellorchestia palawakani* sp. nov., male, 17 mm, AM P.99327, Macintyres Beach, 10 km south of Falmouth, Tasmania. Scales 2 mm.

rotting kelp (mostly bladders) between large granite boulders, 29 January 1991, coll. A. Richardson (TA277); 7 specimens (3 males, 4 females), AM P.96870, Fortescue Bay, Tasman Peninsula (43°08′37″S, 147°57′29″E), burrowing in sand at mid-water mark, coll. A. Richardson (TA72); 12 specimens (3 males, 9 females), AM P.97003, Fortescue Bay, Tasman Peninsula (43°08′33″S, 147°57′29″E), on sand at mid-water mark at night, just above breaking waves, 11 April 1992, coll. A. Richardson

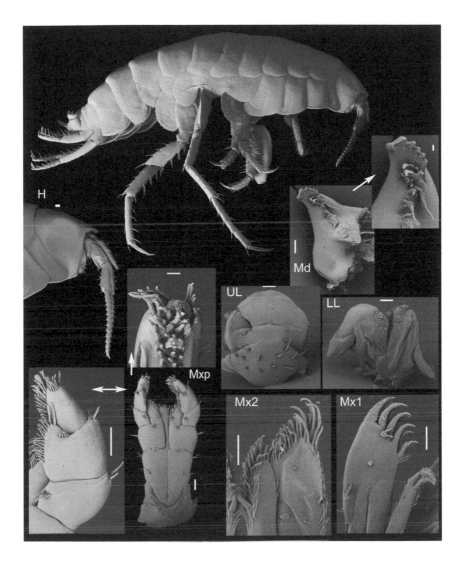

Figure 5.20 *Bellorchestia palawakani* sp. nov., male, 13 mm, AM P.97658, SEM image, Macintyres Beach, 10 km south of Falmouth, Tasmania. Scales H, Ul, LL, Md, Mx1, Mx2, Mxp, 100 μm; Md incisor, 20 μm.

(TA307); 11 specimens (9 males, 2 females), AM P.96846, Two Mile Beach, Forestier Peninsula (42°52′38″S, 147°56′03″E), on sand surface, 2 September 1984, coll. A. Richardson (TA3); 8 specimens (3 males, 5 females/juveniles), AM P.96928, South Cape Bay (43°36′04″S, 146°47′04″E), from burrows in 1 m sand cliff at extreme high water spring, 22 August 1988, coll. R. Swain, J. Tupascz, and A. Richardson (TA195); 20+ specimens, AM P.96929, South Cape Bay (43°36′04″S, 146°47′04″E), from burrows in 1 m sand cliff at extreme high water spring, 22 August 1988, coll. R. Swain, J. Tupascz, and A. Richardson (TA195); 1 male, AM P.99328, South Cape Bay

Figure 5.21　*Bellorchestia palawakani* sp. nov., male, 13 mm, AM P.97658, SEM image, Macintyres Beach, 10 km south of Falmouth, Tasmania. Scales Gn1, Gn2, 200 μm; A1, A2, Gn in part, 100 μm; Gn1 palm, 20 μm.

(43°36′04″S 146°47′04″E), on vegetation at night (feeding), on ridge immediately above extreme high water spring, 21 August 1988, coll. A. Richardson and J. Tupascz (TA197); 23 specimens (3 males, 20 females), AM P.96981, Denison Beach, outflow of Denison River, 41°49′00″S, 148°15′47″E, burrows in sand below high water extending to mid-water mark, 30 January 1991, coll. A. Richardson (TA283); 6 specimens (3 males, 3 females), AM P.96985, Ocean Beach, Strahan, 42°08′47″S, 145°15′39″E, toward high water, under fresh cast strapweed, macroalga cf *Ecklonia* sp., 30 January 1991, coll. M. Moore and K. Heiden (TA286); 6 specimens (2 males, 4 females/

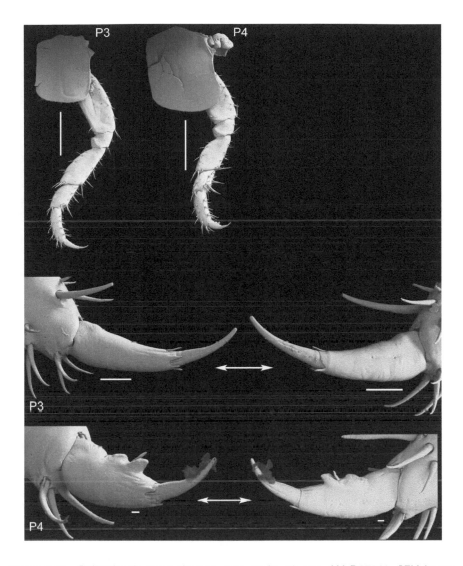

Figure 5.22   *Bellorchestia palawakani* sp. nov., male, 13 mm, AM P.97658, SEM image, Macintyres Beach, 10 km south of Falmouth, Tasmania. Scales P3 dactylus, 100 μm; P4 dactylus, 20 μm; P3, P4, 1 mm.

juveniles), AM P.97035, Shipwreck Point, Perkins Island, 40°45'12"S, 145°02'43"E, burrows in dune slack with cast seagrass, some *Salicornia* sp., 29 October 1994, coll. A. Richardson (TA367); 4 specimens (2 male, 2 female), AM P.97036, North Shore, Perkins Island, 40°47'11"S, 145°05'31"E, sandcliff burrows on actively eroding dune 29 October 1994, coll. A. Richardson (TA368); 20+ specimens, AM P.96942, Hibbs Lagoon, 42°34'26"S, 145°18'05"E, fore dune pitfall 1 (A106), 10 January 1987, coll. S.J. Smith (TA234); 4 specimens (2 males, 2 females with 2 male, 1 female genetic voucher specimens), AM P.99399, South Arm, 45°1'17"S, 147°30'7"E, under kelp at

CRUSTACEAN BIODIVERSITY AND CONSERVATION

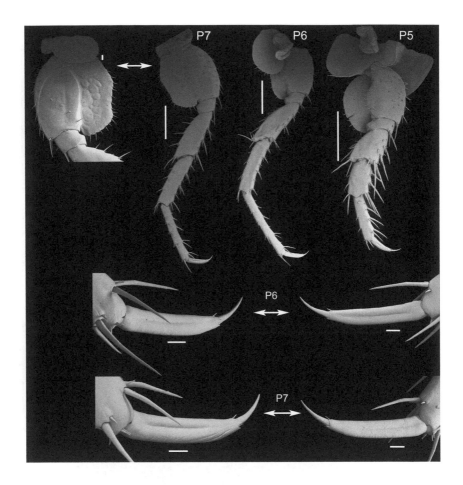

Figure 5.23   *Bellorchestia palawakani* sp. nov., male, 13 mm, AM P.97658, SEM image, Macintyres Beach, 10 km south of Falmouth, Tasmania. Scales P7 dactylus, 100 μm; P5, P6, P7, 1 mm.

highwater, 16 September 2015, coll. A. Richardson; 1 male specimen, TMAG G1334, Marion Bay, Tasmania (-42.81, 147.89), 6 January 1971, coll. Elizabeth Turner and Alan Dartnall (AGD66); 12 specimens, TMAG G8248, Marion Bay, Tasmania (-42.81, 147.89), 6 January 1971, coll. Elizabeth Turner and Alan Dartnall (AGD66); 100 specimens, TMAG G8245, Hobart, Tasmania (-42.88, 147.32), 23 June 1979, coll. J. Lim (AGD66); 2 specimens, AM P.99842, Tiddys Beach (42°12′35″S, 145°11′24″E), 19 June 1997, coll. ARTP; many specimens, AM P.99832, Randalls Bay (43°14′S, 147°07′E), 28 September 1994, coll. ARTP; 1 specimen, AM P.99809, Coningham Beach (43°04′47″S, 147°17′06″E), 28 September 1994, coll. ARTP; many specimens, AM P.99830, Pollys Bay (41°11′S, 144°41′E), 20 May 1997, coll. ARTP; many specimens, AM P.99795, Sandspit Point (42°17′52″S, 148°14′58″E), 28 April 1995, coll. ARTP; 2 specimens, AM P.99797, Redbill Beach (41°52′04″S,

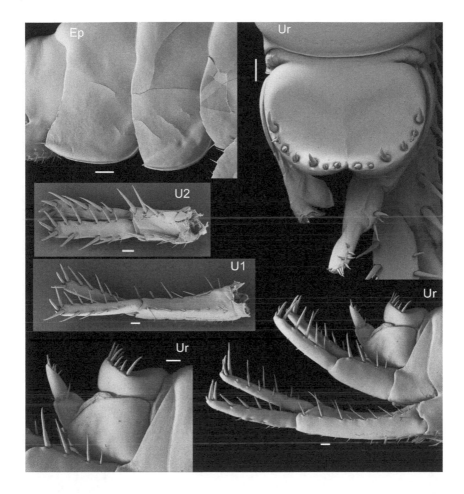

Figure 5.24 *Bellorchestia palawakani* sp. nov., male, 13 mm, AM P.97658, SEM image, Macintyres Beach, 10 km south of Falmouth, Tasmania. Scales Ep, 200 μm; Ur, Ur in part, Ur dorsal view, Ur1, Ur2, 100 μm.

148°17′09″E), 21 January 1995, coll. ARTP; many specimens, AM P.99798, Two Mile Beach, Forestier Peninsula (42°52′31″S, 147°55′50″E), 2 February 1995, coll. ARTP; many specimens, AM P.99804, Bryans Beach (42°15′35″S, 148°17′07″E), 28 April 1995, coll. ARTP; many specimens, AM P.99827, Ocean Beach, Strahan (42°08′47″S, 145°15′39″E), 19 June 1997, coll. ARTP; 3 specimens, AM P.99810, Cooks Beach (42°13′26″S, 148°16′07″E), 28 April 1995, coll. ARTP; many specimens, AM P.99799, Wrinklers Beach (41°26′44″S, 148°16′08″E), 18 April 1995, coll. ARTP; 3 specimens, AM P.99779, Endeavour Bay (42°39′S, 145°21′E), 11 April 1997, coll. ARTP; 3 specimens, AM P.99821, Little Beach (41°37′35″S, 148°18′53″E), 21 January 1995, coll. ARTP; many specimens, AM P.99801, Abbotsbury (41°02′08″S, 148°17′48″E), 17 April 1995, coll. ARTP; many specimens, AM P.99782, Four Mile

Figure 5.25   *Bellorchestia palawakani* sp. nov., female, 15.7 mm, AM P.97659, SEM image, Macintyres Beach, 10 km south of Falmouth, Tasmania. Scales H, Gn1, Gn2, 200 μm; Gn1 in part, Gn2 in part, Ur, 100 μm; Gn1 palm, 30 μm.

Creek, south of Scamander (41°33′32″S, 148°17′21″E), 21 January 1995, coll. ARTP; many specimens, AM P.99746, Fortescue Bay (43°08′14″S, 147°57′15″E), 26 April 1995, coll. ARTP; 5 specimens, AM P.99747, Freemans Beach (42°12′55″S, 148°02′09″E), 22 March 1995, coll. ARTP; many specimens, AM P.99748, Friendly Beach (41°58′04″S, 148°17′46″E), 13 May 1997, coll. ARTP; many specimens, AM P.99749, Front Beach (42°16′S, 148°01′E), 12 April 1995, coll. ARTP; many specimens, AM P.99750, Lighthouse Beach, south of Marrawah (40°56′43″S, 144°37′23″E), 20 May 1997, coll. ARTP; many specimens, AM P.99751, South Neck Beach

(43°18′23″S, 147°19′14″E), 27 February 1995, coll. ARTP; 1 specimen, AM P.99752, Nelson Bay (41°08′S, 144°40′E), 20 May 1997, coll. ARTP; 3 specimens, AM P.99753, Nine Mile Beach, King Island (39°44′S, 144°06′E), 24 March 1997, coll. ARTP; 6 specimens, AM P.99754, Pilot Beach (42°12′S, 145°12′E), 19 June 1997, coll. ARTP; many specimens, AM P.99775, Deephole Bay (43°27′23″S, 146°58′09″E), 20 January 1995, coll. ARTP; many specimens, AM P.99794, Primrose Beach (42°52′51″S, 147°39′26″E), 19 October 1994, coll. ARTP; many specimens, AM P.99783, Friendly Beach (41°58′04″S, 148°17′46″E), 13 May 1997, coll. ARTP; many specimens, AM P.99784, Grassy, King Island (40°02′S, 144°03′E), 22 March 1997, coll. ARTP; 3 specimens, AM P.99787, Kingston Beach (42°58′53″S, 147°19′30″E), 15 September 1994, coll. ARTP; many specimens, AM P.99785, Hardwicke Bay (41°40′S, 144°55′E), 12 April 1997, coll. ARTP; many specimens, AM P.99786, Hope Beach (43°02′16″S, 147°26′31″E), 4 October 1994, coll. ARTP; many specimens, AM P.99824, South Marion Beach (42°50′11″S, 147°52′21″E), 2 February 1995, coll. ARTP; 7 specimens, AM P.99829, Perkins Island (40°46′12″S, 145°03′00″E), 1 June 1997, coll. ARTP; many specimens, AM P.99831, Prion Beach (43°32′S, 146°34′E), 27 June 1997, coll. ARTP; many specimens, AM P.99769, Arthur Beach (41°04′48″S, 144°40′11″E), 20 May 1997, coll. ARTP; many specimens, AM P.99806, Carlton Beach (42°52′16″S, 147°37′02″E), 19 October 1994, coll. ARTP; many specimens, AM P.99796, Stephens Bay (43°23′S, 145°58′E), 27 June 1997, coll. ARTP; 5 specimens, AM P.99802, Adventure Bay (43°22′04″S, 147°19′60″E), 6 March 1995, coll. ARTP; many specimens, AM P.99812, Crockets Bay (42°17′54″S, 148°17′09″E), 28 April 1995, coll. ARTP; many specimens, AM P.99790, Maurouard Beach (41°17′43″S, 148°20′16″E), 18 April 1995, coll. ARTP; many specimens, AM P.99788, Little Lagoon (43°32′05″S, 146°55′55″E), 5 May 1995, coll. ARTP; many specimens, AM P.99789, Mariposa Beach (41°31′36″S, 148°16′40″E), 18 April 1995, coll. ARTP; 1 specimen, AM P.99399, under kelp at high water, South Arm (45°01′17″S, 147°30′07″E), 16 September 2015, coll. A.M.M. Richardson Talitrid Project; many specimens, AM P.99815, Grindstone Beach (42°26′20″S, 147°59′45″E), 22 March 1995, coll. ARTP; many specimens, AM P.99803, Binalong Bay (41°14′37″S, 148°17′37″E), 18 April 1995, coll. ARTP; many specimens, AM P.99770, Banwell Beach (42°21′54″S, 148°00′29″E), 12 April 1995, coll. ARTP; many specimens, AM P.99763, West Beach, Swan Island (40°44′06″S, 148°06′27″E), 1 September 1996, coll. ARTP; 1 specimen, AM P.99764, White Beach, King Island (39°36′29″S, 144°00′45″E), 17 March 1995, coll. ARTP; 6 specimens, AM P.99765, White Beach, King Island (39°36′29″S, 144°00′45″E), 24 March 1997, coll. ARTP; many specimens, AM P.99766, Window Pane Bay (43°27′35″S, 146°00′35″E), 27 June 1997, coll. ARTP; many specimens, AM P.99780, Finns Beach (43°32′41″S, 146°53′28″E), 20 January 1995, coll. ARTP; many specimens, AM P.99781, Four Mile Creek, south of Pieman Heads (41°43′09″S, 144°57′28″E), 12 April 1997, coll. ARTP; 3 specimens, AM P.99791, Nye Bay (43°03′S, 145°40′E), 11 April 1997, coll. ARTP; many specimens, AM P.99811, Crescent Beach (43°12′04″S, 147°51′55″E), 17 March 1995, coll. ARTP; many specimens, AM P.99776, Denison Beach (41°49′59″S, 148°15′52″E), 21 January 1995, coll. ARTP; 2 specimens, AM P.99768, Nine Mile Beach west, north of Swansea (42°05′33″S, 148°05′14″E), 22 March 1995, coll. ARTP; 2 specimens, AM

P.99836, Shoal Bay (42°40'21"S, 148°03'55"E), 31 March 1995, coll. ARTP; many specimens, AM P.99839, Steels Beach (41°28'22"S, 148°15'53"E), 18 April 1995, coll. ARTP; many specimens, AM P.99826, Mickeys Bay (43°25'S, 147°11'E), 24 March 1995, coll. ARTP; 5 specimens, AM P.99825, Mawson Bay (41°00'S, 144°37'E), 20 May 1997, coll. ARTP; many specimens, AM P.99835, Roaring Beach (43°17'36"S, 147°05'11"E), 20 January 1995, coll. ARTP; 3 specimens, AM P.99838, South Simpsons Bay (43°17'48"S, 147°18'47"E), 27 February 1995, coll. ARTP; many specimens, AM P.99818, Little Jetty Beach, Bruny Island (43°27'33"S, 147°09'21"E), 6 March 1995, coll. ARTP; many specimens, AM P.99823, North Marion Beach (42°46'04"S, 147°52'48"E), 2 February 1995, coll. ARTP; many specimens, AM P.99843, between Sandy Cape and Pieman Heads (41°33'26"S, 144°52'01"E), 12 April 1997, coll. ARTP; 9 specimens, AM P.99808, Cloudy Bay, Bruny Island (43°26'21"S, 147°12'14"E), 17 May 1997, coll. ARTP; 7 specimens, AM P.99819, Johnsons Bay (41°28'S, 144°48'E), 12 April 1997, coll. ARTP; 1 specimen, AM P.99816, Hazards Beach (42°10'59"S, 148°17'01"E), 10 March 1995, coll. ARTP; many specimens, AM P.99813, Four Mile Creek, Maria Island (42°37'08"S, 148°02'16"E), 31 March 1995, coll. ARTP; many specimens, AM P.99805, Cape Naturaliste (40°50'34"S, 148°12'02"E), 17 April 1995, coll. ARTP; many specimens, AM P.99807, Bruny Island, Cloudy Bay (43°27'35"S, 147°13'12"E), 17 May 1997, coll. ARTP; many specimens, AM P.99817, Howrah Beach (42°53'02"S, 147°23'53"E), 15 September 1994, coll. ARTP; many specimens, AM P.99774, Calverts Beach (43°02'15"S, 147°29'37"E), 4 October 1994, coll. ARTP; many specimens, AM P.99800, Yellow Rock Beach, King Island (39°40'11"S, 143°54'36"E), 24 March 1997, coll. ARTP; many specimens, AM P.99792, Ocean Beach, Strahan (42°08'47"S, 145°15'39"E), 19 June 1997, coll. ARTP; many specimens, AM P.99793, Plain Place Beach (42°29'03"S, 147°59'34"E), 29 April 1995, coll. ARTP; many specimens, AM P.99820, Lisdillon Beach (42°17'18"S, 148°00'46"E), 12 April 1995, coll. ARTP; 4 specimens, AM P.99755, Planter Beach (43°34'24"S, 146°54'26"E), 22 January 1995, coll. ARTP; many specimens, AM P.99756, Red Ochre Beach (42°51'54"S, 147°36'35"E), 19 October 1994, coll. ARTP; many specimens, AM P.99757, Sandy Cape Beach (41°22'S, 144°46'E), 12 April 1997, coll. ARTP; many specimens, AM P.99771, Beaumaris Beach (41°24'44"S, 148°16'49"E), 18 April 1995, coll. ARTP; many specimens, AM P.99777, Dora Point, Georges Bay (41°16'42"S, 148°19'44"E), 18 April 1995, coll. ARTP; many specimens, AM P.99778, Elliott Bay (42°58'47"S, 145°33'35"E), 11 April 1997, coll. ARTP; 6 specimens, AM P.99814 Green Point (40°54'28"S, 144°40'53"E), 20 May 1997, coll. ARTP; many specimens, AM P.99773, Broken Arm Beach, King Island (40°04'S, 144°02'E), 23 March 1997, coll. ARTP; many specimens, AM P.99828, Peggs Beach (40°50'24"S, 145°19'48"E), 31 January 1997, coll. ARTP; many specimens, AM P.99833, Riedle Bay (42°40'08"S, 148°04'26"E), 31 March 1995, coll. ARTP; many specimens, AM P.99834, Roaring Beach (43°17'36"S, 147°05'11"E), 20 January 1995, coll. ARTP; 5 specimens, AM P.99735, Nine Mile Beach east, north of Swansea (42°05'18"S, 148°05'14"E), 22 March 1995, coll. ARTP; many specimens, AM P.99736, Bellerive Beach (42°52'46"S, 147°22'24"E), 15 September 1994, coll. ARTP; 1 specimen, AM P.99737, Black River Beach (40°49'S, 145°17'E), 1 June 1997, coll. ARTP; 5 specimens, AM P.99738,

Blackmans Bay Beach (43°00′21″S, 147°19′31″E), 14 September 1994, coll. ARTP; many specimens, AM P.99739, Cloudy Bay, Bruny Island (43°27′22″S, 147°15′12″E), 24 March 1995, coll. ARTP; 10 specimens, AM P.99740, Cloudy Bay, Bruny Island (43°27′22″S, 147°15′12″E), 17 May 1997, coll. ARTP; many specimens, AM P.99741, Cox Bight (43°31′S, 146°14′E), 27 June 1997, coll. ARTP; 9 specimens, AM P.99742, Darlington Bay, Maria Island (42°34′54″S, 148°03′46″E), 31 March 1995, coll. ARTP; many specimens, AM P.99743, Dianas Beach (41°22′48″S, 148°17′23″E), 18 April 1995, coll. ARTP; many specimens, AM P.99744, Discovery Beach (42°23′S, 145°14′E), 11 April 1997, coll. ARTP; 1 specimen, AM P.99745, South Foochow Beach, Flinders Island (39°55′S, 148°09′E), 7 June 1996, coll. ARTP; 4 specimens, AM P.99840, Stumpys Bay (40°52′33″S, 148°13′29″E), 17 April 1995, coll. ARTP; many specimens, AM P.99758, Sloop Beach (41°12′S, 148°16′E), 11 April 1997, coll. ARTP; 5 specimens, AM P.99759, South Cape Rivulet (43°34′11″S, 146°46′12″E), 23 February 1997, coll. ARTP; many specimens, AM P.99760, Spain Bay (43°22′S, 145°58′E), 27 June 1997, coll. ARTP; many specimens, AM P.99761, Spring Beach (42°34′46″S, 147°54′28″E), 16 January 1995, coll. ARTP; 7 specimens, AM P.99762, Telegraph Bay, Swan Island (40°44′24″S, 148°05′23″E), 1 September 1996, coll. ARTP; 4 specimens, AM P.99767, Wineglass Bay, Freycinet Peninsula (42°10′29″S, 148°18′01″E), 10 March 1995, coll. ARTP; many specimens, AM P.99822, Louisa Bay (43°31′12″S, 146°20′23″E), 27 June 1997, coll. ARTP; many specimens, AM P.99841, Taylors Beach (41°11′49″S, 148°16′12″E), 18 April 1995, coll. ARTP; 8 specimens, AM P.99772, Boltons Beach (42°24′11″S, 147°58′55″E), 16 January 1995, coll. ARTP; many specimens, AM P.99837, North Simpsons Bay (43°16′S, 147°18′E), 27 February 1995, coll. ARTP. Victoria. 2 specimens (1 male and 1 female), AM P99045, Venus Bay Beach, 38°39′51″S, 145°42′4″E, pitfall trap, October 2015, coll. L. Carracher.

Type locality. Macintyres Beach, 10 km south of Falmouth, Tasmania, Australia, 41°34′23″S, 148°18′31″E.

Etymology. Named from Palawa Kani, a revival composite language, which brings together words from many of the original Tasmanian Aboriginal languages.

Description. Based on type material. Head: eye medium size (greater than 1/5 to 1/3 head length). Antenna 1: short, not reaching midpoint of peduncular article 5 of antenna 2. Antenna 2: less than half body length (0.30 × length); peduncular articles slender, with many large robust setae; article 5 longer than article 4; flagellar articles final article large, cone-shaped forming a *virgula divina*. Labrum: with apical setal patch; epistome with many robust setae. Mandible: left lacinia mobilis 4-dentate. Labium: distolateral setal tuft present, distomedial setal tuft present; without inner plates. Maxilla 1: with 1-articulate small palp. Maxilliped: palp article 2 distomedial lobe well developed; article 4 fused with article 3.

Pereon. Gnathopod 1: sexually dimorphic; subchelate; coxa subequal to coxa 2; carpus longer than propodus, 1.8 as long as propodus; propodus subrectangular, 1.8 × as long as broad, anterior margin with 4 groups of robust setae, posterior margin with palmate lobe, palm transverse to slightly oblique; simplidactylate, dactylus overreaching palm, without anterodistal denticular patch. Gnathopod 2: sexually dimorphic; subchelate; basis slender; ischium anterior margin with lateral and medial rounded lobes asymmetrical; carpus triangular, reduced, enclosed by merus and propodus; propodus 1.2 × as long as wide, palm subacute, 65% along posterior margin, palm margin entire, smooth, lined with robust setae; dactylus curved, subequal in length to palm, apically acute, closing into dactylar socket. Pereopods 3–7: asymmetrically tricuspidate laterally, dactylus without anterodistal denticular patch. Pereopods 3–4: coxae deeper than wide. Pereopod 3: carpus length 1.8 × width. Pereopod 4: significantly shorter than pereopod 3; carpus significantly shorter than carpus of pereopod 3, length 1.1 × width; dactylus posterior margin thickened with 1 proximal and 1 distal projections. Pereopod 5: merus length 1.1 × width; carpus length 1.8 × width; propodus distinctly longer than carpus, length 4.5 × width; dactylus long, slender, length 4 × width. Pereopods 6–7: not sexually dimorphic. Pereopod 6: longer than pereopod 7; coxa posterior lobe posteroventral corner rounded, posterior margin perpendicular to ventral margin; merus not expanded, length twice width; carpus not expanded, length 4 × width; propodus length 8.1 × width, distinctly longer than carpus. Pereopod 7: basis expanded, produced posteriorly, posterior margin convex, lined with small robust setae; merus length twice width; carpus length 3 × width; propodus length 5.5 × width, distinctly longer than carpus.

Pleon. Pleopods 1–3: well developed, rami segmented. Epimeron 1– 2: posterior margin weakly convex, broad minute serrations; posterodistal corner with small tooth. Epimeron 3: posteroventral corner produced, margin with broad minute serration; posterodistal corner with small tooth. Uropod 1: peduncle with 15+ robust setae, without apical spear-shaped setae; rami positioned directly on top of each other; inner ramus subequal in length to outer ramus, with 10 marginal robust setae in 2 rows; outer ramus with 5 marginal robust setae in 1 row. Uropod 2: peduncle with 9 robust setae; inner ramus subequal in length to outer ramus, with 9 marginal robust setae in 2 rows; outer ramus with 4 marginal robust setae in 1 row. Urosomite 3: subrectangular, not enclosing around telson, broader than deep, 1.2 × as broad as deep. Uropod 3: vestigial; peduncle not dorsally concave, length 1.1 × depth, with 3 robust setae; ramus linear, fused to peduncle (when viewed dorsally), subequal to peduncle, 2.1 × as long as broad, with 5 marginal setae and cluster of apical setae. Telson: subovate, broader than long, lateral margins straight to convex, apical margins convex, incised apically, dorsal midline weakly present, with single row of around 6–7 marginal and apical robust setae per lobe; length shorter than uropod 3 peduncle.

Female (sexually dimorphic characters). Gnathopod 1: simple; propodus posterior margin without palmate lobe; propodus palm near transverse. Gnathopod 2: mitten-shaped; basis expanded anteromedially; ischium without lobe; carpus well developed (not enclosed by merus and propodus), length 2.5 × width; propodus twice width, palm obtuse; dactylus shorter than palm.

Variation. There is little morphological variation with growth stage in males.

Remarks. The entire, smooth palmar margin of the male gnathopod 2 readily separates *Bellorchestia palawakani* sp. nov. from other *Bellorchestia* species. *Bellorchestia palawakani* sp. nov. is known to cooccur with *B. needwonnee* sp. nov. and *B. lutruwita* sp. nov.

Ecology. Burrowing in sand.

Distribution. Tasmania: Perkins Island, Strahan, Falmouth, Denison Beach, South Cape Bay, Hibbs Lagoon, Forestier Peninsula, Tasman Peninsula, Marion Bay, Hobart, Hibbs Laboon, Tiddys Beach, Randalls Bay, Coningham, Pollys Bay, Sandspit Point, Redbill Beach, Wrinklers Beach, Endeavour Bay, Little Beach, Abbotbury, Four Mile Creek, Fortescue Bay, Freemans Beach, Friendly Beach, Front Beach, Lighthouse Beach (Marrawah), South Neck Beach, Nelson Bay, King Island, Deephole Bay, Primrose Beach, Kingston Beach, Hope Beach, South Marion Beach, Perkins Island, Prion Beach, Arthur Beach, Adventure Bay, Crockets Bay, Maurouard Beach, Little Lagoon, Mariposa Beach, South Arm, Grindstone Beach, Binalong Bay, Swan Island, Window Pane Bay, Finns Beach, Four Mile Creek, Nye Bay, Crescent Beach, Denison Beach, Nine Mile Beach, Shoal Bay, Steels Beach, Mickeys Bay, Mawson Bay, Roaring Beach, South Simpsons Bay, Bruny Island, North Marion Beach, Cloudy Bay, Howrah Beach, Calverts Beach, Strahan, Plain Place Beach, Lisdillon Beach, Planter Beach, Red Ochre Beach, Sandy Cape Beach, Beaumaris Beach, Georges Bay, Elliott Bay, Green Point, Black River Beach, Blackmans Bay Beach, Cox Bight, Maria Island, Dianas Beach, Discovery Beach, Flinders Island, South Cape Rivulet, Sloop BEacy Stumpys Bay, Spain Bay, Spring Beach, Boltons Beach, North Simpsons Bay (current study). Victoria: Venus Bay (current study).

Bellorchestia pravidactyla Haswell 1880
   Talorchestia pravidactyla Haswell 1880: 100, pl. 5, Fig. 5.5. – Stebbing 1906: 546. – Stebbing 1910: 645. – Springthorpe and Lowry 1994: 129. – Lowry and Stoddart 2003: 275.
   ? *Talorchestia marmorata* – Springthorpe and Lowry 1994: 24 (part, AM P.3412 only; not *Talorchestia marmorata* Haswell 1880).
   Bellorchestia pravidactyla – Serejo and Lowry 2008: 169. Hughes and Ahyong 2017: 54–60, Figs. 5.1–5.10.
   Bellorchestia richardsoni Serejo and Lowry 2008: 170–174, Figs. 5.5–5.8. – Lowry and Springthorpe 2009: 899. – Lowry 2012: 7.
   Bellorchestia mariae Lowry 2012: 3–6, Figs. 5.1–5.3.

Type material. Probable syntypes, 2 specimens (1 male, 1 female), specimens not dissected, B.M. (N.H.) 1928.12.1.2428–2429, Tasmania, no further locality data, Stebbing Collection; 2 female specimens, B.M.(N.H.) 1895.11.14:119–1120, Tasmania, no further locality data (outer jar label Australian Museum (E).
   Material examined. New South Wales: 13 specimens (7 male, 6 female), B.M. (N.H.) 1935.6.26.9–14. Bowen Island, Jervis Bay, New South Wales, coll. Dr. R.A. Rodway, no date. Tasmania: 30 specimens, TMAG G84, Eaglehawk Neck (43°0′23″S, 147°55′12″E), March 1941, coll. D. Pearse (GDA94); 3 specimens, TMAG G1092, Eaglehawk Neck (43°0′23″S, 147°55′12″E), no further details (GDA94); 4 specimens, TMAG G85, Seven Mile Beach (42°50′24″S, 147°33′0″E), 8 May 1952, coll. W. Radford (GDA94); 6 specimens, TMAG G419, Seven Mile Beach (42°50′24″S, 147°33′0″E), 27 July 1962, coll. Julia Greenhill (GDA94); 3 specimens, TMAG G339, South Arm, Hope Beach (43°2′24″S, 147°26′24″E), 19 August 1957, coll.

Barbara Nielsen (AGD66); 12 specimens, TMAG G1275, Granville Harbour (41°47′60″S, 145°1′12″E), May 1967, coll. Alan Dartnall (GDA94); 1 specimen, TMAG G2439, Arthur River, 300 m upstream from mouth of river (41°3′36″S, 145°11′24″E), July 1977, coll. T. Walker (AGD66); 1 specimen, TMAG G2440, Dru Point, Margate (43°1′48″S, 147°16′48″E), March 1977, coll. T. Walker (AGD66); 100 specimens, TMAG G2441, Hobart (42°52′48″S, 14719′12″E), 23 June 1979, coll. J. Lim (AGD66); 2 specimens, TMAG G2515, Darlington beach, Maria Island (42°35′24″S, 148°3′36″E), 16 April 1969, coll. Elizabeth Turner and Alan Dartnall (AGD66); 12 specimens, TMAG G2516, Marion Bay (42°48′36″S, 147°53′24″E), 6 January 1971, coll. Elizabeth Turner and Alan Dartnall (AGD66); 2 specimens, TMAG G2522, Sandspit Point, Schouten Island (42°17′24″S, 148°15′36″S), 6 June 1981, coll. Roberta Barnett (GDA94); few specimens, AM P.99646, Roaring Beach (43°17′36″S, 147°05′11″E), 20 January 1995, coll. ARTP; few specimens, AM P.99651, Tam O'Shanter Bay (41°00′S, 147°04′E), 2 May 1997, coll. ARTP; few specimens, AM P.99609, Buttons Beach (41°09′S, 146°11′E), 7 December 1996, coll. ARTP; many specimens, AM P.99642, Pollys Bay (41°11′S, 144°41′E), 20 May 1997, coll. ARTP; 2 specimens, AM P.99686, South Neck Beach (43°18′23″S, 147°19′14″E), 27 February 1995, coll. ARTP; few specimens, AM P.99547, East Circular (no GPS data) 31 January 1997, coll. ARTP; many specimens, AM P.99694, Sisters Beach (40°55′S, 145°34′E), 16 February 1997, coll. ARTP; few specimens, AM P.99643, Porky Beach, King Island (39°50′S, 143°52′E), 22 March 1997, coll. ARTP; many specimens, AM P.99690, Purdon Bay (40°57′32″S, 148°18′04″E), 17 April 1995, coll. ARTP; few specimens, AM P.99685, Nebraska Beach (43°05′S, 147°20′E), 6 March 1995, coll. ARTP; 6 specimens, AM P.99612, Deephole Bay (43°27′23″S, 146°58′09″E), 20 January 1995, coll. ARTP; many specimens, AM P.99611, Coningham Beach (43°04′47″S, 147°17′06″E), 28 September 1994, coll. ARTP; many specimens, AM P.99701, Wineglass Bay, Freycinet Peninsula (42°10′29″S, 148°18′01″E), 10 March 1995, coll. ARTP; 3 specimens, AM P.99641, Plain Place Beach (42°29′03″S, 147°59′34″E), 29 April 1995, coll. ARTP; many specimens, AM P.99669, Crockets Bay (42°17′54″S, 148°17′09″E), 28 April 1995, coll. ARTP; few specimens, AM P.99697, Tatlows Beach (40°46′12″S, 145°16′47″E), 31 January 1997, coll. ARTP; few specimens, AM P.99613, Discovery Beach (42°23′S, 145°14′E), 11 April 1997, coll. ARTP; many specimens, AM P.99394, Top Camp, Cape Naturaliste (40°50′31″S, 148°12′02″E), 1 September 1994, coll., A.M.M. Richardson; 4 specimens, AM P.99549, Denison Beach (41°49′59″S, 148°15′52″E), 21 January 1995, coll. ARTP; 4 specimens, AM P.99645, Roches Beach (42°54′22″S, 147°29′46″E), 4 October 1994, coll. ARTP; few specimens, AM P.99675, Kelvedon Beach (42°11′56″S, 148°03′14″E), 16 January 1995, coll. ARTP; many specimens, AM P.99673, Friendly Beach (41°58′04″S, 148°17′46″E), 13 May 1997, coll. ARTP; many specimens, AM P.99604, Yellow Rock Beach, King Island (39°40′11″S, 143°54′36″E), 24 March 1997, coll. ARTP; many specimens, AM P.99700, White Beach, King Island (39°36′29″S, 144°00′45″E), 17 March 1995, coll. ARTP; many specimens, AM P.99662, Broken Arm Beach, King Island (40°04′S, 144°02′E), 23 March 1997, coll. ARTP; few specimens, AM P.99622, Fraser Beach, King Island (39°53′S, 144°06′E), 23 March 1997, coll. ARTP; 6 specimens, AM P.99634, Nine Mile Beach King Island (39°44′S,

144°06'E), 24 March 1997, coll. ARTP; few specimens, AM P.99650, Surprise Bay, King Island (40°07'S, 143°54'E), 22 March 1997, coll. ARTP; few specimens, AM P.99548, Colliers Beach, King Island (40°06'S, 143°59'E), 23 February 1997, coll. ARTP; few specimens, AM P.99561, White Beach, King Island (39°36'29"S, 144°00'45"E), 24 March 1997, coll. ARTP; many specimens, AM P.99661, British Admiral Beach, King Island (39°57'S, 143°52'E), 22 March 1997, coll. ARTP; few specimens, AM P.99551, Grassy, King Island (40°02'S, 144°03'E), 22 March 1997, coll. ARTP; many specimens, AM P.99670, Fitzmaurice Bay, King Island (40°03'S, 143°52'E), 23 March 1997, coll. ARTP; many specimens, AM P.99647, Sandfly Beach, King Island (39°58'S, 143°53'E), 23 March 1997, coll. ARTP; 2 specimens, AM P.99582, Lavinia Beach, King Island (39°39'S, 144°05'E), 24 March 1997, coll. ARTP; 2 specimens, AM P.99676, Kingston Beach (42°58'53"S, 147°19'30"E), 15 September 1994, coll. ARTP; 6 specimens, AM P.99682, South Marion Beach (42°50'11"S, 147°52'21"E), 2 February 1995, coll. ARTP; many specimens, AM P.99618, Five Mile Beach (42°49'45"S, 147°35'14"E), 19 October 1994, coll. ARTP; 3 specimens, AM P.99614, Dora Point, Georges Bay (41°16'42"S, 148°19'44"E), 18 April 1995, coll. ARTP; many specimens, AM P.99616, East Wynyard Beach (40°59'S, 145°44'E), 17 January 1997, coll. ARTP; many specimens, AM P.99617, Endeavour Bay (42°39'S, 145°21'E), 11 April 1997, coll. ARTP; many specimens, AM P.99637, Orford Beach (42°34'11"S, 147°52'56"E), 16 January 1995, coll. ARTP; 1 specimen, AM P.99610, Cod Bay (40°55'10"S, 148°16'15"E), 17 April 1995, coll. ARTP; few specimens, AM P.99631, Lisdillon Beach (42°17'18"S, 148°00'46"E), 12 April 1995, coll. ARTP; 5 specimens, AM P.99663, Carlton Beach (42°52'16"S, 147°37'02"E), 19 October 1994, coll. ARTP; 4 specimens, AM P.99625 Godfreys Beach (40°45'S, 145°17'E), 31 January 1997, coll. ARTP; many specimens, AM P.99626, Greens Beach (41°05'S, 146°45'E), 23 November 1996, coll. ARTP; 4 specimens, AM P.99632, Little Lagoon (43°32'05"S, 146°55'55"E), 5 May 1995, coll. ARTP; many specimens, AM P.99665, West Circular (no GPS data), 1 June 1997, coll. ARTP; 2 specimens, AM P.99623, Friendly Beach (41°58'04"S, 148°17'46"E), 13 May 1997, coll. ARTP; 2 specimens, AM P.99605, Banwell Beach (42°21'54"S, 148°00'29"E), 12 April 1995, coll. ARTP; many specimens, AM P.99680, Lighthouse Beach, south of Marrawah (40°56'43"S, 144°37'23"E), 20 May 1997, coll. ARTP; many specimens, AM P.99672, Forwards Beach (40°52'S, 145°28'E), 31 January 1997, coll. ARTP; 4 specimens, AM P.99649, Spring Beach (42°34'46"S, 147°54'28"E), 16 January 1995, coll. ARTP; many specimens, AM P.99684, Muirs Beach (42°06'53"S, 148°16'12"E), 10 March 1995, coll. ARTP; many specimens, AM P.99659, Big Lagoon Beach (43°30'15"S, 146°57'59"E), 20 January 1995, coll. ARTP; many specimens, AM P.99629, Jetty Beach, Swan Island (40°44'05"S, 148°07'09"E), 1 September 1996, coll. ARTP; many specimens, AM P.99630, Landing Down, Swan Island (40°43'43"S, 148°07'01"E), 1 September 1996, coll. ARTP; few specimens, AM P.99698, Swan Island (40°44'03"S, 148°05'53"E), 1 September 1996, coll. ARTP; few specimens, AM P.99554, Littlewest, Swan Island (40°44'03"S, 148°05'53"E), 1 September 1996, coll. ARTP; many specimens, AM P.99699, West Beach, Swan Island (40°44'06"S, 148°06'27"E), 1 September 1996, coll. ARTP; 8 specimens, AM P.99599, Telegraph Bay, Swan Island (40°44'24"S, 148°05'23"E), 1

September 1996, coll. ARTP; many specimens, AM P.99693, Richardsons Beach (42°07′38″S, 148°17′40″E), 10 March 1995, coll. ARTP; 1 specimen, AM P.99689, Porpoise Hole (42°50′11″S, 147°52′08″E), 2 February 1995, coll. ARTP; 4 specimens, AM P.99695, Sloop Beach (41°12′S, 148°16′E), 11 April 1997, coll. ARTP; many specimens, AM P.99696, Sloping Main Beach (42°59′26″S, 147°40′46″E), 24 January 1995, coll. ARTP; 2 specimens, AM P.99691, Red Ochre Beach (42°51′54″S, 147°36′35″E), 19 October 1994, coll. ARTP; many specimens, AM P.99692, North Rheban Beach (42°38′16″S, 147°56′04″E), 16 January 1995, coll. ARTP; many specimens, AM P.99658, Adventure Bay (43°22′04″S, 147°19′60″E), 6 March 1995, coll. ARTP; few specimens, AM P.99687, Okines Beach (42°50′49″S, 147°37′09″E), 2 December 1994, coll. ARTP; few specimens, AM P.99652, Templestowe Beach (41°43′29″S, 148°16′58″E), 21 January 1995, coll. ARTP; few specimens, AM P.99667, Cloudy Bay, Bruny Island (43°26′21″S, 147°12′14″E), 17 May 1997, coll. ARTP; many specimens, AM P.99668, Cressy Beach (42°10′15″S, 148°04′39″E), 22 March 1995, coll. ARTP; many specimens, AM P.99678, Lagoons Beach (42°57′17″S, 147°40′19″E), 21 January 1995, coll. ARTP; many specimens, AM P.99640, Pirates Bay (43°01′30″S, 147°55′36″E), 24 January 1995, coll. ARTP; few specimens, AM P.99628, Hazards Beach (42°10′59″S, 148°17′01″E), 10 March 1995, coll. ARTP; many specimens, AM P.99627, Hellyer Beach (40°51′S, 145°25′E), 31 January 1997, coll. ARTP; many specimens, AM P.99638, Piermont Beach (42°09′26″S, 148°04′38″E), 22 March 1995, coll. ARTP; many specimens, AM P.99620, Four Mile Creek, south of Pieman Heads (41°43′09″S, 144°57′28″E), 12 April 1997, coll. ARTP; many specimens, AM P.99660, Blackmans Bay Beach (43°00′21″S, 147°19′31″E), 14 September 1994, coll. ARTP; few specimens, AM P.99636, Nye Bay (43°03′S, 145°40′E), 11 April 1997, coll. ARTP; 2 specimens, AM P.99666, Cloudy Bay, Bruny Island (43°27′22″S, 147°15′12″E), 17 May 1997, coll. ARTP; 1 specimen, AM P.99656, Window Pane Bay (43°27′35″S, 146°00′35″E), 27 June 1997, coll. ARTP; many specimens, AM P.99677, Lagoon Bay, Forestier Peninsula (42°57′05″S, 147°55′06″E), 4 February 1995, coll. ARTP; 9 specimens, AM P.99608, Boobyalla Beach (40°51′S, 147°54′E), 9 May 1997, coll. ARTP; many specimens, AM P.99633, Nelson Bay (41°08′S, 144°40′E), 20 May 1997, coll. ARTP; 2 specimens, AM P.99653, between Sandy Cape and Pieman Heads (41°33′26″S, 144°52′01″E), 12 April 1997, coll. ARTP; few specimens, AM P.99543, Nine Mile Beach, west north of Swansea (42°05′33″S, 148°05′14″E), 22 March 1995, coll. ARTP; few specimens, AM P.99544, Arthur Beach (41°04′48″S, 144°40′11″E), 20 May 1997, coll. ARTP; 5 specimens, AM P.99545, Binalong Bay (41°14′37″S, 148°17′37″E), 18 April 1995, coll. ARTP; few specimens, AM P.99644, Redbill Beach (41°52′04″S, 148°17′09″E), 21 January 1995, coll. ARTP; 1 specimen, AM P.99639, Pilot Beach (42°12′S, 145°12′E), 19 June 1997, coll. ARTP; 4 specimens, AM P.99546, Cape Naturaliste (40°50′34″S, 148°12′02″E), 17 April 1995, coll. ARTP; few specimens, AM P.99550, Friendly Beach (41°58′04″S, 148°17′46″E), 13 May 1997, coll. ARTP; few specimens, AM P.99552, Grindstone Beach (42°26′20″S, 147°59′45″E), 22 March 1995, coll. ARTP; few specimens, AM P.99553, Howrah Beach (42°53′02″S, 147°23′53″E), 15 September 1994, coll. ARTP; 5 specimens, AM P.99555, Mariposa Beach (41°31′36″S, 148°16′40″E), 18 April 1995, coll. ARTP; 3

specimens, AM P.99556, Maurouard Beach (41°17′43″S, 148°20′16″E), 18 April 1995, coll. ARTP; few specimens, AM P.99557, Riedle Bay (42°40′08″S, 148°04′26″E), 31 March 1995, coll. ARTP; few specimens, AM P.99558, Seymour Beach (41°45′19″S, 148°17′13″E), 21 January 1995, coll. ARTP; few specimens, AM P.99560 Varna Bay (42°30′S, 145°15′E), 11 April 1997, coll. ARTP; few specimens, AM P.99559, South Simpsons Bay (43°17′48″S, 147°18′47″E), 27 February 1995, coll. ARTP; 6 specimens, AM P.99562, Wrinklers Beach (41°26′44″S, 148°16′08″E), 18 April 1995, coll. ARTP; 5 specimens, AM P.99674, Johnsons Bay (41°28′S, 144°48′E), 12 April 1997, coll. ARTP; many specimens, AM P.99664, Nelson Bay Highway (41°09′10″S, 144°40′39″E), 14 January 1997, coll. ARTP; few specimens, AM P.99563, Nine Mile Beach east, north of Swansea (42°05′18″S, 148°05′14″E), 22 March 1995, coll. ARTP; few specimens, AM P.99564, Bakers Beach, Asbestos Range National Park (41°08′41″S, 146°35′50″E), 23 November 1996, coll. ARTP; 3 specimens, AM P.99565, Bellerive Beach (42°52′46″S, 147°22′24″E), 15 September 1994, coll. ARTP; few specimens, AM P.99566 Boat Harbour (40°55′44″S, 145°36′59″E), 16 February 1997, coll. ARTP; few specimens, AM P.99567, Boltons Beach (42°24′11″S, 147°58′55″E), 16 January 1995, coll. ARTP; few specimens, AM P.99568, Boobyalla Beach (40°51′S, 147°54′E), 9 May 1997, coll. ARTP; few specimens, AM P.99569, Bryans Beach (42°15′35″S, 148°17′07″E), 28 April 1995, coll. ARTP; 2 specimens, AM P.99570, Calverts Beach (43°02′15″S, 147°29′37″E), 4 October 1994, coll. ARTP; few specimens, AM P.99571, Cooks Beach (42°13′26″S, 148°16′07″E), 28 April 1995, coll. ARTP; 6 specimens, AM P.99572, Crayfish Creek Beach (40°51′S, 145°23′E), 31 January 1997, coll. ARTP; 1 specimen, AM P.99573, Crescent Beach (43°12′04″S, 147°51′55″E), 17 March 1995, coll. ARTP; many specimens, AM P.99574, Darlington Bay, Maria Island (42°34′54″S, 148°03′46″E), 31 March 1995, coll. ARTP; many specimens, AM P.99575 Eddystone Point (40°59′37″S, 148°19′54″E), 17 April 1995, coll. ARTP; 5 specimens, AM P.99576, Fortescue Bay (43°08′14″S, 147°57′15″E), 26 April 1995, coll. ARTP; 5 specimens, AM P.99577, Freemans Beach (42°12′55″S, 148°02′09″E), 22 March 1995, coll. ARTP; 4 specimens, AM P.99578, Green Point (40°54′28″S, 144°40′53″E), 20 May 1997, coll. ARTP; 5 specimens, AM P.99688, Picnic Point, Ulverstone (41°08′40″S, 146°09′49″E), 7 December 1996, coll. ARTP; few specimens, AM P.99579, Halfmoon Bay (43°01′08″S, 147°24′49″E), 4 October 1994, coll. ARTP; 2 specimens, AM P.99580, Hardwicke Bay (41°40′S, 144°55′E), 12 April 1997, coll. ARTP; 5 specimens, AM P.99654, Waterhouse Beach (40°54′35″S, 147°33′E), 22 April 1997, coll. ARTP; 4 specimens, AM P.99635, Noland Bay (40°59′S, 147°10′E), 2 May 1997, coll. ARTP; 4 specimens, AM P.99607, Black River Beach (40°49′S, 145°17′E), 1 June 1997, coll. ARTP; many specimens, AM P.99581, Little Jetty Beach, Bruny Island (43°27′33″S, 147°09′21″E), 6 March 1995, coll. ARTP; few specimens, AM P.99583, Mawson Bay (41°00′S, 144°37′E), 20 May 1997, coll. ARTP; 6 specimens, AM P.99584, North Neck Beach (43°14′39″S, 147°23′57″E), 27 February 1995, coll. ARTP; 2 specimens, AM P.99585, Ocean Beach, Strahan (42°08′47″S, 145°15′39″E), 19 June 1997, coll. ARTP; few specimens, AM P.99586, Peggs Beach (40°50′24″S, 145°19′48″E), 31 January 1997, coll. ARTP; 6 specimens, AM P.99587, Perkins Island (40°46′12″S, 145°03′00″E), 1 June 1997, coll. ARTP; few specimens, AM P.99588, Planter Beach

(43°34'24"S, 146°54'26"E), 22 January 1995, coll. ARTP; few specimens, AM P.99589, South Rheban Beach (42°39'52"S, 147°57'15"E), 16 January 1965, coll. ARTP; few specimens, AM P.99590, Roaring Beach (43°17'36"S, 147°05'11"E), 20 January 1995, coll. ARTP; few specimens, AM P.99591, Sandspit Point (42°17'52"S, 148°14'58"E), 28 April 1995, coll. ARTP; few specimens, AM P.99648, Somerset Beach (41°02'S, 145°49'E), 17 January 1997, coll. ARTP; many specimens, AM P.99681, Little Beach (41°37'35"S, 148°18'53"E), 21 January 1995, coll. ARTP; few specimens, AM P.99592, Sandy Cape Beach (41°22'S, 144°46'E), 12 April 1997, coll. ARTP; 3 specimens, AM P.99593, Settlement Beach, Flinders Island (40°01'S, 147°53'E), 9 June 1996, coll. ARTP; 1 specimen, AM P.99655, Whitemark Beach, Flinders Island (40°07'11"S, 148°00'35"E), 9 June 1996, coll. ARTP; many specimens, AM P.99671, South Foochow Beach, Flinders Island (39°55'S, 148°09'E), 7 June 1996, coll. ARTP; 5 specimens, AM P.99603, Watering Beach, Flinders Island (40°15'S, 148°10'E), 7 June 1996, coll. ARTP; few specimens, AM P.99619, North Foochow Beach, Flinders Island (39°46'S, 147°59'E), 8 June 1996, coll. ARTP; few specimens, AM P.99594, Shoal Bay (42°40'21"S, 148°03'55"E), 31 March 1995, coll. ARTP; few specimens, AM P.99679, Lemons Beach (40°46'S, 147°57'E), 9 May 1997, coll. ARTP; few specimens, AM P.99624, Front Beach (42°16'S, 148°01'E), 12 April 1995, coll. ARTP; few specimens, AM P.99595, South Clifton Beach (42°59'32"S, 147°31'30"E), 8 September 1994, coll. ARTP; many specimens, AM P.99596, Stumpys Bay (40°52'33"S, 148°13'29"E), 17 April 1995, coll. ARTP; 1 specimen, AM P.99597, St Albans Bay (40°57'S, 147°18'E), 2 May 1997, coll. ARTP; 3 specimens, AM P.99598, Taylors Beach (41°11'49"S, 148°16'12"E), 18 April 1995, coll. ARTP; few specimens, AM P.99600, The Shank (42°54'S, 145°27'E), 11 April 1997, coll. ARTP; many specimens, AM P.99601, Tiddys Beach (42°12'35"S, 145°11'24"E), 19 June 1997, coll. ARTP; many specimens, AM P.99602, Two Mile Beach, Forestier Peninsula (42°52'31"S, 147°55'50"E), 2 February 1995, coll. ARTP; many specimens, AM P.99621, Four Mile Creek, south of Scamander (41°33'32"S, 148°17'21"E), 21 January 1995, coll. ARTP; 2 specimens, AM P.99606, Beaumaris Beach (41°24'44"S, 148°16'49"E), 18 April 1995, coll. ARTP; many specimens, AM P.99657, Yellow Beach, Hunter Island (40°29'24"S, 144°46'47"E), 7 June 1996, coll. ARTP; few specimens, AM P.99683, Moreys Bay, Schouten Island (42°18'04"S, 148°16'53"E), 28 April 1995, coll. ARTP; many specimens, AM P.99615, East Beach (41°04'S, 146°48'E), 11 March 1997, coll. ARTP.

Victoria: 1 specimen, AM P.99396, Betka Beach near Mallacoota (37°34'22"S, 149°45'44"E), under weed on sand at high tide mark, under boat ramp, 21 September 2002, coll., M. Miller, J.K. Lowry, J.H. Peart, and R.A. Peart; 1 specimen, AM P.99395, near Devlins Inlet, near Mallacoota (37°33'55"S, 149°45'35"E), under washed-up weed on sand, Australia, 20 September 2002, coll., J.K. Lowry, J.H. Peart, R.A. Peart, and M. Miller.

Remarks. The material identified from Natural History Museum, London, and TMAG provides additional historic and distribution information, with new records from multiple Tasmanian locations. The Natural History Museum, London, material includes two separate collection depositions, both of which are considered contemporary with the type of material in the Australian Museum. Material registered in

1895 (B.M. [N.H.] registration No.) presented as a gift from the Australian Museum are contemporary with Haswell depositing material across natural history institutions at the time of the species descriptions. The male and female specimen from the Stebbing Amphipod Collection deposited with the B.M. (N.H.) in the early 20th century would also have provenance with Haswell distributing material to colleagues in the late 1800s.

Distribution. Tasmania: Eagle Hawk Neck, Seven Mile Beach, Hope Beach, Granville Harbour, Arthur River, Margate, Hobart, Maria Island, Marion Bay, Schouten Island, Bowen Island, Tam O'Shanter Bay, Pollys Bay, South Neck Beach, East Circular, West Circular, Sisters Beach, King Island, Purdon Bay, Nebraska Beach, Deephole Bay, Conningham Beach, Freycinet Peninsula, Plain Place Beach, Crockets Bay, Tatlows Beach, Discovery Beach, Cape Naturaliste, Denison Beach, Roches Beach, Kelvedon Beach, Kingston Beach, Five Mile Beach, Georges Bay, East Wynyard Beach, Endeavour Bay, Orford Beach, Lisdillon Beach Carlton Beach, Greens Beach, Little Lagoon, Forwards Beach, Spring Beach, Muirs Beach, Big Lagoon Beach, Swan Island, Porpoise Hole, Sloop Beach, Sloping Main Beach, Red Ochre Beach, North Rheban Beach, Adventure Bay, Hazards Beach, Hellyer Beach, Piermont Beach, Four Mile Creek, Blackmans Bay Beach, Nye Bay, Bruny Island, Nelson Bay, Nine Mile Beach, Arthur Beach, Binalong Bay, Friendly Beach, Grindstone Beach, Howrah Beach, Mariposa Beach, Maurouard Beach, Riedle Bay, Seymour Beach, Varna Bay, South Simpsons Bay, Wrinklers Beach, Johnsons Bay, Nelson Bay, Bakers Beach, Bellerive Beach, Boat Harbour, Boltons Beach, Boobyalla, Bryans Beach, Calverts Beach, Cooks Beach, Crayfish Creek Beach, Crescent Beach, Eddystone Point, Fortescue Bay, Freemans Beach, Green Point, Ulverstone, Halfmoon Bay, Hardwicke Bay, Waterhouse Beeach, Noland Bay, Black River Beach, Little Jetty Beach, Mayson Bay North Neck Beach, Strahn Peggs Beach, Perkins Island, Planter Beach, Roaring Beach, Sanspit Point Somerset Beach, Little Beach, Sandy Cape Beach, Flinders Island, Lemons Beach, Front Beach, South Clifton Beach, Stumpys Bay, St Albans Bay, Taylors Beach, The Shank, Tiddys Beach, Two Mile Beach, Four Mile Creek, Ceaumaris Beach, Hundter Island, East Beach, Three Hummocks Island, Pirates Bay, Red Rock Point, Marsh Creek, Tasman Peninsula, Mulcahy Bay, Asbestos Range National Park. Victoria: Mallacoota, Point Ricardo Betka Beach, Devlins Inlet. New South Wales: Port Jackson, Jervis Bay, Narrawallee, Ulladulla, Moruya, Twofold Bay (current study; Serejo and Lowry 2008; Lowry 2012; Hughes and Ahyong 2017).

Bellorchestia reliquia sp. nov. (Figs. 5.26–5.35)

Type material. Holotype 1 c male, 19 mm, 3 slides, AM P.100158, Window Pane Bay, Tasmania (43°27′35″S, 146°35″E), 27 June 1997, coll. A. Richardson and party. Paratypes: 1 a male, 14 mm, 4 slides, AM P.100159, Window Pane Bay, Tasmania (43°27′35″S, 146°35″E), 27 June 1997, coll. A. Richardson and party; 1 b male, 16 mm, 4 slides, AM P.100160, Window Pane Bay, Tasmania (43°27′35″S, 146°35″E), 27 June 1997, coll. A. Richardson and party; 1 d female, 17.5 mm, 4 slides, AM P.100161, Window Pane Bay, Tasmania (43°27′35″S, 146°35″E), 27 June 1997, coll. A. Richardson and party; 1 e male, 22 mm, for SEM, AM P.99364, Window

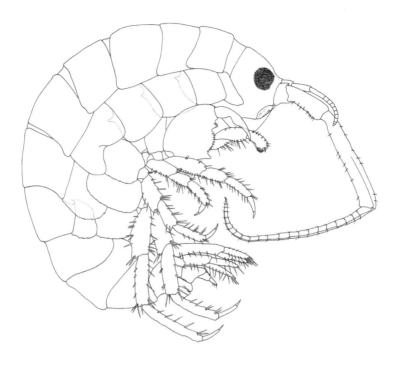

Figure 5.26   *Bellorchestia reliquia* sp. nov. paratype f male 20 mm, AM P.100162, Window Pane Bay, Tasmania.

Pane Bay, Tasmania (43°27′35″S, 146°35″E), 27 June 1997, coll. A. Richardson and party; 1 f male, 20 mm, not dissected, AM P.100162, Window Pane Bay, Tasmania (43°27′35″S, 146°35″E), 27 June 1997, coll. A. Richardson and party; 1 g female, 20 mm, for SEM, AM P.99365, Window Pane Bay, Tasmania (43°27′35″S, 146°35″E), 27 June 1997, coll. A. Richardson and party; 21 specimens, AM P.100163, Window Pane Bay, Tasmania (43°27′35″S, 146°35″E), 27 June 1997, coll. A. Richardson and party.

Additional material examined. 20+ specimens, AM P.100164, Cox Bight (43°31′S, 146°14′E), 27 June 1997, coll. A. Richardson and party; 9 specimens, AM P.100219, Louisa Bay (43°31′12″S, 146°20′23″E), 27 June 1997, coll. A. Richardson and party.

Type locality. Window Pane Bay, Tasmania (43°27′35″S 146°35″E), Tasmania, Australia.

Etymology. Named from the Latin *reliquia* for relic, in reference to the significance of the southwest Tasmania Wilderness region where this species occurs.

Description. Based on type material. Head: eye medium size (greater than 1/5 to 1/3 head length). Antenna 1: short, not reaching midpoint of peduncular article 5 of antenna 2. Antenna 2: more than half body length (0.7 × length); peduncular articles slender, with many large robust setae; article 5 longer than article 4; flagellar articles final article large, cone-shaped forming a *virgula divina*. Labrum: with apical setal patch; epistome with many robust setae. Mandible: left lacinia mobilis 5-dentate. Labium: distolateral setal tuft present, distomedial setal tuft present; without inner

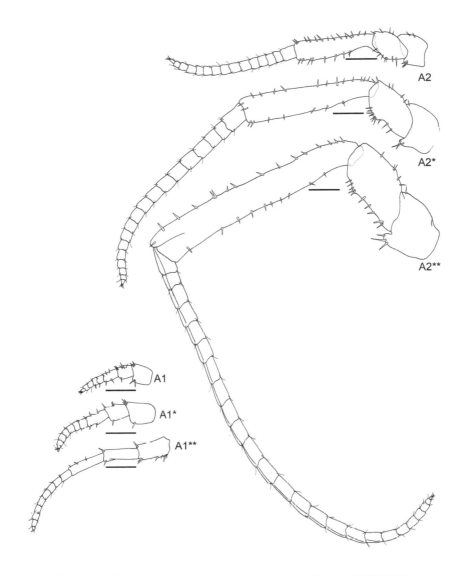

Figure 5.27   *Bellorchestia reliquia* sp. nov. paratype a male 14 mm, AM P.100159; paratype
b male 16 mm, AM P.100160; holotype c male, 19 mm, AM P.100158, Window
Pane Bay, Tasmania. Scales 2 mm.

plates. Maxilla 1: with 1-articulate small palp. Maxilliped: palp article 2 distomedial
lobe well developed; article 4 fused with article 3.

Pereon. Gnathopod 1: sexually dimorphic; subchelate; coxa subequal to coxa 2;
carpus longer than propodus, 1.8 as long as propodus; propodus subrectangular, 2.4
× as long as broad, anterior margin with 5 groups of robust setae, posterior margin
with palmate lobe, palm transverse; simplidactylate, dactylus overreaching palm,
without anterodistal denticular patch. Gnathopod 2: sexually dimorphic; subchelate;

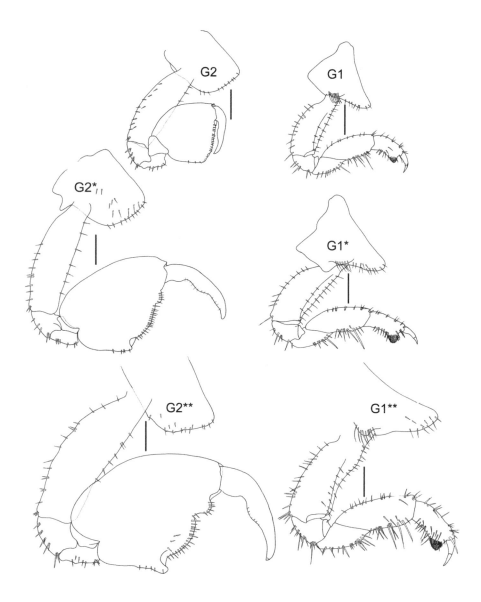

Figure 5.28 *Bellorchestia reliquia* sp. nov. paratype a male 14 mm, AM P.100159; paratype
b male 16 mm, AM P.100160; holotype c male, 19 mm, AM P.100158, Window
Pane Bay, Tasmania. Scales 2 mm.

basis slender; ischium anterior margin with lateral and medial rounded lobes asymmetrical; carpus triangular, reduced, enclosed by merus and propodus; propodus 1.3
× as long as wide, palm subacute, 65% along posterior margin, palm margin convex with posterodistal shelf, lined with robust setae and defined by rounded corner
with tooth; dactylus curved, subequal in length to palm, apically acute, posterior
mid-margin with hump, closing into dactylar socket. Pereopods 3–7: bicuspidate,

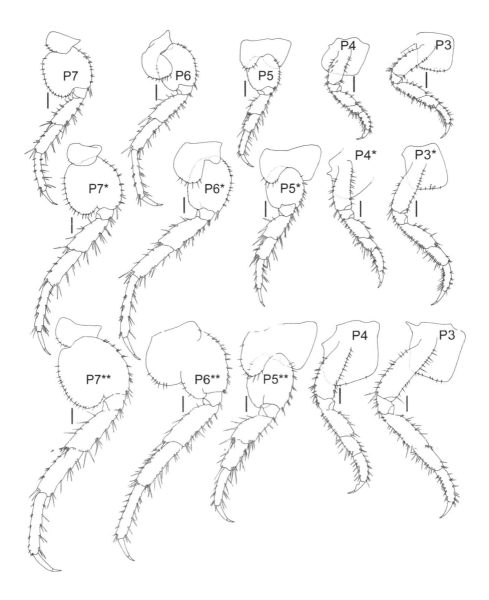

Figure 5.29   *Bellorchestia reliquia* sp. nov. paratype a male 14 mm, AM P.100159; paratype b male 16 mm, AM P.100160; holotype c male, 19 mm, AM P.100158, Window Pane Bay, Tasmania. Scales 2 mm.

dactylus without anterodistal denticular patch. Pereopods 3–4: coxae deeper than wide. Pereopod 3: carpus length twice width. Pereopod 4: significantly shorter than pereopod 3; carpus significantly shorter than carpus of pereopod 3, length 1.5 × width; dactylus posterior margin thickened with 1 proximal and 1 distal projections. Pereopod 5: merus length 1.6 × width; carpus length twice width; propodus longer than carpus, length 4.5 × width; dactylus long, slender, length 3.1 × width. Pereopods

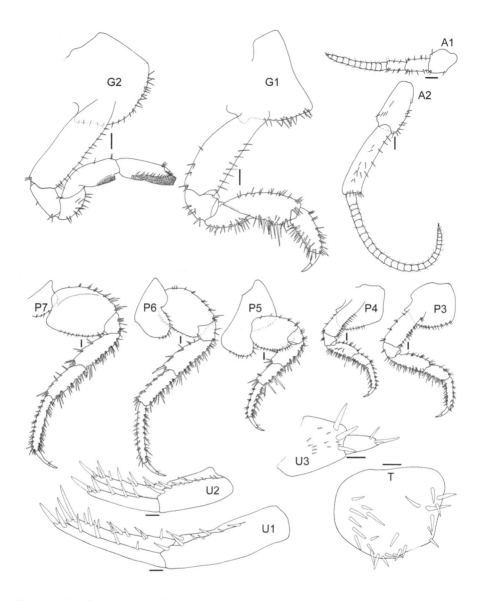

Figure 5.30   *Bellorchestia reliquia* sp. nov. paratype female 17.5 mm, AM P.100161, Window Pane Bay, Tasmania. Scales 2 mm.

6–7: not sexually dimorphic. Pereopod 6: longer than pereopod 7; coxa posterior lobe posteroventral corner rounded, posterior margin perpendicular to ventral margin; merus not expanded, length 2.6 × width; carpus not expanded, length 4 × width; propodus length 6 × width, longer than carpus. Pereopod 7: basis expanded, produced posteriorly, posterior margin convex, lined with small robust setae; merus length 2.1

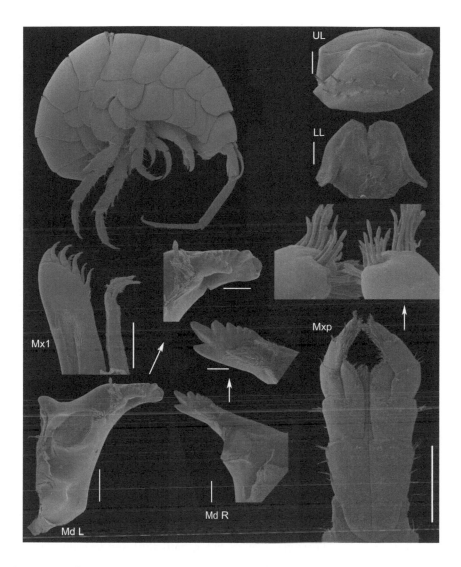

Figure 5.31 *Bellorchestia reliquia* sp. nov., male, 22 mm, AM P.99364, SEM image, Window Pane Bay, Tasmania. Scale UL, LL, Md, Mx1, 100 µm; Mxp 200 µm.

× width; carpus length 3.5 × width; propodus length 5.5 × width, distinctly longer than carpus.

Pleon. Pleopods 1–3: well developed, rami segmented. *Epimeron 1* posterior margin convex, broad minute serrations. Epimeron 2–3: posteroventral corner weakly convex to straight, margin with broad minute serration. Uropod 1: peduncle with 15+ robust setae, without apical spear-shaped setae; rami positioned directly on top of each other; inner ramus subequal in length to outer ramus, with 13+ marginal

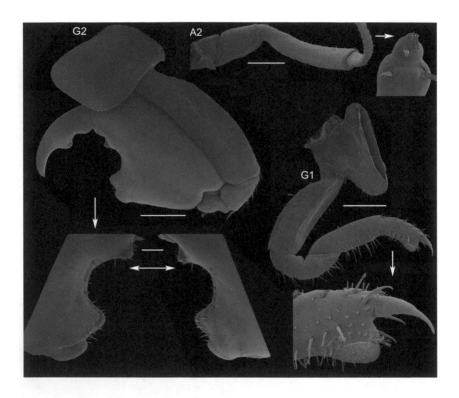

Figure 5.32   *Bellorchestia reliquia* sp. nov., male, 22 mm, AM P.99364, SEM image, Window Pane Bay, Tasmania. Scale G1, G2, 200 μm; A1, 200 μm; A2, 1 mm.

robust setae in 2 rows; outer ramus with 5 marginal robust setae in 1 row. Uropod 2: peduncle with 12 robust setae; inner ramus subequal in length to outer ramus, with 14+ marginal robust setae in 2 rows; outer ramus with 5 marginal robust setae in 1 row. Urosomite 3: subrectangular, not enclosing around telson, broader than deep, 1.2 × as broad as deep. Uropod 3: vestigial; peduncle not dorsally concave, length 1.1 × depth, with 3 robust setae; ramus linear, partially fused to peduncle (when viewed dorsally), longer than peduncle, twice as long as broad, with 11 marginal setae and cluster of apical setae. Telson: subovate, broader than long, lateral margins straight to convex, apical margins convex, incised apically, dorsal midline weakly present, with single row of around 17–19 marginal and apical robust setae per lobe; longer than uropod 3 peduncle.

Female (sexually dimorphic characters). Gnathopod 1: simple; propodus posterior margin without palmate lobe; propodus palm near transverse. Gnathopod 2: mitten-shaped; basis expanded anteromedially; ischium without lobe; carpus well developed (not enclosed by merus and propodus), length 2.5 × width; propodus twice width, palm obtuse; dactylus shorter than palm.

Variation. In more juvenile males, the gnathopod 2 palm is smooth and entire, developing a more tooth and sinusoidal characters with increasing body size. The

Figure 5.33   *Bellorchestia reliquia* sp. nov., male, 22 mm, AM P.99364, SEM image, Window Pane Bay, Tasmania. Scale P, 500 μm; Dactyl, 100 μm.

female specimen used for SEM (AM P. 99365, 20 mm) appears to be aberrant with no robust setae on the upper lip, while other examined had robust setae.

Remarks. The male gnathopod 2 sinsusoidal palm with distal tooth and palm-defining tooth readily separate *Bellorchestia reliquia* sp. nov. from other Australian *Bellorchestia* species. The distribution of this species appears to be highly restricted, having only been collected from three closely located beaches in the southwest Tasmanian Wilderness World Heritage Area. *Bellorchestia reliquia* sp. nov. is known to cooccur at sites with *B. lutriwita*, *B. pravidactyla*, and *B. palawakani*.

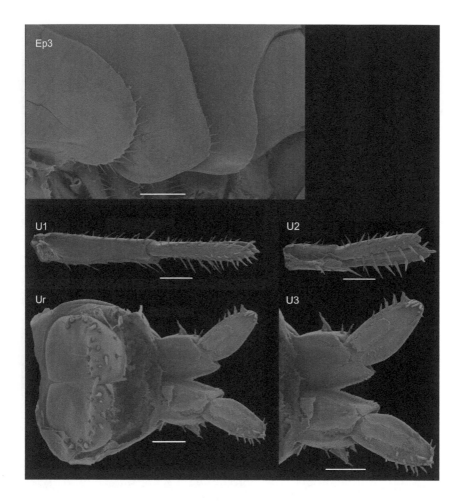

Figure 5.34 *Bellorchestia reliquia* sp. nov., male, 22 mm, AM P.99364, SEM image, Window Pane Bay, Tasmania. Scale Ur, 200 µm; Ep, U1, U2, 500 µm; T, 100 µm.

Distribution. Tasmania: Window Pane Bay, Cox Bight, Louisa Bay (current study).

5.4 KEY TO MALE AUSTRALIAN *BELLORCHESTIA* SPECIES

1 Gnathopod 2 propodus palm with rounded sinus or sculptured 2

1′ Gnathopod 2 propodus palm entire, smooth .... *B. palawakani* sp. nov.

2(1) Gnathopod 2 propodus palm with midpalmar sinus 3

2′ Gnathopod 2 propodus palm convex .... *B. reliquia* sp. nov.

3(2) Gnathopod 2 propodus palm defined by corner with tooth 4

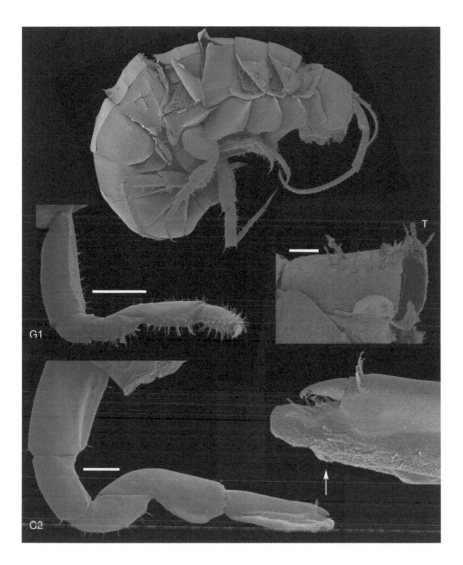

Figure 5.35   *Bellorchestia reliquia* sp. nov., female, 20 mm, AM P.99365, SEM image, Window Pane Bay, Tasmania. Scale G1, 500 μm; G2, 200 μm; T, 100 μm.

3′ Gnathopod 2 propodus palm corner weakly
   defined, subacute... *B. needwonnee* sp. nov.

4(3) Gnathopod 2 propodus palm sinus broad, dactylus with
   proximal hump ....*B. lutruwita* sp. nov.
4′ Gnathopod 2 propodus palm with three teeth,
   dactylus convex with distal hump ....*B. pravidactyla* (Haswell 1880)

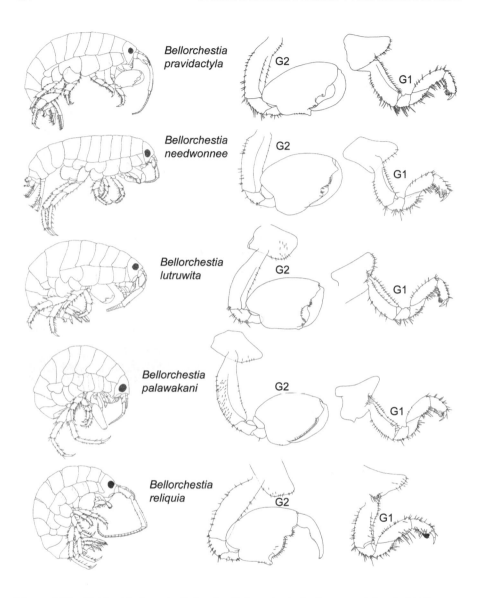

Figure 5.36 Field guide to male *Bellorchestia* from Australia (Lowry and Serejo 2008).

ACKNOWLEDGMENTS

We are greatly indebted to Dr. A.M.M. Richardson who made this collection available for study. Funding for this research is from ABRS grant RF 215–51.

REFERENCES

Dana, J.D. 1853. Crustacea. Part II. United States Exploring Expedition during the years 1838, 1839, 1840, 1841, 1842 under the command of Charles Wilkes. *U.S. Navy* 14:689–1618.

Haswell, W.A. 1880. On some new amphipods from Australia and Tasmania. *Proceedings of the Linnean Society of New South Wales* 5:97–105.

Hughes, L.E. and Ahyong, S.T. 2017. The identity of the Australian sand-hopper *Talorchestia pravidactyla* Haswell, 1880 (Amphipoda: Talitridae). *Journal of Crustacean Biology* 37:53–62.

Hughes, L.E. and Lowry, J.K. 2017. *Hermesorchestia alastairi* gen. et sp. nov. from Australia (Talitridae: Senticaudata: Amphipoda: Crustacea). *Zootaxa* 4311:491–506.

Hurley, D.E. 1956. Studies on the New Zealand amphipodan fauna, No. 13. Sandhoppers of the genus *Talorchestia. Transactions of the Royal Society of New Zealand* 84:359–89.

Lowry, J.K. 2012. Talitrid amphipods from ocean beaches along the New South Wales coast of Australia (Amphipoda, Talitridae). *Zootaxa* 3575:1–26.

Lowry, J.K. and Springthorpe, R.T. 2009. *Talorchestia brucei* sp. nov. (Amphipoda, Talitridae), the first talitrid from the Northern Territory, Australia. *Crustaceana* 82:897–912.

Myers, A.A. and Lowry, J.K. 2020. A phylogeny and classification of the Talitroidea (Amphipoda, Senticaudata) based on interpretation of morphological synapomorphies and homoplasies. *Zootaxa, 4778*(2), 281–310.

Rafinesque, C.S., Ed. 1815. *Analyse de la nature ou tableau de l'univers.* Palerme, Aux dépens de l'auteur.

Richardson, A.M.M. and Araujo P.B. 2015. Lifestyles of terrestrial crustaceans. In *The Life Styles and Feeding Biology of the Crustacea*, eds. M. Thiel, and L. Watling, 299–336. Oxford: Oxford University Press.

Richardson, A.M.M., Swain, R. and Smith, S.J. 1991. Local distributions of sandhoppers and landhoppers (Crustacea: Amphipoda: Talitridae) in the coastal zone of western Tasmania. *Hydrobiologia* 223:127–40.

Richardson A.M.M., Swain, R. and Wong, V. 1997a. The crustacean and molluscan fauna of Tasmanian salt marshes. *Papers and Proceedings of the Royal Society of Tasmania* 131:21–30.

Richardson A.M.M., Swain, R. and Wong, V. 1997b. Translittoral Talitridae (Crustacea: Amphipoda) and the need to reserve transitional habitats: examples from Tasmanian saltmarshes and other coastal sites. *Memoirs of Museum Victoria* 56:521–29.

Serejo, C.S. and Lowry, J.K. 2008. The coastal Talitridae (Amphipoda: Talitroidea) of Southern and Western Australia, with comments on *Platorchestia platensis* (Kroyer, 1845). *Records of the Australian Museum* 60:161–206.

Springthorpe, R.T. and Lowry, J.K. 1994. Catalogue of crustacean type specimens in the Australian Museum: Malacostraca. *Technical Reports of the Australian Museum* 11:1–134.

Stebbing, T.R.R. 1906. Amphipoda. I. Gammaridea. In *Das Tierreich*, in Friedlander and
    Son. 21:1–806.
Stebbing, T.R.R. 1910. Scientific results of the trawling expedition of the H.M.C.S. "Thetis."
    Crustacea. Part V. *Australian Museum Memoir* 4:565–658.
Stephensen, K. 1948. Amphipods from Curaçao, Bonaire, Aruba and Margarita. *Studies on
    the Fauna of Curaçao, Aruba, Bonaire and the Venezuelan Islands* 3, 11:1–20.

CHAPTER **6**

Freshwater Malacostraca of the Mediterranean Islands – Diversity, Origin, and Conservation Perspectives

Kamil Hupało, Fabio Stoch, Ioannis Karaouzas, Anna Wysocka,
Tomasz Rewicz, Tomasz Mamos, and Michał Grabowski

CONTENTS

6.1 INTRODUCTION

Malacostraca exhibit a huge variety of body forms and ecological adaptations, allowing them to inhabit marine, freshwater, and terrestrial habitats. Of the estimated 26,000 malacostracan species described, about 6,000 inhabit freshwater (Balian et al. 2008). In freshwaters, malacostracans adapted to colonize the plethora of available ecosystems, including springs, rivers, estuaries, lakes, caves, all kinds of groundwater habitats, and even ephemeral water bodies (Fig. 6.1) (Balian et al. 2008, Figueroa et al. 2013). The highest number of freshwater malacostracan species has been (2,165, ca. 35%) recorded from the Palearctic (Balian et al. 2008). Freshwater malacostracan diversity in the Mediterranean Basin is still largely unknown, particularly due to the lack of the reliable published information on isopods. According to various estimates, 25%–40% of the Palearctic fauna occurs in this area (Balian et al. 2008, Figueroa et al. 2013).

The Mediterranean Basin has been recognized as one of the 25 most important biodiversity and endemism hotspots worldwide (Myers et al. 2000, Woodward 2009, Blondel et al. 2010). Even though the basin represents only 1.6% of the Earth's surface, it hosts nearly 10% of global flora, 3% of vertebrates, and over 7% of marine biodiversity, with a high level of endemism (Medail and Quezel 1997, Myers et al. 2000, Coll et al. 2010). Although the freshwater fauna is still largely understudied, it is estimated that the region supports about 35% of all Palearctic species. This corresponds to more than 6% of the world freshwater species. At least 43% of the freshwater Mediterranean fauna is considered to be endemic with the majority inhabiting islands (Figueroa et al. 2013).

Figure 6.1 Examples of Mediterranean island habitats inhabited by freshwater malacostracans. (A) River – Rio Alcantara, Sicily. (B) Lake – Limni Kournas, Crete. (C) Captured springs – Crete. (D) River – Rio Salso, Sicily (E) Karstic springs – Sardinia. (F) Phreatic cave – Grotta di Nettuno, Sardinia. (G) Artificial gallery – Sicily. (H) Phreatic waters accessed through wells – Euboea (Evia).

The Mediterranean Sea is a semienclosed basin, covering an area of approximately 2.5 million km². The sea connects three continents, being bordered on the north by Europe, on the south by Africa, and on the east by Asia. It has two narrow connections with other waters. The Strait of Bosphorus connects with the Black Sea and the Strait of Gibraltar opens the basin to the Atlantic. The Mediterranean Sea is divided by the Siculo-Tunisian Strait (aka Strait of Sicily), a biogeographical barrier between Cape Bon (Tunisia) and Mazara del Vallo (Sicily), into two sub-basins: the Western Mediterranean with Sardinia, Corsica, the Tuscan Archipelago as well as the Balearic Islands, and the Eastern Mediterranean with Sicily, the Maltese Archipelago, the Pelagie Islands, the Adriatic, Ionian, Aegean Islands along with Crete, Cyprus and the islands of the Marmara Sea (Bianchi and Morri 2000). More than 5,000 islands in the Mediterranean Basin cover about 100,000 km², i.e., more than 5% of the surface of the entire region (Hopkins 2002, Vogiatzakis et al. 2008). The islands can be either of oceanic or continental origin with some being oceanic plate wedges, formed by volcanic activity, created by tectonic plate fragmentation, or by sea-level changes (Schüle 1993, Whittaker and Fernández-Palacios 2007).

The Mediterranean Sea originated from the Tethys Ocean, which once separated the supercontinents of Laurasia and Gondwana. Over millions of years, due to continental plate shifts, Tethys gradually closed and at the end of the Eocene, a major reorganization of Tethys took place, resulting in its division into two residuals: Paratethys and the circum-Mediterranean Basin. At the end of the Oligocene, the latter became the proto-Mediterranean Sea (Rögl 1999, Popov et al. 2004). Some islands, like Cyprus and the Adriatic Islands, emerged before the formation of the circum-Mediterranean Basin (Harland et al. 1982), although the majority of the Mediterranean islands emerged or were isolated from the continental plates during the Oligocene and Miocene (Fig. 6.2). These include Corsica, Sardinia (Speranza et al. 2002; Advokaat et al. 2014), Sicily (Catalano et al. 1994), the Aegean Islands with Crete (Meulenkamp 1971, Poulakakis et al. 2015) as well as the Maltese (Savona Ventura 1975) and the Tuscan Archipelagos (Barbato et al. 2018). One of the most crucial events in the history of the Mediterranean Basin was the closure of the Strait of Gibraltar, which resulted in the subsequent desiccation of the basin – the so-called Messinian Salinity Crisis (MSC, 5.96–5.33 Ma) (Boccaletti et al. 1990, Krijgsman et al. 1999). During that time, nearly all the islands had temporal land connections with the mainland, resulting in substantial faunal exchange (e.g., Ketmaier and Caccone 2013, Chueca et al. 2015, Poulakakis et al. 2015). For several islands (e.g., Sicily, many Aegean Islands), this situation repeated in the Pleistocene, due to recurrent glaciations and associated eustatic sea-level oscillations, e.g., during the last glacial maximum, the global sea level dropped down by ca. 120 m (e.g., Arias et al. 1980, Perissoratis and Conispoliatis 2003, Marra 2009, Muscarella and Baragona 2017). The geologically youngest islands that emerged only during Pliocene and Pleistocene are the Pelagie Islands and the Ionian Islands (Sakellariou and Galanidou 2015, Muscarella and Baragona 2017) (Fig. 6.2).

Due to their isolation as well as climatic, topographic, and geological heterogeneity, the Mediterranean islands are natural evolutionary laboratories with a high number of local endemics (Schüle 1993, Hopkins 2002, Vogiatzakis et al.

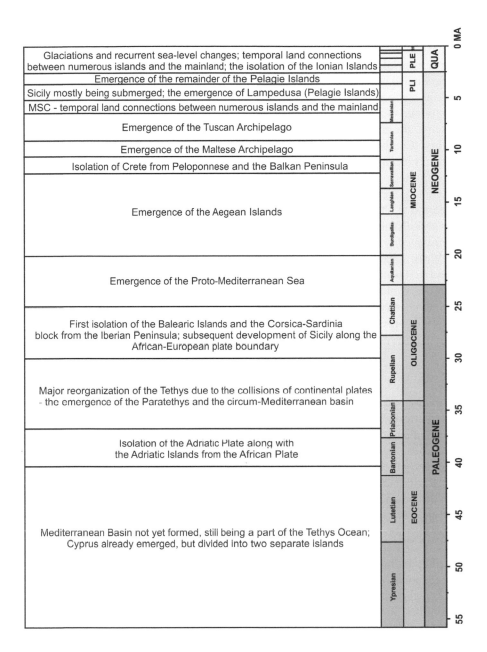

Figure 6.2 The stratigraphic table highlighting the major steps in the history of the Mediterranean Sea and the Mediterranean islands.

2008, Blondel et al. 2010). So far, however, the majority of biodiversity studies and Mediterranean insular fauna origin studies have focused mostly on terrestrial, semiterrestrial, or marine coastal species, leaving the freshwater fauna relatively understudied (Figueroa et al. 2013). Freshwater biota extinction rates are estimated to be as much as five times higher as terrestrial biota (Dudgeon et al. 2006), mainly due to human activity (Cuttelod et al. 2008, Darwall et al. 2009). Freshwater ecosystems are critical to sustaining life and establishing civilizations throughout history (Combes 2003). Given a long history of civilization development in the Mediterranean Basin, combined with the limited local freshwater resources and consequences of increasing tourist pressures, it is one of the most anthropogenically altered regions in the world with the island fauna among the most threatened (Myers et al. 2000, Hopkins 2002, Mittermeier et al. 2004). Thus, gathering diversity and distribution data on freshwater species is of paramount importance for planning conservation strategies.

We (i) assembled all published records, including gray literature on Mediterranean island freshwater Malacostraca; (ii) reviewed information on species biogeographical affinities, evolutionary histories, and colonization of insular freshwater ecosystems; (iii) summarized data on alien malacostracan species, and; (iv) discussed possible risks and conservation perspectives for the local malacostracan fauna.

6.2 METHODS

We focused on the Mediterranean insular malacostracan taxa that complete their life cycle in freshwaters, including both epigean and hypogean biota. We also cover taxa of anchialine caves and aquifers. However, we excluded coastal brackish water malacostracans, i.e., inhabiting marine caves, lagoons, and euryhaline coastal lakes and estuaries, which cannot be considered freshwater. The selection of excluded taxa was necessarily arbitrary where complete environmental data were not available. Only taxa described before the end of 2018 were counted, including legitimate subspecies, with the taxonomic nomenclature following the World Register of Marine Species (WoRMS) database (as of November 2018).

We used updated and exhaustive literature to include information on presence, distribution, biogeographical affinities, evolutionary history, and, where available, conservation. We consulted general reviews (i.e., Balian et al. 2008, Figueroa et al. 2013) and checklists (i.e., Ruffo and Stoch 2006, Stoch et al. 2006) of freshwater malacostracans from Mediterranean islands, web databases and repositories (e.g., WoRMS, WADA), and specialist articles and books (reported in the each section of the chapter). Our review is divided into sections covering all major Mediterranean islands and archipelagos: the Balearic Islands, Corsica and Sardinia, Sicily, the Maltese and Pelagie Islands, the Adriatic Islands, Crete, the Aegean and Ionian Islands, the islands of the Marmara Sea and Cyprus. The data in each section was compiled by individual authors, summarized in Tables 6.1 and 6.2 and illustrated on a map (Fig. 6.3). A detailed and updated checklist of all freshwater Mediterranean island malacostracans is provided in Appendix 1.

Table 6.1 Summary of Described Mediterranean Freshwater Insular Malacostracan Biodiversity, Divided by Island

	Islands													Total*
	BAL	COR	SAR	TUS	SIC	PEL	MAL	ADR	ION	CRE	AEG	MAR	CYP	
Number of species + subspecies	32	20	36	16	21	2	12	29	20	19	27	2	9	182
Proportion of the Med. malacostracan insular fauna	18%	11%	20%	9%	12%	1%	7%	16%	11%	10%	15%	1%	5%	100%
Number of genera	17	15	21	13	14	2	9	15	11	11	12	1	8	51
Number of families	13	13	17	11	12	2	7	12	11	6	11	1	8	27
Number of endemics (proportion)	20 (63%)	4 (20%)	21 (58%)	4 (20%)	9 (57%)	0 (0%)	0 (0%)	6 (21%)	7 (35%)	12 (63%)	6 (22%)	0 (0%)	3 (33%)	95 52%
Number of hypogean species (proportion)	20 (63%)	12 (60%)	23 (64%)	12 (73%)	8 (38%)	1 (50%)	0 (0%)	21 (72%)	9 (45%)	10 (52%)	12 (44%)	0 (0%)	2 (22%)	112 (62%)
Number of alien, introduced species (proportion)	1 (3%)	2 (10%)	2 (6%)	0 (0%)	3 (14%)	0 (0%)	5 (42%)	0 (0%)	0 (0%)	1 (5%)	0 (0%)	0 (0%)	2 (22%)	9 (5%)

SAR – Sardinia and circum-Sardinian Islands; COR – Corsica; TUS – Tuscan Archipelago; SIC – Sicily; PEL – Pelagie Islands; MAL – Maltese Islands; ADR – Adriatic Islands; ION – Ionian Islands; AEG – Aegean Islands; CRE – Crete; MAR – Marmara Islands; CYP – Cyprus.
*Numbers from Total column is not a sum of the former columns due to the possibility of species overlap between the islands.

CRUSTACEAN BIODIVERSITY AND CONSERVATION

Table 6.2 Summary of Reported Species from Each of the Malacostraca Order, Divided by Island

Malacostracan orders (proportion of endemics/aliens)	Islands													Total* (proportion)
	BAL	COR	SAR	TUS	SIC	PEL	MAL	ADR	ION	CRE	AEG	MAR	CYP	
Amphipoda	22 (64/0)	12 (25/0)	18 (61/0)	11 (27/0)	11 (82/0)	2 (0/0)	4 (0/0)	20 (25/0)	10 (33/0)	12 (67/0)	11 (36/0)	2 (0/0)	3 (33/0)	106 – 58 (58/0)
Isopoda	7 (71/0)	4 (25/0)	10 (70/0)	4 (0/0)	4 (50/25)	0	1 (0/0)	6 (0/0)	7 (57/0)	3 (67/33)	8 (25/0)	0	3 (67/0)	47 – 26 (53/4)
Thermosbaenacea	1 (100/0)	0	0	0	1 (100/0)	0	0	1 (0/0)	0	0	0	0	0	3 – 2 (67/0)
Bathynellacea	1 (0/0)	1 (0/0)	3 (67/0)	0	0	0	0	0	0	0	0	0	0	4 – 2 (50/0)
Decapoda	1 (0/100)	3 (0/67)	4 (0/50)	0	5 (0/40)	0	7 (0/71)	2 (50/0)	3 (0/0)	4 (50/0)	8 (0/0)	0	3 (0/67)	21 – 12 (14/33)

SAR – Sardinia and circum-Sardinian Islands; COR – Corsica; TUS – Tuscan Archipelago; SIC – Sicily; PEL – Pelagie Islands; MAL – Maltese Islands; ADR – Adriatic Islands; ION – Ionian Islands; AEG – Aegean Islands; CRE – Crete; MAR – Marmara Islands; CYP – Cyprus.
*Numbers from Total column is not a sum of the former columns due to the possibility of species overlap between the islands

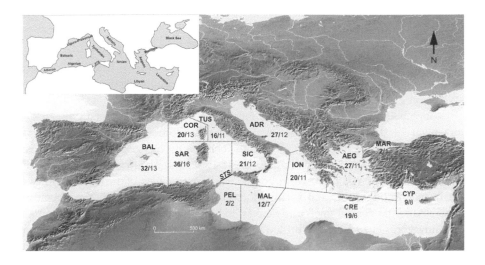

Figure 6.3 Distribution map of the Mediterranean islands freshwater malacostracans. BAL – Balearic Islands; SAR – Sardinia and circum-Sardinian Islands; COR – Corsica; TUS – Tuscan Archipelago; SIC – Sicily; PEL – Pelagie Islands; MAL – Maltese Islands; ADR – Adriatic Islands; ION – Ionian Islands; AEG – Aegean Islands; CRE – Crete; MAR – Marmara Islands; CYP – Cyprus. Numbers in bold indicate the number of species, the regular number is the number of families per region. STS refers to Siculo-Tunisian Strait, dividing the Mediterranean Sea into western and eastern basin. The smaller map in the left upper corner indicates the major sea basin in the Mediterranean region.

6.3 RESULTS

6.3.1 The Balearic Islands

6.3.1.1 Geographical Setting

The Balearic Islands are the most isolated Mediterranean archipelago, located off the Iberian Peninsula coast. It covers an area of approximately 4,992 km². The archipelago comprises four large islands: Mallorca, Menorca, Ibiza (in Catalan: Eivissa), and Formentera, plus minor islands and islets such as Cabrera and Dragonera. Mallorca, Menorca, and Ibiza are 6th (with 3,640 km²), 11th (702 km²), and 14th (572 km²) largest islands in the Mediterranean, respectively (Cherry and Leppard 2018). The Baleares are divided into two island units or "subarchipelagos": the western set, Pityusic, with Ibiza, Formentera, and nearly 60 islets, and the eastern set, Gymnesic, with Mallorca, Menorca, Cabrera, Dragonera, and nearly 30 islets. While the two Pityusic islands are separated only by a narrow, shallow channel, the two main Gymnesic subarchipelago islands Mallorca and Menorca are more isolated and have different paleogeographical history (Bover et al. 2008, also see below). Physiographically, the Balearic Islands are low-lying, highly karstic, and characterized by lack of surface waters, except for the Serra de Tramuntana mountain range on Mallorca, which reaches more than 1,000 m (Palmer et al. 1999).

6.3.1.2 Malacostracan Fauna

Thirty-two species are known from the archipelago: 29 from the Gymnesic Islands (72% from Mallorca, 31% – Menorca, 25% – Cabrera, 10% – Dragonera) and only 6 from the Pityusic Islands. Most are endemic with the majority being stygobionts (63%).

Amphipoda is represented by five families, associated with interstitial and subterranean karstic waters. Metacrangonyctidae is represented by *Metacrangonyx longipes* Chevreux, 1909, an endemic stygobiont known from both Mallorca and Menorca (Pons, Jaume, and Damians 1995). It occurs in various types of groundwater habitats on Mallorca, from coastal anchialine caves with raised salinity to freshwater wells, caves, and springs located inland. On Menorca, this species is restricted to the coastal anchialine caves and wells and is absent from fresh inland groundwaters. The species is limited to lowlands and is absent from potentially suitable habitats located at the elevations higher than 125 m above sea level on both islands. The stygobiont Bogidiellidae is represented on Mallorca by two species of *Bogidiella* (Hertzog 1933): *B. torrenticola* (Pretus and Stock 1990), an interstitial species known only from a stream hyporheic zone on the west coast of Mallorca, and *B. balearica* (Dancau 1973), a cave-dwelling anchialine species, present also on the neighboring Cabrera (Pretus and Stock 1990; Pretus 1991, Pons et al. 1995, Jaume et al. 2007). Recently, a new genus and species of bogidiellid was described, *Racovella birramea* (Jaume et al. 2007), an endemic stygobiont discovered in a Mallorcan anchialine cave (Jaume et al. 2007; remark: in the abstract of the paper the invalid species name, *Racovella uniramea*, appears). According to Pretus (1991), the Balearic groundwaters are inhabited by at least three other, still undescribed, bogidiellid taxa. *Psammogammarus burri* (Jaume and Garcia 1992) is the sole representative of Eriopisidae. It is a Balearic endemic, so far known only from Cova des Burrí, an anchialine cave on Cabrera (Jaume and Garcia 1992).

Gammaridae is species-rich, especially on Mallorca and Menorca. Among five species of *Echinogammarus* (Stebbing 1899), two are endemic: *E. pungens minoricensis* (Margalef 1952), inhabiting the springs on Menorca, and *E. ebusitanus* (Margalef 1952) occurring on Mallorca and Ibiza (Pretus 1991, Pons et al. 1995). However, recent molecular studies (Hou et al. 2013; A. Wysocka pers. comm) suggest that *E. pungens minoricensis* is a separate species. A similar situation concerns the local *E. ebusitanus.*

The remaining nonendemic gammarid taxa are *Echinogammarus eisentrauti* (Schellenberg 1937) and *E. monomerus* (Stock 1978) from Mallorca (Pretus 1991) as well as *E. pungens* sensu lato, a species complex common in southern Europe and recorded in the permanent coastal lagoons on Menorca (Lucena-Moya et al. 2010).

Rhipidogammarus is represented by two species. The nonendemic *R. rhipidophorus* (Catta 1878) is widely distributed in the Mediterranean Basin. It is known from karstic springs close to the sea, on the northern slope of Serra de Tramuntana, as well as from freshwaters of Menorca and Ibiza (Pretus 1991, Pons et al. 1995). On Mallorca it cooccurrs with the endemic *Rhipidogammarus variicauda* (Stock 1978). Pseudoniphargidae is represented in cave environments by seven species, with most species endemic. *Pseudoniphargus racovitzai* (Pretus and Stock 1990) is known from one locality on Mallorca, and *P. pityusensis* (Pretus and Stock 1990), along with *P. pedrerea* (Pretus and Stock 1990), are endemic to the Pityusic Islands (Ibiza, Formentera) (Pretus and Stock 1991, Pons et al. 1995). *Pseudoniphargus daviui* (Jaume 1991), a stygobitic species, endemic to the Balearic Islands (Mallorca, Cabrera) is sympatric with another Cabreran endemic, *P. triasi* (Jaume 1991) (Pons et al. 1995). The other species, *P. mercadali* (Pretus 1988) and *P. adriaticus* (Karaman 1955), are known from the groundwaters of Mallorca and Menorca, but their distribution extends much beyond the Balearic Islands (Pretus 1991, Pons et al. 1995).

The stygobiontic Salentinellidae is represented by two species. *Salentinella angelieri* (Ruffo and Delamare-Deboutteville 1952) displaying a wide circum-Mediterranenan distribution (including the Ionian and Aegean archipelagos) is one of the most common macroinvertebrate in the anchialine caves of Mallorca. *Salentinella formenterae* (Platvoet 1984) is endemic to Formentera, where it cooccurs with *P. pedrerea* (Pesce et al. 1979, Pesce 1985, Pretus 1991, Pons et al. 1995).

Isopods include four families: Asellidae, Cirolanidae, Janiridae, and Lepidocharontidae, including both epigean and stygobitic species. Asellidae is represented by one, widely distributed taxon *Proasellus coxalis gabriellae* (Margalef 1950). It inhabits the groundwaters (wells) as well as channels near agricultural areas (Pons et al. 1995) on the Gymnesic (Mallorca, Menorca, Cabrera, Dragonera) and the Pityusic (Ibiza, Formentera) Islands. Margalef (1954) questioned the taxonomic identity of this taxon and considered it to be *Proasellus coxalis banyulensis* (Racovitza 1919); the problem remains unresolved so far (F. Stoch, unpublished). Two endemic cirolanid species are known: *Typhlocirolana moraguesi* (Racovitza 1905) and *Metacirolana ponsi* (Jaume and Garcia 1992) (see Pons et al. 1995). The first is widely distributed in freshwater and anchialine ecosystems of Mallorca, Menorca, Cabrera, and Dragonera (Palmer et al. 1999), whereas the second is found only in an anchialine cave on Cabrera (Jaume and Garcia 1992). The epigean freshwater habitats on the islands are prone to colonization by euryhaline marine species, which can migrate upstream and settle in the pure freshwater springs and brooks (Jaume and Garcia 1988, Jaume and Queinnec 2007). The isolated populations of *Jaera nordmanni brevicaudata* (Jaume and Garcia 1990) and *J. balearica* (Jaume and Garcia 1990), from the generally marine and brackishwater Janiridae, are found under stones in freshwater streams, up to 500 m above sea level in the Serra de Tramuntana mountain range on Mallorca (Margalef 1952, Jaume and Garcia 1988).

The stygobitic Lepidocharonitidae is represented by two species: the endemic *Microcharon comasi* (Coineau 1968) (Mallorca) and the nonendemic *M. marinus* (Chappuis and Delamare-Deboutteville 1954) (Mallorca and Menorca) (Pretus 1991).

Only one endemic thermosbaenacean, the stygobitic *Tethysbaena scabra* (Pretus 1991) of Thermosbaenidae, has been reported from the Balearic Islands. It is known from numerous localities on the Gymnesic Islands (Mallorca, Menorca, Cabrera, and Dragonera) (Palmer et al. 1999, Jaume 2008). It occurs in anchialine waters of coastal caves and wells and never has been found in fresh inland groundwaters (Pretus 1991). The Bathynellacea are represented only by *Paraiberobathynella fagei* (Delamare-Deboutteville and Angelier 1950) of the Parabathynellidae; a karstic, interstitial, subterranean freshwater species, known also from the Iberian Peninsula (Serban 1977, Guil and Camacho 2001, Camacho 2006, Camacho et al. 2014). This bathynellid was reported first from the Cueva de Genova on Mallorca (Margalef 1952), followed by a few other records in the Serra de Tramuntana range (Pretus 1991; National Biodiversity Data Portal, http://registros.gbif.es).

Only one decapod has been reported from Mallorca: *Procambarus clarkii* (Girard 1852) of the Cambaridae (Hobbs 1942) – an invasive freshwater species (Barbaresi et al. 2007, Pinya et al. 2008, Souty-Grosset et al. 2016). This crayfish prefers warm waters but occasionally is found also in small mountain streams. It tolerates low oxygen concentrations, high temperatures, and salinity (Oscoz et al. 2011).

6.3.1.3 Geological History and Malacostracan Biogeography

The geological history of the Balearic Islands is a combination of complex interactions between the orogenic processes and the extensional tectonic activity observed in the entire Western Mediterranean. The Balearic Islands were affected by extensional tectonics: the Liguro-Provençal Basin (to the west), the Tyrrhenian Basin (to the east), and the Algerian Basin (to the south), accompanied by the orogenesis in the Rif (Morocco) and the Baetic (southern Spain) mountain chains, isolated by the Gibraltar Strait (Rosenbaum et al. 2002). The Balearic archipelago possibly came into existence during the Oligocene (30–25 Ma); the internal parts of the Rif-Baetic belt (the Alboran Domain) may have been located near the present-day Balearic Islands (Rosenbaum et al. 2002, Rosenbaum and Lister 2004).

The Balearic Islands seem to have been connected to the mainland during the Miocene, in the Langhian-Serravallian age (16 to 11.6 Mya), via the Baetic–Balearic corridor (Fontbote et al. 1990). This land connection ceased in the Tortonian (11.3 Ma), due to the transgression of the epicontinental sea, that flooded Mallorca and Menorca and reduced their size to the area roughly corresponding to their present uplands (Pomar 1991). The Messinian Salinity Crisis (MSC) (5.97 to 5.33 Mya) created new temporal connections with the mainland, which allowed the faunal dispersal from Mallorca to Menorca and Ibiza (Hsü et al. 1977, Chueca et al. 2015). Such migration between islands is well documented for numerous taxa (e.g., Bover et al. 2008, Chueca et al. 2015), including the groundwater amphipod, *Metacrangonyx longipes* (see Bauzà-Ribot et al. 2012). It is hypothesized that during the MSC, colonization of the Balearic Islands from the mainland also took place (Delicado et al. 2014).

The final isolation of the Balearic Islands, at the end of the MSC, allowed allopatric speciation of endemic taxa on each of the islands (Chueca et al. 2015). The islands have remained isolated from the continent and from the rest of the Balearic Archipelago since the beginning of the Pliocene and were connected with a marine transgression–regression cycle (Sosdian and Rosenthal 2009, Bauzà-Ribot et al. 2011). The upper Pliocene transgression in the Western Mediterranean triggered the sea-level rise up to 100 m above the current sea level (Sosdian and Rosenthal 2009, Bauzà-Ribot et al. 2011). Fluctuating sea levels have modified the shape and size of the islands, allowing temporal connections between Mallorca (including the Cabrera archipelago) and Menorca, and between Ibiza, Formentera, and the Ses Bledes archipelago. These connections may have allowed species to migrate between islands within each of the Gymnesic and Pityusic islands of the Baleares (Bover et al. 2008, Brown et al. 2008). However, the sea-level oscillations neither reconnected the western and eastern part of the Balearic Islands (separated by 640 m depth) nor linked the Baleares to the Iberian Peninsula (more than 800 m depth) (Chueca et al. 2015).

The paleogeographical history of the Gymnesic and Pityusic Islands triggered the diversification of local endemic fauna as well as affected the biogeographical affinities across the islands within subarchipelagos (Palmer et al. 1999). Islands sharing their palaeogeographical history (e.g., temporal land connections) have similar endemic faunal patterns. Malacostracan fauna on Mallorca is the most species-rich among all the Balearic Islands. It is most similar to the Menorcan fauna, with the presence of three shared species: *Typhlocirolana moraguesi*, *Tethysbaena scabra*, and *Metacrangonyx longipes*, all endemic to the two islands (Pretus 1991, Pons et al. 1995). Similarities within the Gymnesic fauna, however, are also observed on Cabrera and Dragonera that share two endemic species – *T. moraguesi* and *T. scabra* (Pretus 1991, Pons et al. 1995). The Pityusic Islands fauna is relatively poor and greatly differs with the Gymnesic subarchipelago (Pretus 1991, Pons et al. 1995). Thus, the faunistic differences between the Gymnesic and the Pityusic Islands are the consequence of the historical events. Based on vertebrate fossils (mammals and birds) from the upper Pleistocene/early Holocene, the palaeoenvironmental conditions (e.g., vegetation and, thus, hydrology) of the Gymnesic Islands might have been different from that on the Pityusic Islands (Palmer et al. 1999). Ibiza and Formentera share only two endemics: *Pseudoniphargus pityusensis* and *P. pedrerea*. The distinct character of Formentera is manifest by the endemic *Salentinella formenterae*, which is closely related to *S. angelieri*, from Mallorca (Platvoet 1984, Pretus 1991).

Procambarus clarkii is native to southeastern North America (Palmer et al. 1999) and was intentionally introduced in 1973 in the Iberian Peninsula (Barbaresi et al. 2007) as a species of commercial interest in Spain. The species is considered as one of the 100 most dangerous alien aquatic invaders worldwide (Oscoz et al. 2011).

6.3.2 Corsica, Sardinia, and Circum-Sardinian Islands

6.3.2.1 Geographical Setting

Corsica and Sardinia are located on the Tyrrhenian Sea. Corsica covers an area of ca. 8,700 km², ca. 90 km from Tuscany (Italy, Apennine Peninsula) and 170 km from France. Sardinia covers ca. 24,000 km² and is separated by 187 km from the Apennine Peninsula and 184 km from Tunisia. Corsica and Sardinia are separated by the Strait of Bonifacio (11 km wide). Several small islands surrounding Sardinia are collectively known as circum-Sardinian islands.

Corsica is made up of two major geological domains: (i) the crystalline western Corsica, which comprises two-thirds of the island, with the highest peaks and a steep coastline; and (ii) the alpine Corsica in the northeast (including cape Corsica), formed mainly by schists. Moreover, a coastal alluvial plain is present in the northwestern part and a small karstic plateau in the southernmost part, around the town of Bonifacio. The island is mainly mountainous with 28 main watersheds. The average annual flow is ca. 105 m³/s in an area of 6,204 km² that makes up 70% of the island's surface (Arrighi et al. 2005). These average values display a great seasonal variability due to the diverse landscape and the Mediterranean climate, e.g., the annual inflow reaches only 7 m³/s during periods of low precipitation with 74% of total precipitations concentrated in the eight main watersheds (Arrighi et al. 2005). The Corsican Mountains are quite rich with lakes, while ponds are predominant in the coastal areas. The subsurface is a fractured environment, which contains alluvial aquifers, small coastal sedimentary aquifers (Santoni et al. 2016), as well as aquifers in granitic and metamorphic rocks (which make up more than 80% of the groundwater resources in Corsica; Arrighi et al. 2005).

In Sardinia, two main geological complexes may be recognized: the Hercynian basement and the post-Hercynian covers and Quaternary deposits (Baiocchi et al. 2016). The Hercynian basement includes metamorphic rocks, which outcrop extensively in the eastern and in the southwestern sectors of the island (approximately 13,000 km²). The karstic areas, hosting huge aquifers, extend for about 2,088 km², corresponding to the 8.7% of the entire island surface (Cassola 1983). Supramonte is the most important karstic massif, located in the central-eastern part of the island and occupying an area of 170 km² (Cabras et al. 2008). Different from Corsica, Sardinia does not have a well-developed surface hydrographic network, although it is rich in springs, fed mainly from the karstic, volcanic, and alluvial aquifers (Baiocchi et al. 2016).

6.3.2.2 Malacostracan Fauna

Forty-five malacostracan species are recorded so far from Corsica, Sardinia, and the circum-Sardinian islands: 20 from Corsica and 35 from Sardinia and neighboring islands, while 10 (i.e., 22%) are common to both islands. Notwithstanding their common origin and proximity, the malacostracan fauna greatly differs between the two islands, with Sardinia being more species-rich. This difference may be due to several reasons: (i) the greater surface area of Sardinia than Corsica (2.8× in areal extension); (ii) 31 out of 45 species (i.e., 69%) are stygobitic, and groundwater dwelling species in these islands are usually strict endemics (Stoch et al. 2006, Ruffo and Stoch 2006); (iii) the presence of complex, extended, and fragmented karstic aquifers in Sardinia (Cassola 1983), where endemic karstic stygobionts are present (Grafitti 2001, Casale et al. 2008), while the extent of carbonatic rocks in Corsica is restricted to its southern part close to Bonifacio (Santoni et al. 2016); and (iv) less research has been conducted in Corsican freshwaters.

Amphipods are present on these islands with eight families. Nine stygobitic species of Bogidiellidae (five endemic to Sardinia, and two to Corsica) are present, in both interstitial and subterranean karstic waters. The only nonendemic species, *Medigidiella chappuisi* (Ruffo 1952), lives in interstitial anchialine waters (Ruffo and Delamare-Deboutteville 1952). The Gammaridae is well represented; however, Corsica is less studied than Sardinia. Two species of *Echinogammarus* along with *Gammarus italicus* Goedmakers and Pinkster, 1977 and the subterranean *Tyrrhenogammarus sardous* (Karaman and Ruffo 1989) are endemic to Sardinia (Ruffo and Stoch 2006). *Rhipidogammarus rhipidiophorus* (Catta 1878) is present in coastal springs of both islands, while *R. karamani* (Stock 1971) is known only from Corsica, in interstitial stream habitats near estuaries. Another interesting subterranean family of marine origin, the Ingolfiellidae (Hansen 1903), is represented by *Ingolfiella cottarellii* (Ruffo and Vigna-Taglianti 1989) – an endemic stygobiont, found only in an anchialine cave on the small karstic island of Tavolara (Ruffo and Vigna-Taglianti 1989). The Ischyroceridae (Stebbing 1899), including almost exclusively marine species, has a representative collected in central Sardinia in thermal, slightly sulphidic waters (54–58°C, Krapp et al. 2010). A single juvenile specimen was found, identified on a slide in the Museum of Verona as *"Thermojassa"*, was reported by Krapp et al. (2010) as *Jassa* sp., and is still awaiting formal description. Only two stygobitic members of Niphargidae are present in Corsica and Sardinia. *Niphargus corsicanus* (Schellenberg 1950) is widely distributed in the springs of Corsica (Stock 1972), springs and caves of northern Sardinia and on some islets (Ruffo and Stoch 2006). *Niphargus* aff. *speziae* (Schellenberg 1936) is present on Corsica and belongs to a species complex endemic to Italy (Stock 1972). Another species, *N. stefanellii* (Ruffo and Vigna-Taglianti 1968), was reported for Sardinia by Karaman (1985); this single specimen was recently reexamined, and turned out to be a juvenile of *N. corsicanus* (F. Stoch, unpublished data). The Pseudoniphargidae are represented by three species: *Pseudoniphargus obritus* (Messouli et al. 2006) and *P. adriaticus* (Karaman 1955) s.l. in Corsica, and *P. mercadali* (Pretus 1988) in the northwestern karstic area near Alghero in Sardinia). The Salentinellidae, represented by a single species complex – *Salentinella angelieri* Delamare-Deboutteville and (Ruffo 1952) s. l., is present on both islands. Similarly, the Talitridae is represented by one species, *Macarorchestia remyi* (Schellenberg 1950), inhabiting two anchialine caves on both islands (Schellenberg 1950, Ruffo 1960).

Isopoda is represented by three families, including both surface and groundwater species. The Asellidae is particularly speciose in Sardinia, where *Proasellus* (Dudich 1925) is present in springs, caves, and interstitial waters (Argano and Campanaro 2004). *Proasellus banyulensis* (Racovitza 1919), widely distributed in Western Europe, is the only common epigean species on both islands (Arcangeli 1942). Three stygobiotic species known from Sardinia and one known from Corsica are all strict endemics (Argano and Campanaro 2004). The stygobitic *Stenasellus* (Dollfus 1897) (Stenasellidae), present in both islands both in interstitial and karstic waters are under taxonomic revision (Ketmaier et al. 2003). Finally, the minute representatives of the stygobitic Lepidocharontidae are represented by four endemic species of *Microcharon* (Karaman 1934), found up to now only in interstitial waters (Coineau 1994).

Bathynellacea is well represented on the islands, with the members of the group being quite common in interstitial waters, as well as in percolating waters in caves (Ruffo 2006a). The four known species belong to four genera. One of them, *Sardobathynella* (Serban 1973) of the Bathynellidae, is endemic to Sardinia.

Decapoda is represented on the two main islands by four families. The species of Atyidae (*Atyaephyra desmarestii* – Millet 1831, sensu Christodoulou et al. 2012) and Palaemonidae (*Palaemon antennarius* H. Milne Edwards 1837, sensu Tzomos and Koukouras 2015) are quite common and autochtonous inhabitants of coastal lakes, as well as rivers, ditches, artificial freshwater lakes, ponds, including even some brackish water ponds (Froglia 2006). In contrast, Astacidae with *Austropotamobius fulcisianus* (Ninni 1886) (sensu Amouret et al. 2015) (as *Austropotamobius italicus meridionalis* sensu Fratini et al. 2005, *nomen nudum*) and Cambaridae (*Procambarus clarkii*) were recently introduced (Arrignon et al. 1999, Morpurgo et al. 2010, Bertocchi et al. 2010, Amouret et al. 2015). The highly invasive red swamp crayfish (*P. clarkii*) is now widespread in some areas of the two islands, and is the greatest threat to surface native malacostracan fauna (Morpurgo et al. 2010).

6.3.2.3 Geological History and Malacostracan Biogeography

Corsica and Sardinia are a 30-km thick crustal block bounded by two areas affected by extensional tectonics: the Liguro-Provençal basin to the West and the Tyrrhenian basin to the East. The Liguro-Provençal basin, located between the Provençal-Catalan coast and the Corsican–Sardinian block, opened during the Oligocene (the process began at least 29–30 Ma; Speranza et al. 2002, Advokaat et al. 2014). Eastward migrations of the Alpine belt, Corsican–Sardinian block, and Calabrian–Peloritan block, accompanied by several microplates, were almost simultaneous (18.3 to 17.5 Ma; Malinverno and Ryan 1986, Edel et al. 2001). The question of coupling or decoupling between Corsica and Sardinia during the drift was resolved by Vigliotti et al. (1990), who showed that these two islands rotated as a single block.

During the Messinian Salinity Crisis (5.7 to 5.3 Ma; Boccaletti et al. 1990), dry land connections between Corsica–Sardinia, northern Italy, and southern France were created. Sea-level oscillations creating land bridges between Sardinia and Corsica continued during the Pliocene and Pleistocene (5.3 to 0.23 Ma; Arias et al. 1980). During the Quaternary, Sardinia may have been in contact with Peninsular Italy via Elba Island (Tuscan Archipelago), and the contemporaneous Corsican–Tuscan connection was possible as well (Cassian regression, 1.8 Ma; Baccetti 1983). During the Last Glacial Maximum (ca. 21,000 years ago), Corsica and Sardinia were connected to each other, but not to the continental landmasses (Antonioli and Vai 2004).

The complex paleogeography of these islands suggested a three-phase model (Baccetti 1983) of faunal colonization (pre-Miocenic, Messinian, and Quaternary; see Ketmaier and Caccone 2013 and bibliography cited therein); Baccetti (1983) suggested adding a fourth phase dealing with human-mediated species introductions.

The first phase would correspond to the detachment of the microplate from the Provençal-Catalan coast. Several of the endemic malacostracan species nowadays distributed in Corsica and/or Sardinia differentiated from ancestors that were supposedly distributed on the single landmass formed by the Corsican–Sardinian block, western Iberian Peninsula, and southern France. The origin of these ancient lineages (sometimes called paleo-endemics; Ruffo and Stoch 2006) is at least 29 Myr old, even though we cannot exclude that cladogenesis predated geological splits.

Isopods, such as *Proasellus* and *Stenasellus*, conform to this model, together with Bathynellacea (Ketmaier et al. 2001, 2003, Camacho 2003).

The second phase took place during the Messinian Salinity Crisis; it allowed a number of species to reach the islands. Following Ketmaier and Caccone (2013), colonization proceeded along two major paths from south and east: thus, Corsica and Sardinia share genera and species complexes with northern Africa and Sicily on one side and Peninsular Italy on the other side. Although molecular phylogenetic studies are underway, it can be hypothesized that *Gammarus* and *Niphargus* followed this colonization pathway from the Apennine Peninsula, while it is still uncertain if African or Sicilian affinities exist within Malacostraca.

The last connections between this insular Corsican–Sardinian system and the adjacent continent (especially the Apennine Peninsula) took place during the Quaternary ice ages. These connections were relatively short-lived, but would have allowed dispersal. The presence of *Proasellus banyulensis* on these islands is probably a good example of such a recent colonization (although transport by humans cannot be excluded; Stoch et al. 1996). Also, the nonendemic representatives of the decapod Atyidae and Palaemonidae, present in brackish water as well as freshwater, may be recent colonizers. Colonization by crayfish may have been also enhanced by waterbird-mediated passive dispersal (Banha and Anastácio 2014).

In the fourth phase, humans introduced species intentionally or accidentally. Introduction of alien species is an ongoing process, and the decapods *Austropotamobius fulcisianus* and *Procambarus clarkii* are the only examples among Malacostraca so far (Arrignon 1996, Amouret et al. 2015).

Considering that stygobitic species can be of ancient freshwater (limnicoid stygobionts) or marine (thalassoid stygobionts) origin (Coineau and Boutin 1992), some of the cited genera and families seem not to conform to the previous four-stage hypothesis, but may have had single or multiple origins directly from marine ancestors. The stygobitic Bogidiellidae, Ingolfiellidae, Ischyroceridae, Pseudoniphargidae, and Microparasellidae are probably thalassoid stygobionts (Stock 1980, Coineau and Boutin 1992, Jurado-Rivera et al. 2017).

6.3.3 Tuscan Archipelago

6.3.3.1 Geographical Setting

The Tuscan Archipelago is located in the northern Tyrrhenian Sea, between Tuscany (Peninsular Italy) and Corsica. It consists of seven main islands (Elba, Giglio, Capraia, Montecristo, Pianosa, Giannutri, and Gorgona) and some islets, with a total area of more than 300 km². The archipelago once included an eighth island, Monte Argentario, now a promontory connected to mainland Tuscany. The carbonatic island of Pianosa is almost completely flat and devoid of a surface hydrographic system, while the other islands (mainly formed by volcanic, granitic, and metamorphic rocks) are mountainous and, with the exception of the small Giannutri, rich in springs and/or temporary streams as well as small and isolated groundwater aquifers. Perennial watercourses are few and restricted to Elba Island.

6.3.3.2 Malacostracan Fauna

Sixteen malacostracan species are known from the Tuscan Archipelago, with 12 (75%) being stygobitic. The scarcity of surface water limits a diversified epigean fauna. Eleven species of amphipods and four isopods are known, as well as a rare, stygobitic thermosbaenacean species, *Tethysbaena argentarii* (Stella 1951), recorded from a single cave of Monte Argentario (Ruffo 2006b).

Two interstitial species, the bogidiellid amphipod *Medigidiella chappuisi* and the lepidocharontid isopod *Microcharon marinus* (Chappuis and Delamare-Deboutteville 1954), are widespread in the Mediterranean and in shallow marine waters. On Elba, both species were collected in coastal freshwater wells (Ruffo and Stoch 2006, Stoch et al. 2006). The janirid isopod *Jaera* cf. *nordmanni* (Rathke 1837) was collected in a freshwater spring on the hills in the northeastern part of the islands (Stoch et al. 2006).

The semiterrestrial talitrid amphipod *Cryptorchestia garbinii* (Ruffo et al. 2014) is common in a subterranean freshwater reservoir on Montecristo Island (Ruffo and Stoch 2006). *Cryptorchestia garbinii* is strictly linked to freshwater springs and streams (Ruffo and Stoch 2006).

The only epigean malacostracans present are: (i) *Proasellus banyulensis*, widespread over Europe, recorded from Elba, Capraia, and Giglio, associated with permanent standing waters (ponds and pools) or slowly flowing brooks (Stoch et al. 2006); (ii) the gammarid amphipod *Rhipidogammarus rhipidiophorus*, which was collected in wells on Pianosa (Messana and Ruffo 2001). This species is also common on the semi-island Monte Argentario (Ruffo and Stoch 2006).

Except the stygobitic *Proasellus acutianus* Argano and Henry, 1972, widespread on Elba and present on the Apennine Peninsula in Tuscany and Latium (Stoch et al. 2006), all other subterranean species are amphipods. Gammaridae is represented by *Ilvanella inexpectata* (Vigna-Taglianti 1971), present in interstitial environments of Elba and Tuscany (Ruffo and Stoch 2006) as well as by *Longigammarus planasiae* (Messana and Ruffo 2001), found only in the wells of Pianosa (Messana and Ruffo 2001). *Metacrangonyx ilvanus* (Stoch 1997) (Metacrangonyctidae) is present in few wells in central Elba, to which it is endemic (Stoch 1997). Niphargidae is represented by two species: *Niphargus longicaudatus* (Costa 1851) s.l., a species complex under revision, is present on the main islands (except Pianosa), including Monte Argentario. Another species complex, *Niphargus speziae* s.l. (Schellenberg 1936), was recorded on Elba, Capraia, Montecristo, and Monte Argentario (Vigna Taglianti 1976, Ruffo and Stoch 2006). Representatives of Pseudoniphargidae are known from springs on Montecristo and artificial wells on Pianosa: *Pseudoniphargus adriaticus* s.l. (Karaman 1955) (another species complex in need of revision, Messouli et al. 2006) and the endemic *Pseudoniphargus planasiae* (Messouli et al. 2006). Finally, *Salentinella angelieri* was recorded from Monte Argentario and from Elba (Ruffo and Stoch 2006).

6.3.3.3 Geological History and Malacostracan Biogeography

The interaction between the Corsica–Sardinia block and the newly forming Apennines caused the emergence of the Tuscan Archipelago, including some islands that became incorporated later into the mainland (the so-called fossil islands; Lanza 1984). Although small and well isolated in the sea, the origin of the Tuscan Archipelago and their geological history differ markedly. Capraia is the oldest island, having emerged from the sea during volcanic eruptions (9 to 5 Ma), while the others emerged between 7 and 5 Ma, except Pianosa, which probably originated about 3 Ma as a consequence of the uplift of a marine ridge (Barbato et al. 2018). Elba has more complex geology, being the largest remaining stretch of land from the ancient tract that once (Cassian regression, 1.8 Ma; Baccetti 1983) connected the Apennine Peninsula with Corsica. Elba is formed of slices of rock, once being a part of the ancient Tethyan seafloor, which emerged through the Alpine and the Apennine orogeny (Marroni et al. 1998).

The biogeographical features of the Tuscan Archipelago fauna are linked to paleo-geographic events that occurred after the disjunction and rotation of the Corsica–Sardinia block (Alvarez 1972), the Messinian Salinity Crisis (Boccaletti et al. 1990), and the Pleistocene sea regressions. According to the most recent paleogeographical reconstructions, the lower sea level during the Pleistocene glacial maxima resulted in the connection of Elba, Pianosa, Giannutri, Giglio, and Monte Argentario to the mainland (Tuscany), while Capraia, Gorgona, and Montecristo remained isolated (Bossio et al. 2000). During the Holocene, Monte Argentario became a promontory, connected with the mainland by two narrow sandy isthmuses (Barbato et al. 2018).

A paleogeographical imprint on malacostracan fauna in the Tuscan Archipelago is evident. Surely the most intriguing species is *Metacrangonys ilvanus*. Stoch (1997) postulated that the ancestor of *M. ilvanus* colonized the brackish coastal waters of the Tethys Sea and survived the Oligo-Miocenic fragmentation of the Western Mediterranean (Briançonnais) microplate that gave an origin to the Corsica–Sardinia block and, successively, caused the emergence of the Tuscan Archipelago. A recent molecular phylogeny of Metacrangonyctidae (Bauzà-Ribot et al. 2012) suggested that this species is part of the so-called insular clade together with *Metacrangonyx longipes* (Balearic Islands), *M. repens* (Stock and Rondé-Brockhuizen 1986) (Canary Islands), and two species from the Caribbean islands. The initial diversi-fication of the insular clade was estimated to have occurred between 60 and 108 Ma; of course, at that time none of the archipelagos hosting the "insular group" nor the Mediterranean Sea existed. Bauzà-Ribot et al. (2012) hypothesized that Paleo-Macaronesian islands of the Tethys Ocean were located close to the present Western Mediterranean, and the existence of these vanished archipelagos suggests the ancestor of the insular lineage was a shallow-water marine species. The origin of *M. ilvanus* dated back to 50–90 Ma, and the species can be considered a thalas-soid stygobiont (*sensu* Coineau and Boutin 1992). Other thalassoid stygobionts are the representatives of *Pseudoniphargus* (see Jurado-Rivera et al. 2017), including *P. planasiae*, and probably *Longigammarus planasiae* (Messana and Ruffo 2001). The colonization of Pianosa, which hosts these two endemics, cannot be older than the age of the island, i.e., 3 Myr (Barbato et al. 2018).

Another ancient, pre-Messinian colonizer is the thermosbaenacean *Tethysbaena argentarii*, known only from the "fossil island" of Monte Argentario, its closest rela-tive being *T. scabra* (Pretus, 1991) from the Baleares (Canovas et al. 2016). The estimated age for the most recent common ancestor of the two *Tethysbaena* species varied, depending on the population model used, from 10.7 to 18.1 Myr, a date compatible with the Oligocenic and early Miocenic vicissitudes of the Western Mediterranean (Canovas et al. 2016).

The Messinian Salinity Crisis and the incidence of the Corsica–Tuscany Pliocene connection may be responsible for the presence of limnicoid stygobionts (like *Niphargus longicaudatus* s.l. and *N. speziae* s.l.) in the most remote islands, like Capraia, Gorgona, and Montecristo, which remained isolated since those ancient paleogeographical events (Vigna Taglianti 1976).

The Pleistocene sea regressions could explain the relationships between the Italian mainland and the Tuscan block of islands that have had contact with the neighboring areas during the last glacial maximum, and this seems to be true at least for *Proasellus banyulensis*, although the presence of the species on the Capraia Islands is unexplained, with a human-mediated introduction being a plausible option. A Pleistocene origin can explain the presence of *Ilvanella inexpextata*, *Salentinella angelieri* s.l., and *Proasellus acutianus* on Elba as well. Of course, this is highly speculative, and molecular research is being conducted to test this idea (F. Stoch, unpublished data).

6.3.4 Sicily, Circum-Sicilian, Maltese, and Pelagie Islands

6.3.4.1 Geographical Setting

Sicily is the largest Mediterranean island (25,710 km^2) and one of the most biodiverse areas in the Mediterranean Basin (Médail and Quézel 1997). About 62% of Sicily's surface is covered with hills, 24% with mountains, and 14% with plains (Barbera and Cullotta 2012). Mount Etna is the highest point on the island with 3,350 m.a.s.l. The northern and northeastern parts are the most heterogeneous in terms of topography, climate, and hydrography due to the mountain ranges of Madonie (limestones, with huge karstic caves), Nebrodi (mainly sandstones and limestones), and Peloritani (mainly igneous and metamorphic rocks) (Barbera and Cullotta 2012), where a perennial surface hydrographic network is well developed. In the southwest, the karstic Hyblean massif hosts some perennial streams, as well as a rich karstic aquifer (Ruggieri and Grasso 2000) and is rich in caves (Cavallaro 1995). Most of the central part of the island is occupied by gypsum formations, with a poor and intermittent surface hydrography, and some well-developed caves hosting subterranean streams (Madonia and La Manna 1986). Ponds and pools are common on the island, while all the lakes are artificial reservoirs for drinking water supply (Marrone et al. 2006).

Sicily is surrounded by smaller islands (isolated or grouped in archipelagos) called circum-Sicilian islands, consisting of the Aeolian Islands, the Island of Ustica (volcanic, on the Tyrrhenian Sea), and the Aegadian Islands (carbonatic, to the west). Surface running waters are absent from all of these small islands; the carbonatic Aegadian Islands host small karstic aquifers, while the groundwaters on the volcanic islands are reduced to shallow freshwater lenses occurring on top of the saltwater. Artificial ponds and rock pools are present (Marrone et al. 2006).

In the Sicilian channel are the isolated Pantelleria (80 km², a volcanic island), the Maltese Islands (316 km²), and the Pelagie Islands (25.5 km²). The Maltese Archipelago consists of three islands, i.e., Malta, Gozo, and Comino, and lies 93 km south of Sicily and 288 km north of the African coast. The Maltese Islands have no large permanent river systems. Inland surface water systems are small and restricted to several dry river valleys, locally called "widien" (Moore and Schembri 1986). Sea-level groundwater bodies developed in the limestones take the form of freshwater lenses floating over seawater, while unconfined perched aquifers sustained in the upper limestone formation overlay, on western part of Malta and on Gozo, the sea-level aquifers (AA.VV. 2015). The Pelagie Islands consist of three small islands (Lampedusa, Linosa, and Lampione) between Malta and Tunisia. Surface hydrography is absent. A small aquifer in the limestones and a spring are present on Lampedusa. The volcanic island of Linosa has only a small lens of fresh groundwaters on saltwater, while no freshwaters are known on the small islet of Lampione.

6.3.4.2 Malacostracan Fauna

Twenty-one malacostracan species are recorded so far from Sicilia and circum-Sicilian islands, 12 from the Maltese Islands and only 2 from the Pelagie Islands. No freshwater malacostracans are known for Pantelleria.

Amphipods are represented by four families, with Gammaridae being the most speciose. Six species of *Echinogammarus*, three on Sicily and three on Maltese Islands, have been recorded (Moore and Schembri 1986, Pinkster 1993, Ruffo and Stoch 2006). Recent molecular analyses (Hupało et al. 2017, 2018b) revealed an extraordinary cryptic diversity in freshwater *Echinogammarus* from Sicily, where 24 Molecular Operational Taxonomic Units (MOTUs) were identified as the neutral and tentative species equivalents (sensu Grabowski et al. 2017a): 2 within *Echinogammarus* aff. *tibaldii* morphospecies, 6 within *E. adipatus*, and 16 within *E. sicilianus*. The majority of the sites were inhabited by one MOTU only, except one spring where five different MOTUs co-occurred. According to these molecular studies, the report of *Echinogammarus veneris* (Heller 1865) from Anapo River (Hyblean area) by Stock (1968) must be an error. Three species are present on Malta: *Echinogammarus klaptoczi* (originally reported from Malta as *E. ebusitanus* by Karaman [1977] and Moore and Schembri [1986]); *E. pungens* (Baldacchino 1973, Moore and Schembri 1986), and *E. sicilianus* (from a well on Malta; Pinkster 1993). Given the large number of cryptic species found in Sicily, this last record is doubtful. Another gammarid, *Rhipidogammarus rhipidiophorus*, was collected in wells and springs in Sicily, the Aegadian Islands (Marettimo and Favignana), the Maltese Islands (Gozo), and the Pelagie Islands (Lampedusa) (Moore and Schembri 1986, Ruffo and Stoch 2006). The only stygobitic gammarid, *Tyrrhenogammarus catacumbae* (Karaman and Ruffo 1977), was recorded from caves, catacombs, wells, and a spring in the karstic area of the Hyblean massif, in southeastern Sicily (Ruffo and Stoch 2006). All gammarid amphipods reported by Moore and Schembri (1986) were included in the Maltese Islands Red Data Book (Schembri and Sultana 1989).

A single species of Ischyroceridae, *Jassa trinacriae* (Krapp et al. 2010), endemic to Sicily, was described from Conza cave (175 m.a.s.l.) (Krapp et al. 2010). The species complex *Niphargus longicaudatus* s.l. (Niphargidae) is present in western Sicily, where it is common in caves and springs (Ruffo and Stoch 2006); however, the family is absent from all the other islands. Finally, Pseudoniphargidae are the most species-rich family in these islands; so far four species are endemic to Sicily, where they never coexist with Niphargidae (Ruffo and Stoch 2006). Another species (reported as *Pseudoniphargus adriaticus* s.l.) is undoubtedly an unknown new species (Messouli et al. 2006, Ruffo and Stoch 2006) present in the groundwaters of Lampedusa and Linosa islands.

Asellidae is present on Sicily with three species of the *Proasellus coxalis* (Dollfus 1892) species complex (Stoch et al. 1996). The endemic *P. montalentii* (Stoch et al. 1996) is widespread on Sicily, where it inhabits springs and streams. Another species, *P. wolfi* (Dudich 1925), is known only from some slow-flowing rivers at the foot of the Hyblean massif. The third species, *P. banyulensis*, is present in a small area around Palermo. This species was also recently found in reservoirs of the Ustica fortress (F. Stoch, personal observation), possibly introduced in historical times (Stoch et al. 1996, Marrone and Naselli Flores 2015). One more species, *P. ragusani* reported by Galletti (2002), was never formally described and remains a "nomen nudum". The members of the *P. coxalis* species complex on the Maltese Islands were reported for the first time for Gozo (as "*Asellus (Proasellus) coxalia*") by Baldacchino (1983); their taxonomic status is still unstudied. Another interesting stygobitic species endemic to southeastern karstic areas of Sicily is the cirolanid *Typhlocirolana* cf. *moraguesi* (see Ruffo and Stoch 2006). The species is morphologically similar to *T. moraguesi* from the Balearic Islands, but distinguished based on molecular data (Baratti et al. 2004).

The Thermosbaenacea is represented by a single, endemic stygobitic species, *Tethysbaena siracusae* (Wagner 1994), having the same distribution area as *Typhlocirolana* (Ruffo 2006b).

Six decapod families are present on Sicily (five species) and on the Maltese Islands (seven species). Apart from Atyiidae, Palaemonidae, and Potamidae, all others are introduced (Deidun et al. 2018). *Palaemon antennarius* is common in the Sicilian brackish waters, especially in the estuaries, as well as in some inland freshwater reservoirs and rivers, while *A. desmarestii* is present in freshwaters (Froglia 2006). *Potamon fluviatile* (Herbst 1785) is present on Sicily and a distinct, endemic subspecies, *P. fluviatile lanfrancoi* (Capolongo and Cilia 1990) was thought to be present in the Maltese Islands (Schembri 2003). However, recent molecular analyses demonstrated that a single taxon, of recent expansion, is present on the Apennine Peninsula, Sicily, Maltese Islands as well as on the Ionian Islands (Vecchioni et al. 2017).

The alien invasive *Procambarus clarkii* was found on Sicily (Marrone and Naselli Flores 2015). Its distribution on Sicily likely derives from multiple deliberate and independent introduction events (Faraone et al. 2017). This species was also reported from Malta by Vella et al. (2017). The presence of another alien species on Sicily, *Cherax destructor* (Clark 1936) (Parastacidae), was reported by Deidun et al. (2018)

with three specimens collected in a stream in close proximity to a crayfish aquaculture concern. However, at the moment there are no data demonstrating the species' reproduction in Sicily (Marrone, personal communication). Most recently, Deidun et al. (2018), reported the occurrence of five other alien decapods (see Appendix 1), important in the pet trade or in the aquaculture, from the freshwater localities on the Maltese Islands. According to the authors, the Malta's Environmental and Resources Authority (ERA) was alerted about the localities of the nonindigenous malacostracan species and a preliminary eradication program has commenced.

6.3.4.3 Geological History and Malacostracan Biogeography

Sicily developed along the African-European plate boundary. During the Oligocene, between 29 and 24 Ma, the central-western Sicily was part of the African Maghrebides–Southern Apennines accretionary wedge (Catalano et al. 1994). The major tectonic units of the island (Broquet 2016) are: (i) the Hyblean foreland (the emergence of the carbonatic Hyblean Plateau is dated to early Miocene, while during the middle Miocene African subduction, ca. 15 Ma, it became the Hyblean-African plateau – Broquet 2016); (ii) the Gela foredeep, which extends from the northern margin of the Hyblean Plateau to southwestern Sicily, which developed during Late Pliocene and filled with limestones, Messinian evaporites, and clays; (iii) the Apenninic-Maghrebian orogen forming the central and western part of Sicily, which represents the fold and thrust belt of the African subduction system; (iv) the Calabrian–Peloritan arc, which, being still attached to Sardinia in the Early Miocene, formed the northeastern part of Sicily (Peloritani Mountains). Some authors (e.g., Ruggieri 1973) hypothesized that at the beginning of Pliocene, there was no emergent land in the Sicilian territory, and that the island began to raise up again from the sea during Pliocene (at ca. 4 Ma). At that time Sicily was divided into two islands: the Madonie-Nebrodi-Peloritani chain in the northern part, and the Hyblean Plateau in the southeastern part (Guglielmo and Marra 2011). Finally, during the Pleistocene glaciations, Sicily had temporal land connections with Calabria, the Maltese Archipelago, and most of the circum-Sicilian islands (Marra 2009). Nowadays all geologists agree that the African connection during Plio-Pleistocene, through the alleged African-Sicilian bridge (Sacchi 1955), never took place (Broquet 2016).

Among the circum-Sicilian islands hosting Malacostraca, the Aegadian Islands (except Marettimo) are a fragment of Sicily, to which they were connected during the eustatic sea-level variations in Pleistocene (Muscarella and Baragona 2017).

The Maltese Islands are geologically associated with the Hyblean Plateau of the southeastern Sicily. They emerged from the sea during the Alpine orogenesis at ca. 10 Ma (Savona Ventura 1975). The sea between the Maltese Islands and Sicily reaches a maximum depth of 200 m (mostly less than 90 m) and since the Messinian Salinity Crisis, these islands were repeatedly connected to Sicily. In contrast, the sea depth between the Maltese Islands and the African coast is considered to be much deeper, making post-Messinian connections improbable (Savona Ventura 1975).

The Pelagie Islands are connected to each other only from a geographical point of view. Lampedusa (together with Lampione) is a part of the African continental shelf, being a stable carbonatic platform, raised up presumably during the late Miocene or the Early Pliocene (Muscarella and Baragona 2017). The connections between Lampedusa and the Tunisian coast were frequent during Pliocene as well as during the Quaternary glaciations (Foglini et al. 2015), while any past connection with Sicily is considered unlikely (Muscarella and Baragona 2017). On the other hand, the volcanic islands, such as Linosa and Pantelleria, emerged during Pliocene and were never connected with the mainland (Muscarella and Baragona 2017).

La Greca (1957) formed the first hypothesis explaining the origin of Sicilian fauna, including the other islands. Following the paleogeographical scenario accepted at that time, La Greca postulated five main categories of colonizers depending on their origin: widely distributed species, Western Mediterranean species, Eastern Mediterranean species, Sicilian-African species, and the endemics. Among the endemics, he distinguished recent Pleistocene endemics from the relictual endemics, derived from the fragmentation of the "Tyrrhenidis", i.e., the post-Oligocenic vicissitudes of the Western Mediterranean, of which Sicily was considered to be part (Furon 1950). More recently, Massa et al. (2011), following the paleogeographic reconstruction by Ruggieri (1973), rejected the hypothesis of the ancient, pre-Pliocenic, endemics. Recent molecular analyses confirm both the existence of ancient endemic species (e.g., *Typhlocirolana* cf. *moraguesi*; Baratti et al. 2004) and the presence of the Quaternary colonizers (e.g., *Potamon fluviatile*; see Vecchioni et al. 2017).

Baratti et al. (2004) considered *T. moraguesi* (Balearic Islands) and *T.* cf. *moraguesi* (Sicily) as sisters of the Moroccan *Typhlocirolana* species. The two clades may have separated when the hydrological continuity was interrupted in the Aquitanian (ca. 20–23 Ma), with the subsequent fragmentation of the "Alboran-Kabylian-Calabrian Plate". Thus, the interruption of gene flow between the populations colonizing inland waters and their coastal, brackish water ancestors was probably a vicariance event during the Tethian regression. Moreover, Baratti et al. (2004) suggested that the divergence of *T.* cf. *moraguesi* from *T. moraguesi* took place about 15 Ma. Although this date is tentative, it suggests that colonization could have taken place well before the alleged re-emersion of Sicily hypothesized by Ruggieri (1973) and Massa et al. (2011). The same scenario could apply to *Tethysbaena siracusae*, considering that the congeners live in the Balearic Islands and on Monte Argentario in Tuscany, well-known Miocenic relict lands (Ruffo 2006b).

In a recent, ongoing molecular study on the epigean *Echinogammarus* of Sicily, Hupało ct al. (2017, 2018b), using both mitochondrial and nuclear markers, postulated an ancient (i.e., when Sicily was isolated) origin for both *E. adipatus* (late Serravallian/early Tortonian) and *E.* cf. *tibaldii*. *Echinogammarus sicilianus*, presenting a formidable level of cryptic diversity, forms a sister group to both the Apennine *Echinogammarus* and the other Sicilian congeners. The divergence from *E. adipatus* was dated at around late Oligocene/early Miocene. Thus, the presence of an ancient, endemic fauna in Sicily is confirmed, as postulated by La Greca (1957). Other molecular studies, aiming to clarify the biogeographical affiliations, are being conducted on the isopod *Proasellus* and on the amphipods *Rhipidogammarus*, *Tyrrhenogammarus*, *Pseudoniphargus*, and *Niphargus* (F. Stoch, unpublished).

The presence of recent colonizers is demonstrated in the decapods *Potamon fluviatile* (see Vecchioni et al. 2017), with eastern Mediterranean affinities, and *Atyaephyra desmarestii* (see Christodoulou et al. 2012), widely distributed in the Western Mediterranean, demonstrating the importance of recent, Quaternary connections between Peninsular Italy and the islands.

No relict species are known so far from the Pelagian Islands with the presence of only two species, *Rhipidogammarus rhipidiophorus* and *Pseudoniphargus adriaticus* s.l. (see Ruffo and Stoch 2006), of a confirmed alleged marine origin; *P. adriaticus* s.l. is also present on Linosa, which was never connected to the mainland. These scarce data are insufficient to hypothesize any biogeographical scenario for this small archipelago.

6.3.5 Adriatic Islands

6.3.5.1 Geographical Setting

The Adriatic Sea is in the northernmost part of the Mediterranean Sea. The basin is mostly shallow, elongated, and enclosed by the mountain ranges of the Dinarides (Dinaric Alps) from the east and the Apennines from the west. After the Black Sea, it is the most continental basin of the Mediterranean region. The Adriatic Islands form the second largest archipelago of the Mediterranean Sea. The archipelago is composed of over 1,300 island and islets, with the majority (over 1,200) located in the northeastern Adriatic and belonging to Croatia. The largest islands are Cres and Krk. These two islands are almost identical in size, each covering ca. 405 km² (Duplančić Leder et al. 2004). The islands located along the Croatian and Montenegrin coasts are of karstic character, composed of carbonate rocks, limestones, and dolomites of Lower and Upper Cretaceous origins. The karstic lotic waters are characterized by water loss along their course or even by complete percolation, as well as frequent seawater intrusions producing brackish conditions (Bonacci 2014). This, combined with Mediterranean climate and relatively small island size, results in a scarcity of epigean freshwater habitats. Only sparse springs, especially where marly arenaceous rocks meet limestones (e.g., on Krk Island; Lončarić et al. 2011) and few temporary karstic lakes, together with a large number of pools and artificial ponds, are present. The biggest natural lake is Lake Vrana (Vransko) on Cres, with surface 5,750 km². It is the only drinking water source for the island (Katalinic et al. 2008). While surface waters are scarce, the islands harbor considerable reserves of groundwater (Trincardi et al. 1996, Bonacci and Roje-Bonacci 2003, Duplančić Leder et al. 2004, McKinney 2007).

6.3.5.2 Malacostracan Fauna

Amphipoda is represented by 20 species. Two gammarids are found in both freshwater and brackish water: *Echinogammarus pungens* and *E. veneris*. The latter recorded from springs on the islands of Rab and Hvar (Karaman 1969, Karaman and Sket 1989), the former from springs on Cres and Hvar as well as streams of Krk, Rab, and Šipan (Karaman 1969, Zganec 2009). *Gammarus balcanicus* (Schäferna 1923) was reported from Krk Island (Sket 1988, Zganec 2009, Mamos 2015). Mamos et al. (2016) analyzed molecular diversity in *G. balcanicus* showing it to be a complex of deeply divergent phylogenetic lineages, possibly representing a plethora of cryptic species. Therefore, the Krk population should be treated as *G. balcanicus* sensu lato until revision. Finally, the gammarid *Rhipidogammarus karamani* is reported from coastal springs in Pag (Sket 1988). The sole resident member of the Crangonyctidae is *Synurella ambulans* (Müller 1846) from Vransko Lake on the island of Cres and from Pag (Grube 1861, Karaman 1974, Sket 1988). Given that *S. ambulans* is broadly distributed (Ponto-Caspian region, Central and Eastern Europe), it is likely a species complex (Sidorov and Palatov 2012 and references therein). The most speciose amphipod family is the hypogean Niphargidae, with 13 species reported from the islands and 5 of them being endemic (Appendix 1, Karaman and Sket 1989, Sket and Karaman 1990, Fišer et al. 2006). This diverse subterranean fauna is attributed to the abundance of karstic groundwater habitats, even on the smaller islands (Bonacci and Roje-Bonacci 2003). Five niphargid species are classified as endangered (Ozimec et al. 2009): *N. hebereri* (Schellenberg 1933), *N. hvarensis* (Karaman 1952), *N. miljeticus* (Straškraba 1959), and *N. pectencoronatae* (Sket and Karaman 1990), with *N. jadranko* (Sket and Karaman 1990) being critically endangered. Other Adriatic Island hypogean amphipods are *Salentinella angelieri* sensu lato (Salentinellidae), a species complex under revision (F. Stoch, pers. comm.), reported from an anchihaline cave on Gangarol island of the Kornati Archipelago (Sket and Karaman 1990, Sket 1994) and *Hadzia fragilis* (Karaman 1932) recorded from several islands and islets of the Kornati archipelago (Sket and Karaman 1990, Gottstein et al. 2007), Krk, Cres, and Lošinj (Sket 1994 and authors cited therein).

Isopods are also well represented. Surface waters are inhabited by the Asellidae, with *Asellus aquaticus* (Linnaeus 1758) present only in Vransko Lake on Cres (Sket 1988) and *Proasellus coxalis* species complex widespread on several islands (Krk, Cres, Rab, Pag, and Molat: Sket 1988). Three stygobitic families are found in groundwaters. Sphaeromatidae are represented by *Monolistra pretneri* (Sket 1964) (Prevorčnik et al. 2010). Cirolanidae are present in the central and southern islands with *Sphaeromides virei virei* (Brian 1923) and *S. virei mediodalmatina* (Sket 1964) (Delić and Sket 2015). Finally, Sket (1988) reported an unknown representative of *Microcharon* (Karaman 1934) (Lepidocharontidae) from the hyporheos of a small brook on Krk.

The sole member of Thermosbaenacea is the stygobitic *Tethysbaena halophila* (Karaman 1953) (Monodellidae), recorded in anchialine caves on Korcula, Lošinj, and the Kornati archipelago (Sket 1988, Wagner 1994).

Finally, Decapoda are present with two families. Small surface streams of Krk and Vransko Lake on Cres islands are inhabited (Sket 1988) by *Austropotamobius* (Astacidae) (Maguire et al. 2003); following Clavero et al. (2016), the populations are attributed to *A. italicus carsicus* (Karaman 1962). Ugljan and Brač karstic groundwaters host a stygobitic atyid, *Troglocaris anophthalmus periadriaticus* (Jugovic et al. 2012) (Jugovic et al. 2012); this subspecies also occurs along the mainland Croatian coastal areas.

6.3.5.3 Geological History and Malacostracan Biogeography

The Adriatic basin topography is a consequence of a long and complex geological history. In the Paleogene (66 to 23 Ma), the Adriatic Plate (aka Apulian Plate) decoupled from the African Plate and moved north, colliding with the Eurasian Plate, contributing to the Alpine orogeny that resulted in the uplift of the Alps, Dinarides, and Hellenides. It also led to closing of the Tethys Ocean and formed the northeastern border of the Adriatic basin. In the Early Neogene (ca. 20 Ma), movement of the Adriatic Plate led to uplift of the Apennine Mountains, establishing the western borders of the Adriatic basin, leaving the basin largely open in the south (Devoti et al. 2002, Pinter et al. 2016 and references therein) until the Messinian (5.7–5.3 Ma). During the Messinian Salinity Crisis, the region equivalent to the present Adriatic basin almost completely dried out, leaving several small hypersaline waterbodies and probably only a sparse hydrological network. Such conditions changed in the Lago Mare episode at the end of Messinian, when the freshwater/brackish water of Paratethys flushed into the proto-Mediterranean Sea, partially refilling the Adriatic basin. At ca. 5 Ma, with the reopening of the Gibraltar Strait, the Messinian ended and marine conditions were reestablished in the Mediterranean Basin, including the Adriatic Sea (Rögl 1999 and references therein, Popov et al. 2004, 2006, Orszag-Sperber 2006). Major glaciation events in the Pleistocene caused recurrent eustatic sea-level regressions in the Mediterranean and a rearrangement of local hydrological networks. For example, there is evidence that during the Last Glacial Maximum, the extended Po River (aka Mega-Po) unified most of the Periadriatic hydrological networks (i.e., eastern Appenines, southern Alps, and western Dinarides) within one vast river system. In the Holocene (ca. 11 Ka), the Mega-Po valley was flooded due to the rise of sea level leading to formation of the current north Adriatic basin and isolation of the Adriatic Islands from the Balkan Peninsula (Colantoni et al. 1979, Stanley and Wezel 1985, Correggiari et al. 1996).

Geological history suggests that inland waters colonization (including those of the future islands) may have started in the Eocene by thalassoid malacostracans of Tethyan and Paratethyan origin. However, so far, studies employing molecular markers show rather younger patterns of colonization and diversification. The most diverse crustacean group and probably one of the first inhabitants of the freshwater systems in the Adriatic region is Niphargidae. Molecular studies revealed that this group colonized the Periadriatic region in late Oligocene (ca. 25 Ma) from northern Europe via land connections and hydrological networks developed during the Alpine orogeny (McInerney et al. 2014). Diversification of this hypogean group led to great diversification of species that survived through the dramatic geological and hydrological changes of the Adriatic region in relatively stable underground habitats. In the Miocene, between 20 Ma and 10 Ma, freshwaters of the eastern Periadriatic were colonized by gammarids of Paratethyan (e.g., *Gammarus balcanicus* complex) and Tethyan (e.g., *Gammarus roeselii* Gervais 1835 complex, *Echinogammarus* spp.) origins (Hou et al. 2013, Hou and Sket 2016, Mamos et al. 2016, Grabowski et al. 2017b). So far, two species of *Echinogammarus*, presumably of Tethyan origin, and only one Paratethyan *G. balcanicus* lineage was found on the islands. The population of *G. balcanicus* on Krk diverged from the continental clade at ca. 7 Ma in the Messinian, possibly due to isolation by hypersaline or arid environments during the salinity crisis (Krijgsman et al. 1999). Changes in this geological period may have influenced isopod diversification based on what is observed in the *A. aquaticus* complex divergence time frame in the western Periadriatic; however no molecular data is available from the insular populations (Sworobowicz et al. 2015). During the Lago Mare episode (5.3 Ma), due to oligohaline environments connecting the hydrological systems on both sides of the Periadriatic, freshwater hydrobionts may possibly have migrated between the continent and the Adriatic Islands. However, molecular data suggest that most colonizations occurred as late as the Pleistocene. In particular, the most recent terrestrialization of almost the entire Periadriatic region and the formation of the Mega-Po river systems of the Balkan and Apennine Peninsulas (ca. 18 ka) created a wide possibility for freshwater fauna dispersal. We can assume that this opportunity would be used by species that show low molecular variability in the region, such as *Austropotamobius* (Trontelj et al. 2005). This dispersal opportunity ended with the Holocene and sea-level rise. So far no invasive or alien species have been found on the Adriatic Islands.

6.3.6 Crete, Aegean, and Ionian Islands

6.3.6.1 Geographical Setting

The Aegean Sea is one of four major basins of the East Mediterranean and covers an area of approximately 240,000 km^2. It stands in the center of the conjunction of three continents, Europe, Asia, and Africa, and it is situated between the coasts of mainland Greece, Crete, and Asia Minor. Around 7,500 islands and islets occur at a variety of isolation levels and topographic features placing the Aegean among the archipelagos with the highest number of the islands worldwide (Triantis and Mylonas 2009, Poulakakis et al. 2015). Geographers and geologists have divided the Aegean Sea in three parts following the morphology of the coasts, the position of the islands, and the formation of the seabed: North, Central, and South Aegean. The South Aegean is considered the most important part of the Aegean Sea, since it comprises the majority of the islands, including the two big island complexes, Cyclades and the Dodecanese.

Crete is the fifth largest Mediterranean island, with a surface area of 8,261 km^2, and it forms the border between the Aegean and Libyan seas. The island extends 260 km from west to east and its width varies from 12 to 60 km. Its coastline has a total length of 1,065 km and consists of both sandy beaches and rocky shores. The island is highly mountainous (52%) and has four great mountain ranges: White Mountains (Lefka Ori), Psiloritis (Ida), Dikti (Lasithian Mountains), Thryptis (Mountains of Siteia), with numerous peaks exceeding 2,000 m.a.s.l. Crete is gifted with a plethora of caves, gorges, and plateaus, and it has been estimated that around 5,200 caves and karst formations, rich in underground waters, exist on the island (Fassoulas et al. 2007, Fassoulas 2017), thus creating ideal habitats for subterranean animals.

The Ionian Islands are located in the Ionian Sea at the westernmost part of Greece and include the islands of Kerkyra (Corfu), Paxoi, Lefkada (Lefkas), Cephalonia (Kefallinia), Ithaki (Ithaca), Zakynthos (Zante), and Kythera. They are traditionally called Heptanese or "the Seven Islands", but the group includes many smaller islands covering a total area of 2,318 km^2. The relief of the majority of the islands is mainly mountainous or semimountainous. The largest of the Ionian Islands is Cephalonia, covering a third of an area with 779.3 km^2.

The climate of the Aegean Sea Islands (including Crete) is typical dry Mediterranean. The average yearly precipitation on Crete is about 900 mm, which correlates to approximately 7,500 hm^3, while for the Aegean Islands the corresponding quantities are 585 mm and 5,192 hm^3, respectively (Gikas and Angelakis 2009). However, less than 15% of the precipitation percolates through the ground in both the regions. Evapotranspiration and surface runoff to the sea accounts for 65% and 21%, respectively, for Crete, and 60% and 26%, respectively, for the Aegean Islands (Gikas and Angelakis 2009). The salinity of rainfall in the Aegean is high, due to airborne sea spray, which has a strong effect on the chemical composition of the groundwater of the islands. During the last decades, Crete and many islands of the Aegean are subjected to increased water demands, due to tourism development and intense irrigation, resulting in surface as well as groundwater degradation and depletion which threaten aquatic biodiversity (Diamantopoulou and Voudouris 2008). As a result, the annual water demands are steadily growing and the groundwater abstraction during summer is estimated to further increase. Precipitation levels are higher in the Ionian Islands with mean annual precipitation of 1,038 mm according to World Clim-Global climate data (http://www.worldclim.org).

6.3.6.2 Malacostracan Fauna

Sixty-five malacostracan species have been recorded from the area: 19 from Crete, 27 from the Aegean Islands, and 20 from the Ionian Islands. The malacostracan fauna differs greatly between the Ionian and Aegean Islands, while Crete shares only a few common species with the Aegean Islands. Out of the 20 Ionian malacostracan species, 9 (45%) are stygobionts and 8 (40%) are endemics; Crete has 9 (47%) stygobionts and 10 (52%) endemics; and the Aegean has 14 (51%) stygobionts and 6 (22%) endemics. Despite the significantly greater number of Aegean Islands, species richness is not markedly higher than in the Ionian Islands and Crete, as would have been expected. This may be due to the Aegean Islands fauna being less examined than the Ionian Islands and Crete. Thus, the number of malacostracan species in the Aegean Islands may be underestimated.

Amphipods are present with seven families. Four species of Bogidiellidae (*Medigidiella minotaurus* (Ruffo 1976) is endemic to Crete) are present, linked both to interstitial and subterranean karstic waters with all being freshwater stygobionts, except *Medigidiella chappuisi* (Ruffo 1952), which lives in interstitial anchialine waters (Ruffo and Delamare-Deboutteville 1952). Gammaridae known from the islands are relatively poor (seven species), compared to other families (e.g., Niphargidae), with one endemic from the Ionian (*Echinogammarus kerkuraios*

Pinkster 1993) and two from Crete (*Echinogammarus platvoeti* Pinkster 1993, *Gammarus plaitisi* Hupało et al. 2018a) (Karaman and Pinkster 1977, Pinkster 1993, Hupało et al. 2018a). Hadziidae is represented in Greece only by *Metahadzia helladis* (Pesce and Argano 1980), endemic to Cephalonia (Pesce and Argano 1980). Ingolfiellidae is represented by one species, *Ingolfiella petkovskii* (Karaman 1957), that has been recorded only from Keramou, Euboea Island (Bou 1970). The Greek subterranean amphipod fauna is rich with many endemics, due to the complex geological history, geography, climate, and hydromorphological conditions. Many taxa are restricted to springs, wells, caves, and groundwater pools. Within Niphargidae, three genera are known from Greece: *Niphargus* (Schiödte 1849), *Exniphargus* (Karaman 2016b), and *Niphargobatoides* (Karaman 2016b); the last two being endemic. *Niphargus* is currently represented by 22 species from continental Greece and the Greek islands (Pesce and Maggi 1983, Karaman 2015, 2016a,b, 2017a,c, Ntakis et al. 2015). Fourteen occur in the Ionian, Aegean Islands, and Crete, of which 75% are endemic. Sket (1990a) described *Niphargobatoides lefkodemonaki* from Lefka Ori in western Crete. Other Cretan niphargids include *Exniphargus tzanisi* (Karaman 2016a) from Lefka Ori, *Niphargus impexus* (Karaman 2016a) from near Heraklion, *N. lakusici* (Karaman 2017a) from Pyrgos, and *N. zarosiensis* (Zettler and Zettler 2017) from Lake Zaros (aka Limni Votomos). The Aegean Islands have five niphargids, three of which are endemic: *Niphargus spasenijae* (Karaman 2015) from Thasos Island, *N. lourensis skiroci* (Karaman 2018) from Skyros, and *N. rhodi* (Karaman 1950) from Rhodes (Karaman 2015, 2016a, 2018), *N. adei* (Karaman 1934) endemic to Samothrace (Samothraki), and *N. jovanovici* (Karaman 1931) recorded from two wells in Amarinthos, Euboea (Karaman 2015, 2017c). Four species are known from the Ionian Islands; *Niphargus denarius* (Karaman 2017) endemic to Cephalonia, *N.* cf. *lourensis* (Fišer et al. 2006), from Cephalonia, *N. skopljensis* (Karaman 1929) from Lefkada, and *N. versluysi* (Karaman, 1950) from Zakynthos (Karaman 2015, 2017d). The stygobitic Salentinellidae is represented by a single species complex: *Salentinella angelieri* sensu lato, present on both Ionian and Aegean archipelagos (Pesce et al. 1979, Pesce 1985). Talitridae is represented by one freshwater species, *Cryptorchestia ruffoi* (Latella and Vonk 2017), endemic to Rhodes, from a freshwater spring on Monte Smith, and in the streams flowing out of the Epta Pyges springs (Davolos et al. 2017).

Isopoda is present with three families, Asellidae, Cirolanidae, and Microparasellidae, including both epigean and groundwater species. Asellidae show high endemism in the Ionian, Aegean Islands, and Crete. *Proasellus* is present in springs, caves, and interstitial waters with eight species, six of which are endemic (Strouhal 1966, Henry 1975, Argano and Pesce 1979, Pesce and Argano 1980). *Proasellus* is relatively speciose in the Ionian Islands with four endemics; *Proasellus coxalis cephallenus* (Strouhal 1942), *P. coxalis corcyraeus* (Strouhal 1942), *P. coxalis leucadius* (Strouhal 1942), and *P. coxalis versluysi* (Strouhal 1966), occurring on the islands of Cephalonia, Kerkyra, Lefkada, and Zakynthos, respectively (Strouhal 1966). In Crete, two subterranean endemics, *P. cretensis* (Pesce and Argano 1980) and *P. minoicus* (Pesce and Argano 1980), are present, while on the Aegean Islands only one stygobiont, *P. sketi* (Henry 1975), has been recorded from Euboea and the epigean *P. coxalis* s.l. from Rhodes (Pesce at al. 1979, Pesce and Argano 1980). Arcangeli (1942) reported *P. c. rhodiensis* nomen nudum from Rhodes. *Asellus aquaticus* has been reported from Crete by Chappuis (1949) and was confirmed in 2015 by Wysocka (unpublished data). Chappuis (1949) suggested that *A. aquaticus* on Crete may be due to anthropogenic introduction in historical times; humans may have played a key role in the island colonization by this isopod. This seems to be plausible in light of the undergoing molecular research (Wysocka et al. unpublished). Recently, Sworobowicz et al. (2015) discovered that *A. aquaticus* on Santorini Island is represented by two divergent lineages, one possibly a local endemic and the other widespread in Thrace and some other parts of the Balkans.

The cirolanid *Turcolana* occurs in the freshwater and brackish groundwater environments around the eastern Mediterranean (Argano and Pesce 1980). Here, only the endemic stygobitic *Turcolana rhodica Botosaneanu* (Boutin and Henry 1985) has been reported in the springs of Gadouras River on Rhodes (Botosaneanu et al. 1985). Finally, the stygobitic Lepidocharontidae is represented by five species of *Microcharon*. Four have been recorded from the Aegean Islands and one from the Ionian Islands; none from Crete (Argano and Pesce 1979, Pesce 1981; Galassi et al. 1994). One species, *Microcharon hellenae* (Chappuis and Delamare-Deboutteville 1954) is endemic to the Greek mainland and Sporades. Within Microparasellidae, two representatives occur in Greece: *Microparasellus hellenicus* (Argano and Pesce 1979) and *Microparasellus puteanus* (Karaman 1933) from Cephalonia (Argano and Pesce 1979).

Decapoda is represented here by Potamidae, Palaemonidae, and Atyidae. The freshwater crabs *Potamon* (Potamidae) are most speciose with six species (Jesse et al. 2011). Genetic studies demonstrate high genetic diversity and the presence of cryptic lineages among previously identified *Potamon* species (Jesse et al. 2010, 2011). *Potamon hippocratis* (Ghigi 1929) inhabits the Aegean Islands of Kos, Samos, Ikaria, and Naxos, but is also present on Crete. *Potamon rhodium* (Parisi 1913) occurs on Rhodes and Tilos and is also known to occur on southern Turkey (Özbek and Ustaoğlu 2006). *Potamon karpathos* (Giavarini 1934) was described from Karpathos, but is distributed all along the southern Anatolian coast. *Potamon kretaion* (Giavarini 1934) is endemic to Crete, whereas *P. ibericum* (de Bieberstein, 1808) occurs in Chios and Lesbos (Jesse et al. 2011). *Potamon ibericum* has a relatively wide distribution range, with its easternmost limit in the southern Caspian Sea drainages, extending west to the northern Aegean Sea (Brandis et al. 2000). Recently, the westernmost populations in the northern Aegean Sea, specifically in Chalikidiki, Thasos, and Samothrace, have been distinguished as a separately evolving lineage representing possibly a new, yet undescribed, species (Jesse et al. 2011). The widespread *P. fluviatile* is found on Euboea and Sporades. On the Ionian Islands, only *P. fluviatile* is present, occurring on Kerkyra and mainland Greece (Jesse et al. 2011). Nine palaemonid shrimp species occur in the circum-Mediterranean area, with three being present here: *Palaemon antennarius* reported from the Ionian Islands of Kerkyra and Zakynthos, *P. minos* (Tzomos and Koukouras 2015) endemic to Crete, and *P. colossus* occurring in Rhodes and the Antalya region of Turkey (Tzomos and Koukouras 2015). One atyid, *Atyaephyra thyamisensis* (Christodoulou et al. 2012), has been recorded on the Ionian islands Kerkyra and Lefkada (Christodoulou et al. 2012).

6.3.6.3 Geological History and Malacostracan Biogeography

The landmass of Aegeis first emerged in the Neogene (23 to 12 Ma) (Meulenkamp 1971). The fragmentation of Aegeis started during the collision of the African tectonic plate with the Eurasian Plate in the middle Miocene (ca. 16 Ma; Krijgsman 2002, Steininger and Rögl 1984) and the formation of the Mid-Aegean Trench (12 to 9 Ma; Dermitzakis and Papanikolaou 1981). The subsequent sea expansion and continent compartmentalization took place, dividing the former landmass into large number of islands of various sizes, which were connected and isolated repeatedly during the Messinian Salinity Crisis (5.96 to 5.33 Ma (Krijgsman et al. 1999). Afterward, some of the connections with the mainland were temporarily reestablished during the Pleistocene (2.58 Ma–11.70 ka) due to eustatic sea-level changes (Perissoratis and Conispoliatis 2003). However, there is strong evidence suggesting that during the Pleistocene, the Cyclades were isolated and a wide sea barrier existed between them, Crete and the Dodecanese Islands (Dermitzakis and De Vos 1987, Dermitzakis 1990). The isolation of Cyclades started at the end of the Pliocene, while the Dodecanese Islands, being much closer to Asia Minor, have been present only for the last few thousand years (Perissoratis and Conispoliatis 2003). Intensive volcanic activity contributed to the formation of several islands, a few of which, such as Milos, Nisyros, and Kimolos, are purely volcanic (Francalanci et al. 2005).

Until the Vallesian period of the late Miocene (9 Ma), Crete was connected to the Balkan Peninsula and Asia Minor, as shown by fossil remains of mainland fauna (van der Geer et al. 2006). The split of the Balkan Peninsula (including Crete and Peloponnesus) from Asia Minor began around 12 Ma. The isolation of Crete from Peloponnesus started about 11 to 8 Ma due to sea-level rise. The desiccation of the Mediterranean Sea during the Messinian Salinity Crisis led to the formation of hypersaline deserts and semi-deserts around Crete and other islands, being the last known land connection between Crete and the mainland (Poulakakis et al. 2015). During the Pliocene, Crete was divided temporarily into at least four islands due to sea-level rise associated with the Zanclean Flood (Sondaar and Dermitzakis 1982). At the end of the Pliocene or in the Early Pleistocene, Crete gained its present configuration (Sondaar et al. 1986).

Because of its complex paleogeographical history, high biodiversity, and endemism, the Aegean archipelago as well as Crete have become key areas for biogeographical studies (Sfenthourakis and Triantis 2017), especially upon the terrestrial taxa (Heller 1976, Sfenthourakis 1996, Fattorini 2002, Poulakakis et al. 2005, Jesse at al. 2011) showing that Aegean species distribution often mirrors palaeogeographical patterns and processes. The phylogeny of several freshwater malacostracan species has been examined over the last decades. For example, the divergence of the recently discovered Cretan species, *Gammarus plaitisi*, from its continental relatives occurred most probably in the late Miocene, around 9.2 Ma, possibly due to the isolation of Crete from the mainland during that time (Hupało et al. 2018a). Although molecular phylogenetic studies are still underway (unpublished data), it may be possible that the amphipod genera *Gammarus* and *Niphargus* followed this colonization pathway from peninsular Greece, as it remains uncertain if Asia Minor affinities exist within Malacostraca of the region. However, the evolutionary history of insular *Echinogammarus* is apparently different, presumably with direct Tethyan origins (Hou et al. 2013, Hou and Sket 2016, Hupało et al. 2018b).

Jesse et al. (2011) showed that *Potamon* has no insular endemics in the Aegean archipelago, with exception of *P. kretaion* on Crete, indicating that the Pleistocene terrestrial connections, with some anthropogenic dispersal in historical times (e.g., in case of *P. hippocratis*), may have facilitated dispersal. Most probably, *Potamon* diversification occurred during the Late Pliocene and Early Pleistocene in the eastern Aegean region. This coincides with climatic fluctuations and increased aridity in Anatolia, thus suggesting a relationship between these climatic oscillations and speciation events (Jesse et al. 2011). Most of the malacostracan distributional patterns in the studied islands are possibly related to the Pleistocene glacial cycles, following similar patterns observed in the other taxa, such as reptiles, amphibians, and beetles, as well as terrestrial malacostracans (Triantis and Mylonas 2009). The Pleistocene climatic fluctuations repeatedly allowed invasions and retreats of numerous species. Several genera endemic to the Aegean either went extinct during glaciations or, at least some of them, were able to survive in extreme habitats such as caves and subalpine mountainous areas with isolated springs (Triantis and Mylonas 2009).

The geological evolution of the Ionian Islands was comparatively simpler compared to the Aegean, with most islands becoming isolated from western continental Greece during the Pleistocene or even in the Holocene (Sakellariou and Galanidou 2015). The Ionian Islands formation took place only in the Quaternary, as a result of intense compressive tectonism and uplift, which started in the Lower Pliocene (Evelpidou 2012). Most of the Ionian islands were connected to each other and to the mainland during the last glacial maximum (ca. 26.5 ka) and the earlier glaciations of the late Pleistocene resulting in lowered sea levels (Lykousis 2009, Sakellariou and Galanidou 2015). Ferentinos et al. (2012), who reconstructed the paleocoastline of this area for the last glacial maximum (sea level at -120 m bpsl), showed that the islands of Cephalonia, Ithaki, and Zakythos were connected to each other and separated from Lefkada by a narrow strait. The latter was connected to the mainland and to most of the smaller islets in the east (i.e., Meganissi, Skorpios, etc.). Lefkada is the only Ionian Island that was separated from the mainland throughout the Pleistocene. Kerkira was suggested to be connected to the mainland in early Holocene, about 8,000 years ago, when the sea level was lower (Perissoratis and Conispoliatis 2003).

While the history of biogeographic studies upon Aegean insular fauna is long, the similar studies on the Ionian Islands are limited, with the two rarely compared. The biota of the Ionian Islands is similar to those of the adjacent mainland, although several endemic taxa occur, most of which live on the larger and more heterogeneous islands, such as Kerkyra and Cephalonia (Strouhal 1966, Pesce and Argano 1980, Pinkster 1993, Karaman 2017b).

6.3.7 Marmara Islands

6.3.7.1 Geographical Setting

The Sea of Marmara is the northeasternmost part of the Mediterranean Basin. It is also the world's smallest inner-continental sea (over 11,000 km^2), entirely located within the borders of Turkey. The Marmara Basin is elongated east–west, with a length of 275 km and width of ca. 80 km. From the east, it is connected with the Aegean Sea through the Strait of the Dardanelles and from the northeast with the Black Sea through the Bosporus strait. The water of the Marmara Sea has a less saline upper water layer coming from the mesohaline Black Sea, while the lower water layer is of Aegean origin with fully marine salinity. These layers do not intermix, but form a halocline at a depth of 25 m, which influences the faunal composition in the Marmara Islands. The literature concerning the Sea of Marmara Islands is scant and typically in Turkish. There are over 40 islands and islets. The Marmara archipelago is the only one from which freshwater malacostracans have been recorded, and it is located in the southwest. The archipelago consists of 21 islands, of which the Marmara Island is the largest, covering 118.9 km^2. It is mostly mountainous with the highest point of 700 m.a.s.l. There are only five villages, with a total human population of 6,000. The climate is dry Mediterranean and the island has poor surface water resources with few small rivers and streams. Vegetation cover consists mostly of Mediterranean forest trees, maquis, and frigana. The other islands of the Marmara archipelago are similar (Beşiktepe et al. 1994, Wong et al. 1995, Görür al. 1997, Akyol et al. 2009, Bulut 2016, Ayfer et al. 2017).

6.3.7.2 Malacostracan Fauna

The marine fauna of the Marmara Islands is scarcely known (e.g., Ayfer et al. 2017), and there is only one study upon its freshwaters. Özbek et al. (2015) reported the occurrence of two gammarid amphipods: *Gammarus uludagi* (Karaman 1975) and *G. pulex* (Linnaeus 1758) on Marmara Islands and on (10 km distance) Kapıdağ Peninsula. Both were previously recorded from the Turkish mainland (Karaman 1975, Özbek 2011). While *G. pulex* is widely distributed in Europe and is probably a complex of cryptic species (Hupało et al. 2018a), *G. uludagi* was reported only from Turkey and from the adjacent Aegean Islands (Karaman and Pinkster 1977, Özbek et al. 2015). Even if some widely distributed malacostracan species, such as *Potamon fluviatile* or *P. ibericum*, occur in the region, there are no records from the Marmara Islands (e.g., Güner 2009). Similarly, there are no data upon the subterranean fauna of the islands.

6.3.7.3 Geological History and Malacostracan Biogeography

In the middle Miocene (ca. 12 Ma), the Marmara region was overflown by fully saline waters of the proto-Mediterranean Sea. In the late Miocene (ca. 8.5 Ma), Paratethys waters prevailed with oligohaline conditions. These conditions dominated the present Marmara Sea until the end of Pliocene (ca. 3 Ma) when another intrusion of waters from the Eastern Mediterranean Basin occurred. In that period, the connection between Paratethys and the Mediterranean ceased and led to the formation of an isolated basin that eventually became the Black Sea. During the Pleistocene, an open connection with the Mediterranean was reestablished and the water salinity fluctuated greatly due to eustatic sea-level changes in the present Marmara Basin (Görür et al. 1997 and references therein, Popov et al. 2004). The Marmara Basin was completely isolated during the last glaciation (ca. 115 to 11.7 ka), both from the Mediterranean and the Black Sea. During this isolation, it transformed into an anoxic brackish lake (Görür et al. 1997). At the end of the glaciation, the Mediterranean Sea intruded into the Marmara and ca. 7 ka broke through the Bosphorus Strait, overflowing the isolated and now hyposaline Black Sea. This eventually led to the dual-flow regime in the Bosphorus Strait and the Marmara Sea (Ryan et al. 1997). The freshwater fauna may have colonized the islands in the brackishwater or lacustrine phases of the Marmara Sea. The Gammaridae recorded on Marmara Island are presumably of Paratethyan origins (Hou et al. 2011). The widely distributed *G. pulex* species complex occupies lowland and upland river systems and may have colonized the Island in the late Miocene. Similarly, *G. uludagi* is known only from Anatolia and Thrace and may be a recent colonizer. However, it is important to state that *G. uludagi* is a sister lineage to *G. pulex* (see Hou et al. 2011) and is difficult to draw conclusions without molecular studies. It is important to emphasize the lack of knowledge on the hydrological regime and the crustaceans from the Marmara Islands.

6.3.8 Cyprus

6.3.8.1 Geographical Setting

Cyprus is located in the eastern Mediterranean Sea and is 75 km south of Turkey and 100 km west of Syria. The island covers an area of 9,251 km², which makes it third largest in the Mediterranean Basin. It stretches 230 km in length and is up to 100 km wide. The shoreline is well developed, rocky, and with cliffs in the north, while the south is more flat with numerous sandy beaches. Total coastline length is about 671 km (World Resource Institute).

Cyprus geomorphologically consists of three main units. The northern part is dominated by the ca. 160-km-long Kyrenia Mountain range. It is primarily limestone with some marble, and the highest peak is Pentadaktylos (1,024 m.a.s.l.). The Troodos Mountains are in the southern and western parts of the island, covering around half of its area. This massif is made of basalts, granites, and gabbros originating from the oceanic crust. The highest peak is Mt. Olympus (1,952 m.a.s.l.). These two mountain ranges are separated by the wide alluvial plain of Mesaoria. Due to deforestation initiated in antiquity, much of the Mesaoria is covered with a hardpan (local name "Kafkalla"), formed by compacted calcium carbonate impervious to water.

Generally, freshwater resources are rather scarce. The surface hydrographic network is extensively developed, though not even one river or stream is permanent its full length. The primary and longest (98 km) watercourse is the Pedieos River, originating in the Troodos Mountains and flows from late autumn to late spring through the Mesaoria to Famagusta Bay. The Troodos Mountains have numerous springs and above 1,000 m.a.s.l. the streams are often permanent. The Mesoaria plain is rich in underground waters; however, they are heavily exploited by agriculture and tourism (Charalambous 2001). Natural flows are often interrupted by dams; their density is one of the highest in Europe (Zogaris et al. 2012).

6.3.8.2 Malacostracan Fauna

Only nine species are known from Cyprus, of which two are stygobitic. Considering the extent of the local hydrological network, the island area, and the local diversity of malacostracans on other Mediterranean islands, the crustacean fauna of Cyprus seems to be either poor or largely understudied. Only Amphipoda, Isopoda, and Decapoda are represented, with three species from each order.

The sole subterranean representative of amphipods is the endemic *Stygogidiella cypria* (Karaman 1989) of Bogidiellidae. It is known only from a spring in Amathus, Limassol, and from a spring in a tunnel in Neofytos, Pafos (Karaman 1989). The other two species are the gammarid *Echinogammarus veneris,* widely distributed around the Mediterranean (probably a species complex) (Hupało et al. 2018b), and the semiterrestrial talitrid *Cryptorchestia cavimana* (Heller 1865), a Cyprian endemic (Ruffo et al. 2014, Davolos et al. 2017, 2018). Interestingly, more recent sampling in streams and springs on the Greek part of Cyprus yields only records of talitrids (F. Stoch, unpublished data).

Isopoda is represented by two families. The epigean asellid *Proasellus coxalis* s.l., widespread in the Mediterranean region, reported from springs in Arothes and Kritou Tera (Sket 1990b), and the endemic *P. coxalis nanus* (Sket 1990) occurs in a spring in Agia Mavri. The sole representative of Lepidocharontidae, *Microcharon luciae* (Sket 1991), was described from a well in the shallow gravel deposits in Amathous.

Freshwater decapods include three families, each with one representative. The crab *Potamon hippocratis* (Ghigi 1929) (Potamidae) was reported by Lewinsohn and Holthuis (1986) (Jesse et al. 2011). This species, used as food source by ancient Greeks, may be present due to anthropogenic activity (Jesse et al. 2011). The two other decapod species are the alien invasive crayfishes *Pacifastacus leniusculus* (Dana 1852) (Astacidae) and *Procambarus clarkii* (Cambaridae), introduced probably, respectively, in the 1980s and in 1990s (Hobbs et al. 1989, Holdich et al. 1999, Lewis 2002). Further monitoring of these species is needed, as the two invaders pose serious threats to the native invertebrate fauna (Morpurgo et al. 2010, Vella et al. 2017, Deidun et al. 2018).

6.3.8.3 Geological History and Malacostracan Biogeography

Cyprus began emerging in the Mesozoic (McCallum and Robertson 1990), between 85 and 92 Ma (Harland et al. 1982). However, Gass (1980) suggested that Mt. Troodos was a volcanic island that had emerged at the break of the Cretaceous and Paleogene, whereas the Kyrenia Mountains were another island or began as a part of the southern Taurus Mountains, originating in the Eocene and subsequently split off (Cavazza and Wezel 2003). Troodos island and the southern Tauruian-Kyrenian Peninsula (present Kyrenia Mountains) had a land connection (Hsü et al. 1977, Cavazza and Wezel 2003) during the Messinian Salinity Crisis (5.96 to 5.3 Ma) (Krijgsman et al. 2002). During this time, Cyprus remained isolated from the continental mainland by hypersaline deserts forming the floor of the Miocene death valleys, ca. 3,000 m below the sea level (Hsü 1983). Now, together with Crete, Cyprus is among the most isolated Mediterranean islands (Moores et al. 1984). After the refilling of the Mediterranean Sea with Atlantic waters during the Zanclean Flood, the new island, composed of two formerly disjunct landmasses, remained isolated from the mainland. It is now bordered on the north by the Adana Strait and from the Levant by the Latakia Basin, as much as 1,000 m deep. Thus, even during the last glacial maximum, when the sea level dropped ca. 120 m, the closest mainland (the now submerged Alexandretta Bay, or Gulf of İskenderun) was still separated from Cyprus by ca. 30 km of sea (Hadjisterkotis 2012, and the citations therein).

The widely distributed *Echinogammarus veneris* is reported from the Adriatic coast, the mainland Greece and Peloponnesus, the islands of Corfu, Zakinthos, and Crete as well as from the Mediterranean coasts of Turkey, Lebanon, and Israel (Pinkster 1993). Ongoing molecular studies demonstrate that it is a complex of deeply divergent and narrowly distributed lineages, which may date before the Messinian Salinity Crisis (Hupało et al. 2018b). The hypogean bogidiellids, represented only by the endemic *Stygogidiella cypria* (Karaman 1989), are probably thalassoid stygobionts of ancient marine origin (Coineau and Boutin 1992, Boutin 1997).

Two or possibly three of the nine Cyprus malacostracan species were introduced by humans, either intentionally or accidentally in historical times. Given the growing economic and tourist activity in the area and favorable climatic conditions, one may expect that other alien malacostracans, particularly those kept for ornamental purposes, will be introduced.

6.4 DISCUSSION

6.4.1 Diversity and Endemism

The freshwaters of the Mediterranean islands contain 182 known species in five different orders (Amphipoda, Isopoda, Thermosbaenacea, Bathynellacea, Decapoda), representing 51 genera and 27 families (Table 6.1, Appendix 1). The Mediterranean insular freshwaters are a freshwater malacostracan biodiversity hotspot, given that the islands cover only about 5% of the Mediterranean region and about 0.2% of the whole Palearctic, but being inhabited by more than 25% of all the malacostracan species reported from the region and more than 8% of the total Palearctic Malacostraca (Balian et al. 2008, Figueroa et al. 2013). The species composition seems to mirror patterns previously observed for the Palaeartic (Balian et al. 2008) with amphipods and isopods being the most speciose groups (58% and 26%, respectively) of all freshwater malacostracans and Thermosbaenacea and Bathynellacea having the lowest numbers of species (Table 6.2).

The highest alpha diversity was recorded on the large islands such as Sicily, Sardinia, Corsica, and Crete as well the largest archipelagos such as the Baleares, the Aegean Islands, the Ionian Islands, and the Adriatic Islands (Table 6.1). This is expected given both the age and large area of those islands and the high number of archipelagos as well as the general correlation between species number and area size (Rosenzweig 1995). Equally expected, the lowest biodiversity was observed on the Marmara Islands and on the Pelagie Islands, being among the smallest Mediterranean archipelagos. Surprising, however, is the low number of species reported from Cyprus, the third largest island in the Mediterranean and on the Maltese Islands. This may reflect the rarity of freshwater biota studies there and also may be why no insular freshwater malacostracans have been recorded from Djerba or the Kerkennah Islands. Conversely, local environmental conditions may simply be unfavorable for freshwater species. Sardinia exhibits the highest biodiversity with 36 species in 21 genera and 17 families, plus the highest number of endemic malacostracan taxa (Table 6.1). Such high diversity levels may be a result of the long and complex geological history of the island (Speranza et al. 2002), combined with the landscape heterogeneity (Vogiatzakis et al. 2008).

The Mediterranean islands are natural evolutionary laboratories, housing a significant number of endemic species (Schüle 1993, Hopkins 2002, Vogiatzakis et al. 2008, Blondel et al. 2010); more than half of all reported freshwater malacostracan species being localized are endemics (Table 6.1). Sardinia, Crete, and the Balearic Islands exhibit the highest rates of endemism, with 50% or more of all malacostracans as narrow range endemics (Table 6.1). This trend has been demonstrated in other freshwater organisms, such as insects, molluscs, and annelids (Gómez-Campo et al. 1984, Stoch 2000, Ruffo and Stoch 2006, Sfenthourakis and Triantis 2017). The proximity to the mainland and the often reestablished land connections, combined with the infrequency of freshwater studies, may explain lower local endemism rates on islands like the Tuscan Archipelago, the Adriatic Islands, and the Maltese Islands (Table 6.1).

Generally, each malacostracan order shows high endemism rates, exceeding 50% of reported taxa, with the exception of decapods (Table 6.2). This is explained by the relatively greater dispersal abilities in this group, enabling some taxa to disperse through terrestrial connections (e.g., Ponniah and Hughes 2004, Jesse et al. 2011). The highest number of endemics recorded was in amphipods with 61 insular endemic species (Appendix 1).

With the exception of the Maltese Islands and Cyprus, amphipods were always the most species-rich order (Table 6.2), with Gammaridae being the most diverse, comprising almost one-third of all epigean freshwater malacostracans (Appendix 1). Gammarid amphipods are keystone species, structuring freshwater epigean macroinvertebrate communities, and are also among the most abundant and biomass-dominant groups in European lotic ecosystems (MacNeil et al. 1997, Kelly et al. 2002). Following the data patterns, Gammaridae is also the most speciose epigean freshwater malacostracan family worldwide (Väinölä et al. 2007, Balian et al. 2008) with *Gammarus* being globally the largest epigean freshwater malacostracan genus. There are 204 *Gammarus* species reported from inland waters (Väinölä et al. 2007) and this number is still growing (e.g., Cannizzaro et al. 2017, Grabowski et al. 2017a, Hou et al. 2018, Hupało et al. 2018a). The emergence and increasing use of DNA-based identification has led to identifying a plethora of cryptic and pseudocryptic species (Fišer et al. 2018). In the Mediterranean, there are numerous cryptic species complexes, e.g., *Gammarus balcanicus* (e.g., Mamos et al. 2014, 2016) or *Asellus aquaticus* (e.g., Sworobowicz et al. 2015), with many more studies currently underway (e.g., *Gammarus italicus*, *Echinogammarus veneris*, *Rhipidogammarus rhipidiophorus* [Stoch et al., unpublished data], and Sicilian *Echinogammarus* [Hupało et al. 2017, 2018b, unpublished data]).

The majority of insular species inhabit subterranean habitats (Table 6.1). This exceeds the estimated level of groundwater diversity, which was estimated to be lower than in surface waters (Sket 1999, Väinölä et al. 2007). At the regional scale, it is often equal to or greater than epigean diversity (Stoch 1995). Again, amphipods lead with more than 60% of hypogean species reported from Mediterranean islands, represented by both members of strictly stygobitic families (such as Bogidiellidae, Niphargidae, Pseudoniphargidae, Salentinellidae) and a few members of Gammaridae or Ingolfiellidae (Appendix 1). The most speciose hypogean amphipod genus here is *Niphargus*, represented by 27 species and subspecies, being also the globally largest known stygobiont freshwater amphipod genus (Väinölä et al. 2007, Fišer et al. 2008), with many endemic insular species (Karaman 2017a, 2018; Zettler and Zettler 2017). Interestingly, the subterranean malacostracan fauna in the Mediterranean islands show an extreme level of endemism with 60% of all recorded stygobionts being endemic, comprising 67 endemics out of 111 malacostracan species recorded from all Mediterranean islands (Appendix 1). This exceeds the already high estimates for all the European groundwaters (Deharveng et al. 2009). These numbers are most probably underestimated for epigean taxa given that more and more cryptic diversity is being detected both within hypogean amphipods (e.g., Jurado-Rivera et al. 2017) and isopods (Ketmaier et al. 2003) with many more ongoing molecular studies on groundwater malacostracans (Stoch et al., unpublished).

6.4.2 Origin of Insular Freshwater Malacostracans

The origin of the recently introduced invasive fauna on the Mediterranean islands is known to be within the last decades (e.g., Chappuis 1949, Holdich et al. 1999). The majority of insular freshwater malacostracans have been present for millions of years, often being on the island since its formation. Even though the islands largely differ in terms of their dates of origin and geological history, the colonization patterns of freshwater malacostracan crustaceans remain similar throughout the islands. They are either of ancient marine origin (thalassoid, Coineau and Boutin 1992), which colonized insular groundwater habitats, or they are of continental origin. The continental malacostracans were either inhabiting the island when it was still a part of the mainland (e.g., Ketmaier et al. 2001, 2003, Chueca et al. 2015) or they migrated using temporary land connections established either by land movement or sea-level fluctuations from the Miocene to the Pleistocene (e.g., Triantis and Mylonas 2009, Ketmaier and Caccone 2013, Delicado et al. 2014). The origin of the Mediterranean insular malacostracans has been one of biogeographical studies focal points in the last decades. In these studies, several authors hypothesized possible biogeographical scenarios based mostly on taxonomic and systematic evidence (e.g., La Greca 1957, Ruggieri 1973, Coineau and Boutin 1992). Now, with DNA-based molecular clock methods (Ho and Duchêne 2014), the growing number of studies on the origin and biogeography of insular faunas are able to test previous hypotheses. Although numerous Mediterranean insular freshwater fauna studies have shed some light on malacostracan biogeography (e.g., Jesse et al. 2011, Ketmaier and Caccone 2013, Poulakakis et al. 2015, Hupało et al. 2018a), more molecular studies are needed to fully and reliably resolve the biogeographical entanglements.

6.4.3 Threats

Freshwater ecosystems are among the world's most species-rich, but are also the most endangered (Cuttelod et al. 2008, Darwall et al. 2009). Extinction rates for freshwater biota are believed to be five times higher than in terrestrial environments (Dudgeon et al. 2006). The Mediterranean region, including the islands, recognized as a biodiversity hotspot, is inhabited by ca. 400 million people and exposed to growing mass tourism, already reaching 150 million visitors a year and projected to double by 2025 (e.g., Vogiatzakis et al. 2008, Benoit and Comeau 2012). Even though tourism brings economic benefits for particular islands, it has a devastating impact on the environment, including freshwaters (Hopkins 2002). Besides the degradation of freshwater ecosystems by pollution, daming, gravel mining, and other habitat conversion, humans are also responsible for introducing alien, often invasive taxa (Hopkins 2002, Vogiatzakis et al. 2008), which often have a devastating effect on local fauna (e.g., Palmer and Pons 1996, Hentonnen and Huner 1999, Geiger et al. 2005). Although the number of alien malacostracan species present in Mediterranean island freshwaters is relatively low (nine species, Table 6.1), it comprises a third of all reported insular freshwater decapods (Table 6.2). Notably, the islands with the highest percentage of alien species present (the Maltese Islands, Cyprus, Sicily; Table 6.1) are also the ones with the highest tourist activity (Hopkins 2002). *Procambarus clarkii* is among the most invasive organisms globally and is one of the most common malacostracan in Mediterranean island freshwaters (Appendix 1). The invasiveness of *P. clarkii* is manifest by its strong impact on entire ecosystems, often resulting in the extinction of coexisting macroinvertebrates (Geiger et al. 2005). These invaders along with extensive human activity pose the largest ecological threat. Even though the freshwater malacostracan biodiversity is relatively high, it may possibly be a remnant of even higher diversity with many recent extinctions, when the *loci typici* of some species, known now only from the museum specimens, are gone due to the anthropogenic environmental changes (personal observations).

6.4.4 Conservation Perspectives

There is a need for planning reasonable conservation strategies for Mediterranean island freshwaters. Although freshwaters are recognized globally as affected by anthropogenic pressures, with an estimated rate of about 20,000 freshwater species already extinct or threatened (Dudgeon et al. 2006, Vörösmarty et al. 2010), conservation policies regarding freshwater biota are poorly addressed. Few freshwater representatives are included in the European Council Directive 92/43/ EEC on the conservation of natural habitats and of wild fauna and flora (1992, also known as the Habitats Directive). Only 0.1% of all European invertebrates arc under protection and freshwater invertebrate fauna is only a small fraction (Balian et al. 2008, Figueroa et al. 2013). Even though Mediterranean island freshwater habitats have been identified as threatened, no island-specific, extensive conservation strategy has been planned or implemented (Hopkins 2002). The first step in protecting Mediterranean island freshwater ecosystems has to be a biodiversity summary. Apart from insular freshwater species assessments, studies on their biology and population structures must follow. Recognizing and establishing "Freshwater Key Biodiversity Areas" in the Mediterranean islands could help formulate long-term national protection plans and protect local freshwater biodiversity hotspots (Darwall et al. 2014). The second step would be to reduce anthropogenic pressure on freshwaters. According to EU Water Framework Directive (and to the Groundwater Directive 2006/118/EC, which was developed in response to the requirements of Article 17 of the Water Framework Directive), the EU Member States must protect, enhance, and restore all bodies of surface and subterranean water, aiming to achieve "satisfactory water status". However, in reality freshwaters are being even more affected not only by habitat degradation, pollution, modification, overexploitation, and introduction of alien fauna, but also by climate change (Dudgeon et al. 2006, Kernan et al. 2011). Thus, perspectives for maintaining the current freshwater malacostracan biodiversity in the Mediterranean islands are not optimistic. Preparing similar biodiversity reviews of other freshwater biota inhabiting the Mediterranean islands and initiating activities will raise public awareness and help protect the remaining insular freshwater habitats.

ACKNOWLEDGMENTS

This chapter was written within the framework of the following projects funded by the Polish National Science Center (projects no. 2014/15/B/NZ8/00266, 2015/17/N/ NZ8/01628 and 2018/28/T/NZ8/00022). We would also like to thank Khaoula Ayati, Sonia Dhaouadi, Murat Özbek, and Krešimir Žganec for providing the valuable literature references.

APPENDIX 1

Islands ('species' name=hypogean; Xe=endemic; Xi=alien,introduced)	Balearic islands	Corsica	Sardinia	Tuscan Archipelago	Sicily	Pelagie islands	Maltese islands	Adriatic islands	Ionian Islands	Crete island	Aegean Islands	Marmara islands	Cyprus
Order: Amphipoda Latreille, 1816													
Family:													
Bogidiellidae Hertzog, 1936													
*Bogidiella balearica Dancau, 1973	Xe												
*Bogidiella calicali G. Karaman, 1988			Xe										
*Bogidiella cyrnensis Hovenkamp, Hovenkamp & Van der Heide, 1983		Xe											
*Bogidiella ichnusae Ruffo & Vigna Taglianti, 1975			Xe										
*Bogidiella longiflagellum S. Karaman, 1959									x	x			
*Bogidiella silverii Pesce, 1981			Xe										
*Bogidiella torrenticola Pretus & Stock, 1990	X	X											

(Continued)

Islands (*species' name=hypogean; Xe=endemic; Xi=alien, introduced)	Balearic islands	Corsica	Sardinia	Tuscan Archipelago	Sicily	Pelagie islands	Maltese islands	Adriatic islands	Ionian Islands	Crete island	Aegean Islands	Marmara islands	Cyprus
*Bogidiella vandeli Coineau, 1968			Xe										
*Medigidiella aquatica (G. Karaman, 1990)										X			
*Medigidiella chappuisi (Ruffo, 1952)			X	X							X		
*Medigidiella minotaurus (Ruffo & Schiecke, 1976)										Xe			
*Medigidiell paolii Hovenkamp, Hovenkamp & Van der Heide, 1983		Xe											
*Medigidiella pescei (Karaman, 1989)			Xe										
*Racovella birramea Jaume, Gràcia & Boxshall, 2007	Xe												
*Stygogidiella cypria (G. Karaman, 1989)													Xe
Family: **Crangonyctidae Bousfield, 1973**													
Synurella ambulans (F. Müller, 1846)								X					

(Continued)

Islands ('species' name=hypogean; Xe=endemic; Xi=alien,introduced)	Balearic islands	Corsica	Sardinia	Tuscan Archipelago	Sicily	Pelagie islands	Maltese islands	Adriatic islands	Ionian Islands	Crete island	Aegean Islands	Marmara islands	Cyprus
Family: Eriopisidae Lowry & Myers, 2013													
Psammogammarus burri Jaume & Garcia, 1992	Xe												
Family: Gammaridae Leach, 1813													
Echinogammarus adipatus Karaman & Tibaldi, 1972					Xe								
Echinogammarus ebusitanus Margalef, 1951	Xe												
Echinogammarus eisentrauti (Schellenberg, 1937)	X												
Echinogammarus kerkuraios Pinkster 1993									Xe				
Echinogammarus klaptoczi Schäferna, 1908							X						
Echinogammarus kretensis Pinkster 1993										X			

(Continued)

Islands ('species' name=hypogean; Xe=endemic; Xi=alien,introduced)	Balearic islands	Corsica	Sardinia	Tuscan Archipelago	Sicily	Pelagie islands	Maltese islands	Adriatic islands	Ionian Islands	Crete island	Aegean Islands	Marmara islands	Cyprus
Echinogammarus monomerus Stock, 1978	X												
Echinogammarus platvoeti Pinkster 1993										Xe			
Echinogammarus pungens s.l. (H. Milne Edwards, 1840)	X						X	X	X				
Echinogammarus pungens minoricensis (Margalef, 1952)	Xe												
Echinogammarus aff. *tibaldii* Pinkster & Stock, 1970					Xe								
Echinogammarus sardus Pinkster, 1993			Xe										
Echinogammarus sp. (= *fluminensis* Pinkster & Stock, 1970)		X											
Echinogammarus tibaldii bolo Karaman & Tibaldi, 1972			Xe										

(Continued)

Islands ('species' name=hypogean; Xe=endemic; Xi=alien,introduced)	Balearic islands	Corsica	Sardinia	Tuscan Archipelago	Sicily	Pelagie islands	Maltese islands	Adriatic islands	Ionian Islands	Crete island	Aegean Islands	Marmara islands	Cyprus
Echinogammarus aff. *sicilianus* Karaman & Tibaldi, 1972	X												X
Echinogammarus sicilianus Karaman & Tibaldi, 1972					Xe								
Echinogammarus veneris s.l. (Heller, 1865)								X	X	X			
Gammarus balcanicus s.l. Schäferna, 1923								X					
Gammarus italicus Goedmakers & Pinkster, 1977			Xe										
Gammarus komareki Schäferna, 1923											X		
Gammarus plaitisi Hupalo, Mamos, Wrzesińska & Grabowski, 2018										Xe			
Gammarus pulex s.l.(Linnaeus, 1758)												X	
Gammarus uludagi G. Karaman, 1975											X	X	
*Ilvanella inexpectata Vigna-Taglianti, 1971				X									

(Continued)

Islands (*species' name=hypogean; Xe=endemic; Xi=alien,introduced)	Balearic islands	Corsica	Sardinia	Tuscan Archipelago	Sicily	Pelagie islands	Maltese islands	Adriatic islands	Ionian Islands	Crete island	Aegean Islands	Marmara islands	Cyprus
*Longigammarus planasiae Messana & Ruffo, 2001				Xe									
Rhipidogammarus rhipidiophorus s.l. (Catta, 1878)	X	X	X	X	X	X	X						
Rhipidogammarus karamani Stock, 1971		X						X					
Rhipidogammarus variicauda Stock, 1978	Xe												
*Tyrrhenogammarus catacumbae (G. Karaman & Ruffo, 1977)					Xe								
*Tyrrhenogammarus sardous Karaman & Ruffo, 1989			Xe										
Sarathrogammarus madeirensis (Dahl, 1958)	Xe												
Family: Hadziidae S. Karaman, 1943													
Hadzia fragilis fragilisS. Karaman, 1932								X					

(Continued)

Islands ('species' name=hypogean; Xe=endemic; Xi=alien,introduced)	Balearic islands	Corsica	Sardinia	Tuscan Archipelago	Sicily	Pelagie islands	Maltese islands	Adriatic islands	Ionian Islands	Crete island	Aegean Islands	Marmara islands	Cyprus
Metahadzia helladis Pesce, 1980									Xe				
Family: Ingolfiellidae Hansen, 1903													
Ingolfiella petkovskii S. Karaman, 1975											X		
*Ingolfiella cottarellii*Ruffo & Vigna-taglianti, 1989			Xe										
Family: Ischyroceridae Stebbing, 1899													
Jassa sp. Leach, 1814			Xe										
*Jassa trinacriae*Krapp, Grasso & Ruffo, 2010					Xe								
Family: Metacran-gonyctidae Boutin & Messouli, 1988													
Metacrangonyx longipes Chevreux, 1909	Xe												
Metacrangonyx ilvanus Stoch, 1997				Xe									

(Continued)

Islands ('species' name=hypogean; Xe=endemic; Xi=alien,introduced)	Balearic islands	Corsica	Sardinia	Tuscan Archipelago	Sicily	Pelagie islands	Maltese islands	Adriatic islands	Ionian Islands	Crete island	Aegean Islands	Marmara islands	Cyprus
Family: Niphargidae Bousfield, 1977													
*Exniphargus tzanisi G. Karaman, 2016										Xe			
*Niphargobatoides lefkodemonaki Sket, 1990										Xe			
*Niphargus adei S.Karaman, 1934											Xe		
*Niphargus aff. aquilex Schiödte, 1855								X					
*Niphargus alphaeus Delić, Švara, Coleman, Trontelj & Fišer, 2017								Xe					
*Niphargus arethusa Delić, Švara, Coleman, Trontelj & Fišer, 2017								Xe					
*Niphargus corsicanus Schellenberg, 1950		X	X										
*Niphargus dalmatinus Schäferna, 1932								X					
*Niphargus denarius G. Karaman, 2017									Xe				

(Continued)

CRUSTACEAN BIODIVERSITY AND CONSERVATION

Islands ('species' name=hypogean; Xe=endemic; Xi=alien,introduced)	Balearic islands	Corsica	Sardinia	Tuscan Archipelago	Sicily	Pelagie islands	Maltese islands	Adriatic islands	Ionian Islands	Crete island	Aegean Islands	Marmara islands	Cyprus
*Niphargus doli Delić, Švara, Coleman, Trontelj & Fišer, 2017								X					
*Niphargus hebereri s.l. Schellenberg, 1933								X					
*Niphargus hvarensis S. Karaman, 1952								X					
*Niphargus impexus G. Karaman, 2016										Xe			
*Niphargus jadranko Sket & G. Karaman, 1990								Xe					
*Niphargus jovanovici S. Karaman, 1931											X		
*Niphargus lakusici G. Karaman, 2017										Xe			
*Niphargus liburnicus G. Karaman & Sket, 1989								X					
*Niphargus longicaudatus s.l. (A. Costa, 1851)				X	X			X					

(Continued)

Islands ('species' name=hypogean; Xe=endemic; Xi=alien,introduced)	Balearic islands	Corsica	Sardinia	Tuscan Archipelago	Sicily	Pelagie islands	Maltese islands	Adriatic islands	Ionian Islands	Crete island	Aegean Islands	Marmara islands	Cyprus
*Niphargus cf. lourensis Fišer, Trontelj & Sket, 2006									X				
*Niphargus lourensis skiroci G. Karaman, 2018											Xe		
*Niphargus miljeticus Straškraba, 1959								Xe					
*Niphargus pectencoronatae Sket & G. Karaman, 1990								Xe					
*Niphargus rhodi S. Karaman, 1950											Xe		
*Niphargus skopljensis S.Karaman, 1929									X				
*Niphargus spasenijae G.Karaman, 2015											Xe		
*Niphargus aff. speziae Schellenberg, 1936				X									
*Niphargus versluysi S.Karaman, 1950									X				
*Niphargus wolfi Schellenberg, 1933								X					

(Continued)

CRUSTACEAN BIODIVERSITY AND CONSERVATION

Islands ('species' name=hypogean; Xe=endemic; Xi=alien,introduced)	Balearic islands	Corsica	Sardinia	Tuscan Archipelago	Sicily	Pelagie islands	Maltese islands	Adriatic islands	Ionian Islands	Crete island	Aegean Islands	Marmara islands	Cyprus
Niphargus zarosiensis M. Zettler & A. Zettler, 2017										Xe			
Family: Pseudoniphargidae Karaman, 1993													
Pseudoniphargus adriaticus s.l. S. Karaman, 1955	X	X		X		X							
Pseudoniphargus daviui Jaume 1991	Xe												
Pseudoniphargus duplus Messana & Yacoubi-Khebiza, 2006					Xe								
Pseudoniphargus inconditus Karaman & Ruffo, 1989					Xe								
Pseudoniphargus italicus Karaman & Ruffo, 1989					Xe								
Pseudoniphargus mercadali Pretus, 1988	X		X										

(Continued)

Islands (*species' name=hyphogean; Xe=endemic; Xi=alien,introduced)	Balearic islands	Corsica	Sardinia	Tuscan Archipelago	Sicily	Pelagie islands	Maltese islands	Adriatic islands	Ionian Islands	Crete island	Aegean Islands	Marmara islands	Cyprus
Pseudoniphargus obritus Messouli, Messana & Yacoubi-Khebiza, 2006		Xe											
Pseudoniphargus planasiae Messouli, Messana & Yacoubi-Khebiza, 2006				Xe									
Pseudoniphargus pedrerea Pretus, 1990	Xe												
Pseudoniphargus pityusensis Pretus, 1990	Xe												
Pseudoniphargus racovitzai Pretus, 1990	Xe												
Pseudoniphargus sodalis Karaman & Ruffo, 1989					Xe								
Pseudoniphargus triasi Jaume 1991	Xe												
Family: Salentinellidae Bousfield, 1977													

(Continued)

Islands ('species' name=hypogean; Xe=endemic; Xi=alien,introduced)	Balearic islands	Corsica	Sardinia	Tuscan Archipelago	Sicily	Pelagie islands	Maltese islands	Adriatic islands	Ionian Islands	Crete island	Aegean Islands	Marmara islands	Cyprus
*Salentinella angelieri s.l. Ruffo & Delamare Debouteville, 1952	X	X	X	X				X	X		X		
*Salentinella formenterae Platvoet, 1984	Xe												
Family: Talitridae Rafinesque, 1815													
Cryptorchestia cavimana (Heller, 1865)													Xe
Cryptorchestia garbinii Ruffo, Tarocco & Latella, 2014				X									
Cryptorchestia ruffoi Latella & Vonk, 2017											X		
Macarorchestia remyi (Schellenberg, 1950)		X	X										
Order: Isopoda Latreille, 1817													
Family: Asellidae Latreille, 1802													
Asellus aquaticus (Linnaeus, 1758)								X		Xi	X		

(Continued)

Islands (*species' name=hypogean; Xe=endemic; Xi=alien,introduced)	Balearic islands	Corsica	Sardinia	Tuscan Archipelago	Sicily	Pelagie islands	Maltese islands	Adriatic islands	Ionian Islands	Crete island	Aegean Islands	Marmara islands	Cyprus
*Proasellus acutianus Argano & Henry, 1972				X									
Proasellus banyulensis Racovitza, 1919		X	X	X	Xi								
*Proasellus beroni Henry & Magniez, 1968		X											
Proasellus coxalis s.l. (Dollfus, 1892)							X	X			X		X
Proasellus coxalis cephallenus (Strouhal, 1942)									Xe				
Proasellus coxalis corcyraeus (Strouhal, 1942)									Xe				
Proasellus coxalis gabriellae (Margalef, 1950)	X												
Proasellus coxalis leucadius (Strouhal, 1942)									Xe				
Proasellus coxalis nanus Sket, 1991													Xe
Proasellus coxalis versluysi (Strouhal, 1966)									Xe				

(Continued)

Islands (*species' name=hypogean; Xe=endemic; XI=alien,introduced)	Balearic islands	Corsica	Sardinia	Tuscan Archipelago	Sicily	Pelagie islands	Maltese islands	Adriatic islands	Ionian Islands	Crete island	Aegean Islands	Marmara islands	Cyprus
*Proasellus cretensis Pesce & Argano, 1980										Xe			
*Proasellus ezzu Argano & Campanaro, 2004			Xe										
*Proasellus minoicus Pesce & Argano, 1980										Xe			
Proasellus montalentii Stoch, Valentino & Volpi, 1996					Xe								
*Proasellus patrizii (Arcangeli, 1952)			Xe										
*Proasellus ruffoi Argano & Campanaro, 2004			Xe										
*Proasellus sketi Henry, 1975											X		
Proasellus wolfi Dudich, 1925					X								
Family: Stenasellidae Dudich, 1924													
*Stenasellus racovitzai s.l. Razzauti, 1925		X	X										

(Continued)

Islands (*species* name=hypogean; Xe=endemic; Xi=alien,introduced)	Balearic islands	Corsica	Sardinia	Tuscan Archipelago	Sicily	Pelagie islands	Maltese islands	Adriatic islands	Ionian Islands	Crete island	Aegean Islands	Marmara islands	Cyprus
*Stenasellus assorgiai Argano, 1968			Xe										
*Stenasellus nuragicus Argano, 1968			Xe										
Family: Sphaeromatidae Latreille, 1825													
*Monolistra (Microlistra) pretneri Sket, 1964								x					
Family: Cirolanidae Dana, 1852													
*Metacirolana ponsi Jaume & Garcia, 1992	Xe												
*Sphaeromides virei virei(Brian, 1923)								x					
*Sphaeromides virei mediodalmatina Sket, 1964								x					
*Typhlocirolana margalefi Pretus, 1986													
*Typhlocirolana moraguesi Racovitza, 1905	Xe												

(Continued)

CRUSTACEAN BIODIVERSITY AND CONSERVATION

Islands ('species' name=hypogean; Xe=endemic; Xi=alien,introduced)	Balearic islands	Corsica	Sardinia	Tuscan Archipelago	Sicily	Pelagie islands	Maltese islands	Adriatic islands	Ionian Islands	Crete island	Aegean Islands	Marmara islands	Cyprus
*Typhlocirolana sp. aff. moraguesi Racovitza, 1905					Xe								
*Turcolana rhodica Botosaneanu, Boutin & Henry, 1985											Xe		
Family: Janiridae G.O. Sars, 1897													
Jaera (Jaera) nordmanni brevicaudata Jaume & Garcia, 1990	Xe												
Jaera (Jaera) nordmanni balearica Jaume & Garcia, 1990	Xe												
Jaera (Jaera) cf. nordmanni (Rathke, 1837)				X									
Family: Lepidocharontidae Galassi & Bruce, 2016													
*Microcharon agripensis Galassi, De Laurentiis & Pesce, 1995											X		

(Continued)

Islands ('species' name=hypogean; Xe=endemic; Xi=alien,introduced)	Balearic islands	Corsica	Sardinia	Tuscan Archipelago	Sicily	Pelagie islands	Maltese islands	Adriatic islands	Ionian Islands	Crete island	Aegean Islands	Marmara islands	Cyprus
*Microcharon comasi Coineau, 1968	Xe												
*Microcharon hellenae Chappuis & Delamare Deboutteville, 1954											Xe		
*Microcharon latus Karaman, 1934									X				
*Microcharon luciae Sket, 1991													Xe
*Microcharon marinus Chappuis & Delamare-Deboutteville, 1954	X		X	X									
*Microcharon nuragicus Pesce & Galassi, 1988			Xe										
*Microcharon prespensis Karaman, 1954											X		
*Microcharon silverii Pesce & Galassi, 1988			Xe										
*Microcharon sisyphus Chappuis & Delamare-Deboutteville, 1954		Xe											

(Continued)

Islands (*species' name=hypogean; Xe=endemic; Xi=alien,introduced)	Balearic islands	Corsica	Sardinia	Tuscan Archipelago	Sicily	Pelagie islands	Maltese islands	Adriatic islands	Ionian Islands	Crete island	Aegean Islands	Marmara islands	Cyprus
*Microcharon ullae Pesce, 1981											X		
*Microcharon sp. Karaman, 1934								X					
Family: Microparasellidae Karaman, 1933													
*Microparasellus hellenicus Argano & Pesce, 1979									X				
*Microparasellus puteanus Karaman, 1933									X				
Order: Thermosbaenacea Monod, 1927													
Family: Monodellidae Taramelli, 1954													
*Tethysbaena argentarii Stella, 1951				Xe									
*Tethysbaena halophila (S.L. Karaman, 1953)								X					
*Tethysbaena scabra (Pretus, 1991)	Xe												

(Continued)

Islands ('species' name=hypogean; Xe=endemic; Xi=alien,introduced)	Balearic islands	Corsica	Sardinia	Tuscan Archipelago	Sicily	Pelagie islands	Maltese islands	Adriatic islands	Ionian Islands	Crete island	Aegean Islands	Marmara islands	Cyprus
*Tethysbaena siracusae Wagner, 1994					Xe								
Order: Bathynellacea Chappuis, 1915													
Family: Bathynellidae Grobben, 1905													
*Hispanobathynella aff. catalanensis (Serban, Coineau & Delamare, 1971)			Xe										
*Sardobathynella cottarellii Serban, 1973			Xe										
Family: Parabathynellidae Noodt, 1965													
*Hexabathynella knoepffleri (Coineau, 1964)		X	X										
*Paraiberobathynella fagei (Delamare Debouteville & Angelier, 1950)	X												
Order: Decapoda Latreille, 1802													

(Continued)

Islands ('species' name=hypogean; Xe=endemic; Xi=alien,introduced)	Balearic islands	Corsica	Sardinia	Tuscan Archipelago	Sicily	Pelagie islands	Maltese islands	Adriatic islands	Ionian Islands	Crete island	Aegean Islands	Marmara islands	Cyprus
Family: Astacidae Latreille, 1802													
Astacus (Pontastacus) leptodactylus (Eschscholtz, 1823)							x						
Austropotamobius fulcisianus (Ninni, 1886)			Xi										
Austropotamobius italicus carsicus (Karaman, 1962)								x					
Austropotamobius pallipes (Lereboullet, 1858)		Xi											
Pacifastacus leniusculus (Dana, 1852)							Xi						Xi
Family: Atyidae De Haan, 1849													
Atyaephyra desmarestii (Millet, 1831)		x	x		x								
Atyaephyra thyamisensis Christodoulou, Antoniou, Magoulas & Koukouras, 2012									x				

(Continued)

Islands (*species' name=hypogean; Xe=endemic; Xi=alien,introduced)	Balearic islands	Corsica	Sardinia	Tuscan Archipelago	Sicily	Pelagie islands	Maltese islands	Adriatic islands	Ionian Islands	Crete island	Aegean Islands	Marmara islands	Cyprus
Atyopsis moluccensis (De Haan, 1849)							Xi						
Troglocaris anophthalmus periadriaticus Jugovic, Jalžić, Prevorčnik & Sket, 2012								Xe					
Family: Cambaridae Hobbs, 1942													
Procambarus clarkii (Girard, 1852)	Xi	Xi	Xi		Xi		Xi						Xi
Procambarus virginalis Lyko, 2017							Xi						
Family: Palaemonidae Rafinesque, 1815													
Palaemon antennarius Milne Edwards, 1837			X		X				X	X	X		
Palaemon colossus Tzomos & Koukouras, 2015											X		
Palaemon minos Tzomos & Koukouras, 2015										Xe			

(Continued)

CRUSTACEAN BIODIVERSITY AND CONSERVATION

Islands (*species' name=hypogean; Xe=endemic; Xi=alien,introduced)	Balearic islands	Corsica	Sardinia	Tuscan Archipelago	Sicily	Pelagie islands	Maltese islands	Adriatic islands	Ionian Islands	Crete island	Aegean Islands	Marmara islands	Cyprus
Family: Parastacidae Huxley, 1879													
Cherax destructor Clark, 1936					Xi								
Cherax quadricarinatus von Martens, 1868							Xi						
Family: Potamidae Ortmann, 1896													
Potamon fluviatile (Herbst, 1785)					X		X		X		X		
Potamon hippocratis Ghigi, 1929										X			X
Potamon ibericum (Bieberstein, 1808)											X		
Potamon karpathos Giavarini, 1934											X		
Potamon kretaion Giavarini, 1934										Xe			
Potamon potamios (Olivier, 1804)											X		
Potamon rhodium Parisi, 1913											X		
TOTAL	32	20	35	16	21	2	12	29	20	19	27	2	9

REFERENCES

AA.VV. 2015. *The 2nd Water Catchment Management Plan for the Malta Water Catchment District 2015–2021.* Italy: Sustainable Energy and Water Conservation Unit, Environment and Resources Authority.

Advokaat, E.L. van Hinsbergen D.J.J. Maffione M. et al. 2014. Eocene rotation of Sardinia, and the paleogeography of the western Mediterranean region. *Earth and Planetary Science Letters* 401:183–95.

Akyol, O., Ceyhan, T. and Ertosluk, O. 2009. Marmara Adasi Kiyi Balikciligi ve Balikcilik Kaynaklari. *Ege Journal of Fisheries and Aquatic Sciences* 26:143–48.

Alvarez, W. 1972. Rotation of the Corsica-Sardinia microplate. *Nature Physics Science* 235:103–5.

Amouret, J., Bertocchi, S., Brusconi, S., et al. 2015. The first record of translocated white-clawed crayfish from the *Austropotamobius pallipes* complex in Sardinia (Italy). *Journal of Limnology* 74:491–500.

Antonioli, F. and Vai G.B. 2004. *Litho-Paleoenvironmental Maps of Italy during the Last Two Climatic Extremes. Climex Maps Italy, ENEA.* Bologna: Museo geologico Giovanni Capellini.

Arcangeli, A. 1942. *Il genere Asellus in Italia con speciale riguardo alla diffusione del genere Proasellus.* Italy: Bollettino del Museo di Zoologia e Anatomia Comparata della Reale Università di Torino.

Argano, R. and Campanaro, A. 2004. Two new *Proasellus* (Crustacea, Isopoda, Asellidae) species from Sardinia: Evidences of an old colonization wave of the island. *Studi Trentini di Scienze Naturali, Acta Biologica* 81:49–52.

Argano, R. and Pesce, G.L. 1979. Microparasellids from phreatic waters of Greece (Isopoda, Asellota). *Crustaceana* 37:173–83.

Argano, R. and Pesce, G.L. 1980. A cirolanid from subterranean waters of Turkey (Crustacea, Isopoda, Flabellifera). *Revue Suisse de Zoologie* 87:439–44.

Arias, C., Azzaraoli, A., Bigazzi, G. and Bonadonna, F.P. 1980. Magnetostratigraphy and Pliocene-Pleistocene boundary in Italy. *Quarternary Research* 13:65–74.

Arrighi, M.E., Khoumeri, B., Ottavi, V., et al. 2005. Report on Corsica. In *Water on Mediterranean Islands: Current Conditions and Prospects for Sustainable Management,* eds. A.A. Donta, M.A. Lange, and Herrmann, A., 97–226. Münster: Centre for Environment Research, University of Muenster.

Arrignon, J. 1996. *Il gambero d'acqua dolce e il suo allevamento.* Bologna: Edagricole.

Arrignon, J.C.V., Martini, V. and Mattei, J. 1999. *Austropotamobius pallipes pallipes* (Lereb.) in Corsica. *Freshwater Crayfish* 12:811–16.

Ayfer, B., Balkis, H., and Mulayim, A. 2017. Decapod crustaceans in the Marmara Island (Marmara Sea) and ecological characteristics of their habitats. *European Journal of Biology* 76:20–5.

Baccetti, B. 1983. Biogeografia sarda venti anni dopo. *Biogeographia* 8:859–70.

Baiocchi, A., Lotti, F. and Piscopo, V. 2016. Occurrence and flow of groundwater in crystalline rocks of Sardinia and Calabria (Italy): an overview of current knowledge. *Italian Journal of Groundwater* AS16-195:7–13.

Baldacchino, A.E. 1983. A preliminary list of freshwater crustaceans from the Maltese islands. *The Central Mediterranean Naturalist* 1:49–50.

Balian, E.V., Lévêque, C., Segers, H., and Martens, K. (eds.) 2008. *Freshwater Animal Diversity Assessment* (Vol. 198). Dordrecht: Springer (eBook), https://www.springer.com/gp/book/9781402082580.

Banha, F. and Anastácio, P.M. 2014. Desiccation survival capacities of two invasive crayfish species. *Knowledge and Management of Aquatic Ecosystems* 413:01.

Baratti, M., Yacoubi Khebiza, M. and Messana, G. 2004. Microevolutionary processes in the stygobitic genus *Typhlocirolana* (Isopoda Flabellifera Cirolanidae) as inferred by partial 12S and 16S rDNA sequences. *Journal of Zoological Systematics and Evolutionary Research* 42:27–32.

Barbaresi, S., Gherardi, F., Mengoni, A. and Souty-Grosset C. 2007. Genetics and invasion biology in fresh waters: A pilot study of *Procambarus clarkii* in Europe. In *Biological Invaders in Inland Waters: Profiles, Distribution, and Threats*, ed. F. Gherardi, 381–400. Dordrecht: Springer.

Barbato, D., Benocci, A. and Manganelli, G. 2018. The biogeography of non-marine molluscs in the Tuscan Archipelago reveals combined effects of current eco-geographical drivers and paleogeography. *Organisms Diversity and Evolution* 18:443–45

Barbera, G. and Cullotta, S. 2012. An inventory approach to the assessment of main traditional landscapes in Sicily (Central Mediterranean Basin). *Landscape Research* 37:539–69.

Barnes, R.S.K. 1988. The faunas of land-locked lagoons: Chance differences and the problems of dispersal. *Estuarine, Coastal and Shelf Science* 26:309–18.

Bauzà-Ribot, M.M., Jaume, D., Fornós, J.J., Juan C. and Pons, J. 2011. Islands beneath islands: Phylogeography of a groundwater amphipod crustacean in the Balearic archipelago. *BMC Evolutionary Biology* 11:221.

Bauzà-Ribot, M. M., Juan, C., Nardi, F., Orom, P., Pons J. and Jaume, D. 2012. Mitogenomic phylogenetic analysis supports continental-scale vicariance in subterranean thalassoid crustaceans. *Current Biology* 22:2069–74.

Benoit, G. and Comeau, A. 2012. *A Sustainable Future for the Mediterranean: The Blue Plan's Environment and Development Outlook*. London, UK: Routledge.

Bertocchi, S., Brusconi, S., Gherardi, F. and Chessa, L. 2010. Prima segnalazione del gambero minacciato *Austropotamobius pallipes* complex in Sardegna. In XIII Congresso Nazionale Associazione Italiana Ittiologi Acque Dolci, Sansepolcro, Arezzo, 12–13 Novembre 2010, 52.

Beşiktepe, Ş.T., Sur, H.İ., Özsoy, E., Latif, M.A., Oğuz, T. and Ünlüata, Ü. 1994. The circulation and hydrography of the Marmara Sea. *Progress in Oceanography* 34:285–334.

Bianchi, C. N. and Morri, C. 2000. Marine biodiversity of the Mediterranean Sea: Situation, problems and prospects for future research. *Marine Pollution Bulletin* 40:367–76.

Blondel, J., Aronson, J., Bodiou J.-Y. and Boeuf, G. 2010. *The Mediterranean Region. Biological Diversity in Space and Time*. Oxford: Oxford University Press.

Boccaletti, M., Ciaranfi, N., Casentino, D. et al. 1990. Palinspastic restoration and palaeogeographic reconstruction of the peri-Tyrrhenian area during the Neogene. *Palaeogeography, Palaeoclimatology, Palaeoecology* 77:41–50.

Bonacci, O. 2014. Karst hydrogeology/hydrology of dinaric chain and isles. *Environmental Earth Sciences* 74:37–55.

Bonacci, O. and Roje-Bonacci, T. 2003. Groundwater on small Adriatic karst islands. *RMZ - Materials and Geoenvironment* 50:41–44.

Bossio, A., Cornamusini G., Ferrandini, J., et al. 2000. Dinamica dal Neogene al Quaternario della Corsica orientale e della Toscana. In *Progetto Interreg II Toscana- Corsica 1997–1999*, 87–95. Pisa: Edizioni ETS.

Botosaneanu, L., Boutin, C. and Henry, J.P. 1985. Deux remarquables cirolanides stygobies nouveaux du Maroc et de Rhodes. *Stygologia* 1:186–207.

Bou, C. 1970. Observations on the Ingolfiellidae (Amphipod crustaceans) of Greece, (in French). *Biologia Gallo-Hellenica* 3:57–70.

Boutin, C. 1997. Stygobiologie et géologie historique: l'émersion des terres de Méditerranée orientale datée à partir des amphipodes Metacrangonyctidae actuels (micro-crustacés souterrains) [Stygobiology and historical geology: Emersion of lands in Eastern Mediterranean Basin dated with present Metacrangonyctid Amphipods (subterranean micro-crustaceans.]. *GEOBIOS*, M.S. 21:67–74.

Bover, P., Quintanab, J. and Alcoverc, J.A. 2008. Three islands, three worlds: Paleogeography and evolution of the vertebrate fauna from the Balearic Islands. *Quaternary International* 182:135–44.

Brandis, D., Storch V. and Türkay, M. 2000. Taxonomy and zoogeography of the freshwater crabs of Europe, North Africa, and the Middle East. *Senckenbergiana Biologica* 80:5–56.

Broquet, P. 2016. Sicily in its Mediterranean geological frame. *Boletín Geológico y Minero* 127:547–62.

Brown, R.P., Terrasa, B., Pérez-Mellado, V., Castro, J. A., Hoskisson, P.A., Picornell, A. and Ramon M. M. 2008. Bayesian estimation of post-Messinian divergence times in Balearic Island lizards. *Molecular Phylogenetics and Evolution* 48:350–58.

Bulut, G. 2016. Medicinal and wild food plants of Marmara Island (Balikesir - Turkey). *Acta Societatis Botanicorum Poloniae* 85:3501.

Cabras, S., De Waele, J. and Sanna L. 2008. Caves and karst aquifer drainage of Supramonte (Sardinia, Italy): A review. *Acta Carsologica* 37:227–40.

Camacho, A.I. 2003. Historical biogeography of *Hexabathynella*, a cosmopolitan genus of groundwater Syncarida (Crustacea, Bathynellacea, Parabathynellidae). *Biological Journal of the Linnean Society* 78:457–66.

Camacho, I.A. 2006. An annotated checklist of the Syncarida (Crustacea, Malacostraca) of the world. *Zootaxa* 1374:1–54

Camacho, A.I., Dorda B.A., and Rey I. 2014. Iberian Peninsula and Balearic Island Bathynellacea (Crustacea, Syncarida) database. *ZooKeys* 386:1–20.

Cannizzaro, A.G., Walters, A.D., and Berg, D.J. 2017. A new species of freshwater *Gammarus* Fabricius, 1775 (Amphipoda: Gammaridae) from a desert spring in Texas, with a key to the species of the genus Gammarus from North America. *The Journal of Crustacean Biology* 37:709–72.

Cánovas, F., Jurado-Rivera, J.A., Cerro-Gálvez, E., Juan, C., Jaume, D. and Pons, J. 2016. DNA barcodes, cryptic diversity and phylogeography of a W Mediterranean assemblage of thermosbaenacean crustaceans. *Zoologica Scripta* 45:659–70.

Casale, A., Graffiti, G., Lana, E., Marcia, P., Molinu, A., Mucedda, M., Onnis, C. and Stoch, F. 2008. La Grotta del Bue Marino: cinquanta anni di ricerche biospeleologiche in Sardegna. *Memorie dell'Istituto Italiano di Speleologia*, s. II, 21:197–209.

Cassola, F. 1983. L'esplorazione naturalistica della Sardegna. *Biogeographia* 8:5–34.

Catalano, R., Di Stefano, P., Nigro, F. and Vitale, F.P. 1994. The Sicily mainland thrust belt. Evolution during the Neogene. *Bollettino di Geofisica Teorica e Applicata* 36:141–4.

Cavallaro, F. 1995. Fenomeni carsici nei Monti Iblei (Sicilia sud - orientale). In Atti Convegno Regionale di Speleologia della Sicilia, Ragusa, 14–16 Dicembre 1990, 2:237–55.

Cavazza, W, and Wezel, F.C. 2003. The Mediterranean region - a geological primer. *Episodes* 26:160–68.

Chappuis, A. 1949. Les Asellides d'Europe et pays limitrophes. *Archives de Zoologie expérimentale et générale* 86:78–94.

Charalambous, C.N. 2001. Water management under drought conditions. *Desalination* 138:3–6.

Cherry, J.F. and Leppard T.P. 2018. The Balearic Paradox: Why were the islands colonized so Late? *Pyrenae* 49:49–70.

Christodoulou, M., Antoniou, A., Magoulas, A. and Koukouras, A. 2012. Revision of the freshwater genus *Atyaephyra* (Crustacea, Decapoda, Atyidae) based on morphological and molecular data. *ZooKeys* 229:53–110.

Chueca, L.J., Madeira, M.J. and Goomez-Moliner, B.J. 2015. Biogeography of the land snail genus *Allognathus* (Helicidae): Middle Miocene colonization of the Balearic Islands. *Journal of Biogeography* 42:1845–57.

Clavero, M., Nores, C., Kubersky-Piredda, S. and Centeno-Cuadros, A. 2016. Interdisciplinarity to reconstruct historical introductions: Solving the status of crypto-genic crayfish. *Biological Reviews* 91:1036–49.

Coineau, N. 1994. Evolutionary biogeography of the Microparasellid isopod *Microcharon* (Crustacea) in the Mediterranean Basin. *Hydrobiologia* 287:77–93.

Coineau, N. and Boutin, C. 1992. Biological processes in space and time. Colonization, evolu-tion and speciation in interstitial stygobionts. In *The Natural History of Biospeleology, Madrid Monografias, 7,* ed. A.I. Camacho, 423–451. Madrid: Museo Nacional de Ciencias Naturales.

Colantoni, P., Gallignani, P. and Lenaz, R. 1979. Late Pleistocene and Holocene evolution of the North Adriatic continental shelf (Italy). *Marine Geology* 33:M41–50.

Coll, M., Piroddi, C., Steenbeek, J., et al. 2010. The biodiversity of the Mediterranean Sea: Estimates, patterns, and threats. *PloS One* 5: doi.org/10.1371/journal.pone.0011842 2010.

Combes, S. 2003. Protecting freshwater ecosystems in the face of global climate change. In *Buying Time: A User's Manual for Building Resistance and Resilience to Climate Change in Natural Systems,* ed. L.J., Hansen, et al., 175–214. Washington, DC: World Wildlife Fund.

Correggiari, A., Roveri, M. and Trincardi. F. 1996. Late Pleistocene and Holocene evolution of the north Adriatic Sea. *Il Quaternario Italian Journal of Quaternary Sciences* 9: 697–704.

Cuttelod, A., García, N., Abdul Malak, D., Temple, H. and Katariya, V. 2008. The Mediterranean: A biodiversity hotspot under threat. In *The 2008 Review of the IUCN Red List of Threatened Species,* ed. J.-C. Vié, C. Hilton-Taylor and S. N. Stuart. Gland, Switzerland: The International Union for Conservation of Nature.

Darwall, W., Carrizo, S., Numa, C., Barrios V., Freyhof J. and Smith K. 2014. *Freshwater Key Biodiversity Areas in the Mediterranean Basin Hotspot: Informing Species Conservation and Development Planning in Freshwater Ecosystems.* Cambridge: IUCN.

Darwall, W., Smith, K., Allen, D., Seddon, M.B., Reid, G.M., Clausnitzer, V. and Kalkman, V.J. 2009. Freshwater biodiversity: A hidden resource under threat. In *Wildlife in a Changing World—An Analysis of the 2008 IUCN Red List of Threatened Species,* eds. J.C. Vié, C. Hilton-Taylor, and S.N. Stuart, 43–54. Gland, Switzerland: IUCN.

Davolos, D., De Matthaeis, E., Latella, L. and Vonk, R. 2017. *Cryptorchestia ruffoi* sp. n. from the island of Rhodes (Greece), revealed by morphological and phylogenetic analy-sis (Crustacea, Amphipoda, Talitridae). *ZooKeys* 652:37–54.

Davolos, D., De Matthaeis, E., Latella, L., Tarocco, M., Özbek, M. and Vonk, R. 2018. On the molecular and morphological evolution of continental and insular *Cryptorchestia* species, with an additional description of *C. garbinii* (Talitridae). *ZooKeys* 783:37–54.

Deharveng, L. Stoch, F. Gibert, J., et al. 2009. Groundwater biodiversity in Europe. *Freshwater Biology* 54:709–26.

Deidun, A., Sciberras, A., Formosa, J., Zava, B., Insacco, G., Corsini-Foka, M. and Crandall K.A. 2018. Invasion by non-indigenous freshwater decapods of Malta and Sicily, central Mediterranean Sea. *Journal of Crustacean Biology* 38:748–59.

Delić, T. and Sket, B., 2015. Found after 60 years: The hows and whys of *Sphaeromides virei montenigrina* (Crustacea: Isopoda: Cirolanidae) rediscovery in Obodska pecina, Montenegro/Najden po 60 letih: razlogi za ponovno odkritje Sphaeromides virei montenigrina (Crustacea: Isopoda: Cirolanidae) v Obodski pecini v Crni gori. *Natura Sloveniae* 17(2):1–59.

Delicado, D., Machordom, A. and Ramos, M.A. 2014. Vicariant versus dispersal processes in the settlement of *Pseudamnicola* (Caenogastropoda, Hydrobiidae) in the Mediterranean Balearic Islands. *Zoological Journal of the Linnean Society* 171:38–71.

Dermitzakis, M.D. 1990. Paleogeography, geodynamic process and event stratigraphy during the late Cenozoic of the Aegean area. *Atti. Convegni Lincei* 85:263–88.

Dermitzakis, M.D. and Papanikolaou, D.J. 1981. Paleogeography and geodynamics of the Aegean region during the Neogene. *Annales Geologiques des pays Helleniques* 30:245–89.

Dermitzakis, M.D. and De Vos, J. 1987. Faunal succession and the evolution of mammals in Crete during the Pleistocene. *Neues Jahrbuch für Geologie und Paläontologie. Abhandlungen* 173:377–408.

Devoti, R. Ferraro, C., Gueguen, E., et al. 2002. Geodetic control on recent tectonic movements in the central Mediterranean area. *Tectonophysics* 346:151–67.

Diamantopoulou, P. and Voudouris, K. 2008. Optimization of water resources management using SWOT analysis: The case of Zakynthos Island, Ionian Sea, Greece. *Environmental Geology* 54:197–211.

Dudgeon, D., Arthington, A.H., Gessner, M.O., et al. 2006. Freshwater biodiversity: Importance, threats, status and conservation challenges. *Biological Reviews* 81:163–82.

Duplančić Leder, T., Ujević, T. and Čala, M. 2004. Coastline lengths and areas of islands in the Croatian part of the Adriatic Sea determined from the topographic maps at the scale of 1: 25 000. *Geoadria* 9:28.

Edel, J.B., Dubois, D., Marchant, R., Hernandez, J. and Cosca, M. 2001. La rotation miocène inférieur du bloc corso-sarde. Nouvelles contraintes paléomagnetiques sur la fin du mouvement. *Bulletin Societé Géologique France* 172:275–83.

Evelpidou, N. 2012. Modelling of erosional processes in the Ionian Islands (Greece). *Geomatics, Natural Hazards and Risk* 3(4): 293–310.

Faraone, F.P., Giacalone, G., Canale, D.E. et al. 2017. Tracking the invasion of the red swamp crayfish *Procambarus clarkii* (Girard, 1852) (Decapoda Cambaridae) in Sicily: A "citizen science" approach. *Biogeographia (The Journal of Integrative Biogeography)* 32:25–29.

Fassoulas, C. 2017. The geological setting of Crete: An overview. In *Minoan Earthquakes: Breaking the Myth through Interdisciplinarity*, eds. S. Jusseret, and M. Sintubin. Leuven: Leuven University Press.

Fassoulas, C., Paragamian, K. and Iliopoulos, G. 2007. Identification and assessment of Cretan geotopes. *Bulletin of the Geological Society of Greece* 40:1780–95.

Fattorini, S., 2002. Biogeography of the tenebrionid beetles (Coleoptera, Tenebrionidae) on the Aegean Islands (Greece). *Journal of Biogeography* 29:49–67.

Ferentinos, G., Gkioni, M., Geraga, M. and Papatheodorou, G. 2012. Early seafaring activity in the southern Ionian Islands, Mediterranean Sea. *Journal of Archaeological Science* 39:2167–76.

Figueroa, J.M.T., López-Rodríguez, M.J., Fenoglio, S., Sánchez-Castillo, P. and Fochetti, R. 2013. Freshwater biodiversity in the rivers of the Mediterranean Basin. *Hydrobiologia* 719:137–86.

Fišer, C., Robinson, C. T. and Malard, F. 2018. Cryptic species as a window into the paradigm shift of the species concept. *Molecular Ecology* 27:613–35.

Fišer, C., Sket, B., and Trontelj, P. 2008. A phylogenetic perspective on 160 years of troubled taxonomy of *Niphargus* (Crustacea: Amphipoda). *Zoologica Scripta* 37:665–80.

Fišer, C., Trontelj, P. and Sket, B. 2006. Phylogenetic analysis of the *Niphargus orcinus* species-aggregate (Crustacea: Amphipoda: Niphargidae) with description of new taxa. *Journal of Natural History* 40:2265–315.

Foglini, F., Prampolini, M., Micallef, A., Angeletti, L., Vandelli, V., Deidun, A., Soldati, M. and Taviani, M. 2015. Late Quaternary coastal landscape morphology and evolution of the Maltese Islands (Mediterranean Sea) reconstructed from high-resolution sea-floor data. In *Geology and Archaeology: Submerged Landscapes of the Continental Shelf*, eds. J. Harff, G. Bailey, and F. Lüth, special publication 411. London: Geological Society, http://doi.org/10.1144/SP411.12.

Fontbote, J.M., Guimera, J., Roca, E., Sabat, F., Santanach, P. and Fernandez-Ortigosa, F. 1990. The Cenozoic geodynamic evolution of the Valencia Trough (western Mediterranean). *Revista de la Sociedad Geologica de Espana* 3:249–59.

Francalanci, L., Vougioukalakis, G.E., Perini, G. and Manetti, P. 2005. A west-east traverse along the magmatism of the South Aegean Volcanic Arc in the light of volcanological, chemical and isotope data. In *The South Aegean Active Volcanic Arc: Present Knowledge and Future Perspectives*, eds. M. Fytikas, and G.E. Vougioukalakis, 65–111. Amsterdam: Elsevier.

Froglia, C. 2006. Crustacea Malacostrca Decapoda. In: *Checklist and Distribution of the Italian Fauna. 10,000 Terrestrial and Inland Water Species*, Memorie del Museo Civico di Storia Naturale di Verona, 2. Serie, Scienze della Vita 17, eds. S. Ruffo, and F. Stoch, 113–4. Verona : Museo Civico di Storia Naturale.

Furon, M. 1950. Les grandes lignes de la Paléogéographie del la Méditerranée (Tertiaire et Quaternaire). *Vie et Milieu* 1:131–62.

Galassi, D.M.P., De Laurentiis, P. and Pesce, G.L. 1994. Some remarks on the genus *Microcharon* Karaman in Greece, and description of *M. agripensis* n. sp. (Crustacea, Isopoda, Microparasellidae). *International Journal of Speleology* 23:133–55.

Galletti, I. 2002. L'attività biospeleologica del CIRS in Sicilia e all'estero. *Speleologia Iblea* 10:155–64.

Gass, I.A. 1980. The Troodos massif: Its role in the unravelling of the ophiolite problem and its significance in the understanding of constructive plate margin processes. In *Proceedings International Ophiolite Symposium in 1979*, ed. A. Panayiotou, 23–35. Nicosia: Geological Survey Department, Cyprus.

Geiger, W., Alcorlo, P., Baltanas, A. and Montes, C. 2005. Impact of an introduced Crustacean on the trophic webs of Mediterranean wetlands. *Biological Invasions* 7:49–73.

Gikas, P. and Angelakis, A. N. 2009. Water resources management in Crete and in the Aegean Islands, with emphasis on the utilization of non-conventional water sources. *Desalination* 248:1049–64.

Gómez-Campo, C., Bermudez-de-Castro, L., Cagiga, M. J. and Sánchez-Yélamo, M.D. 1984. Endemism in the Iberian Peninsula and Balearic Islands. *Webbia* 38:709–14.

Gottstein, S., Ivković, M., Ternjej, I., Jalžić, B. and Kerovec, M. 2007. Environmental features and crustacean community of anchihaline hypogean waters on the Kornati islands, Croatia. *Marine Ecology* 28:24–30.

Görür, N., Cagatay, M.N., Sakinc, M., Sümengen, M., Sentürk, K., Yaltirak, C. and Tchapalyga, A. 1997. Origin of the sea of marmara as deduced from neogene to quaternary paleogeographic evolution of its frame. *International Geology Review* 39:342–52.

Grabowski, M., Mamos, T., Bącela-Spychalska, K., Rewicz, T. and Wattier, R. A. 2017b. Neogene paleogeography provides context for understanding the origin and spatial distribution of cryptic diversity in a widespread Balkan freshwater amphipod. *PeerJ* 5: e3016.

Grabowski, M., Wysocka A., and Mamos, T. 2017a. Molecular species delimitation methods provide new insight into taxonomy of the endemic gammarid species flock from the ancient Lake Ohrid. *Zoological Journal of the Linnean Society* 181:272–85.

Grafitti, G. 2001. Osservazioni sulla fauna cavernicola della Sardegna. In *Atti Convegno "Biospeleologia dei sistemi carsici della Sardegna", Cagliari, 10 giugno 2000*, eds. G. Piras, and F. Randaccio, 13–33.

Grube, A.E. 1861. *Ein Ausflug nach Triest und dem Quarnero. Beitrage zur Kenntniss der Thierwelt dieses Gebietes.* Berlin: Nicolaische Verlagsbuchhandlung.

Guglielmo, M. and Marra, A.C. 2011. Le due Sicilie dle Pleistocene Medio: osservazioni paleobiogeografiche. *Biogeographia* 30:11–25.

Guil N. and Camacho, A.I. 2001. Historical biogeography of *Iberobathynella* (Crustacea, Syncarida, Bathynellacea), an aquatic subterranean genus of Parabathynellids, endemic to the Iberian Peninsula. *Global Ecology and Biogeography* 10:487–501.

Güner, U. 2009. Distribution of freshwater crab (*Potamon* sp.) in Turkish Thrace. *Trakya Universitesi Fen Bilimleri Dergisi* 10:69–74.

Hadjisterkotis, E. 2012. The arrival of elephants on the island of Cyprus and their subsequent accumulation in fossil sites. In *Elephants: Ecology, Behavior and Conservation*, ed. M. Aranovich, and O. Dufresne, 49–75. Hauppauge, NY: Nova Science Publishers, Inc.

Harland, W.B., Cox, A.V., Llewellyn, P.G., Pickton, C.A., Smith, A.G. and Walters, R. 1982. *A Geological Time Scale.* Cambridge: Cambridge University Press.

Heller, J., 1976. The biogeography of enid landsnails on the Aegean Islands. *Journal of Biogeography* 3:281–92.

Henry, J.P. 1975. Données sur les Asellides de Grèce et description de *Proasellus sketi* n. sp. (Crustacea, Isopoda, Asellota). *Biologia Gallo-Hellenica* 6:139–44.

Hentonnen, P. and Huner, J.V. 1999. The introduction of alien species of crayfish in Europe. In *Crayfish in Europe as Alien Species. How to Make the Best of a Bad Situation?* eds. F. Gherardi, and D.M. Holdich, 13–22. Brookfiled, Rotterdam: AA Balkema.

Ho, S.Y. and Duchêne, S. 2014. Molecular clock methods for estimating evolutionary rates and timescales. *Molecular Ecology* 23:5947–65.

Hobbs, Jr. H.H., Jass, J.P. and Huner, J.V. 1989. A review of global crayfish introductions with particular emphasis on two North American species (Decapoda: Cambaridae). *Crustaceana* 56:299–316.

Holdich, D.M., Ackefors, H., Gherardi, F., Rogers, W.D. and Skurdal, J. 1999. Native and alien crayfish in Europe, some conclusions. In *Crayfish in Europe as Alien Species. How to Make the Best of a Bad Situation?* eds. F. Gherardi, and D.M. Holdich, 281–91. Brookfield, Rotterdam: AA Balkema.

Hopkins, L. 2002. *IUCN and the Mediterranean Islands: Opportunities for Biodiversity Conservation and Sustainable Use.* 63. Gland, Switzerland: International Union for Conservation of Nature.

Hou, Z. and Sket, B. 2016. A review of Gammaridae (Crustacea: Amphipoda): the family extent, its evolutionary history, and taxonomic redefinition of genera. *Zoological Journal of the Linnean Society* 176:323–48.

Hou, Z., Sket, B., Fišer, C., and Li, S. 2011. Eocene habitat shift from saline to freshwater promoted Tethyan amphipod diversification. *Proceedings of the National Academy of Sciences* 108:14533–38.

Hou, Z., Sket, B. and Li, S. 2013. Phylogenetic analyses of Gammaridae crustacean reveal different diversification patterns among sister lineages in the Tethyan region. *Cladistics* 30:352–65.

Hou, Z., Zhao, S. and Li, S. 2018. Seven new freshwater species of *Gammarus* from southern China (Crustacea, Amphipoda, Gammaridae). *ZooKeys* 749:1–79.

Hsü, K.J. 1983. *The Meditterranean Was a Desert. A Voyage of the Glomar Challenger.* Princeton: Princeton University Press.

Hsü, K.J., Montadert, L., Bernoulli, D., et al. 1977. History of the Mediterranean salinity crisis. *Nature* 267:399–403.

Hupało, K., Rewicz T., Mamos, T., Boulaaba, S. and Grabowski, M. 2017. Diversity and origin of freshwater gammarids from Sicily (Italy): Preliminary results. *Biodiversity Journal* 8:515–16.

Hupało, K., Mamos, T., Wrzesińska, W. and Grabowski, M. 2018a. First endemic freshwater *Gammarus* from Crete and its evolutionary history—An integrative taxonomy approach. *PeerJ* 6: e4457.

Hupało, K., Stoch, F., Mamos, T., Rewicz T., Boulaaba S., Flot, J.F. and Grabowski, M. 2018b. The more the merrier—The extraordinary cryptic diversity in freshwater *Echinogammarus* (Gammaridae, Amphipoda) from Sicily. In *3rd Central European Symposium for Aquatic Macroinvertebrate Research, 8-13 July 2018*, 42. Łódź, Poland: Book of Abstracts.

Jaume, D. 2008. Global diversity of spelaeogriphaceans and thermosbaenaceans (Crustacea; Spelaeogriphacea and Thermosbaenacea) in freshwater. *Hydrobiologia* 595:219–24.

Jaume, D. and Garcia, L. 1988. Revision de la especies politipica Jaera nordmanni (Rathke, 1837) (Isopoda, Asellota, Janiridae) de les aguas dulces de Mallorca. *Miscellanea Zoologica (Barcelona)* 12:79–88.

Jaume, D. and Garcia, L. 1992. A new *Metacirolana* (Crustacea: Isopoda: Cirolanidae) from anchihialine cave lake on Cabrera (Balearic Islands). *Stygologia* 7:179–86.

Jaume, D. and Queinnec, E. 2007. A new species of freshwater isopod (Sphaeromatidea: Sphaeromatidae) from an inland karstic stream on Espíritu Santo Island, Vanuatu, southwestern Pacific. *Zootaxa* 1653:41–55.

Jaume, D., Garcia, Lado, F. and Boxshall, G. 2007. New genera of Bogidiellidae (Amphipoda: Gammaridea) from SW Pacific and Mediterranean marine caves. *Journal of Natural History* 41:419–44.

Jesse, R., Schubart, C.D. and Klaus, S. 2010. Identification of a cryptic lineage within *Potamon fluviatile* (Crustacea: Brachyura: Potamidae). *Invertebrate Systematics* 24:348–56.

Jesse, R., Grudinski, M., Klaus, S., Streit, B. and Pfenninger, M. 2011. Evolution of freshwater crab diversity in the Aegean region (Crustacea: Brachyura: Potamidae). *Molecular Phylogenetics and Evolution* 59:23–33.

Jugovic, J., Jalžić, B., Prevorčnik, S. and Sket, B. 2012. Cave shrimps *Troglocaris* s. str. (Dormitzer, 1853), taxonomic revision and description of new taxa after phylogenetic and morphometric studies. *Zootaxa* 3421:1–31.

Jurado-Rivera, J.A., Álvarez, G., Caro, J.A., Juan, C., Pons, J. and Jaume, D. 2017. Molecular systematics of *Haploginglymus*, a genus of subterranean amphipods endemic to the Iberian Peninsula (Amphipoda: Niphargidae). *Contributions to Zoology* 86:239–60.

Karaman, G.S. 1969. 27. Beitrag zur Kenntnis der Amphipoden- Arten der Genera *Echinogammarus* Stebb. und Chaetogammarus Mart. an der Jugoslawischer Adriaküste. *Glasnik Republičkog Zavoda Za Zaštitu Prirode i Prirodnjačke Zbirke* 2:59–84.

Karaman, G.S. 1974. 58Contribution to the knowledge of the Apmhipoda. Genus *Synurella* Wrzes. In Yugoslavia with remarks on its all World known species, their synonomy, bibliography and distribution (fam. Gammaridae). *Poljoprivreda i šumarstvo* 20:49–60.

Karaman, G.S. 1975. 56Contribution to the knowledge of the Amphipoda. Several new and interesting *Gammarus* species from Asia Minor (fam. Gammaridae). *Bollettino del Museo civico di storia naturale di Verona* 1:311–43.

Karaman, G.S. 1977. Contribution to the Knowledge of the Amphipoda 84. One interesting member of the genus *Echinogammarus* Stebb. from Malta island, E. ebusitanus (Marg. 1951) (fam. Gammaridae). *Poljoprivreda i Šumarstvo* 23:29–38.

Karaman, G.S. 1985. New data on genus *Niphargus* (fam. Niphargidae) in Italy and adjacent regions. *Bollettino del Museo civico di Storia naturale di Verona* 12:209–28.

Karaman, G.S. 1989. 190. Contribution to the knowledge of Amphipoda. *Bogidiella cypria*, new species of the family Bogidiellidae from Cyprus Island in the Mediterranean Sea. *The Montenegrin Academy of Sciences and Arts, Glasnik of the Section of Natural Sciences* 7:7–23.

Karaman, G. 2015. New data of genus *Niphargus* Schiödte, 1849 (Fam. Niphargidae) from Greece (Contribution to the knowledge of the Amphipoda 284). *Agriculture and Forestry* 61:43–60.

Karaman, G. 2016a. On two new or interesting species of the family Niphargidae from Greece and Croatia. (Contribution to the knowledge of the Amphipoda 286). *Ecologica Montenegrina* 5:1–17.

Karaman, G. 2016b. Two new genera of the family Niphargidae from Greece (Contribution to the Knowledge of the Amphipoda 287). *Agriculture and Forestry* 62:7–27.

Karaman, G.S. 2017a. A new member of the genus *Niphargus* Schiödte, 1849 (Amphioda Gammaridea, fam. Niphargidae) from Crete Island, Greece (contribution to the knowledge of the Amphipoda 293). *Ecologica Montenegrina* 10:1–10.

Karaman, G.S. 2017b. New subterranean species of the family Niphargidae from Greece, *Niphargus denarius*, sp. n. (contribution to the knowledge of the Amphipoda 295). *Agriculture and Forestry* 63:337–56.

Karaman, G.S. 2017c. *Niphargus cymbalus*, new species and *N. jovanovici* S. Kar. 1931 in Greece (contribution to the knowledge of the Amphipoda 298). *Agriculture and Forestry* 63:263–79.

Karaman, G.S. 2017d. On the endemic subterranean amphipod *Niphargus versluysi* S. Karaman, 1950 (Fam. Niphargidae) in Greece (Contribution to the Knowledge of the Amphipoda 297). *Biologia Serbica* 39:52–67.

Karaman, G.S. 2018. Further discoof new or partially known taxa of the genus *Niphargus* Schiödte, 1849 (Fam. Niphargidae) in Greece (contribution to the knowledge of the Amphipoda 302). *Agriculture and Forestry* 64:5–31.

Karaman, G. and Pinkster, S. 1977. Freshwater *Gammarus* species from Europe, North Africa and adjacent regions of Asia (Crustacea-Amphipoda). Part I. *Gammarus pulex*-group and related species. *Bijdragen tot de Dierkunde* 47:1–97.

Karaman, G.S. and Sket, B. 1989. *Niphargus* species (Crustacea: Amphipoda) of the Kvarner-Velebit Islands (NW Adriatic, Yugoslavia). *Biološki vestnik* 37:19–36.

Karaman, S.L. 1931. Beitrag zur Kenntnis der Süsswasseramphiopden. *Bulletin de la Société Scientifique de Skoplje* IX:93–107.

Katalinic, A., Rubinic, J. and Buselic, G. 2008. Hydrology of two coastal karst cryptodepressions in Croatia: Vrana lake vs Vrana lake. In Proceedings of Taal 2007: The 12th World Lake Conference, 732:743.

Kelly, D.W., Dick, J.T. and Montgomery, W.I. 2002. The functional role of *Gammarus* (Crustacea, Amphipoda): shredders, predators, or both? *Hydrobiologia* 485:199–203.

Kernan, M., Battarbee, R.W. and Moss, B.R. (eds.). 2011. *Climate Change Impacts on Freshwater Ecosystems.* West Sussex: Wiley.

Ketmaier, V. and Caccone, A. 2013. Chapter 4. Twenty years of molecular biogeography in the west Mediterranean islands of Corsica and Sardinia: Lessons learnt and future prospects. In *Current Progress in Biological Research*, ed. M. Silva-Opps, 71–93. London: IntechOpen.

Ketmaier, V., Argano, R., Cobolli, M., De Matthaeis, E. and Messana, G. 2001. A systematic and biogeographical study of hypogean and epigean populations of the *Proasellus* species group from Sardinia, Central Italy and Jordan: Allozyme insights. *Journal Zoological Systematic Evolutionary Research* 39:53–61.

Ketmaier, V., Argano, R. and Caccone, G. 2003. Phylogeography and molecular rates of subterranenan aquatic Stenasellid Isopod within a peri-Tyrrhenian distribution. *Molecular Ecology* 12:547–55.

Krapp, T., Grasso, R. and Ruffo, S. 2010. New data on the genus *Jassa* Leach (Amphipoda, Ischyroceridae). *Zoologica Baetica* 21:85–100.

Krijgsman, W. 2002. The Mediterranean: Mare nostrum of earth sciences. *Earth and Planetary Science Letters* 205:1–12.

Krijgsman, W., Hilgen, F.J., Raffi, I., Sierro, F.J. and Wilson, D.S. 1999. Chronology, causes and progression of the Messinian salinity crisis. *Nature* 400:652.

Krijgsman, W., Blanc-Valleron, M.M., Flecker, R., Hilgen, F.J., Kouwenhoven, T.J., Orszag-Sperber, F. and Rouchy, J.M. 2002. The onset of the Messinian salinity crisis in the eastern Mediterranean (Pissouri basin, Cyprus). *Earth and Planet Science Letters* 194:299–310.

La Greca, M. 1957. Considerazioni sull'origine della fauna siciliana. *Italian Journal of Zoology* 24:593–631.

Lanza, B. 1984. Sul significato biogeografico delle isole fossili con particolare riferimento all'arcipelago pliocenico della Toscana. *Atti Società Italiana Scienze Naturali, Museo Civico Storia Naturale Milano* 125:145–58.

Lewinsohn, C. and Holthuis, L.B. 1986. The Crustacea Decapoda of Cyprus. *Zoologische Verhandelingen, Leiden* 230:1–64.

Lewis S.D. 2002 *Pacifastacus*. In *Biology of Freshwater Crayfish*, ed. D.M. Holdich, 511–540. Oxford: Blackwell Scientific.

Lončarić, R., Magaš D. and Surić M. 2011. The influence of water availability on the historical, demographic and economic development of the Kvarner Islands (Croatia). *Annales, Series Historica et Sociologica* 21:1–12.

Lucena-Moya P., Abraín R., Pardo I., Hermida B. and Domínguez M. 2010. Invertebrate species list of coastal lagoons in the Balearic Islands. *Transitional Waters Bulletin* 4 1:1–11.

Lykousis, V. 2009. Sea-level changes and shelf break prograding sequences during the last 400 ka in the Aegean margins: Subsidence rates and palaeogeographic implications. *Continental Shelf Research* 29:2037–44.

MacNeil, C., Dick, J.T. and Elwood, R.W. 1997. The trophic ecology of freshwater *Gammarus* spp. (Crustacea: Amphipoda): problems and perspectives concerning the functional feeding group concept. *Biological Reviews of the Cambridge Philosophical Society* 72:349–64.

Madonia, P. and La Manna, M. 1986. Fenomeni carsici ipogei nelle evaporiti in Sicilia. *Le Grotte d'Italia* 13:163–89.

Maguire, I., Klobucar, G.I.V., Matocec, S.G. and Erben, R. 2003. Distribution of *Austropotamobius pallipes* (Lereboullet) in Croatia and notes on its morphology. *Bulletin Francais De La Peche Et De La Pisciculture* (Abstract for poster, https://www.bib.irb.hr/135544?rad=135544).

Malinverno, A. and Ryan, W.B.F. 1986. Extension in the Tyrrhenian Sea and shortening in the Apennines as result of arc migration driven by sinking of the lithosphere. *Tectonics* 5:227–45.

Mamos, T. 2015. Phylogeography and cryptic diversity of *Gammarus balcanicus* Schäferna, 1922 in Europe. PhD thesis, Faculty of Biology and Environmental Protection, University of Lodz.

Mamos, T., Wattier, R., Majda, A., Sket, B., and Grabowski, M. 2014. Morphological vs. molecular delineation of taxa across montane regions in Europe: The case study of *Gammarus balcanicus* Schäferna, (Crustacea: Amphipoda). *Journal of Zoological Systematics and Evolutionary Research* 52:237–48.

Mamos, T., Wattier, R., Burzyński, A., and Grabowski, M. 2016. The legacy of a vanished sea: A high level of diversification within a European freshwater amphipod species complex driven by 15 My of Paratethys regression. *Molecular Ecology* 25:795–810.

Margalef, R. 1952. Une *Jaera* dans les eaux douces des BalCares, *Jaera balearica* n. sp. (Isopoda, Asellota). *Hydrobiologia* 4:209–13.

Margalef, R. 1954. Algunos crustáceos de agua dulce y salobre de la Romagna (Colección Zarrgheri). *Bolletino Societa Entomolologica Italiana* 34:146–50.

Marra, A.C. 2009. Pleistocene mammal faunas of Calabria (Southern Italy): Biochronology and palaeobiogeography. *Bollettino della Società Paleontologica Italiana* 48:113–22.

Marrone, F. and Naselli Flores, L. 2015. A review on the animal xenodiversity in Sicilian inland waters (Italy). *Advances in Oceanography and Limnology* 6:2–12.

Marrone, F., Barone, R. and Naselli Flores, L. 2006. Cladocera (Branchiopoda: Anomopoda, Ctenopoda, and Onychopoda) from Sicilian inland aaters: An updated inventory. *Crustaceana* 78:1025–39.

Marroni, M., Molli G., Montanini A. and Tribuzio R. 1998. The association of continental crust rocks with ophiolites in the Northern Apennines (Italy): Implications for the continent-ocean transition in the Western Tethys. *Tectonophysics* 292:43–66.

Massa, B., Sbordoni, V. and Vigna Taglianti. A. 2011. La biogeografia della sicilia: considerazioni conclusive sul XXXVII Congresso della Società Italiana di Biogeografia. *Biogeographia* 30:685–94.

McCallum, J.E. and Robertson, A.H.F. 1990. Pulsed uplift of the Troodos massif-evidence from the Plio-Pleistocene Mesaoria basin. Ophiolites: Crustal Analogues. In Proceedings of the International Symposium 'Troodos 1987', eds. E.M. Moores, J. Malpas, A. Panayiotou, and C. Xenophontos, 217–230. Nicosia: Cyprus Geological Survey Department, Cyprus.

McInerney, C.E., Maurice, L., Robertson, A.L., et al. 2014. The ancient Britons: Groundwater fauna survived extreme climate change over tens of millions of years across NW Europe. *Molecular Ecology* 23:1153–66.

McKinney, F.K. 2007. *The Northern Adriatic Ecosystem: Deep Time in a Shallow Sea.* New York: Columbia University Press.

Medail, F. and Quezel, P. 1997. Hot-spots analysis for conservation of plant biodiversity in the Mediterranean Basin. *Annals of the Missouri Botanical Garden* 84:112–27.

Messana, G. and Ruffo, S. 2001. A new species of *Longigammarus* (Crustacea Amphipoda, Gammaridae) from the Pianosa Island (Tuscany Archipelago). *Italian Journal of Zoology* 68:161–64.

Messouli, M., Messana G. and Yacoubi-Kebiza, M. 2006. Three new species of *Pseudoniphargus* (Amphipoda), from the groundwater of three Mediterranean islands, with notes on the Ps. Adriaticus. *Subterranean Biology* 4:79–101.

Meulenkamp, J.E. 1971. The Neogene in the southern Aegean area. *Opera Botanica* 30:5–12.

Mittermeier, R.A., Robles, Gil P., Hoffmann, M., et al. 2004. *Hotspots Revisited: Earth's Biologically Richest and Most Endangered Ecoregions, 2nd ed*. Mexico City: Univerrsity of Chicago Press.

Moore, P.G. and Schembri P.J. 1986. Notes concerning the semi-terrestrial and freshwater amphipods (Crustacea: Peracarida) of the Maltese Islands. *Animalia* 13:65–75.

Moores, E.M., Robinson, P.T., Malpas, J. and Xenophontos, C. 1984. A model for the origin of the troodos massif, Cyprus, and other mideast ophiolites. *Geology* 12:500–03.

Morpurgo, M., Aquiloni, L., Bertocchi, S., Brusconi, S., Tricarico, E. and Gherardi, F. 2010. Distribuzione dei gamberi d'acqua dolce in Italia. *Studi Trentini di Scienze Naturali* 87:125–32.

Muscarella, C. and Baragona, A. 2017. The endemic fauna of the Sicilian islands. *Biodiversity Journal* 8:249–78.

Myers, N., Mittermeier, R.A., Mittermeier, C.G., Da Fonseca, G.A. and Kent, J. 2000. Biodiversity hotspots for conservation priorities. *Nature* 403:853.

Ntakis A., Anastasiadou C., Zakšek V. and Fišer C. 2015. Phylogeny and biogeography of three new species of *Niphargus* (Crustacea: Amphipoda) from Greece. *Zoologischer Anzeiger* 255:32–46.

Orszag-Sperber, F. 2006. Changing perspectives in the concept of "Lago-Mare" in Mediterranean Late Miocene evolution. *Sedimentary Geology* 188–189:259–77.

Oscoz J., Galicia D. and Miranda R. (eds.) 2011. *Identification Guide of Freshwater Macroinvertebrates of Spain*, 1–153. Dordrecht: Springer, (eBook), https://www.spr inger.com/gp/book/9789400715530.

Özbek, M. 2011. An overview of *Gammarus* species distributed in Turkey, with an updated check-list and additional records. *Zoology in the Middle East* 53:71–8.

Özbek, M. and Ustaoğlu, M.R. 2006. Check-list of malacostraca (crustacea) species of Turkish inland waters. *EU Journal of Fisheries and Aquatic Sciences* 23:229–34.

Özbek, M., Öztürk, H.H. and Özkan, N. 2015. Marmara ve Paşalimanı adaları ile Kapıdağ Yarımadası içsularının Gammaridae (Amphipoda) faunası. *Ege Journal of Fisheries and Aquatic Sciences* 32:213–16.

Ozimec R., Bedek J., Gottstein S., et al. 2009. *Crvena knjiga špiljske faune Hrvatske*, 372. Ministarstvo kulture, Državni zavod za zaštitu prirode, Croatia.

Palmer, M. and Pons, G. X. 1996. Diversity in Western Mediterranean islets: Effects of rat presence on a beetle guild. *Acta Oecologica* 17:297–305.

Palmer M., Pons G.X., Cambefort I. and Alcover J.A. 1999. Historical processes and environ-mental factors as determinants of inter-island differences in endemic faunas: The case of the Balearic Islands. *Journal of Biogeography* 26:813–23.

Perissoratis, C. and Conispoliatis, N. 2003. The impacts of sea-level changes during latest Pleistocene and Holocene times on the morphology of the Ionian and Aegean seas (SE Alpine Europe). *Marine Geology* 196:145–56.

Pesce, G.L. 1981. *Microcharon ullae* n. sp., a microparasellid from subterranean waters of Rhodes, Greece (Isopoda: Asellota). *Fragm, Balcanica* 11:57–62.

Pesce, G.L. 1985. New records for *Salentinella* Ruffo (Crustacea Amphipoda) from phreatic waters of Italy and Greece. *International Journal of Speleology* 14:19–29.

Pesce, G. L. and Argano, R. 1980. Nouvelles données sur les Asellides de la Grèce continentale et insulaire (Crustacea, Isopoda). *Bulletin Zoologisch Museum* 7:49–54.

Pesce, G.L. and Maggi, D. 1983. Ricerche faunistiche in acque sotterranee freatiche della Grecia Meridionale ed insulare e stato attuale delle conoscenze sulla stygofauna di Grecia. *Natura, Milano* 74:15–73.

Pesce G. L., Maggi, D., Ciocca, A. and Argano, R. 1979. Biological researches on the subterranean phreatic waters of Northern Greece. *Biologica Gallo-Helen* 8:109–26.

Pinkster, S. 1993. *A Revision of the Genus Echinogammarus Stebbing, 1899 with Some Notes on Related Genera (Crustacea, Amphipoda)*. Memorie del Museo Civico di Storia Naturale di Verona, 2. serie, Scienze della Vita, 19, 1–185. Verona : Museo Civico di Storia Naturale.

Pinter N., Grenerczy G., Weber J., Stein S. and Mcdak D. (eds.). 2016. *The Adria Microplate: GPS Geodesy, Tectonics and Hazards*. Nato Science Series: *IV*: 61. Netherlands: Springer.

Pinya, S., Guijarro, B., Alberti, R., Amorós, L. and Muñoz, A. 2008. Sobre la presència del cranc de riu americà, Procambarus clarkii, a l'illa de Mallorca. In Jornades de Medi de les Illes, Balears.

Platvoet, D. 1984. Observations on the genus *Salentinella* (Crustacea, Amphipoda) with description of *Salentinella formenterae* n. sp. *Bijdragen lot de Dierkunde* 54:178–84.

Pomar, L. 1991. Reef geometries, erosion surfaces and high-frequency sea-level changes, upper miocene Reef Complex, Mallorca, Spain. *Sedimentology* 38:243–69.

Ponniah, M. and Hughes, J. M. 2004. The evolution of Queensland spiny mountain crayfish of the genus *Euastacus*. I. Testing vicariance and dispersal with interspecific mitochondrial DNA. *Evolution* 58:1073–85.

Pons, G. X., Jaume D. and Damians J. 1995. Fauna cavernicola de Mallorca. *Bolletí de la Societat d'Història Natural de les Balears* 3:125–43.

Popov, S.V., Rögl, F., Rozanov, A.Y., Steininger, F.F., Shcherba, I.G. and Kovac, M. 2004. Lithological-paleogeographic maps of Paratethys-10 maps Late Eocene to Pliocene. *Courier Forschungsinstitut Senckenberg* 250:1–46.

Popov, S.V., Shcherba, I.G., Ilyina, L.B., Nevesskaya, L.A., Paramonova, N.P., Khondkarian, S.O. and Magyar, I. 2006. Late Miocene to Pliocene palaeogeography of the Paratethys and its relation to the Mediterranean. *Palaeogeography, Palaeoclimatology, Palaeoecology* 238:91–106.

Poulakakis, N., Lymberakis, P., Valakos, E., Zouros, E. and Mylonas, M., 2005. Phylogenetic relationships and biogeography of *Podarcis* species from the Balkan Peninsula, by Bayesian and maximum likelihood analyses of mitochondrial DNA sequences. *Molecular Phylogenetics and Evolution* 37:845–57.

Poulakakis, N., Kapli, P., Lymberakis, P., Trichas, A., Vardinoyiannis, K., Sfenthourakis, S. and Mylonas, M. 2015. A review of phylogeographic analyses of animal taxa from the Aegean and surrounding regions. *Journal of Zoological Systematics and Evolutionary Research* 53:18–32.

Pretus, J. L. 1991. *Estudio taxonómico biogeográfico y ecológico de los crustáceos epigeos e hipogeos de las Baleares (Branchiopoda, Copepoda, Mystacocarida y Malacostraca)*. Universitat de Barcelona.

Pretus, J.L. and Stock J.H. 1990. A new hyporheic *Bogidiella* (Crustacea, Amphipoda) from Mallorca. *Endins* 16:47–51.

Prevorčnik, S., Verovnik, R., Zagmajster, M. and Sket, B., 2010. Biogeography and phylogenetic relations within the Dinaric subgenus *Monolistra* (Microlistra)(Crustacea: Isopoda: Sphaeromatidae), with a description of two new species. *Zoological Journal of the Linnean Society*, 159(1):1–21.

Rögl, F. 1999. Mediterranean and Paratethys. Facts and hypotheses of an Oligocene to Miocene paleogeography (short overview). *Geologica Carpathica* 50:339–49.

Rosenbaum, G. and Lister, G.S. 2004. Formation of arcuate orogenic belts in the western Mediterranean region. *Geological Society of America Special Papers* 383:41–56.

Rosenbaum, G., Lister, G.S. and Duboz, C. 2002. Reconstruction of the tectonic evolution of the western Mediterranean since the Oligocene. *Journal of the Virtual Explorer* 8:107–30.

Rosenzweig, M.L. 1995. *Species Diversity in Space and Time*. Cambridge: Cambridge University Press.

Ruffo, S. 1960. Studies on Crustacean amphipods. A contribution to the knowledge of Crustacean amhpipods of subterranean waters of Sardinia and Baleari Islands. *Atti dell'Istituto Veneto di Scienze, Lettere ed Arti* 118:168–80.

Ruffo, S. 2006a. Crustacea Malacostraca Bathynellacea. In *Checklist and Distribution of the Italian Fauna. 10,000 Terrestrial and Inland Water Species*. Memorie del Museo Civico di Storia Naturale di Verona, 2. serie, Scienze della Vita 17, eds. S. Ruffo, and F. Stoch, 101. Verona : Museo Civico di Storia Naturale.

Ruffo, S. 2006b. Crustacea Malacostraca Thermosbaenacea. In *Checklist and Distribution of the Italian Fauna. 10,000 Terrestrial and Inland Water Species*. Memorie del Museo Civico di Storia Naturale di Verona, 2. serie, Scienze della Vita 17, eds. S. Ruffo, and F. Stoch, 103. Verona : Museo Civico di Storia Naturale.

Ruffo, S. and Delamare Deboutteville, C. 1952. Deux nouveaux Amphipodes souterrains de France, *Salentinella angelieri* n. sp. et *Bogidiella chappuisi* n. sp. *Comptes Rendus de l'Académie des Sciences, Paris* 234:1636–38.

Ruffo, S. and Stoch F. 2006. Crustacea Malacostraca Amphipoda. In *Checklist and Distribution of the Italian Fauna. 10,000 Terrestrial and Inland Water Species*. Memorie del Museo Civico di Storia Naturale di Verona, 2. serie, Scienze della Vita 17, eds. S. Ruffo, and F. Stoch, 109–11. Verona : Museo Civico di Storia Naturale.

Ruffo, S. and Vigna-Taglianti, A. 1989. Description of a new cavernicolous *Ingolfiella* species from Sardinia, with remarks on the systematics of the genus (Crustacea, Amphipoda, Ingolfiellidae). *Annali Museo civico di Storia naturale di Genova* 87:237–61.

Ruffo, S., Tarocco, M. and Latella, L. 2014. *Cryptorchestia garbinii* n. sp. (Amphipoda: Talitridae) from Lake Garda (Northern Italy), previously referred to as *Orchestia cavimana* Heller, 1865, and notes on the distribution of the two species. *Italian Journal of Zoology* 81:92–9.

Ruggieri, G. 1973. Due parole sulla paleogeografia delle isole minori a Ovest e a Nord della Sicilia. *Lavori della Società italiana di Biogeografia* 3:5–12.

Ruggieri, R. and Grasso, M. 2000. Caratteristiche stratigrafiche e strutturali dell'Altipiano Ibleo Ragusano e sue implicazioni sulla morfologia carsica. *Speleologia Iblea* 8:19–35.

Ryan, W.B.F., Pitman, W.C., Major, C.O., et al. 1997. An abrupt drowning of the Black Sea shelf. *Marine Geology* 138:119–26.

Sacchi, C.F. 1955. Il contributo dei Molluschi terrestri alle ipotesi del "Ponte Siciliano". *Archivio Zoologico Italiano* 40:49–181.

Sakellariou, D. and Galanidou, N. 2015. Pleistocene submerged landscapes and Palaeolithic archaeology in the tectonically active Aegean region. In *Geology and Archaeology: Submerged Landscapes of the Continental Shelf*. Special Publication 411, eds. J. Harff, G. Bailey, and F. Lüth, 145–178. London: Geological Society.

Santoni, S., Huneau F., Garel E., Vergnaud-Ayraud V., Labasque, Aquilina, L., Jaunat, J. and Celle-Jeanton, H. 2016. Residence time, mineralization processes and groundwater origin within a carbonate coastal aquifer with a thick unsaturated zone. *Journal of Hydrology* 540:50–63.

Savona Ventura, C. 1975. The geographical evolution of the Maltese Archipelago. *The Maltese Naturalist* 2:9–12.

Schellenberg, A. 1950. Subterrane Amphipoden korsikanischer Biotope. *Archiv für Hydrobiologie, (Supplement* 2) 44:325–32.

Schembri, P.J. 2003. Current state of knowledge of the Maltese non-marine fauna. *Malta Environment and Planning Authority Annual Report and Accounts* 2003:33–65.

Schembri, J. and Sultana, J. 1989. *Red Data Book of the Maltese Islands*. Department of Information, Ministry of Education, Malta.

Schüle, W. 1993. Mammals, vegetation and the initial human settlement of the Mediterranean islands: A palaeoecological approach. *Journal of Biogeography* 20:399–411.

Serban, E. 1977. Sur les péréiopodes VIII mâles de Iberobathynella cf. fagei de Majorque (Bathynellacea, Parabathynellidae). *Crustaceana* 33:1–16.

Sfenthourakis, S. 1996. A biogeographical analysis of terrestrial isopods (Isopoda, Oniscidea) from the central Aegean islands (Greece). *Journal of Biogeography* 23:687–98.

Sfenthourakis, S, and Triantis, K.A. 2017. The Aegean archipelago: A natural laboratory of evolution, ecology and civilisations, *Journal of Biological Research-Thessaloniki* 24(4):1–13.

Sidorov, D.A. and Palatov, D. 2012. Taxonomy of the spring dwelling amphipod *Synurella ambulans* (Crustacea: Crangonyctidae) in West Russia: With notes on its distribution and ecology. *European Journal of Taxonomy* 23:1–19.

Sket, B. 1988. Zoogeografija sladkovodnih i somornih rakov (Crustacea) v kvarnersko- vele-bitskem območju. *Biološki vestnik* 36:63–76.

Sket, B. 1990a. Is *Niphargobates lefkodemonaki* sp. n. (Crustacea: Amphipoda) from Kriti (Greece) a Zoogeographical Enigma? *Zoologische Jahrbücher, Abteilung für Systematik*, 117:1–10.

Sket, B. 1990b. Isopoda (Crustacea: Isopoda: *Microcharon, Jaera, Proasellus*) and other fauna in hypogean waters of southern Cyprus. *International Journal of Speleology* 19:39–50.

Sket, B. 1994. Distribution patterns of some subterranean Crustacea in the territory of the former Yugoslavia. *Hydrobiologia* 287:65–75.

Sket, B., 1999. The nature of biodiversity in hypogean waters and how it is endangered. *Biodiversity and Conservation* 8:1319–38.

Sket, B. and Karaman, G. 1990. *Niphargus rejici* (Amphipoda), its relatives in the Adriatic islands, and its possible relations to S.W. Asian taxa. *Stygologia* 5:153–72.

Sondaar, P.Y. and Dermitzakis, M.D. 1982. Relation Migration Landvertebrates, Paleogeography and Tectonics. In International Symposium on the Hellenic Arc and Trench (H.E.A.T.), Athens, April 8–10, 1981, eds. X.L. Pichon, S.S. Augustidis, and J. Mascle, 283–308.

Sondaar, P.Y., De Vos, J. and Dermitzakis, M.D. 1986. Late Cenozoic faunal evolution and palaeogeography of the South Aegean island arc. *Modern Geology* 10:249–59.

Sosdian, S. and Rosenthal, Y. 2009. Deep-sea temperature and ice volume changes across the Pliocene-Pleistocene climate transitions. *Science* 325:306–10.

Souty-Grosset, C., Anastacio, P. M., Aquiloni, L., Banha, F., Choquer, J., Chucholl, C., and Tricarico, E. 2016. The red swamp crayfish *Procambarus clarkii* in Europe: Impacts on aquatic ecosystems and human well-being. *Limnologica* 58:78–93.

Speranza, F., Villa, I.M., Sagnotti, L., Florindo, F., Cosentino, D., Cipollari, P. and Mattei, M. 2002. Age of the Corsica-Sardinian rotation and Liguro-Provençal Basin spreading: New paleomagnetic and Ar/Ar evidence. *Tectonophysics* 347:231–51.

Stanley, D.J. and Wezel F-C 1985. *Geological Evolution of the Mediterranean Basin.* New York: Springer.

Steininger, F.F., and Rögl, F. 1984. Paleogeography and palinspastic reconstruction of the Neogene of the Mediterranean and Paratethys. *Geological Society, London, Special Publications* 17:659–68.

Stoch, F. 1995. Diversity in groundwaters, or: Why are there so many. *Mémoires de Biospéologie* 22:139–60.

Stoch, F. 1997. *Metacrangonyx ilvanus* n. sp., the first Italian representative of the family Metacrangonyctidae (Crustacea: Amphipoda). *Annales Limnologie* 33:255–62.

Stoch, F. 2000. How many endemic species? Species richness assessment and conservation priorities in Italy. *Belgian Journal of Entomology* 2:125–33.

Stoch, F., Valentino, F. and Volpi, E. 1996. Taxonomic and biogeographic analysis of the Proasellus coxalis-group (Crustacea, Isopoda, Asellidae) in Sicily, with description of *Proasellus montalentii* n. sp. *Hydrobiologia* 317:247–58.

Stoch, F., Argano, R. and Campanaro, A. 2006. Crustacea Malacostraca Isopoda. In *Checklist and Distribution of the Italian Fauna. 10,000 Terrestrial and Inland Water Species.* Memorie del Museo Civico di Storia Naturale di Verona, 2. serie, Scienze della Vita 17, eds. S. Ruffo, and F. Stoch, 107–108. Verona : Museo Civico di Storia Naturale.

Stock, J. 1968. A revision of the European species of the *Echinogammarus pungens*-group (Crustacea, Amphipoda). *Beaufortia* 16:13–78.

Stock, J.H. 1972. Les Gammarides (Crustacés Amphipodes) des eaux douces et saumâtres de Corse. *Bulletin Zoologisch Museum Universiteit van Amsterdam* 2:197–220.

Stock, J.H. 1980. Regression model evolution as exemplified by the genus Pseudoniphargus (Amphipoda). *Bijdragen tot der Dierkunde* 50:105–44.

Strouhal, H. 1966. Ein weiterer Beitrag zur Süßwasser-und Landasselfauna Korfus. In *Ein weiterer Beitrag zur Süßwasser-und Landasselfauna Korfus*, 257–315. Berlin, Heidelberg: Springer.

Sworobowicz, L., Grabowski, M., Mamos, T., Burzyński, A., Kilikowska, A., Sell, J. and Wysocka, A. 2015. Revisiting the phylogeography of *Asellus aquaticus* in Europe: Insights into cryptic diversity and spatiotemporal diversification. *Freshwater Biology* 60:1824–40.

Triantis, K.A. and Mylonas, M. 2009. Greek Islands, Biology. In *Encyclopedia of Islands*, eds. R.G. Gillespie, and D.A. Ciague, 388–92.

Trincardi, F., Cattaneo, A., Asioli, A., Correggiari, A. and Langone, L. 1996. Stratigraphy of the late–Quaternary deposits in the central Adriatic basin and the record of short-term climatic events. *Memorie dell'Istituto Italiano di Idrobiologia* 55:39–70.

Trontelj, P., Machino, Y. and Sket, B. 2005. Phylogenetic and phylogeographic relationships in the crayfish genus *Austropotamobius* inferred from mitochondrial COI gene sequences. *Molecular Phylogenetics and Evolution* 34: 212–26.

Tzomos, T., and Koukouras, A. 2015. Redescription of *Palaemon antennarius* H. Milne Edwards, 1837 and *Palaemon migratorius* (Heller, 1862) (Crustacea, Decapoda, Palaemonidae) and description of two new species of the genus from the circum-Mediterranean area. *Zootaxa* 3905:27–51.

van der Geer, A., Dermitzakis, M. and de Vos, J. 2006. Crete before the Cretans: The reign of dwarfs. *Pharos (journal from The Netherlands Institute at Athens)* 13:119–30.

Väinölä, R., Witt, J. D. S., Grabowski, M., Bradbury, J. H., Jazdzewski, K. and Sket, B. 2007. Global diversity of amphipods (Amphipoda; Crustacea) in freshwater. In *Freshwater Animal Diversity Assessment*, eds. E.V. Balian, C. Lévêque, H. Segers, and K. Martens, 241–55. Dordrecht: Springer.

Vecchioni, L., Deidun, A., Sciberras, J., Sciberras, A., Marrone, F. and Arculeo, M. 2017. The late Pleistocene origin of the Italian and Maltese populations of *Potamon fluviatile* (Malacostraca: Decapoda): Insights from an expanded sampling of molecular data. *The European Zoological Journal* 84:575–82.

Vella, N., Vella, A. and Mifsud, C.M. 2017. First scientific records of the invasive Red Swamp Crayfish, *Procambarus clarkii* (Girard, 1852) (Crustacea: Cambaridae) in Malta, a threat to fragile freshwater habitats. *Natural and Engineering Sciences* 2: 58–66.

Vigliotti, L., Alvarez, W. and McWilliams, M. 1990. No relative motion detected between Corsica and Sardinia. *Earth Planetary Science Letters* 98:313–18.

Vigna Taglianti, A. 1976. Gli anfipodi sotterranei dell'Arcipelago Toscano. *Biogeographia-The Journal of Integrative Biogeography* 5:375–83.

Vogiatzakis, I.N., Pungetti, G. and Mannion, A.M. (eds.). 2008. *Mediterranean Island Landscapes: Natural and Cultural Approaches*, Vol. 9. Dordrecht: Springer.

Vörösmarty, C.J., McIntyre, P.B., Gessner, M.O., et al. 2010. Global threats to human water security and river biodiversity. *Nature* 467:555.

Wagner, H.P. 1994. A monographic review of the Thermosbaenacea (Crustacea: Peracarida). *Zoologische Verhandelingen, Leiden* 291:1–338.

Whittaker, R.J. and Fernández-Palacios, J.M. 2007. *Island Biogeography: Ecology, Evolution, and Conservation*. Oxford: Oxford University Press.

Wong, H.K., Lüdmann, T., Ulug, A. and Görür, N. 1995. The Sea of Marmara: A plate boundary sea in an escape tectonic regime. *Tectonophysics* 244:231–50.

Woodward, J. (ed.). 2009. *The Physical Geography of the Mediterranean*, Vol. 8. Oxford: Oxford University Press.

Zettler, M. L., and Zettler, A. 2017. A new species of *Niphargus* (Amphipoda, Niphargidae) from Crete (Greece). *Crustaceana* 90:1415–26.

Zganec K. 2009. Rasprostranjenost i ekologija nadzemnih rakušaca (Amphipoda: Gammaroidea) slatkih i bočatih voda Hrvatske. Doktorska disertacija, Prirodoslovno matematički fakultet, Zagreb. 214.

Zogaris, S., Chatzinikolaou, Y., Koutsikos, N., et al. 2012. Freshwater fish assemblages in Cyprus with emphasis on the effects of dams. *Acta Ichthyologica et Piscatoria* 42:165–75.

CHAPTER 7

Conservation Status of the Large Branchiopods (Branchiopoda: Anostraca, Notostraca, Laevicaudata, Spinicaudata, Cyclestherida)

D. Christopher Rogers

CONTENTS

7.1 INTRODUCTION

Branchiopods are allotriocaridan crustaceans with most taxa occurring in seasonally astatic aquatic habitats. The large branchiopods are an artificial subgroup of Class Branchiopoda, representing all of the Branchiopoda clade except the Diplostracan Suborder Cladocera, or "water fleas", which are not considered here. The large branchiopods consist of two orders (Anostraca and Notostraca) and portions of Order Diplostraca (suborders Laevicaudata, Spinicaudata, and Cyclestherida) (Brendonck et al. 2008, Rogers 2009, Schwentner et al. 2018).

Large branchiopods are predominately planktonic or semibenthic, and are used as ecosystem health indicators in temporary and saline inland aquatic habitats, providing impact data regarding land-use practices and management (Rogers 2009). Very few species are legally protected, while some others are economically important in aquaculture, salt production, or for human consumption. Most taxa are primary consumers, and as obligate, passively dispersed aquatic macroinvertebrates, large branchiopods are important in the ephemeral wetland food chain (Brendonck et al. 2008, Rogers 2009).

Seasonally astatic aquatic habitats have distinctive wet and dry phases, and while most aquatic insects readily fly between perennial aquatic habitats and seasonally dry ones, large branchiopods cannot actively disperse, relying on a resident egg bank in a given habitat for population persistence between wet seasons (Rogers 2009, 2014a, c, Wang and Chou 2015). These animals are then directly influenced by and reactive to impacts and affects on their individual habitats in both the wet and dry seasons, as well as the surrounding water shed into their habitats (Rogers 2014a, b, 2015).

At a larger scale, large branchiopod habitats are important arid and dryland, highly localized water resources, generally necessary for aquifer recharge and are important ecologically, municipally, and agriculturally (e.g., Jain 1976, Ebert and Balko 1984, Jain and Moyle 1984, Williams 1987, Zedler 1987, Ikeda and Schlising 1990, Witham et al. 1996, Brown and Jung 2005, Comer et al. 2005, Schlising and Alexander 2007). The branchiopods themselves are vital to the diet of migratory water birds (e.g., Cottam 1939, Proctor 1964, Horne 1966, Krapu 1974, Swanson et al. 1974, MacDonald 1980, Donald 1983, Hurlbert et al. 1984, Camara and De Medeiros Rocha 1987, Saunders III et al. 1993, Brochet et al. 2010, Rogers 2014c), as well as food for a variety of amphibians, other crustaceans, and insects, which are in turn all vital to their dispersal (Rogers 2014c, 2015). Furthermore, the variety of seasonally astatic aquatic habitats, such as playas, pans, vernal pools, gnammas, rock pools, tinajas, et cetera, generates an equal variety of biodiversity, and are excellent model systems for ecological and evolutionary research (Jocque et al. 2010)

Conservation of these organisms requires an understanding of the ecology of the habitats where these animals occur, their dispersal vectors, the distribution of the various taxa, and the phylogenetic and functional diversity of the organisms themselves. Surveys, distribution maps, and records tend to show where these crustaceans are, but do not emphasize where they are not. Compounding these difficulties, records are typically a reflection of where surveys have been conducted, rather than any actual distribution delineations. This makes it difficult to establish baseline conditions for species across their ranges. Consistent data reporting would help prevent ambiguous interpretation or mischaracterization of species conservation needs. The best way to assess the conservation status of the large branchiopod species is to compile what we do know and then identify data needs.

7.2 LARGE BRANCHIOPODS

There are no complete and up-to-date catalogues of all large branchiopod taxa, although this is rapidly being rectified in the WoRMS database (http://www.marine-species.org/). The earliest comprehensive monographs of large branchiopods were produced by Daday (1910a, b, 1915, 1923, 1925, 1926, 1927) who treated all groups except the Notostraca. Belk provided a brief overview in 1982, then in 1996, and with a later supplement in 1997; Belk and Brtek provided a very useful checklist of the Anostraca. Each species was annotated with the locality of the type specimens, generalized taxon distribution, and an extensive introduction and methods section, which explained in detail the decision processes leading to which names were used and why. A few names were changed without justification, and were eventually rectified (Rogers, 2003, 2006). However, Brtek (Belk pers. comm.; Brtek, pers. comm.) disagreed on some of the names used and their placements. Therefore, Brtek created his own separate checklist (1997), including all extant Branchiopoda known at that time, as well as synonyms. Unfortunately, the text has many problems and the checklist created confusion. The English is poor, and many genera previously synonymized based on quantified analyses were resurrected without any justification, and with little if any mention of the previous analyses. Other errors include some species presented with valid names in more than one genus simultaneously, plus poor editing. Brtek cited Brehm's (1958) illustration of a *Branchipodopsis* collected from southern Africa. The illustration is of poor quality, and labeled "*Branchipodopsis* nov. spec.?" The specimens were searched for by Hamer and Appleton (1996) but could not be found. Brtek named this "species" *Branchipodopsis brehmi* with no description, no diagnosis, no designation of type material, and no figures. There is no evidence that any material was even examined. Therefore, *Branchipodopsis brehmi* is a nomen nudum, according to Articles 12 and 13 of the ICZN (1999) and is not an acceptable name.

Dumont and Negrea (2002) provided the only modern comprehensive introduction to the class Branchiopoda. The work generally is a good introduction and should be used as a source to find primary literature, but it should not be relied upon by itself. There are many factual as well as grammatical errors. Keys are provided to family level for all branchiopods (p. 249); however, the keys are sometimes confusing and use characters that are not exclusive; thus, many taxa may be identified incorrectly. For example, in couplet 1 of the Anostraca keys, the suborder Artemiina is separated from suborder Anostracina by three characters, the third being that the male second antennae are "only slightly fused at their base". This character is only true for one of the two families included in this suborder. The second character in the couplet, "Brood pouch ... tending towards the development of lateral lobes" is also applicable to some members of the Anostracina. The other half of the couplet (which leads to the suborder Anostracina and the remaining anostracan families) also has difficulties. The second character states "Brood pouch variously shaped, but not bilobed"; however, a bilobed brood pouch is present in at least two genera in this suborder (Brendonck 1997, Rogers 2002a). The third character of this same couplet is also in error, stating: "Male antennae variously shaped ... but not with two small outgrowths on inner side of median article [sic]". Structures that could be described as "outgrowths" are present on the second antennal medial surface in many families in the Anostracina.

Examples of other errors in Dumont and Negrea (2002) include: "In the Conchostraca ... the carapace encloses only the trunk and its appendages, leaving the head free (Laevicaudata)" (p. 29, repeated on p. 256) (the head may be enclosed in all clam shrimp); or "The Spinicaudata have one pair of appendages per segment (in all up to 32 pairs), while the Cyclestherida have 16 pairs" (p. 59) (obviously, they have only one pair per segment). The book is useful, but any facts gleaned must be verified.

Brendonck et al. (2008) provided good diversity summaries for each zoogeographic region based on the information available at the time and stressed the limitations in the knowledge and understanding for each of the major orders and suborders. Brendonck et al. (2008) also discussed the conservation issues surrounding these animals at a global scale. Rogers (2009) published a slightly updated discussion on diversity and distribution, focusing more on general biology, ecology, and morphology, but also discussed conservation issues related to the large branchiopods. Martin et al. (2016) provided basic information on collecting and culturing large branchiopod crustaceans.

Molecular and morphological studies very strongly support the monophyly of the class and the extant subclasses, orders, and suborders (Schwentner et al. 2018). The primary taxonomic groups Anostraca and Notostraca have been considered stable for quite some time; however, the relationships among the diplostracan suborders have only been recently resolved. Up until 1980, the clam shrimp were regarded as a single order (Conchostraca) and the water fleas as a separate order (Cladocera) based on morphological studies (Belk 1982). However, numerous additional morphological studies (Linder 1945, Olesen 1998, 2007, 2009), molecular studies using increasingly more powerful analyses (Brabrand et al. 2002, DeWaard et al. 2006, Stenderup et al. 2006, Richter et al. 2007, Regier et al. 2010, Sun et al. 2011, Fritsch et al. 2013), and combined analyses (Richter et al. 2007) resolved the relationships between the Diplostracan suborders Spinicaudata, Cyclestherida, and Cladocera. The placement of Laevicaudata within Diplostraca (as opposed to sister to Notostraca) was not fully resolved until the analyses of Schwentner et al. (2018). Thus, there is no support for the alternative classification schemes proposed by Naganawa and Orgiljanova (2001) and Naganawa (2001) dividing the large branchiopods and the Cladocera each into a separate subclass, or the subordinal arrangement of the spinicaudatan families.

Although there are numerous treatments of large branchiopod diversity for many parts of the world, gaps are still present where no studies have been undertaken, or the most recent studies are decades old. This includes large areas such as China, central Asia, as well as northern and central Africa. Examples of recent regional treatments for large branchiopods include: southern Africa (Day et al. 1999, Nhiwatiwa et al. 2014, Tuytens et al. 2015); Australia (Timms 2012, 2015, 2018); Austria (Eder and Hödl 2002); the former Czechoslovakia (Šrámek-Hušek et al. 1962); France (Cart and Rabet 2003); Iberian Peninsula (Alonso 1996); India (Rogers and Padhye 2015, Padhye and Lazo-Wasem 2018); Mexico (Maeda-Martínez et al. 2002a, b, c); Morocco (van den Broeck et al. 2015); the Nearctic region (Rogers and Hann 2016); the Neotropical region (Rogers et al. 2020a, b); the Palaearctic region (Rogers et al. 2019); Poland (Gołdyn et al. 2013); Portugal (Machado et al. 2017); eastern Russia (Vekhov 1992); southeast Asia (Rogers et al. 2013a, 2016a, b), and Tunisia (Marrone et al. 2016).

7.2.1 Order Anostraca

Anostraca (fairy shrimp and brine shrimp) are reasonably well known taxonomically (Rogers 2013), although there is no slowing in the pace of new species descriptions. Strong sexual selection has resulted in most anostracan species being well defined morphologically due to extravagant adaptations of secondary sexual characteristics (Rogers 2002a, 2009). Male genital morphology defines the anostracan genera (Belk 1997, Brendonck 1997, Brendonck and Belk 1997, Rogers 2003, 2005, 2006), and this is well supported molecularly (Remigio and Hebert 2000, Weekers et al. 2002, DeWaard et al. 2006).

Several catalogues of anostracan species have been published (Belk and Brtek 1995, 1997, Brtek 1997, 2002, Rogers 2013). The most recent (Rogers 2013) is the most up to date so far, and provides a history of anostracan taxonomy; however, it is already out of date as many more new species have been described. Molecular phylogenetic studies have been conducted on anostracans, primarily defining the families and genera (Remigio and Hebert 2000, Weekers et al. 2002).

Molecular phylogenies for families and genera are few, with only Parartemiidae (Remigio et al. 2001), Streptocephalidae (Daniels et al. 2004), Branchinectidae (Rogers and Aguillar 2020), and the thamnocephalid genus *Branchinella* (Remigio et al. 2003, Pinceel et al. 2013a, b) having been examined. Recent important revisions include: branchipodid genera (Brendonck 1997), *Metabranchipus* (Rogers and Hamer 2012), *Streptocephalus* (for Africa: Brendonck and Coomans 1994a, b, Hamer et al. 1994a, b; Asia: Rogers and Padhye 2014; North America: Maeda-Martínez et al. 1995), thamnocephalid genera (Rogers 2006), *Branchinella* (Australia: Timms 2015; elsewhere: Rogers et al. 2013b), branchinectid genera (Rogers and Coronel 2011), *Parartemia* (Timms 2014), and some chirocephalid genera (Rogers 2002b). More work is still needed in the other families and the genera, most notably the Chirocephalidae, especially *Chirocephalus*.

Anostracan regional conservation assessments include Algeria (Samraoui and Dumont 2002), Australia (Pinceel et al. 2013a, b), Botswana (Brendonck and Riddoch 1997), Chile (Rogers et al. 2008), the Iberian Peninsula (García-de-Lomas et al. 2015a, b, c), India (Padhye et al. 2017), Mexico (Maeda-Martínez et al. 2002a), Morocco (van den Broeck et al.2015), South Africa (Hamer and Brendonck 1997, Hamer and Martens 1998, De Roeck et al. 2007), United States (California: Eng et al. 1990), and Zimbabwe (Nhiwatiwa et al. 2014). The first large branchiopod preserve was established in Austria in 1982 to protect *Chirocephalus shadini* (Smirnov, 1928). One anostracan species is protected under law in the United Kingdom, five species in the United States, one in Australia, and one in Brazil. *Artemia franciscana* Kellogg, 1906 native to the Americas, is invasive in Eurasia, Africa, and Australia (Amat et al. 2005, Timms 2015).

Rogers (2015) separated the anostracan genera into three completely artificial categories: "major genera", "intermediate genera", and "minor genera" as part of a biogeographical model. The 4 major genera have more than 45 species each (62.6% of total species), the 4 intermediate genera have 15–25 species (22.0% of species), while the 24 minor genera have fewer than 7 species each (15.4% of species). More than half the anostracan species are in the four major genera. In contrast, 56.2% of anostracan species are known from ten or fewer localities and 28.7% from the type locality alone (Belk and Brek 1995, 1997, Rogers 2013). Thus, the extinction potential among anostracans at both species and genus levels from stochastic effects alone is quite high.

Several anostracan species are important economically. Brine shrimp in the genus *Artemia* are important globally in aquaculture (e.g., Browne et al. 1991, Sriputhorn and Sanoamuang 2011) and other species are being examined for similar use in other countries (e.g., Plodsomboon et al. 2012). Anostracans are model organisms for a variety of ecological, genetic, evolutionary, and developmental studies, and have even been used in experiments in outer space (e.g., De Bell 1992, Nunes et al. 2006).

7.2.2 Order Notostraca

The Notostraca (tadpole shrimp, shield shrimp, helmet shrimp) contains one family with two genera. The order was taxonomically confused until the monographic revision by Longhurst (1955), who accepted very few of the previously described taxa, lumping them considerably, and recognizing very few, broadly distributed species based on quantifiable characters. Longhurst's (1955) species concepts for the notostracan taxa were necessarily conservative due to the technical and methodological limitations of the time. Recent molecular phylogenies demonstrate that much of Longhurst's (1955) concepts were correct; however, cryptic taxa, some undescribed, do exist (Korn and Hundsdoerfer 2006, Vanschoenwinkel et al. 2012, Korn et al. 2013).

Notostracan diversity is just being understood and described. *Lepidurus* has 13 described taxa, which are reasonably well defined morphologically, and the genus has strong support for monophyly (Vanschoenwinkel et al. 2012, Korn et al. 2013). The Nearctic species have been revised (Rogers 2001), but additional regions need revisions as well. The genus *Triops* has less support and may actually represent two genera (Vanschoenwinkel et al. 2012) with *Triops cancriformis* sensu lato possibly representing a separate genus. Molecular and morphological analyses demonstrate the presence of many cryptic *Triops* species (Korn and Hundsdoerfer 2006, 2016, Korn et al. 2006, 2010, 2013).

In the United States, *Lepidurus packardi* Simon, 1886 is protected under law and in the UK *Triops* are protected legally as well. In the Republic of Korea (South Korea), nonnative Nearctic populations were legally protected from 2004 until 2012.

7.2.3 Order Diplostraca, Suborder Laevicaudata

These unique clam shrimp encompass a single family (Lynceidae) with 3 genera and 39 species. Molecular and morphological analyses place the Laevicaudata as a basal diplostracan lineage (Schwentner et al. 2018). The group was first treated in toto by Daday (1927). A recent catalogue of the family (Rogers and Olesen 2016) provides a diagnosis for the order, synapomorphies, and a list of all taxa. Rogers and Padhye

(2015) and Rogers et al. (2016a) reviewed the southern and southeast Asian species, respectively, and Timms (2013) revised the Australian taxa. Additional revisionary work of the genera and species using morphological and molecular analyses are in progress. No species are specifically protected under law, although many narrow range endemic species may be at risk.

7.2.4 Order Diplostraca, Suborder Spinicaudata

Spinicaudatan taxonomy is particularly confused. Relationships within the higher taxonomic levels have only recently been resolved, and more work needs to be done. Daday (1915, 1923, 1925, 1926) provided the first comprehensive monographs of the suborder, creating most of the genera we use today. The great polymorphy exhibited by the species in this group yielded a myriad of dubious species and a few dubious genera. Straškraba (1962, 1965a,b, 1966) cleared up much of this confusion demonstrating the range of variability in the species and the Cyzicidae genera. Naganawa (2001) proposed a classification scheme with Spinicaudata as an order composed of three suborders: Cyclostraca (for the Cyclestheriidae, see next section), Spinirostria (for the Leptestheriidae and Cyzicidae), and Procephalida (for the Limnadiidae). This classification ignored data from the fossil taxa as well as numerous molecular and morphological phylogenetic studies, and has not been accepted. Rogers et al. (2012) revised the Limnadiidae genera. Astrop and Hegna (2015) put recent and fossil clam shrimp into a phylogenetic framework, clarifying relationships among the groups. Molecular phylogenies have been conducted by Weeks et al. (2009) for Limnadiidae, and Schwentner et al. (2009) and Sun et al. (2011) for all families. Morphological and molecular revisions have been published for *Australimnadia* (see Timms and Schwentner 2017), *Eocyzicus* (Rogers et al. 2017, Tippelt and Schwentner 2018), the Australian species of *Eulimnadia* (see Timms 2016a), *Limnadopsis* (see Schwentner et al. 2012a, b), and *Paralimnadia* (see Timms 2016b). Most recently, Schwentner et al. (2020) produced the largest molecular phylogeny of the group, demonstrating the presence of a cryptic family, and Rogers (2020) presented the first comprehensive catalogue of the spinicaudatan species. Much more needs to be done and additional revisions are in progress.

There are around 215 described spinicaudatan species (Brendonck et al. 2008, Rogers 2009, 2020); however, many are probably invalid and there may be just as many undescribed species as well. Species, genus, and family definitions need to be redefined. The Spinicaudata has four families – Cyzicidae, Leptestheriidae, Eocyzicidae, and Limnadiidae – with members on all continents, except Antarctica. No species are specifically protected under law, although many narrow range endemic species may be at risk. Two species are economically important in Mexico (Martínez-Pantoja et al. 2002).

7.2.5 Order Diplostraca, Suborder Cyclestherida

This group is morphologically intermediate between the Spinicaudata and the Cladocera. Traditionally treated as a single, pantropical species (*Cyclestheria hislopii*; Baird, 1859), Schwentner et al. (2013) demonstrated that the taxon is composed of numerous cryptic species. Additional collecting and revisionary work is needed. No species are protected under law, and some localized forms may be at risk; however, we really have no quantitative information on distribution and diversity for this suborder.

7.3 DISCUSSION

Large branchiopods are certainly important organisms. Their ecological relevance to seasonally astatic aquatic habitats and hypersaline lakes and wetlands is well understood, as is their use as indicator taxa, umbrella species, flagship species, model organisms for various scientific studies, their important role in the aquaculture industry, and ecotoxicological assessments (summarized in Eder and Hödl 1996, Jocque et al. 2010). These animals have in the past been (Henrikson et al. 1998) and are currently (Creaser 1931, Sriputhorn and Sanoamuang 2011) used as human food, and their importance as food for a great diversity of wildlife makes them important in the food chain for waterfowl hunted for sport (Rogers 2014c).

Conservation assessments of branchiopods must start with an understanding of their biodiversity; we cannot protect what we do not know we have. Large branchiopod crustacean biodiversity is reasonably well understood for the Anostraca and Laevicaudata, such that IUCN Red List assessments are possible for these groups. Biodiversity within the Notostraca is less well understood, in that there appear to be many cryptic species in certain lineages, but again, IUCN Red List assessments are possible. The biodiversity of the Spinicaudata and Cyclestherida are very poorly resolved and it is not possible to provide conservation assessments for these groups at this time.

The uncoupling of large branchiopod habitat definitions from standard water quality parameters will greatly help our understanding of ecological preferences for these taxa. The fact that substrate geochemistry is an important driver of species distribution (Rogers 2014b, d) has great ramifications, in that now soil maps can be used to estimate potential suitable habitat distribution for various branchiopod taxa, allowing us to determine possible distribution limits as well as identify habitat creation opportunities. This, plus the recently developed biogeographical model (Rogers 2015), provides us with a greater understanding of large branchiopod dispersal, colonization, evolution, and habitat suitability. These tools will allow us to better assess the conservation needs and limitations for large branchiopod species. However, the very high number of narrow range endemic species coupled with the increasing rate of new species discovery (Rogers et al. 2015), suggests that most large branchiopod taxa are at some level of risk already, at least from stochastic events.

Currently, wild large branchiopod populations are protected under law in Austria and the United Kingdom, with specific taxa protected in other nations: Australia and Brazil (one anostracan species each) and the United States (five anostracan species and one notostracan species). Conservation is occurring; however, there is some pushback from some stakeholders, especially related to development. To develop acceptable conservation strategies, we need a greater understanding and assessment of large branchiopod biodiversity, we need proper IUCN Red List conservation assessments of all described species (plus future assessments as new species are described), as well as the education and inclusion of all potential stakeholders in local and regional conservation efforts.

REFERENCES

Alonso, M. 1996. *Crustacea Branchiopoda. Fauna Iberica*. Vol. 7. Madrid: Museo Nacional de Ciencias Naturales Consejo Superior de Investigaciones Científicas.

Amat, F., Hontoria, F., Ruiz, O., Green, A.J., Sanchez, M.I., Figuerola, J. and Hortas, F. 2005. The American brine shrimp as an exotic invasive species in the western Mediterranean. *Biological Invasions* 7:37–47.

Astrop, T.I. and Hegna, T.A. 2015. Phylogenetic relationships between living and fossil spinicaudatan taxa (Branchiopoda Spinicaudata): Reconsidering the evidence. *Journal of Crustacean Biology* 35:339–54.

Baird, W. 1859. Description of some new recent Entomostracan from Nagpur, collected by the Rev. S. Hislop. *Proceedings of the Zoological Society of London* 27:231–34.

Belk, D. 1982. *Branchiopoda. Synopsis and Classification of Living Things*. New York: McGraw Hill.

Belk, D. 1996. Was sind "Urzeitkrebse"? *Stapfia* 42:15–9.

Belk, D. 1997. Uncovering the Laurasian roots of Eubranchipus. *Hydrobiologia* 298:241–43.

Belk, D. and Brtek, J. 1995. Checklist of the Anostraca. *Hydrobiologia* 298:315–53.

Belk, D. and Brtek, J. 1997. Supplement to 'checklist of the Anostraca'. *Hydrobiologia* 359:243–45.

Braband, A., Richter, S. Hiesel, R. and Scholtz, G. 2002. Phylogenetic relationships within the Phyllopoda (Crustacea, Branchiopoda) based on mitochondrial and nuclear markers. *Molecular Phylogenetics and Evolution* 25:229–44.

Brehm, V. 1958. Crustacea: Phyllopoda und Copepoda Calanoida. In: Hanström, B., Brinck, P. and Ruderbeck, G. (eds.) *South African Animal Life* 5:10–39.

Brendonck, L. 1997. An updated diagnosis of the branchipodid genera (Branchiopoda: Anostraca Branchipodidae) with reflections on the genus concept by Dubois (1988) and the importance of genital morphology in anostracan taxonomy. *Archiv für Hydrobiologie, Supplement* 107:149–86.

Brendonck, L. and Belk, D. 1997. On potentials and relevance of the use of copulatory structures in anostracan taxonomy. *Hydrobiologia* 359:83–92.

Brendonck, L. and Coomans, A. 1994a. Egg morphology in African Streptocephalidae (Crustacea: Branchiopoda: Anostraca) Part 1: South of Zambezi and Kunene rivers. *Archiv für Hydrobiologie, Supplement* 99:313–34.

Brendonck, L. and Coomans, A. 1994b. Egg morphology in African Streptocephalidae (Crustacea: Branchiopoda: Anostraca) Part 2: North of Zambezi and Kunene rivers, and Madagascar. *Archiv für Hydrobiologie, Supplement* 99:335–56.

Brendonck, L. and Riddoch, B. 1997. Anostracans (Branchiopoda) of Botswana: Morphology, distribution, diversity, and endemicity. *Journal of Crustacean Biology* 17:111–34.

Brendonck, L., Rogers, D.C. Olesen, J. Weeks, S. and Hoeh, R. 2008. Global diversity of large branchiopods (Crustacea: Branchiopoda) in freshwater. *Hydrobiologia* 595:167–76.

Brochet, A.L. Gauthier-Clerc, M., Guillemain, M., Fritz, H., Waterkeyn, A., Baltanás, Á. and Green, A.J. 2010. Field evidence of dispersal of branchiopods, ostracods and bryozoans by teal (*Anas crecca*) in the Camargue (southern France). *Hydrobiologia* 637:255–61.

Brown, L.J. and Jung R.E. 2005. *An Introduction to Mid-Atlantic Seasonal Pools*, EPA/903/B-05/001. Meade, MD: US EPA, Mid-Atlantic Integrated Assessment, FT.

Browne, R.A., Sorgeloos, P. and Trotman, C.N.A. 1991. *Artemia Biology*. Boca Raton, FL: CRC Press/Taylor & Francis.

Brtek, J. 1997. Checklist of the valid and invalid names of the "large branchiopods" (Anostraca, Notostraca, Spinicaudata and Laevicaudata), with a survey of the taxonomy of all Branchiopoda. *Zborník Slovenského Národného Múzea, Prírodné Vedy* 43:3–66.

Brtek, J. 2002. Taxonomical survey of the Anostraca, Notostraca Cyclestherida, Spinicaudata and Laevicaudata. *Zbornik Slovenského Národneho Músea* 48:49–59.

Camara, M.R. and De Medeiros Rocha R. 1987. *Artemia* culture in Brazil: An overview. In *Artemia, Research and Applications*. Vol. III, ed. P. Sorgeloos, D.A. Bengsten, W. Decleir, and E. Jaspers, 195–200. Wetteren: Universa Press.

Cart, J.F. and Rabet, N. 2003 Les grandes branchiopodes. In *Les listes rouges de la nature menacée en Alsace*, ed. par J.-P. Vacher, N. Rabet, J.-F. Cart, and G. Godinat, 186–95. Strasbourg: Collection Conservation.

Comer, P., Goodwin, K., Tomaino, A., Hammerson, G., Kittel, G., Menard, S., Nordman, C., Pyne, M., Reid, M., Sneddon, L. and Snow, K. 2005. *Biodiversity Values of Geographically Isolated Wetlands in the United States*. Arlington, VA: NatureServe.

Cottam, C. 1939. Food habits of North American diving ducks. *US Department of Agriculture Technical Bulletin* 643:1–140.

Creaser, E.P. 1931. North American phyllopods. *Science* 74:267–68.

Daday, E. 1910a. Monographic systématique des Phyllopodes anostracés. *Annales des Sciences Naturelles, Zoologiel, Series* 11:91–489.

Daday, E. 1910b. Quelques phyllopodes anostraces nouveaux. *Annales des Sciences Naturelles, Zoologiel* 9 12:241–64.

Daday, E. 1915. Monographie systématique des Phyllopodes Conchostracés. *Annales des Sciences Naturelles, Zoologie, Series* 9:39–330.

Daday, E. 1923. Monographie systématique des Phyllopodes Conchostracés. II. Leptestheriidae. *Annales des Sciences Naturelles, Zoologie* 6:255–390.

Daday, E. 1925. Monographie systématique des Phyllopodes Conchostracés. III. Limnadiidae. *Annales des Sciences Naturelles, Zoologie* 8:143–84.

Daday, E. 1926. Monographie systématique des Phyllopodous Conchostracés. III. Limnadiidae (suite). *Annales des Sciences Naturelles, Zoologie* 9:1–81.

Daday, E. 1927. Monographie systématique des Phyllopodes Conchostracós Troisiéme partie (fin). *Annales des Sciences Naturelles, Zoologie* 10:1–112.

Daniels, S.R., Hamer, M. and Rogers, D.C. 2004. Molecular evidence suggests an ancient radiation for the fairy shrimp genus *Streptocephalus* (Branchiopoda: Anostraca). *Biological Journal of the Linnean Society* 82:313–27.

Day, J.A., Stewart, B.A., de Moor, I.J. and Louw, A.E. (eds.) 1999. *Guides to the Freshwater Invertebrates of Southern Africa. Crustacea I: Notostraca, Anostraca, Conchostraca, and Cladocera*. South Africa: Water Research Commission Report TT121/00.

De Bell, L. 1992. Scanning electron microscope observations of brine shrimp larvae from space shuttle experiments. *Scanning Microscopy* 6:1129–35.

De Roeck, E.R., Vanschoenwinkel, B.J. Day, J.A. Xu, Y., Raitt, L. and Brendonck, L. 2007. Conservation status of large branchiopods in the Western Cape, South Africa. *Wetlands* 27:162–73.

De Waard, J.A., Sacherova, V. Cristescu, M.E.A., Remigio, E.A., Crease, T.J. and Hebert, P.D.N. 2006. Probing the relationships of the branchiopod crustaceans. *Molecular Phylogenetics and Evolution* 39:491–502.

Donald, D.B. 1983. Erratic occurrence of anostracans in a temporary pool: colonization and extinction or adaptation to variations in annual weather? *Canadian Journal of Zoology* 61:1492–98.

Dumont, H.J. and Negrea, S.V. 2002. *Branchiopoda*. In *Guides to the Identification of the Microinvertebrates of the Continental Waters of the World 19*. Leiden: Backhuys Publishers.

Ebert, T.A. and Balko, L.M. 1984. Vernal pools as islands in space and time. In *Vernal Pools and Intermittent Streams*, eds. S. Jain, and P. Moyle, 1–280. California: University of California Institute for Ecology Publications.

Eder, E. and Hödl, W. 1996. Wozu "Urzeitkrebse"? Praktische Bedeutung der Groß-Branchiopoden für Wirtschaft, Naturschutz und Wissenschaft. *Stapfia* 100:149–58.

Eder, E. and W. Hödl, 2002. Large freshwater branchiopods in Austria: Diversity, threats, conservational status. In *Modern Approaches to the Study of Crustacea*, ed. E. Escobar-Briones, and F. Alvarez, 281–89. New York: Kluwer Academic: Plenum Publishers.

Eng, L., Belk, D. and Ericksen, C. 1990. Californian Anostraca: Distribution, habitat, and status. *Journal of Crustacean Biology* 10:247–77.

Fritsch, F., Bininda-Emonds, O.R.P. and Richter, S. 2013. Unraveling the origin of Cladocera by identifying heterochrony in the developmental sequences of Branchiopoda. *Frontiers in Zoology* 10:35 54.

García-de-Lomas, J., Sala, J., García, C.M. and Alonso, M. 2015a. Clase Branchiopoda Orden Anostraca. *Revista IDE@-SEA* 67:1–12.

García-de-Lomas, J., Sala, J. and Alonso, M. 2015b. Clase Branchiopoda Orden Notostraca. *Revista IDE@-SEA*, 71:1–10.

García-de-Lomas, J., Sala, J. and Alonso, M. 2015c. Clase Branchiopoda Orden Spinicaudata. *Revista IDE@-SEA* 68:1–11.

Gołdyn, B., Bernard, R., Czyż, M.J. and Jankowiak, A. 2013. Diversity and conservation status of large branchiopods (Crustacea) in ponds of western Poland. *Limnologica* 42:264–70.

Hamer, M.L. and Appleton, C.C. 1996. The genus *Branchipodopsis* (Crustacea, Branchiopoda, Anostraca) in Southern Africa: Morphology, distribution, relationships, and the description of five new species. *South African Museum* 104:311–77.

Hamer, M.L. and Brendonck, L. 1997. Distribution, diversity and conservation of Anostraca (Crustacea: Branchiopoda) in southern Africa. *Hydrobiologia* 359:1–12.

Hamer M. and Martens, K. 1998. The large Branchiopoda (Crustacea) from temporary habitats of the Drakensberg region, South Africa. *Hydrobiologia* 384:151–65.

Hamer, M.L., Brendonck, L., Coomans, A. and Appleton, C. 1994a. A review of African Streptocephalidae (Crustacea: Branchiopoda: Anostraca) Part 1: South of Zambezi and Kunene rivers. *Archiv für Hydrobiologie, Supplement* 99:235–77.

Hamer, M.L., Brendonck, L., Coomans, A. and Appleton, C. 1994b. A review of African Streptocephalidae (Crustacea: Branchiopoda: Anostraca) Part 2: North of Zambezi and Kunene rivers, and Madagascar. *Archiv für Hydrobiologie, Supplement* 99:279–311.

Henrikson, L.S., Yohe II, R.M., Newman, M.E. and Druss, M. 1998. Freshwater crustaceans as an aboriginal food resource in the northern Great Basin. *Journal of California and Great Basin Anthropology* 20:72–87.

Horne, F.R. 1966. The effect of digestive enzymes on the hatchability of *Artemia salina* eggs. *Transactions of the American Microscopical Society* 85:271–74.

Hurlbert, S.H., Lopez, M. and Keith, J.O. 1984. Wilson's Phalarope in the Central Andes and its interaction with the Chilean Flamingo. *Revista Chilena de Historia Natural* 57:47–57.

ICZN (International Code of Zoological Nomenclature). 1999. *The International Trust for Zoological Nomenclature* 1999. 4th edition. London, UK: The Natural History Museum, http://siamensis.org/sites/default/files/iczn1999_code.pdf (download on 11st December 2019).

Ikeda, D.H. and Schlising, R.A. 1990. Vernal pool plants. *Studies from the Herbarium, California State University* 8:1–178.

Jain, S. (ed.). 1976. *Vernal Pools, Their Ecology and Conservation*. California: University of California Institute for Ecology Publications.

Jain, S. and Moyle, P. (eds.). 1984. *Vernal Pools and Intermittent Streams*. California: University of California Institute for Ecology Publications.

Jocque, M., Vanschoenwinkel, B. and Brendonck, L. 2010. Freshwater rock pools: A review of habitat characteristics, faunal diversity and conservation value. *Freshwater Biology* 55:1587–602.

Kellogg, V.L. 1906. A new *Artemia* and its life conditions. *Science* 24:594–96.

Korn, M. and Hundsdoerfer, A.K. 2006. Evidence for cryptic species in the tadpole shrimp *Triops granarius* (Lucas, 1864) (Crustacea: Notostraca). *Zootaxa* 1257:57–68.

Korn, M. and Hundsdoerfer, A.K. 2016. Molecular phylogeny, morphology and taxonomy of Moroccan *Triops granarius* (Lucas, 1864) (Crustacea: Notostraca), with the description of two new species. *Zootaxa* 4178:328–46.

Korn, M., Marrone, F., Pérez-Bote, J.L., Machado, M., Cristo, M., Cancela da Fonseca, L. and Hundsdoerfer, A.K. 2006. Sister species within the *Triops cancriformis* lineage (Crustacea, Notostraca). *Zoologica Scripta* 35:301–22.

Korn, M., Green, A.J., Machado, M., García-de-Lomas, J., Cristo, M., Cancela da Fonseca, L., Frisch, D., Pérez-Bote, J.L. and Hundsdoerfer, A.K. 2010. Phylogeny, molecular ecology and taxonomy of southern Iberian lineages of *Triops mauritanicus* (Crustacea: Notostraca). *Organisms Diversity and Evolution* 10:409–40.

Korn, M.N., Rabet, H.V., Ghate, F., Marrone, F. and Hundsdoerfer, A.K. 2013. Molecular phylogeny of the Notostraca. *Molecular Phylogenetics and Evolution* 69:1159–71.

Krapu, G.L. 1974. Foods of breeding pintails in North Dakota. *The Journal of Wildlife Management* 38:408–17.

Linder, F. 1945. Affinities within the Branchiopoda, with notes on some dubious fossils. *Arkiv för Zoologi* 37:1–28.

Longhurst, A.R. 1955. A review of the Notostraca. *Bulletin of the British Museum (Natural History)* 3:1–57.

MacDonald, G.H. 1980. The use of *Artemia* cysts as food by the flamingo (*Phoenicopterus ruber roseus*) and the shelduck (*Tadorna tadorna*). In *Ecology, Culturing, Use in Aquaculture*. The Brine Shrimp *Artemia*. Vol. 3, ed. G. Persoone, P. Sorgeloos, O. Roels, and E. Jaspers, 191–200. Wettern: Universa Press.

Machado, M., Cancela da Fonseca, L. and Cristo, M. 2017. Freshwater large branchiopods in Portugal: An update of their distribution. *Limnetica* 36:567–84.

Maeda-Martínez, A., Belk, D., Obregón-Barboza, H. and Dumont, H.J. 1995. Diagnosis and phylogeny of the New World Streptocephalidae (Branchiopoda: Anostraca). *Hydrobiologia* 298:15–44.

Maeda-Martínez, A.M., Obregón-Barboza, H., García-Velazco, H. and Prieto-Salazar, M.A. 2002a. 14. Branchiopoda: Anostraca. In *Biodiversidad, taxonomía y biogeografía de artrópodos de México*. Volumen III, ed. J.L. Bousquets, and J.J. Morrone, 305–22. Mexico City: Universidas Nacional Autónoma de México.

Maeda-Martínez, A.M., Obregón-Barboza, H. and García-Velazco, H. 2002b. 15. Branchiopoda: Cyclestherida, Laevicaudata, and Spinicaudata. In *Biodiversidad, taxonomía y biogeografía de artrópodos de México*. Volumen III, ed. J.L. Bousquets, and J.J. Morrone, 323–32. Mexico City: Universidas Nacional Autónoma de México.

Maeda-Martínez, A.M., Obregón-Barboza, H., García-Velazco, H. and Murugan, G. 2002c. 16. Branchiopoda: Notostraca. In *Biodiversidad, taxonomía y biogeografía de artrópodos de México*. Volume III, ed. J.L. Bousquets, and J.J. Morrone, 333–40. Mexico City: Universidas Nacional Autónoma de México.

Marrone, F., Korn, F., Stoch, F., Naselli Flores, F. and Turk, S. 2016. Updated checklist and distribution of large branchiopods (Branchiopoda: Anostraca, Notostraca, Spinicaudata) in Tunisia. *Biogeographia* 31:27–53.

Martin, J.W., Rogers, D.C. and Olesen, J. 2016. Collecting and processing branchiopods. *Journal of Crustacean Biology* 36:396–401.

Martínez-Pantoja, M.A., Alcocer, J. and Maeda-Martínez, A.M. 2002. On the Spinicaudata (Branchiopoda) from Lake Cuitzeo, Michoacán, México: First report of a clam shrimp fishery. *Hydrobiologia* 486:207–13.

Naganawa, H. 2001. Current prospect of the recent large branchiopodan fauna of East Asia: 3. revised classification of the recent Spinicaudata. *Aquabiology* 23:291–99.

Naganawa, H. and Orgiljanova, T.I. 2001. Current prospect of the recent large branchiopodan fauna of East Asia: 2. Order Anostraca. *Aquabiology* 23:186–94.

Nhiwatiwa, T., Waterkeyn, A., Riddoch, B.J. and Brendonck, L. 2014. A hotspot of large branchiopod diversity in south-eastern Zimbabwe. *African Journal of Aquatic Science* 39:57–65.

Nunes, B.S., Carvalho, F.D. Guilhermino, L.M. and Van Stappen, G. 2006. Use of the genus *Artemia* in ecotoxicity testing. *Environmental Pollution* 144:453–62.

Olesen, J. 1998. A phylogenetic analysis of the Conchostraca and Cladocera (Crustacea, Branchiopoda, Dipostraca). *Zoological Journal of the Linnean Society, London* 122:491–536.

Olesen, J. 2007. Monophyly and phylogeny of Branchiopoda, with focus on morphology and homologies of branchiopod phyllopodous limbs. *Journal of Crustacean Biology* 27:165–83.

Olesen, J. 2009. Phylogeny of Branchiopoda (Crustacea) – character evolution and contribution of uniquely preserved fossils. *Arthropod Systematics and Phylogeny* 67:3–39.

Padhye, S.M. and Lazo-Wasem, E.A. 2018. An updated and detailed taxonomical account of the large Branchiopoda (Crustacea: Branchiopoda: Anostraca, Notostraca, Spinicaudata) from the Yale North India Expedition deposited in the Yale Peabody Natural History Museum. *Zootaxa* 4394:207–18.

Padhye, S.M., Kulkarni, M.A. and Dumont, H.J. 2017. Diversity and zoogeography of the fairy shrimps (Branchiopoda: Anostraca) on the Indian subcontinent. *Hydrobiologia* 801:117–28.

Pinceel, T., Brendonck, L., Larmuseau, M.H.D., Vanhove, M.P.M., Timms, B.V. and Vanschoenwinkel, B. 2013a. Environmental change as a driver of diversification in temporary aquatic habitats: does the genetic structure of extant fairy shrimp populations reflect historic aridification? *Freshwater Biology* 58:1–17.

Pinceel, T., Vanschoenwinkel, B., Waterkeyn, A., Vanhove, M.P.M. Pinder, A., Timms, B.V. and Brendonck, L. 2013b. Fairy shrimps in distress: a molecular taxonomic review of the diverse fairy shrimp genus *Branchinella* (Anostraca: Thamnocephalidae) in Australia in the light of ongoing environmental change. *Hydrobiologia* 700:313–27.

Plodsomboon, S., Maeda-Martínez, A.M., Obregón-Barboza, H. and Sanoamuang, I. 2012. Reproductive cycle and genitalia of the fairy shrimp *Branchinella thailandensis* (Branchiopoda: Anostraca). *Journal of Crustacean Biology* 32:711–26.

Proctor, V.W. 1964. Viability of crustacean eggs recovered from ducks. *Ecology* 45:656–58.

Regier, J.C., Shultz, J.W., Zwick, A., Hussey, A., Ball, B., Wetzer, R., Martin, J.W. and Cunningham, C.W. 2010. Arthropod relationships revealed by phylogenomic analysis of nuclear protein-coding sequences. *Nature* 463:1029–84.

Remigio, E.A. and Hebert, P.D.N. 2000. Affinities among anostracan (Crustacea: Branchiopoda) families inferred from phylogenetic analyses of multiple gene sequences. *Molecular Phylogenetics and Evolution* 17:117–28.

Remigio, E.A., Hebert, P.D.N. and Savage, A. 2001. Phylogenetic relationships and remarkable radiation in *Parartemia* (Crustacea: Anostraca), the endemic brine shrimp of Australia: evidence from mitochondrial DNA sequences. *The Biological Journal of the Linnean Society* 74:59–71.

Remigio, E.A., Timms, B.V. and Hebert, P.D.N. 2003. Phylogenetic systematics of the Australian fairy shrimp genus *Branchinella* based on mitochondrial DNA sequences. *Journal of Crustacean Biology* 23:436–42.

Richter, S., Olesen, J. and Wheeler, W.C. 2007. Phylogeny of Branchiopoda (Crustacea) based on a combined analysis of morphological data and six molecular loci. *Cladistics* 23:301–36.

Rogers, D.C. 2001. Revision of the Nearctic *Lepidurus* (Notostraca). *Journal of Crustacean Biology* 21:991–1006.

Rogers, D.C. 2002a. The amplexial morphology of selected Anostraca. *Hydrobiologia* 486:1–18.

Rogers, D.C. 2002b. A morphological re-evaluation of the anostracan families Linderiellidae and Polyartemiidae, with a redescription of the linderiellid *Dexteria floridana* (Dexter 1956) (Crustacea: Branchiopoda). *Hydrobiologia* 486:57–61.

Rogers, D.C. 2003. Revision of the thamnocephalid genus *Phallocryptus* (Crustacea: Branchiopoda: Anostraca). *Zootaxa* 257:1–14.

Rogers, D.C. 2005. A new genus and species of chirocephalid fairy shrimp (Crustacea: Branchiopoda: Anostraca) from Mongolia. *Zootaxa* 997:1–10.

Rogers, D.C. 2006. A genus level revision of the Thamnocephalidae (Crustacea: Branchiopoda: Anostraca). *Zootaxa* 1260:1–25.

Rogers, D. C. 2009. Branchiopoda (Anostraca, Notostraca, Laevicaudata, Spinicaudata, Cyclestherida). In *Encyclopedia of inland waters*. Vol. 2, ed. G.F. Likens, 242–49. New York: Academic Press.

Rogers, D.C. 2013. Anostraca Catalogus. *The Raffles Bulletin of Zoology* 61:525–46.

Rogers, D.C. 2014a. Anostracan (Crustacea: Branchiopoda) Biogeography I. North American Bioregions. *Zootaxa* 3838:251–75.

Rogers, D.C. 2014b. Anostracan (Crustacea: Branchiopoda) Biogeography II. Relating distribution to geochemical substrate properties in the USA. *Zootaxa* 3856:1–49.

Rogers, D.C. 2014c. Larger hatching fractions in avian dispersed anostracan eggs (Branchiopoda). *Journal of Crustacean Biology* 34:135–43.

Rogers, D.C. 2014d. Two new cryptic anostracan (Branchiopoda: Streptocephalidae, Chirocephalidae) species. *Journal of Crustacean Biology* 34:862–74.

Rogers, D.C. 2015. A conceptual model for anostracan biogeography. *Journal of Crustacean Biology* 35:686–99.

Rogers, D.C. 2020. Spinicaudata catalogus (Crustacea: Branchiopoda). *Zoological Studies* 59:45.

Rogers, D.C. and Aguillar, A. 2020. Molecular evaluation of the fairy shrimp family Branchinectidae (Crustacea: Anostraca) supports peripatric speciation and complex divergence patterns. *Zoological Studies* 59:14. doi:10.6620/ZS.2020.59-14.

Rogers, D.C. and Coronel, J.S. 2011. A redescription of *Branchinecta pollicifera* Harding, 1940, and its placement in a new genus (Branchiopoda: Anostraca: Branchinectidae). *Journal of Crustacean Biology* 31:717–24.

Rogers, D.C. and Hamer, M. 2012. Two new species of *Metabranchipus* Masi, 1925 (Anostraca: Branchipodidae). *Journal of Crustacean Biology* 32:972–80.

Rogers, D.C. and Hann, B.J. 2016. *Class Branchiopoda (in Chapter 16, Phylum Arthropoda). Thorp and Covich's freshwater invertebrates*. 4th edition. Volume II. Keys to the Nearctic fauna, eds. J.H. Thorp, and D.C. Rogers. Boston: Academic Press.

Rogers, D.C. and Olesen, J. 2016. *Laevicaudata catalogus* (Crustacea: Branchiopoda): an overview of diversity and terminology. *Arthropod Systematics and Phylogeny* 74:221–40.

Rogers, D.C. and Padhye, S.M. 2014. A new species of *Streptocephalus* (Crustacea: Anostraca: Streptocephalidae) from the Western Ghats, India, with a key to the Asian species. *Zootaxa* 3802:75–84.

Rogers, D.C. and Padhye, S.M. 2015. Review of the large branchiopod crustacean fauna of the Indian Subcontinent (Anostraca,Notostraca, Laevicaudata, Spinicaudata, Cyclestherida). *Journal of Crustacean Biology* 35:392–406.

Rogers, D.C., de los Ríos, P. and Zúñiga, O. 2008. Fairy shrimp (Branchiopoda: Anostraca) of Chile. *Journal of Crustacean Biology* 28:543–50.

Rogers, D.C., Rabet, M. and Weeks, S.C. 2012. Revision of the extant genera of Limnadiidae (Branchiopoda: Spinicaudata). *Journal of Crustacean Biology* 32:827–42.

Rogers, D.C., Thaimuangphol, W., Saengphan, N. and Sanoamuang, L. 2013a. Current knowledge of the South East Asian large branchiopod Crustacea (Anostraca, Notostraca, Laevicaudata, Spinicaudata, Cyclestherida). *Journal of Limnology* 72:69–80.

Rogers, D.C., Shu, S. and Yang, J. 2013b. The identity of *Branchinella yunnanensis* Shen, 1949, with a brief review of the subgenus *Branchinellites* (Branchiopoda: Anostraca: Thamnocephalidae). *Journal of Crustacean Biology* 33:576–81.

Rogers, D.C., Schwentner, M., Olesen, J. and Richter, S. 2015. Evolution, classification and global diversity of large Branchiopoda. *Journal of Crustacean Biology* 35:297–300.

Rogers, D.C., Saengphan, N., Thaimuangphol, W. and Sanoamuang, L. 2016a. The lynceid clam shrimps (Branchiopoda: Laevicaudata) of Thailand, with keys to the Eurasian species. *Journal of Crustacean Biology* 36:384–92.

Rogers, D.C., Dadseepai, P. and Sanoamuang, L. 2016b. The spinicaudatan clam shrimps (Branchiopoda: Diplostraca) of Thailand. *Journal of Crustacean Biology* 36:567–75.

Rogers, D.C., Chang, T.C. and Wang, Y.-C. 2017. A new *Eocyzicus* (Branchiopoda: Spinicaudata) from Taiwan, with a review of the genus. *Zootaxa* 4318:254–70.

Rogers, D.C., Kotov, A.A. Sinev, A.Y. Glagolev, S.M. Korovchinsky, N.M. Smirnov, N.N. and Bekker, E.I. 2019. *Class Branchiopoda. Thorp and Covich's Freshwater Invertebrates*. 4th edition. Volume V. Keys to the Palaearctic fauna, eds. D.C. Rogers, and J.H. Thorp. Boston: Academic Press.

Rogers, D.C., Severo-Neto, F., Vieira Volcan, M., De los Ríos, P., Epele, L.B., Ferreira, A.O. and Rabet, N. 2020a. Comments and records on the large branchiopod Crustacea (Anostraca, Notostraca, Laevicaudata, Spinicaudata, Cyclestherida) of the Neotropical and Antarctic bioregions. *Studies on Neotropical Fauna and Environment*. doi: 10.1080/01650521.2020.1728879.

Rogers, D.C., Cohen, R.G. and Hann, B.J. 2020b. *Class Branchiopoda. Damborenea, Thorp and Covich's Freshwater Invertebrates.* 4th edition. Volume IV. Keys to the Neotropical fauna, eds. C.D. Rogers, and Thorp, J.H. Boston: Academic Press.

Samraoui, B. and Dumont, H.J. 2002. The large branchiopods (Anostraca, Notostraca and Spinicaudata) of Numidia (Algeria). *Hydrobiologia* 486:119–23.

Saunders III, J.F., Belk, D. and Dufford, R.1993. Persistence of *Branchinecta paludosa* (Anostraca) in southern Wyoming, with notes on zoogeography. *Journal of Crustacean Biology* 13:184–89.

Schlising, R.A. and Alexander, D.G. (eds.) 2007. *Vernal Pool Landscapes,* 14. California: Studies from the Herbarium, California State University.

Schwentner, M., Timms, B.V. Bastrop, R. and Richter, S. 2009. Phylogeny of *Spinicaudata* (Branchiopoda, Crustacea) based on three molecular markers—An Australian origin for Limnadopsis. *Molecular Phylogenetics and Evolution* 53:716–25.

Schwentner, M., Timms, B.V. and Richter, S. 2012a. Description of four new species of *Limnadopsis* from Australia (Crustacea: Branchiopoda: Spinicaudata). *Zootaxa* 3315:42–64.

Schwentner, M., Timms, B.V. and Richter, S. 2012b. Flying with the birds? Recent large-area dispersal of four Australian *Limnadopsis* species (Crustacea: Branchiopoda: Spinicaudata). *Ecology and Evolution* 2:1605–26.

Schwentner, M, Clavier, S., Fritsch, M., Olesen, J., Padhye, S., Timms, B.V. and Richter, S. 2013. *Cyclestheria hislopi* (Crustacea: Branchiopoda): A group of morphologically cryptic species with origins in the Cretaceous. *Molecular Phylogenetics and Evolution* 66:800–10.

Schwentner, M., Richter, S., Rogers, D.C. and Giribet, G. 2018. Tetraconatan phylogeny with special focus on Malacostraca and Branchiopoda. *Proceedings of Royal Society, Biological Science* 285:15–24.

Schwentner, M., Rabet, N., Richter, S., Giribet, G., Padhye, S., Cart, J.-F., Bonillo, C., and Rogers, D.C. 2020. Phylogeny and biogeography of spinicaudata (crustacea: branchiopoda). *Zoological Studies* 59:44.

Smirnov, S.S. 1928. To the phyllopod fauna of the environs of Murom town. *Raboty Okskoi Biologischen Stantsii* 5:117–24.

Šrámek-Hušek, R., Straškraba, M. and Brtek, J. 1962. *Fauna ČSSR, Lupenonožci– Branchiopoda, svazek 16.* Praha: Československé Akademie Věd.

Sriputhorn, K. and Sanoamuang, L. 2011. Fairy shrimp (*Streptocephalus sirindhor- nae*) as live feed improve growth and carotenoid contents of giant freshwater prawn *Macrobrachium rosenbergii. International Journal of Zoological Research* 7:138–46.

Stenderup, J.T., Olesen, J. and Glenner, H. 2006. Molecular phylogeny of the Branchiopoda (Crustacea)–multiple approaches suggest a 'diplostracan' ancestry of the Notostraca. *Molecular Phylogenetics and Evolution* 41:182–94.

Straškraba, M. 1962. 3.rad Conchostraca-Skeblovk. In *Lupenonozci-Branchiopoda. Fauna CSSR. Svazek 16*, eds. R. Srámek-husek, M. Straskraba, and J. Brtek, 150–174. Praha: Nakladatelství Ceskosolvenské Adademie Ved.

Straškraba, M. 1965a. Taxonomic studies on Czechoslovak Conchostraca, 1. Family Limnadiidae. *Crustaceana* 9:263–73.

Straškraba, M. 1965b. Taxonomical studies on Czechoslovak Conchostraca II. Families Lynceidae and Cyzicidae. *Věstník Československé Zoologické Společnosti v Praze* 29:205–14.

Straškraba, M. 1966. Taxonomical studies on Czechoslovak Conchostraca III. Family Leptestheriidae, with some remarkson the variability and distribution of Conchostraca and a key to the Middle-European species. *Hydrobiologia* 27:571–89.

Sun, X., Xia, X. and Yang, Q. 2011. Phylogeny of Conchostraca (Crustacea: Branchiopoda) based on 28S rDNA D1-D2 and partial 16S rDNA sequences. *Acta Micropalaeontologica Sinica* 28:370–80.

Swanson, G.A., Meyer, M.I. and Serie, J.R. 1974. Feeding ecology of breeding blue-winged teals. *The Journal of Wildlife Management* 38:396–407.

Timms, B.V. 2012. An appraisal of the diversity and distribution of large branchiopods (Branchiopoda: Anostraca, Laevicaudata, Spinicaudata, Cyclestherida, Notostraca) in Australia. *Journal of Crustacean Biology* 32:615–23.

Timms, B.V. 2013. A revision of the Australian species of *Lynceus* Müller, 1776 (Crustacea: Branchiopoda: Laevicaudata, Lynceidae). *Zootaxa* 3702:501–33.

Timms, B.V. 2014. A review of the biology of Australian halophilic anostracans (Branchiopoda: Anostraca). *Journal of Biological Research* 21:1–8.

Timms, B.V. 2015. A revised identification guide to the fairy shrimps (Crustacea: Anostraca: Anostracina) of Australia. *Museum of Victoria Science Report* 20:1–25.

Timms, B.V. 2016a. A partial revision of the Australian Eulimnadia Packard, 1874 (Branchiopoda: Spinicaudata: Limnadiidae). *Zootaxa* 4066:351–89.

Timms, B.V. 2016b. A review of the Australian endemic clam shrimp, Paralimnadia Sars 1896 (Crustacea: Branchiopoda: Spinicaudata). *Zootaxa* 4161:451–508.

Timms, B.V. 2018. Key to the Australian clam shrimps (Crustacea: Branchiopoda: Laevicaudata, Spinicaudata, Cyclestherida). *Museum of Victoria Science Report* 19:1–44.

Timms, B.V. and Schwentner, M. 2017. A revision of the clam shrimp *Australimnadia* Timms and Schwentner, 2012 (Crustacea: Spinicaudata: Limnadiidae) with two new species from Western Australia. *Zootaxa* 4291:81–98.

Tippelt, L. and Schwentner, M. 2018. Taxonomic assessment of Australian *Eocyzicus* species (Crustacea: Branchiopoda: Spinicaudata). *Zootaxa* 4410:401–52.

Tuytens, K., Vanschoenwinkel, B., Clegg, B., Nhiwatiwa, T. and Brendonck, L. 2015. Exploring links between geology, hydroperiod, and diversity and distribution patterns of anostracans and notostracans (Branchiopoda) in a tropical savannah habitat in SE Zimbabwe. *Journal of Crustacean Biology* 35:309–18.

van den Broeck, M., Waterkeyn, A. Rhazi, L. and Brendonck, L. 2015. Distribution, coexistence, and decline of Moroccan large branchiopods. *Journal of Crustacean Biology* 35:355–65.

Vanschoenwinkel, B., Pinceel, T., Vanhove, M.P.M., Denis, C., Jocque, M., Timms, B.V. and Brendonck, L. 2012. Toward a global phylogeny of the "living fossil" crustacean order of the Notostraca. *PLoS One* 7:e34998.

Vekhov, N.V. 1992. The fauna of fairy shrimps (Anostraca) and tadpole shrimps (Notostraca) of Siberia and Far East. 2. Rare and disappearing species for Red Data Books. *Sibirskii Biologicheskii Zhurnal* 1992:45–50.

Wang, C.-C. and Chou, L.-S. 2015. Terminating dormancy: Hatching phenology of sympatric large branchiopods in Siangtian Pond, a temporary wetland in Taiwan. *Journal of Crustacean Biology* 35:301–8.

Weekers, P.H.H., Murugan, G., Vanfleteren, J.R., Belk, D. and Dumont, H.J. 2002. Phylogenetic analysis of anostracans (Branchiopoda: Anostraca) inferred from nuclear 18S ribosomal DNA (18S rDNA) sequences. *Molecular Phylogenetics and Evolution* 25:535–44.

Weeks, S.C., Chapman, E.G., Rogers, D.C., Senyo, D.M. and Hoeh, W.R. 2009. Evolutionary transitions among dioecy, androdioecy and hermaphroditism in limnadiid clam shrimp (Branchiopoda: Spinicaudata). *Journal of Evolutionary Biology* 22:1781–99.

Williams, D.D. 1987. *The Ecology of Temporary Waters.* Portland, Oregon: Timber Press.
Witham, C.W., Bauder, E.T., Belk, D., Ferrin Jr., W.R. and Orduff R. (eds.). 1996. *Ecology, Conservation, and Management of Vernal Pool Ecosystems—Proceedings from a 1996 Conference.* Sacramento, CA: California Native Plant Society
Zedler, P.H. 1987. *The Ecology of Southern California Vernal Pools: A Community Profile. Biological Report 85.* California: U.S. Fish and Wildlife Service.

CHAPTER **8**

Faunal Patterns in the Cladocera (Crustacea: Branchiopoda) on the Indian Subcontinent with Special Emphasis on Their Body Size Distribution

Sameer M. Padhye and Henri J. Dumont

CONTENTS

8.1   INTRODUCTION

Cladocera are an important group of freshwater zooplankton found in all types of aquatic environments (Dumont and Negrea 2002). The earlier notion of a cosmopolitan distribution concept in Cladocera has now given way to regional endemism with species having more restricted distributions than previously thought (Frey 1987, Van Damme and Dumont 2008, Xu et al. 2009, Popova et al. 2016). Distinctive patterns of distribution are observed in cladocerans on local and regional scales in response to factors such as water temperature and local environmental conditions, latitudinal and altitudinal extent, climatic conditions, as well as invasive species (Dumont 1980, Fernando 1980, Chengalath 1987, Gillooly and Dodson 2000, Dumont and Negrea 2002, Havel and Medley 2006, Van Damme and Eggermont 2011).

Research on zoogeography in this group has received unequal attention, with patterns relatively well known in specific groups (e.g., Genus *Daphnia* Müller 1785 – Fernando 1980, Popova and Kotov 2013) or specific regions (e.g., Africa: Dumont 1980, Chiambeng and Dumont 2005). A positive association of species richness with increasing latitude in Cladocera was suggested by Fernando (1980) with the general consensus being "tropics having lesser species than the temperate regions". This was later challenged by Dumont (1994). Korovchinsky (2006) put forth the "ejected relicts" hypothesis for cladoceran distribution wherein current cladoceran zoogeography patterns are thought to be due to their old, relict distributions and several extinctions.

Body size as a trait shapes spatial distributional patterns; it follows a unimodal trend as a function of latitudinal increase with temperate regions having the biggest sizes with a decrease toward the poles (Gillooly and Dodson 2000). Species composition in cladocerans shifts in terms of body size based on predation and/or competition (Brooks and Dodson 1965) affecting their local and/or regional diversity. *Daphnia* species, with some of the largest among cladocerans, show a distinct zoogeography with respect to latitude as well as altitude, especially in the tropics in case of the latter (Green 1995, Benzie 2005, Van Damme and Eggermont 2011, Popova and Kotov 2013).

The Indian subcontinent has an interesting formation and, as a result of this, an interesting extant geography (Mani 1974). A large altitudinal gradient is observed from the high altitudinal Himalayas in the north-northeast to the Gangetic plains that lie on its foothills and a desert in the northwest (Mani 1974). This has resulted in varied climatic conditions within a small region, giving way to diverse eco-regions. The effect of these factors on zooplankton diversity has been described (but not analyzed) earlier given the available data at the time (Fernando 1980, Fernando and Kanduru 1984). Since then, new and interesting discoveries have been made within the subcontinent (Hudec 1987, 1991, Kotov 2000, Manca et al. 2006, Chatterjee et al. 2013, Padhye and Dumont 2014a).

Various local and regional cladoceran checklists exist for the subcontinent, yet most of them are unreliable and, with the exception of some (Michael and Sharma 1988, Sharma and Sharma 2008, 2014, Chatterjee et al. 2013, Padhye and Dumont 2014b), fail to comment on wider regional zoogeographical patterns. To our knowledge, few studies address the overall cladoceran zoogeographic patterns of this region.

Given this background, we here present a regional distribution of the Cladocera on the Indian subcontinent. We specifically address within the Indian subcontinent: (a) trends in faunal composition on a regional scale, (b) differences among cladoceran assemblages, (c) body size variation with respect to latitudinal shift, and (d) relationships between body size and beta diversity patterns. We also highlight some characteristic faunal elements of the Indian subcontinent.

8.2 MATERIAL AND METHODS

Cladoceran occurrence data were collected and compiled from the literature of selected regions of the Indian subcontinent (refer to Fig. 8.1 for selected regions).

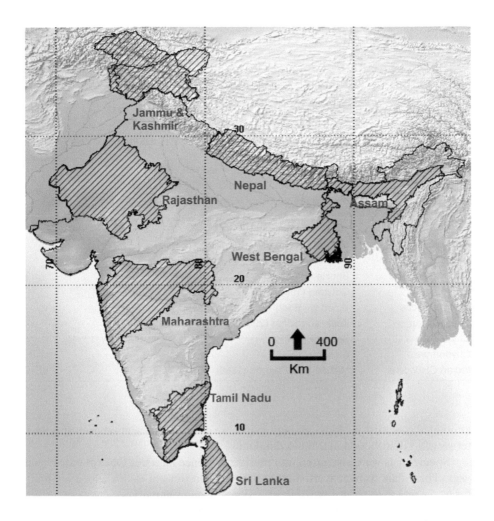

Figure 8.1 The map of the Indian subcontinent showing the regions selected for the study (Map shapefiles are old and hence do not show the current union territories of Jammu & Kashmir and Ladakh)

We used only reliable reports among the many unreliable checklists available. We generated cladoceran species lists for the following regions: Sri Lanka, Tamil Nadu, Maharashtra, Rajasthan, Assam, West Bengal, Nepal, and Jammu and Kashmir (J&K henceforth). These regions were selected for two reasons: (1) they have a distinct geography resulting in different environmental conditions (e.g., J&K has large glacial lakes, Rajasthan has a desert, Assam has large patches of evergreen forests), and (2) reliable species reports were of better quality than in other regions (the literature references used for specific regions are in Table 8.1). Maps were made using shape files of the region available online in DIVA-GIS (v 7.5c). Species names and reports were cross-checked with the annotated cladoceran checklist of India by Chatterjee et al. (2013), edited and updated wherever necessary. Since a detailed

Table 8.1   **References Used for Obtaining the Species Occurrence and Additional Body Size Data**

Region	References used
Assam	Michael and Sharma (1988), Sharma and Sharma (2008), Sharma and Sharma (2014)
Jammu and Kashmir	Michael and Sharma (1988), Raina and Vass (1993)
Maharashtra	Rane (2013), Padhye and Dumont (2014b)
Nepal	Dumont and Van de Velde (1977), Swar and Fernando (1979), Manca et al. (2006)
Rajasthan	Biswas (1971), Venkataraman (1990, 1992)
Sri Lanka	Fernando (1980), Rajapaksa and Fernando (1982)
Tamil Nadu	Michael (1973), Michael and Sharma (1988), Venkataraman (1999)
West Bengal	Michael and Sharma (1988), Venkataraman and Das (1993), Venkataraman et al. (2000)
Body size	**References used**
Sididae	Korovchinsky (1992)
Daphniidae	Benzie (2005)
Moinidae	Goulden (1968)
Macrothricidae	Smirnov (1992)
Ilyocryptidae	Kotov and Štifter (2006)
Chydoridae	Smirnov (1971)
Chydorinae	Smirnov (1996)
Indian Cladocera	Michael and Sharma (1988)

annotated checklist for India already exists, we did not comment on each species in terms of its distribution to avoid redundancy.

Species lists were converted into presence/absence matrices for explorative analyses. These data were converted into relative proportions of each cladoceran family for every region to reduce the high variation in species numbers.

Beta diversity between the different regions was calculated using Jaccard's index of similarity (Real 1999). This index was specifically used since we could not verify true absences in the fauna of these regions. We used UPGMA (Unweighted Pair-Group Method using Arithmetic averages) hierarchical clustering to check regional clustering patterns using the Jaccard's index values (see Padhye et al. 2017 for details). Clusters obtained were named based on the regions defined by Padhye et al. (2017) for Anostraca, viz., Northern Zone and the rest of the subcontinent. We used the "vegan" package (Oksanen et al. 2013, https://CRAN.R-project.org/package=vegan) in R (3.5.0) to calculate beta diversity, which provides dissimilarity values. Bootstrap values for the cluster nodes were calculated using the "fpc" package (Hennig 2018). Bootstrapping was carried out 999 times and only the nodes having a value of 90 or above are reported.

For the mean body size (MBL), values given by Gillooly and Dodson (2000) and Rizo et al. (2017) were used. Additional data on some species were obtained from standard faunal identification references (Table 8.1: Body size references). In most instances, the literature had the range of size for a species, from which an average was calculated. Due to high variation in species number per region, which could influence the average body size per region, we randomly selected 25 species per region and further generated 50 such lists to obtain an average body length (with Standard Error) per region.

In order to check body size–based differences in beta diversity patterns, species were categorized into three size categories: "big" (>1 mm), "medium" (1 mm > x > 0.5 mm), and "small" (<0.5 mm). The Jaccard's index was calculated between all regions for these three size classes. Statistical significance between the regions was studied using the nonparametric Kruskal–Wallis test after testing the data for normality with Shapiro–Wilk test using R.

8.3 RESULTS

At family level, Chydoridae was the most species-rich, followed by Daphniidae. The ratio of Chydoridae to Daphniidae species decreases with increasing latitude (Fig. 8.2).

Two main beta diversity value clusters were observed with high bootstrap support (Fig. 8.3A), one being the higher latitude/higher altitude regions "NZ" (Fig. 8.3A "red" colored regions) and the second of more tropical/subtropical zones "RS" (Fig. 8.3A "black" colored regions).

Body size increases gradually along the latitudinal gradient (Fig. 8.4A). The highest MBL observed was 1.07 (mean ± 0.034 mm S.E.) in Jammu and Kashmir (J&K), while the minimum was from Assam (mean = 0.66 ± 0.016 S.E.). The Northern Zone group mean size was larger (0.99 ± 0.08 mm) than that of the rest of the continent (0.76 ± 0.02 mm).

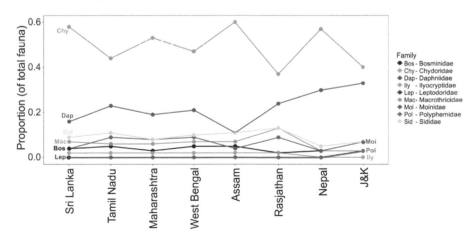

Figure 8.2   Plot showing the composition of cladoceran fauna across the regions of the Indian subcontinent. Each line represents the proportion of species of that respective family across the regions.

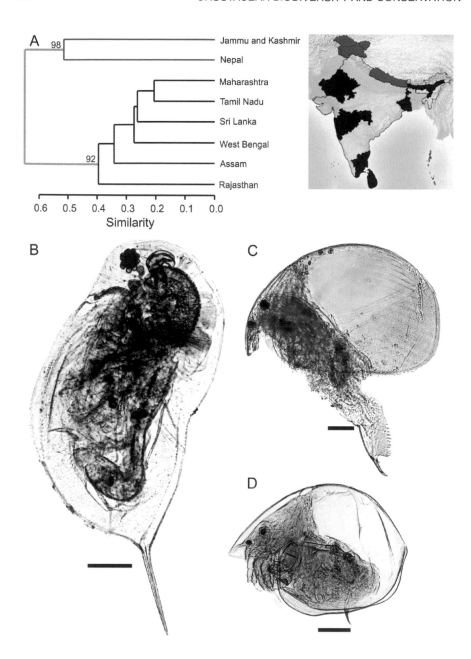

Figure 8.3   (A) Dendrogram showing the clustering pattern of the regions using the Jaccard's index values (red-colored cluster represents the "Northern Zone" also represented by the regions with the same color in the adjacent map; black-colored cluster represents the "Rest of the Subcontinent" represented by the regions with the same color); values at the nodes are the bootstrap values. (B–D) A few characteristic species found in the subcontinent. (B) *Daphnia similoides*. (C) *Flavalona cheni*. (D) *Indialona ganapati*. (Map shapefiles are old and hence do not show the current union territories of Jammu & Kashmir and Ladakh)

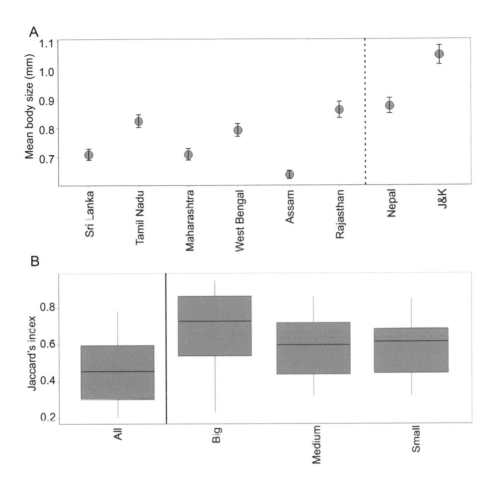

Figure 8.4   (A) Average body size per study region. Each point is an average (+S.E.) of 50 lists of 25 randomly selected species from the species list of each region. The vertical line corresponds with the regional clustering pattern observed. (B) Boxplots showing the range and median values of Jaccard's index calculated between the regions for all species followed by considering only big-, medium-, and small-sized species, respectively, with a vertical line separating the size categories from the "All" (all species) category.

The beta diversity median and range values were higher for big species (median: 0.72; range: 0.23–0.95) as compared to the medium- (median: 0.62; range: 0.32–0.85) and small-sized species (median: 0.62; range: 0.32–0.86) (Fig. 8.4B), though the difference between the three groups was not significant (Kruskal–Wallis $\chi^2 = 4$; p = 0.135).

8.3.1 Comments on Peculiar Fauna

Species Showing an Amphi-Pacific Disjunct Distribution

The genus *Leydigiopsis* (reported as *L. curvirostris* Sars, 1901) was reported in India originally from the NE state of Assam (Sharma and Sharma 2007). It was suggested to be the SE Asian endemic species, *L. pulchra* Van Damme and Sinev, 2013, extending the range of this species from SE Asia and South China into NE India (Fig. 9 in Van Damme and Sinev 2013). Congeneric forms of *Leydigiopsis* live more than 15,000 km away in the Neotropics, where at least three more species are known to occur (Sinev 2004, Van Damme and Sinev 2013).

Padhye and Dumont (2014a) described *Moina hemanti*, which closely resembles the Neotropical/Central American *M. dumonti* Kotov, Elías-Gutiérrez, and Granados-Ramírez, 2005 (Fig. 8.5A). The overall morphology of the two species is similar with only the finest details such as the gnathobase structure of the second trunk limb different between the two species (Kotov et al. 2005, Padhye and Dumont 2014a).

An explanation for this AmphiPacific distribution pattern is provided by Van Damme and Sinev (2013) where they stated that this might result from a boreotropical migration, wherein the species were widespread in the Cenozoic tropics until the Oligocene–Miocene, after which the distribution became fragmented and finally disjunct.

8.3.2 Subcontinental Endemics

Few species are restricted to the Indian subcontinent of which the taxonomy is stable. These are *Latona tiwari* Biswas, 1964; *Moina oryzae* Judec, 1987, *Indialona ganapati* Petkovski, 1966 (Fig. 8.3D), and *M. hemanti* Padhye and Dumont, 2014. *Indialona ganapati* has a distribution over a wide geographical area (Fig. 8.5B), while the other three have restricted distributions. *Indialona ganapati*, a sole member of the tribe Indialonini, is one of few chydorids adapted to open waters (Kotov 2000). In one of the author's collections (SP), *I. ganapati* was consistently observed in the limnetic zones of large water reservoirs in high abundances along with similar sized *B. longirostris* and *B. deitersi*. *Moina oryzae*, which closely resembles *M. reticulata*, is from the *Moina* species group having an ocellus, with the latter species reported from the Neotropics until its discovery in Africa (Hudec 1987). There is an additional species known from the region, viz., *B. tripurae*, but the exact status of this species is still unclear (sensu Chatterjee et al. 2013).

8.3.3 Species Showing a Latitudinal Restricted Distribution

The predatory *Leptodora kindtii* (Focke 1844) s. lat. and *Polyphemus pediculus* (Linnaeus 1761) s. lat. have been reported from some Himalayan glacial lakes of J&K (Michael and Sharma 1988). Xu et al. (2009) suggested that this *Leptodora* might either be a relict, older even than the Himalayan orogenic event, or it could be a recent introduction by migratory birds, which visit J&K in winter. Not much is known about the systematics of *P. pediculus* from the region.

Daphnia shows the most striking distribution among the cladocerans on the ISC with distinct segregation of species with respect to latitude/altitude. Padhye et al. (2016) provided a detailed account of its zoogeography and categorized the species into three groups, separated based on the combination of high altitude and lower

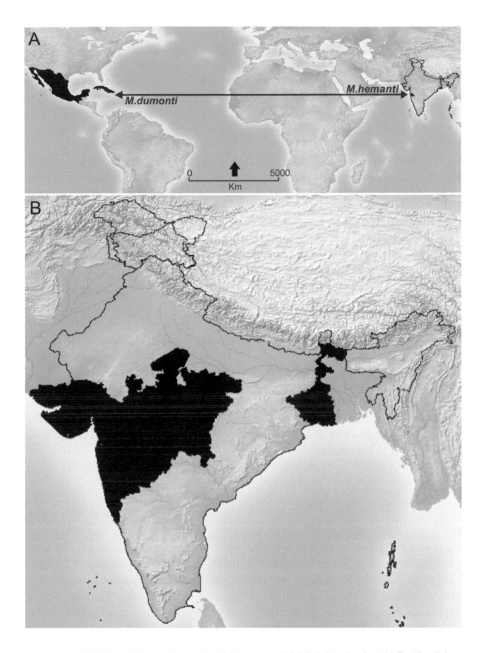

Figure 8.5   (A) Map of *Moina dumonti* and *M. hemanti* highlighting the Amphi-Pacific disjunct distribution. Arrowheads indicate the region (and not the type locality) of their occurrence. (B) Distribution map of *Indialona ganapati*. (Please note that map shapefiles are old and hence do not show the current union territories of Jammu and Kashmir and Ladakh,)

mean temperature. They are: (1) altitudinal endemics; (2) widely distributed species (complexes), and (3) lower latitude (higher temperature) taxa, favoring large *Ctenodaphnia*. *Ctenodaphnia* besides *Daphnia lumholtzi* Sars, 1885 are restricted to small temporary and fishless water bodies at altitudes >1,000 m.a.s.l. in the Western Maharashtra region where they are abundant (SP pers. obs.)

8.3.4 Oriental Endemics

Many "cosmopolitan" species have been reexamined from the Oriental region and described as separate species, and that list is increasing (Korovchinsky 2013). Oriental forms of species complexes such as *Daphnia similis* Claus, 1876, *Pseudochydorus globosus* (Baird 1843), *Alona quadrangularis* (Müller 1776), *Anthalona verrucosa* (Sars 1901), *Karualona karua* (King 1853), and *Alona costata* Sars 1862 have all been redescribed, mostly from the SE Asian region (Hudec 1991, Sinev 1999, Van Damme et al. 2011, 2013, Sinev 2012, Sinev et al. 2016). Some of these species are also reported from parts of the Indian subcontinent (Chatterjee et al. 2013). *Pseudochydorus bopingii* Sinev, Garibian and Gu, 2016 and *Camptocercus vietnamensis* Sinev, 2011 described from SE Asia (Sinev 2011, Sinev et al. 2016) are also found in the Western Ghats (SP pers. obs). *Salinalona* Van Damme and Maiphae, 2013 is an Oriental endemic genus reported from SE Asia as well as India (in India as *A. taraporevalae* Shirgur and Naik 1977), though the authors suggested it could be different until closer examination (Van Damme and Maiphae 2013).

Some of the earlier Indian subcontinental endemics are also reported from SE Asia. *Ilyocryptus bhardwaji* Battish, 1981, *Daphnia similoides* Hudec, 1991 (Fig. 8.3B), and *Flavalona cheni* (Sinev, 1999) (Fig. 8.3C), all described from India (Hudec 1991, Sinev 1999, Kotov and Štifter 2006) have been reported from SE Asia as well (sensu Korovchinsky 2013).

8.4 DISCUSSION

The Indian subcontinent cladoceran fauna is most similar to SE Asia (Padhye and Dumont 2014b) and its family-level species composition is comparable to that of other tropical regions (Chiambeng and Dumont 2005). The proportion of Chydoridae and Daphniidae broadly follows worldwide patterns, where Chydoridae make up the bulk of species (Forró et al. 2008). Occurrence of four orders (Onychopoda and Haplopoda) as opposed to two in SE Asia probably reflects lower mean temperatures prevailing in the parts of the subcontinent where they occur (see Mani 1974 for details on the climatic conditions of the Indian subcontinent). Based on the fauna, the Indian subcontinent can be compared to some African (e.g., Ethiopia: Van Damme and Eggermont 2011) and Neotropical regions (e.g., Chile: Kotov et al. 2010) (not considering species numbers but composition) having a high altitudinal gradient similar to the Indian subcontinent as a whole on a smaller geographical scale.

Environmental filters in species composition change according to the spatial scale under consideration. On a continental scale, factors such as dispersal, temperature, and precipitation govern distribution (Heino et al. 2009). The two biogeographic groups (Northern Zone and the rest of the continent) formed in the clustering have different temperature and precipitation patterns resulting in different eco-regions sets (Padhye et al. 2017). This split into two major zones is analogous to the distribution of a related group, viz., fairy shrimps in the Indian subcontinent, with similar latitudinal restriction observed in the species.

An increase in body size as a function of latitude agrees with the observation made by Fernando (1980) and the model put forth by Gillooly and Dodson (2000) for Cladocera in the western hemisphere. This probably is due to the increase in average surface temperature of the water bodies with decreasing latitude. The average temperature of latitudinal zone of the remaining southern portion of the subcontinent (6–24°N) is 7° higher (25.03°C) than the Northern Zone (25°–37°N, 18.05°C) (Padhye et al. 2016). Body size increases with latitude due to the negative association of "rearing temperature and body size" (Chown and Klok 2003 and references therein). Besides temperature, the pattern of fish occurrence in the tropics in relation to their diversity, habits, and physiology also limits large-bodied plankton in the tropics (via size-selective predation) (Fernando 1994). Removal of the large-bodied competitors from the habitats (by predation) leads to a substitution by small-bodied littoral species (Brooks and Dodson 1965). Meerhof et al. (2007) demonstrated that predator avoidance is inefficient for subtropical zooplankton and that the benefits of macrophytes as a protective measure for zooplankton are less strong in warm lakes. The subtropical and tropical regions within the Indian subcontinent are also devoid of high altitudinal aquatic refuges ("Löffler Islands" *sensu* Van Damme and Eggermont 2011) for sustaining large-bodied *Daphnia* species (Mani 1974). The combined effect of these factors, therefore, would make the beta diversity patterns more evident in the larger sized species than the smaller ones across the Indian subcontinent.

We used body size range data available in literature, though this functional trait varies with environment and shows intraspecific variation (Hebert 1995). Also factors like sampling regime and number of specimens measured need to be considered for making predictive relationships of body size with temperatures in terms of zoogeography. We therefore refrained from making any predictions and/or extrapolations from the collected data.

Apparent contrasts in species richness between the tropics and temperate regions have been explained by differences in habitat types (especially presence of lakes and their continuity), food availability, fish predation, and as well as temperature (Fernando 1980). Fernando (1980) stated that the genus *Simocephalus* Schödler, 1858 was rare in the tropics, but at least five species were found on the subcontinent so far (Orlova-Bienkowskaja 2001, Chatterjee et al. 2013), and some were abundant in the littoral zone of many habitat types (SP pers. obs). Dumont (1994) stated that the paucity is not with respect to species richness but with the quality of the studies and taxonomists. In the last few years, intensive studies in SE Asian regions have led to several new discoveries, increasing the taxa richness, demonstrating it is comparable to temperate regions (Maiphae et al. 2008, Van Damme et al. 2013). Similar studies need to be carried out on the Indian subcontinent fauna.

The distribution ranges of all the cladoceran faunal elements presented here are completely based on the available literature and will change given more surveys and taxonomic studies. A shift from a "cosmopolitan" species perspective to a regional oriental endemic taxocene will help (Van Damme et al. 2013, Padhye and Dumont 2014b, Sharma and Sharma 2014, Sinev et al. 2016, Neretina and Kotov 2017, Xu et al. 2018, Alonso et al. 2019, Elías-Gutiérrez et al. 2019). Such findings will alter the observed beta diversity patterns within the different regions as well. There has been an increase in the faunal checklists of "zooplankton", including cladocerans from all over the subcontinent. The most striking problem with such inventories are the dubious identifications (no images or diagnoses) leading to possible artificial inflation in species numbers. This also leads to misleading zoogeographical patterns; for example, *Daphnia pulex* has been commonly reported from South and Western India (e.g., Pai and Berde 2005, Shahzan and Ambore 2014) but with no accompanying, additional information on species identity.

Therefore, our study is exploratory and our results conservative. Even with these limitations, we still present a preliminary zoogeography of these crustaceans on the Indian subcontinent classifying the region into a "Rest of the subcontinent" and a "High altitudinal temperate zone", which until recently was still looked upon as a region of "cosmopolitan" species.

ACKNOWLEDGMENTS

SP acknowledges Belgian Science Policy (BELSPO) for providing the postdoctoral scholarship. He thanks Sanjay Molur Zoo Outreach Organization, Coimbatore, for their support. He also thanks Shweta Purushe for her comments on the data visualization and analysis.

REFERENCES

Alonso, M., Neretina, A.N., Sanoamuang, L.O., Saengphan, N. and Kotov, A.A. 2019. A new species of Moina Baird, 1850 (Cladocera: Moinidae) from Thailand. *Zootaxa* 4554:199–218.

Benzie, J.A.H. 2005. The genus *Daphnia* (including *Daphniopsis*) (Anomopoda: Daphniidae). In *Guides to the Identification of the Microinvertebrates of the Continental Waters of the World*, ed. H.J. Dumont, 1–383. Hague: SPB Academic Publishing.

Biswas, S. 1971. Fauna of Rajasthan, India. Part II. Crustacea: Cladocera. *Records of Zoological Survey of India* 63:95–140.

Brooks, J.L. and Dodson, S.I. 1965. Predation, body size, and composition of plankton. *Science* 150:28–35.

Chatterjee, T., Kotov, A.A, Van Damme, K., Chandrasekhar, S.V.A. and Padhye, S. 2013. An annotated checklist of the Cladocera (Crustacea: Branchiopoda) from India. *Zootaxa* 3667:1–89.

Chengalath, R. 1987. The distribution of chydorid Cladocera in Canada. *Hydrobiologia* 145:151–57.

Chiambeng, G.Y. and Dumont, H.J. 2005. The Branchiopoda (Crustacea: Anomopoda, Ctenopoda and Cyclestherida) of the rain forests of Cameroon, West Africa: Low abundances, few endemics and a boreal–tropical disjunction. *Journal of Biogeography* 32:1611–20.

Chown, S.L. and Klok, C.J. 2003. Altitudinal body size clines: Latitudinal effects associated with changing seasonality. *Ecography* 26:445–55.

Dumont, H.J. 1980. Zooplankton and the science of biogeography: The example of Africa. In *Evolution and Ecology of Zooplankton Communities*, ed. W.C. Kerfoot, 685–96. New England: University Press of New England.

Dumont, H.J. 1994. On the diversity of the Cladocera in the tropics. *Hydrobiologia* 272:27–38.

Dumont, H.J. and Negrea, S.V. 2002. *Introduction to the Class Branchiopoda. Guides to the Identification of the Microcrustaceans of the Continental Waters of the World 19.* Hague: SPB Academic Publishing.

Dumont, H.J. and Van de Velde, I. 1977. Report on collection of Cladocera and Copepoda from Nepal. *Hydrobiologia* 53:55–65.

Elías-Gutiérrez, M., Juračka, P.J., Montoliu-Elena, L., Miracle, M.R., Petrusek, A. and Kořínek, V. 2019. Who is *Moina micrura*? Redescription of one of the most confusing cladocerans from terra typica, based on integrative taxonomy. *Limnetica* 38:227–52.

Fernando, C.H. 1980. The freshwater zooplankton of Sri-Lanka, with a discussion of tropical freshwater zooplankton composition. *International Review of the Entire Hydrobiology and Hydrography* 65:85–125.

Fernando, C.H. 1994. Zooplankton, fish and fisheries in tropical freshwaters. In *Studies on the Ecology of Tropical Zooplankton*, eds. H.J. Dumont, J. Green and H. Masundire, 105–23. Dordrecht: Springer.

Fernando, C.H. and Kanduru, A. 1984. Some remarks on the latitudinal distribution of Cladocera on the Indian subcontinent. *Hydrobiologia* 113:69–76.

Forró, L., Korovchinsky, N.M., Kotov, A.A. and Petrusek, A. 2008. Global diversity of cladocerans (Cladocera; Crustacea) in freshwater. *Hydrobiologia* 595:177–84.

Frey, D.G. 1987. The taxonomy and biogeography of the Cladocera. *Hydrobiologia* 145:5–17.

Gillooly, J.F. and Dodson, S.I. 2000. Latitudinal patterns in the size distribution and seasonal dynamics of new world, freshwater cladocerans. *Limnology and Oceanography* 45:22–30.

Goulden, C.E. 1968. The systematics and evolution of the Moinidae. *Transactions of the American Philosophical Society* 58:1–101.

Green, J. 1995. Altitudinal distribution of tropical planktonic Cladocera. In *Cladocera as Model Organisms in Biology*, eds. P. Larrson and L.J. Weider, 75–84. Dordrecht: Springer.

Havel, J.E. and Medley, K.A. 2006. Biological invasions across spatial scales: Intercontinental, regional, and local dispersal of cladoceran zooplankton. *Biological Invasions* 8:459–73.

Hebert, P.M. 1995. The *Daphnia* of North America-an illustrated fauna. In *Digital Wisdom*, CD-Rom. Ontario, Canada: University of Guelph.

Heino, J., Virkkala, R. and Toivonen, H. 2009. Climate change and freshwater biodiversity: Detected patterns, future trends and adaptations in northern regions. *Biological Reviews* 84:39–54.

Hennig, C. 2018. fpc: Flexible procedures for clustering. R package version 2.1-11. https://CRAN.R-project.org/package=fpc.

Hudec, I. 1987. *Moina oryzae* n. sp. (Cladocera, Moinidae) from Tamil Nadu (South India). *Hydrobiologia* 145:147–50.

Hudec, I, 1991. A comparison of populations from the *Daphnia similis* group (Cladocera: Daphniidae). *Hydrobiologia* 225:9–22.

Korovchinsky, N.M. 1992. Sididae and Holopediidae. In *Guides to the Identification of the Microinvertebrates of the Continental Waters of the World*, ed. H.J. Dumont, 1–82. Hague: SPB Academic Publications.

Korovchinsky, N.M. 2006. The Cladocera (Crustacea: Branchiopoda) as a relict group. *Zoological Journal of the Linnean Society* 147:109–24.

Korovchinsky, N.M. 2013. Cladocera (Crustacea: Branchiopoda) of South East Asia: History of exploration, taxon richness and notes on zoogeography. *Journal of Limnology* 72:109–24.

Kotov, A.A. 2000. Re-description and assignment of the chydorid *Indialona ganapati* Petkovski, 1966 (Branchiopoda: Anomopoda: Aloninae) to Indialonini, new tribus. *Hydrobiologia* 439:161–78.

Kotov, A.A. and Štifter, P. 2006. Ilyocryptidae of the world. In *Guides to the Identification of the Microinvertebrates of the Continental Waters of the World*, ed. H.J. Dumont, 1–172. Hauge: SPB Academic Publishing.

Kotov, A.A., Elías-Gutiérrez, M. and Granados-Ramírez, J.G. 2005. *Moina dumonti* sp. nov. (Cladocera, Anomopoda, Moinidae) from southern Mexico and Cuba, with comments on moinid limbs. *Crustaceana* 78:41–57.

Kotov, A.A., Sinev, A.Y. and Berrios, V.L. 2010. The Cladocera (Crustacea: Branchiopoda) of six high altitude water bodies in the North Chilean Andes, with discussion of Andean endemism. *Zootaxa* 2430:1–66.

Maiphae, S., Pholpunthin, P. and Dumont, H.J. 2008. Taxon richness and biogeography of the Cladocera (Crustacea: Ctenopoda, Anomopoda) of Thailand. *Annales de Limnologie-International Journal of Limnology* 44:33–43.

Manca, M., Martin, P., Carolina, D.P. and Benzie, J.A.H. 2006. Re-description of *Daphnia* (*Ctenodaphnia*) from lakes in the Khumbu Region, Nepalese Himalayas, with the erection of a new species, *Daphnia himalaya*, and a note on an intersex individual. *Journal of Limnology* 65:132–40.

Mani, M.S., 1974. *Ecology and Biogeography in India*. Dordrecht: Springer, e-book.

Meerhoff, M., Iglesias, C., De Mello, F.T., Clemente, J.M., Jensen, E., Lauridsen, T.L. and Jeppesen E. 2007. Effects of habitat complexity on community structure and predator avoidance behaviour of littoral zooplankton in temperate versus subtropical shallow lakes. *Freshwater Biology* 52:1009–21.

Michael, R.G. 1973. Cladocera. In A guide to the study of freshwater organisms. *Journal of Madurai University* 1:71–85.

Michael, R.G. and Sharma, B.K. 1988. *Fauna of India and Adjacent Countries, Indian Cladocera (Crustacea: Branchiopoda: Cladocera)*. Kolkata: Zoological Survey of India.

Neretina, A.N. and Kotov, A.A. 2017. Diversity and distribution of the *Macrothrix paulensis* species group (Crustacea: Cladocera: Macrothricidae) in the tropics: What can we learn from the morphological data? *Annales de Limnologie-International Journal of Limnology* 53:425–65.

Oksanen, J., Blanchet, F.G., Kindt, R., et al. 2013. Vegan: Community Ecology Package. R package Version 2.4-3. https://CRAN.R-project.org/package=vegan

Orlova-Bienkowskaja, M.Y. 2001. Daphniidae: Genus *Simocephalus*. In *Guides to the Identification of the Microinvertebrates of the Continental Waters of the World*, ed. H.J. Dumont, 1–130. Hauge: Backhuys Publications.

Padhye, S.M. and Dumont, H.J. 2014a. *Moina hemanti* sp. nov., a new species of the genus *Moina* sl (Branchiopoda: Anomopoda) from Pune, India. *Zootaxa* 3860:561–70.

Padhye, S.M. and Dumont, H.J. 2014b. Species richness of Cladocera (Crustacea: Branchiopoda) in the Western Ghats of Maharashtra and Goa (India), with biogeographical comments. *Journal of Limnology* 74:182–91.

Padhye, S.M., Kotov, A.A., Dahanukar, N. and Dumont, H. 2016. Biogeography of the 'water flea' *Daphnia* O.F. Müller (Crustacea: Branchiopoda: Anomopoda) on the Indian subcontinent. *Journal of Limnology* 75:571–80.

Padhye, S.M., Kulkarni, M.R. and Dumont, H.J. 2017. Diversity and zoogeography of the fairy shrimps (Branchiopoda: Anostraca) on the Indian subcontinent. *Hydrobiologia* 801:117–28.

Pai, I.K. and Berde, V.B. 2005. Comparative studies on limnology of freshwater bodies located in coastal and high altitude of Goa and Maharashtra. *Journal of Aquatic Biology* 20:95–100.

Popova, E.Y. and Kotov, A.A. 2013. Latitudal patterns in the diversity of two subgenera of the genus *Daphnia* O.F. Müller (Crustacea: Cladocera: Daphniidae). *Zootaxa* 3736:59–174.

Popova, E.V., Petrusek, A., Kořínek, V., et al. 2016. Revision of the old World *Daphnia* (*Ctenodaphnia*) *similis* group (Cladocera: Daphniidae). *Zootaxa* 4161:1–40.

Raina, H.S. and Vass, K.K. 1993. Distribution and species composition of zooplankton in Himalayan ecosystems. *Internationale Revue der gesamten Hydrobiologie und Hydrographie* 78:295–307.

Rajapaksa, R. and Fernando, H. 1982. The Cladocera of Sri Lanka (Ceylon), with remarks on some species. *Hydrobiologia* 94:49–69.

Rane, P. 2013. Cladocera. In *Fauna of Maharashtra*, State Fauna Series, ed. Z.S.I. Kolkata, 1–673. West Bengal, India: Zoological Survey of India,.

Real R. 1999. Tables of significant values of Jaccard's index of similarity. *Miscel·lània Zoològica* 22:21–40.

Rizo, E.Z.C., Gu, Y., Papa, R.D.S., Dumont, H.J. and Han, B.P. 2017. Identifying functional groups and ecological roles of tropical and subtropical freshwater Cladocera in Asia. *Hydrobiologia* 799:83–99.

Shahzan, T.S. and Ambore, N.E. 2014. Diversity of Zooplankton at Barul Dam, Nanded. Maharashtra. *Knowledge Scholar* 1:1–7.

Sharma, B.K. and Sharma, S. 2007. New records of two interesting chydorid cladocerans (Branchiopoda: Cladocera: Chydoridae) from flood plain lakes of Assam, India. *Zoo's Print Journal* 22:2799–801.

Sharma, B.K. and Sharma, S.2008. Zooplankton diversity in floodplain lakes of Assam. In *Records of Zoological Survey of India*, Occasional Paper 290, ed. B.K. Sharma and S. Sharma, 141–307. West Bengal, India: Zoological Survey of India.

Sharma, B.K. and Sharma, S. 2014. Faunal diversity of Cladocera (Crustacea: Branchiopoda) in wetlands of Majuli (the largest river island), Assam, northeast India. *Opuscula Zoologica (Budapest)* 45:83–94.

Sinev, A.Y. 1999. *Alona costata* Sars, 1862 versus related palaeotropical species: The first example of close relations between species with a different number of main head pores among Chydoridae (Crustacea: Anomopoda). *Arthropoda Selecta* 8:131–48.

Sinev, A.Y. 2004. Re-description of two species of the genus *Leydigiopsis* Sars, 1901 (Branchiopoda, Anomopoda, Chydoridae). *Invertebrate Zoology* 1:75–92.

Sinev, A.Y. 2011. Re-description of the rheophilous Cladocera *Camptocercus vietnamensis* Than, 1980 (Cladocera: Anomopoda: Chydoridae). *Zootaxa* 2934:53–60.

Sinev, A.Y. 2012. *Alona kotovi* sp. nov., a new species of Aloninae (Cladocera: Anomopoda: Chydoridae) from South Vietnam. *Zootaxa* 3475:45–54.

Sinev, A.Y., Garibian, P.G. and Gu, Y. 2016. A new species of *Pseudochydorus* Fryer, 1968 (Cladocera: Anomopoda: Chydoridae) from South-East Asia. *Zootaxa* 4079:129–39.

Smirnov, N.N. 1971. *Chydoridae fauny mira*. Fauna USSR. Rakoobraznie, 1. Leningrad [English translation: Chydoridae of the world. Israel Program for Scientific Translations, Jerusalem, 1974]

Smirnov, N.N. 1992. The Macrothricidae of the world. In *Guides to the Identification of the Microinvertebrates of the Continental Waters of the World*, ed. H.J., Dumont, 1–143. Hauge: SPB Academic Publications.

Smirnov, N.N. 1996. Cladocera: The Chydorinae and Sayciinae (Chydoridae) of the world. In *Guides to the Identification of the Microinvertebrates of the Continental Waters of the World*, ed. H.J. Dumont, 1–197. Hauge: SPB Academic Publications.

Swar, D.B. and Fernando, C.H. 1979. Cladocera from Pokhara Valley, Nepal with notes on distribution. *Hydrobiologia* 66:113–28.

Van Damme, K. and Dumont, H.J. 2008. The 'true' genus *Alona* Baird, 1843 (Crustacea: Cladocera: Anomopoda): Characters of the *A. quadrangularis* group and description of a new species from Democratic Republic Congo. *Zootaxa* 1945:1–25.

Van Damme, K. and Eggermont, H. 2011. The Afromontane Cladocera (Crustacea: Branchiopoda) of the Rwenzori (Uganda-D.R. Congo): ecology, biogeography and taxonomy including the description of *Alona sphagnophila* n. sp. *Hydrobiologia* 676:57–100.

Van Damme, K. and Maiphae, S. 2013. *Salinalona* gen. nov., an euryhaline chydorid lineage (Crustacea: Branchiopoda: Cladocera: Anomopoda) from the Oriental region. *Journal of Limnology* 72:142–73.

Van Damme, K. and Sinev, A.Y. 2013. Tropical amphi-pacific disjunctions in the Cladocera (Crustacea: Branchiopoda). *Journal of Limnology* 72:209–44.

Van Damme, K., Bekker, E.I., and Kotov, A.A. 2013. Endemism in the Cladocera (Crustacea: Branchiopoda) of Southern Africa. *Journal of Limnology* 72:440–63.

Van Damme, K., Sinev, A.Y. and Dumont, H.G. 2011. Separation of *Anthalona* gen.n. from *Alona* Baird, 1843 (Branchiopoda: Cladocera: Anomopoda): morphology and evolution of scraping stenothermic alonines. *Zootaxa* 2875:1–64.

Venkataraman, K. 1990. New records of Cladocera of Keoladeo National Park, Bharatpur. *Journal of the Bombay Natural History Society* 87:166–68.

Venkataraman, K. 1992. Cladocera of Keoladeo National Park, Bharatpur and its environs. *Journal of the Bombay Natural History Society* 89:17–26.

Venkataraman, K. 1999. Freshwater Cladocera (Crustacea) of Southern Tamil Nadu. *Journal of the Bombay Natural History Society* 96:268–80.

Venkataraman, K. and Das, S.R. 1993. Freshwater Cladocera (Crustacea: Branchiopoda) of southern West Bengal. *Journal of Andaman Science Association* 9:19–24.

Venkataraman, K., Das, S.R. and Nandi N.C. 2000. Zooplankton diversity in freshwaterwetlands of Haora district, West Bengal. *Journal of Aquatic Biology* 15:19–25.

Xu, S., Hebert, P.D.N., Kotov, A.A. and Cristescu, M.E. 2009. The noncosmopolitanism paradigm of freshwater zooplankton: Insights from the global phylogeography of the predatory cladoceran *Polyphemus pediculus* (Linnaeus, 1761) (Crustacea, Onychopoda). *Molecular Ecology* 18:5161–79.

Xu, L., Lin, Q., Xu, S., et al. 2018. *Daphnia* diversity on the Tibetan Plateau measured by DNA taxonomy. *Ecology and Evolution* 8:5069–78.

CHAPTER 9

Freshwater and Brackish Water Planktonic Copepods (Crustacea: Copepoda) of Sakhalin Island (Far East Asia)
Diversity, Ecology, and Zoogeography

Denis S. Zavarzin

CONTENTS

9.1 INTRODUCTION

The species composition of freshwater and brackish water copepods of the Russian Far East is much less understood than in western Russia. Sakhalin, the largest island in the northern Pacific Ocean, belongs to one of the most unexplored areas of Russia. Despite the large number of papers on plankton and benthos of freshwater and brackish waters of Sakhalin Island, there is no comprehensive review on the Copepoda. Some of the literature data is outdated.

The first plankton copepod studies from north Sakhalin were published by Rylov (1932b). At about the same time, Japanese researchers Masuzou Ueno (1935a,b, 1936) and Kenzou Kikuchi (1930, 1936) were studying zooplankton in the south (Karafuto prefecture) as part of fisheries research.

The M.V. Lomonosov Moscow State University and the Sakhalin fish hatchery department studied the inland waters in the 1950s. These studies examined acclimatization in the southern lagoons and contained a description of the zooplankton, including copepods (Borutskii and Bogoslovskii 1964).

As part of the marine hydrobiological expedition of the Zoological Institute of the USSR Academy of Sciences in 1963, plankton samples were collected in coastal areas of the south as well as Busse Lagoon. The works form the basis of Kos's (1985) work on the Calanoida of this area.

Zooplankton was sporadically studied in southern Sakhalin by SakhTINRO (the old name of SakhNIRO) (Sabitov and Chernysheva 1976, Chernysheva and Sabitov 1980, 1981, Usova et al. 1980). The integrated hydrobiological expeditions of the Sakhalin State University student scientific group collected zooplankton samples from various reservoirs from 1993 to 1999. Some of the results were published (Zavarzin and Safronov 2001). However, some data were not used and remained unpublished.

From 2000 to 2013, comprehensive lakes and lagoon surveys were conducted by SakhNIRO in collaboration with other organizations. The copepod data were partially published within many papers (Labay et al. 2002, 2010, 2012, 2013, 2014, 2015, 2016, Samatov et al. 2002, Zavarzin 2003, 2004, 2005, 2007, 2011a,b, Nemchinova 2006, 2011, Vinogradov and Zavarzin 2013, Zavarzin and Atamanova 2014). Again, much of the gathered data remained unpublished.

Here, we summarize and update the knowledge on pelagic freshwater and brackish water copepods of Sakhalin Island (including meroplanktonic parasitic species) from the literature and from our personal collections, and the unpublished data from the above-mentioned studies.

9.2 MATERIAL AND METHODS

We examined zooplankton samples from the water bodies of Sakhalin Island collected by the author and colleagues, during the expeditions of Sakhalin State University (1993–1999) and SakhNIRO (2000–2013), the author's personal collections, as well as the scientific literature. We present the unpublished data from the Sakhalin State University (1993–1999) and SakhNIRO (2000–2013) expeditions for the first time. The primary water bodies mentioned in this review are shown in Fig. 9.1. Samples were collected with Juday plankton nets (mouth diameter 18 cm, mesh size 82 μm) and were fixed in a 5% buffered formalin solution. Photographs were taken using an Olympus BX51 microscope with an Olympus DP25 camera or a ScopeTek MDC560 camera, and Olympus CellSens Standard software. Most photos were made using focus stacking. Drawings were made from photographs and/or with a drawing tube.

Based on the concepts of critical salinity (Khlebovich 1974, 1989) and zones of barrier salinity (Kinne 1971, Aladin and Plotnikov 2013), "freshwater copepods" in this chapter are defined as those that occur predominantly in range between 0‰ and upper bound δ-horohalinicum (salinity 2‰), and "brackish water copepods" as living in a range between lower bound δ-horohalinicum (0.5‰) and upper bound α-horohalinicum (8‰), which is the main barrier between freshwater and marine fauna (Kinne 1971, Khlebovich 1989). It may be noted that the expansion of the upper

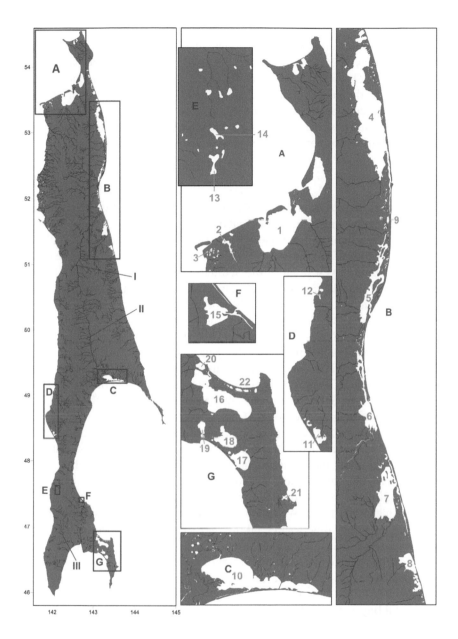

Figure 9.1 Main Sakhalin water bodies mentioned in this study. I – Tym' River, II – Poronai River, III – Lyutoga River; 1 – Baikal Bay, 2 – Lake Sladkoe, 3 – Lake Svetloe, 4 – Pil'tun Bay, 5 – Chaivo Bay, 6 – Nyiskii Bay, 7 – Nabil'skii Bay, 8 – Lun'skii Bay (4–8 – lagoon bays of the northeast), 9 – Lake Panitu, 10 – Lake Nevskoe, 11 – Lake Ainskoe, 12 – Lake Protochnoe, 13 – Lake Mokhovoe, 14 – Lake Osochnoe (13, 14 and around – Spamberg Mountain lakes), 15 – Lake Lebyazh'e, 16 – Lake Tunaicha, 17 – Busse Lagoon, 18 – Vavaiskie Lakes, 19 – Chibisanskie Lakes (18, 19 – Vavaiskie lake system), 20 – Lake Izmenchivoe, 21 – Lake Ptich'e, 22 – Okhotskie freshwater lake group.

limit to 18‰ (category of "true estuarine" zooplankton; Jeffries 1967, Sakaguchi et al. 2011) does not change the species composition. Marine neritic species (*Acartia hudsonica* Pinhey 1926, *Pseudocalanus neumani* Frost 1989, *Centropages abdominalis* Sato 1913, *Oithona similis* Klaus 1866, etc.) that penetrate into the desalinated areas as well as the meiobenthic Cyclopoida and Harpacticoida that fall into plankton nets are not considered in this paper.

The taxonomic position of each taxon was mainly aligned according to the World Register of Marine Species (WoRMS) (http://www.marinespecies.org.).

9.2.1 Geographical Conditions

Sakhalin is the largest island in the North Pacific. It is located off the east coast of Asia, stretching 948 km in the meridional direction between 54°24′ (Cape Elizabeth) and 45°54′ (Cape Crillon) north latitude. The area of the island is 76,400,000 km². The average width of Sakhalin is about 100 km, the maximum width reaches 157 km, and at the narrowest point of the Schmidt Peninsula is only 6 km. The northwestern part of the island is almost closely adjacent to the mainland (Russia) at the mouth of the Amur River. The island is separated from the mainland by the Amur Liman and the shallow Nevelskoy Strait, the narrowest width of which is 7.5 km. The southern part of Sakhalin is separated from Hokkaido (Japan) by the La Pérouse Strait, 40 km wide (Sakhalin region 1994, Bogatov et al. 2007).

On Sakhalin, there are 16,120 water bodies (lagoons and lakes) with a total area of 1,004 km². Approximately 16,000 km² of them have an area of less than 0.4 km² (Onishchenko 1987, Sakhalin region 1994, Matjushkov et al. 2014). Small tundra lakes, most numerous in the north of the island, often are of thermokarst origin. The oxbow lakes are located in floodplains of large rivers in central and southern Sakhalin. In the mountains are landslide or barrier lakes. The largest water bodies are of lagoon origin – these are true lagoons and lagoon lakes. Shoreline lakes created by estuary blockage are located close to these lagoons.

Labay (2011) proposed to divide the island's reservoirs on the basis of critical salinity (Khlebovich 1989). We used this classification, specifying the boundary of freshwater and oligohaline waters, taking into account the concept of a multiplicity of barrier salinity zones (Aladin and Plotnikov 2013). It was proposed to divide the reservoirs into saline or marine – with a salinity above 22‰ (lower limit 22–26‰ – β-horohaline zone) – Baikal Bay, Busse Lagoon, Lake Ptich'e, Lake Izmenchivoe; brackish water – from 22‰ to 5‰ (lower limit 5–8‰ – α-horohaline zone) – Chaivo Bay, Nyiskii Bay, part of Pil'tun Bay, Lake Lebyazh'e; oligohaline – from 5 to 0.5‰ (lower limit 0.5–2‰ – δ-horohaline zone) – part of the Pil'tun Bay, Lake Tunaicha, Lake Nevskoe, Lake Ainskoe and freshwater – less than 0.5‰ – Vavaiskie Lakes, Chibisanskie Lakes, the Okhotskie freshwater lake group, and many small tundra lakes and oxbow lakes.

The island's climate is monsoonal. The average annual air temperature on Sakhalin varies from -2.0°C to -2.7°C in the north and from 4.0°C to 4.5°C in the south. The January amplitude between the southern and northern temperatures is more than 15°C, while in summer it varies only 5°C. The warm Tsushima Current flows to the island from the south, entering the Sea of Japan through the Korea Strait. One of its branches passes through the La Pérouse Strait into the Sea of Okhotsk, the other follows along the western coast. The cold East-Sakhalin Current flows along the eastern coast of the island from north to south, penetrates into the southern part of Sakhalin, and has a strong cooling effect on the entire eastern part (Bogatov et al. 2007).

9.3 ANNOTATED LIST OF SAKHALIN FRESHWATER AND BRACKISH WATER PLANKTONIC COPEPODS

Subclass Copepoda (Milne Edwards 1840)
  Infraclass Neocopepoda (Huys and Boxshall 1991)
  Superorder: Gymnoplea (Giesbrecht 1882)
  Order: Calanoida (Sars 1903)
  Centropagidae (Giesbrecht 1893)
  Boeckella (Guerne and Richard 1889)
  Boeckella triarticulata (Thomson 1883)

Taxonomic notes: Chernysheva (1974) noted that specimens do not differ morpho logically from those described by Rylov (1932a, as *B. orientalis*) from the Ussuri region (mainland Russian Far East).

Distribution: Gondwanan: in Australasia, South America, Eastern Mongolia, Manchuria, the Russian Far East (Borutskii et al. 1991 (as *B. orientalis*), Bayly 1992, Dussart and Defaye 2002). Recorded in Europe as an invasive species imported with Chinese carp (Ferrari et al. 1991, Ferrari and Rossetti 2006).

Sakhalin records: Reported once in August 1972 from the pelagial of Lake Russkoe (Okhotskie freshwater lake group), where it was probably introduced with acclimatized Amur carp (*Cyprinus rubrofuscus* Lacépède 1803) from Khabarovsk Krai (Chernysheva 1974 – as *B. orientalis*) (Fig. 9.2).

Ecology: Warm freshwater pelagic species. The wide salinity tolerance (Nielsen et al. 2003), and resting (diapausing) eggs allow the species to colonize new areas, especially newly formed habitats (Maly 1984, 1991, Alfonso and Belmonte 2008).

Sinocalanus (Burckhardt 1913)
  Sinocalanus tenellus (Kikuchi 1928)
  Taxonomic notes: Our material has no noticeable differences from those described by Kos (1985) (Fig. 9.3).

Distribution: Korea, Japan, China, Russia (Amur River mouth and Suifun River, Kuril Islands, Sakhalin), Philippines, Thailand (Borutskii 1991, Chang 2014). Pacific Asian lower boreal-subtropical species.

Sakhalin records: Reported in lakes on the sand spit near Moskal'vo village (northwest of the island) (Rylov 1932b, as *S. tenellus* var. *sachalinensis*), further south along the west coast in Baikal Bay (Nemchinova 2011), in small lakes, connected to the sea near Cape Pogibi, lakes Protochnoe (new record), Ainskoe (Ueno

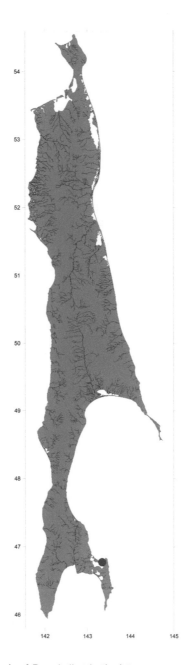

Figure 9.2   Sakhalin records of *Boeckella triarticulata*.

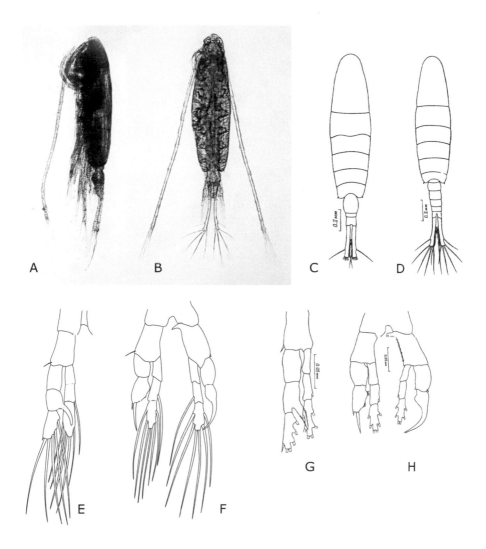

Figure 9.3　*Sinocalanus tenellus* from Sakhalin. Female: A – habitus lateral, B and C – dorsal, E and G – P5; Male: D – habitus dorsal, F and H – P5 (A, B, E, F – from Lake Tunaicha, original; C, D, G, H – from southwest sea coast – Kos 1985).

1935a, b, Borutskii and Bogoslovskii 1964), nearshore of southwest Sakhalin (Kos 1985). Along Okhotsk Sea coast: lagoon bays of the northeast (Labay et al. 2002, 2016, Kafanov et al. 2003), Nevskoe lake and Mereya Lake (Labay et al. 2016), Ptich'e Lake (Zavarzin and Atamanova 2014), Lebyazh'e Lake, Vyselkovoe Lake (Borutskii and Bogoslovskii 1964), Tunaicha Lake (Usova et al. 1980, Samatov et al. 2002, Zavarzin 2003, 2004, 2005, Labay et al. 2013, 2016), the Lyutoga River estuary (Labay et al. 2015), and Busse Lagoon (SakhNIRO unpublished) (Fig. 9.4).

Ecology: Occurring frequently in lagoons, estuaries, brackish water lakes, and coastal marshes. Sometimes occurs in tidal zone, nearly pure freshwater in rivers (Chang

Figure 9.4   Sakhalin records of *Sinocalanus tenellus*.

2014). It is found also in marine coastal areas (Kos 1985). It is mostly found in the open pelagic zone. It is often one of the dominant zooplankton taxa in oligohaline reservoirs (Lake Tunaicha, Lake Nevskoe, Lake Ainskoe and others). Surveys in Lake Tunaicha from 2002 to 2003, demonstrated the presence of *S. tenellus* year round. Three annual nauplii abundance peaks occur: end of May, end of June, and August–early September.

A similar situation was also noted for Lake Abashiri (Hokkaido Island) (Asami and Ito 2003, Asami 2004). The maximum nauplii abundance depends on water temperature and falls on the second (2002) or third (2003) peaks (Labay et al. 2013). The population reached a maximum abundance, up to 70,000 specimens/m^3 in August–September of 2003 under the highest water temperatures.

Pseudodiaptomidae (Sars 1902)

    Pseudodiaptomus (Herrick 1884)

    Pseudodiaptomus japonicus (Kikuchi 1928)

    Taxonomic notes: Previously considered a junior synonym of *P. inopinus* (Burckhardt 1913). The taxon "*P. inopinus*" is now considered a complex of species (Sakaguchi and Ueda 2018). *Pseudodiaptomus* from Sakhalin reservoirs belongs to the "short-process" form described by Sakaguchi and Ueda (2011, 2018), based on the structure of the genital operculum (Zavarzin 2020) (Fig. 9.5E). Sakaguchi and Ueda (2011, 2018) resurrected this form as *P. japonicus* (Kikuchi 1928, Ueda and Sakaguchi 2019).

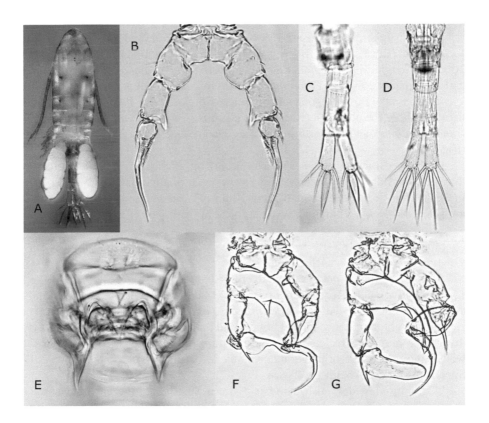

Figure 9.5 *Pseudodiaptomus japonicus* from Sakhalin (A – Pil'tun Bay, B–G – Lake Tunaicha). Female: A – habitus dorsal, B – P5, C – urosome, dorsal view, with thick setae form, D – urosome with thin setae form, E – genital operculum; Male: F – thumb-shaped form P5, G – paddle-shaped form P5 (original).

Two female morphotypes occur. The first are typical specimens with spiniform setae on caudal rami (Fig. 9.5C), whereas the second form has thin setae (Fig. 9.5D) (a similar thin setae form was reported from the Yukiura River in Japan – Sakaguchi and Ueda 2010). In Sakhalin lagoon lakes Tunaicha (2002–2003) (Labay et al. 2016) and Ptich'e (2012–2013) (SakhNIRO unpublished) typical females were found from June to October, and atypical females were found from September to November in Lake Tunaycha, and from October to May in Lake Ptich'e. We also found both forms in the lagoon bays of northeastern Sakhalin.

Two male morphotypes occur: with either a thumb-shaped (Fig. 9.5F) or a paddle-shaped P5 (Fig. 9.5G) (Burckhardt 1913). The latter was previously described as the subspecies *P. inopinus saccupodus* (Shen and Tai 1962b). However, the ITS1 and mtCOI sequence analyses demonstrate that two morphs belong to one species (Soh et al. 2012 (as *P. koreanus*)). Males of both morphotypes were found in the reservoirs of the island (Zavarzin 2020) without notable seasonal patterns.

Distribution: Widespread in northern East Asia–coastal Japan (except the warm Kuroshio Current coasts of the middle and western mainland) (Ueda and Sakaguchi 2019), Korea (Chang 2014), the Japan Sea coast of mainland Russia (Smirnov 1929), and Sakhalin Island (Zavarzin 2020).

Sakhalin records: In seaside reservoirs around the island – lakes Lebyazh'e, Vyselkovoe (Borutskii and Bogoslovskii 1964), Tunaicha (Zavarzin 2003, 2004, 2005 – as *Schmackeria inopina*, Labay et al. 2016 – as *P. inopinus*), Ptich'e (Zavarzin and Atamanova 2014 – as *P. inopinus*); in Lyutoga River estuary (Labay et al. 2015 – as *P. inopinus*), Busse Lagoon and the sea coastal area (Kos 1985 – as *P. inopinus*), lagoon bays of the northeast (Kafanov et al. 2003 – as *S. inopina*, Labay et al. 2016 – as *P. inopinus*), and the northwest of the island (Nemchinova 2011 – as *S. inopina*) (Fig. 9.6).

Ecology: Brackish water species, occurring in estuaries, lagoons, coastal marshes, and tidal embankments. This species is euryhaline and is sometimes found in freshwaters near the seashore (Chang 2014 as *P. koreanus*), also in the sea coastal area (Kos 1985). On Sakhalin it was not found in freshwater (Zavarzin 2020).

Diaptomidae (Baird 1850)

Acanthodiaptomus (Kiefer 1932)

Acanthodiaptomus pacificus s. lato (Burckhardt 1913)

Taxonomic notes: Recent genetic studies (Makino and Tanabe 2009, Makino et al. 2018) demonstrate that *Acanthodiaptomus pacificus* is a complex of cryptic species with slight morphological differences and parapatric distribution in Japan. This divergence may be associated with marine transgression and regression events, as well as introgressive hybridization during the Miocene (Ohtsuka and Nishida 2017). Specimens from Sakhalin are shown (Fig. 9.7).

Distribution: Widely distributed in Eastern Siberia and the northern Far East (Korea, Japan, China – Manchuria; Russia – Siberia, Kamchatka, Sakhalin, Kuril Islands) (Borutskii et al. 1991, Chang 2014).

Sakhalin records: Mountain lakes (800 m above sea level) (Spamberg Mountain lakes), Nikolskii Pass Lake (600 m above sea level), and in tundra lakes (almost at

Figure 9.6   Sakhalin records of *Pseudodiaptomus japonicus.*

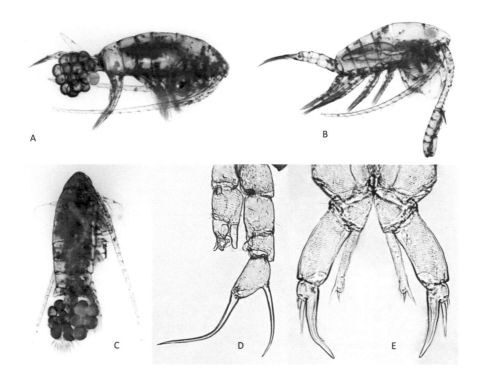

Figure 9.7 *Acanthodiaptomus pacificus* s. lato from Sakhalin (Spamberg Mountain lakes). Female: A – habitus lateral, C – dorsal, E – P5; Male: B – habitus lateral D – P5 (original).

sea level) on the northwestern coast (Rylov 1932b, as *Diaptomus yamanacensis*) and on spits separating lagoons from the sea in the northeast (Zavarzin 2011b) (Fig. 9.8).

Ecology: Occurring in various freshwaters, such as reservoirs, bogs, swamps, and rice fields, especially in winter (Chang 2009). In Sakhalin, the maximum abundance is recorded in the open pelagial, where this species is often one of the dominants.

Neutrodiaptomus (Kiefer 1937)

    Neutrodiaptomus ostroumovi (Stepanova 1981)

    Taxonomic notes: Ueno reported this species as *Eudiaptomus pachypoditus* (Rylov 1925) for the Vavaiskie and Chibisanskie lakes (Ueno, 1935a,b), and as *Diaptomus amurensis* Rylov, 1918 from the tundra marsh lake (bog-pond) not far from the Lake Nevskoe (Ueno 1936a). Borutskii and Bogoslovskii (1964) reported this species from the Okhotskie freshwater lake group and the Vavaiskie lake system (Vavaiskie and Chibisanskie lakes). Stepanova (1981) described *Neutrodiaptomus ostroumovi* from Kamchatka Lake (as *Ligulodiaptomus*). We reported specimens from Sakhalin reservoirs (Zavarzin 2007) Fig. 9.9A–D and J–K). The illustrations from Ueno (1936) (Fig. 9.9E and F) are much closer to ours (Fig. 9.9J and K) and

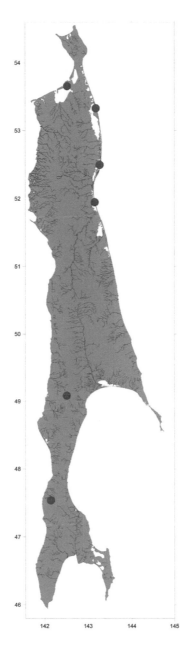

Figure 9.8 Sakhalin records of *Acanthodiaptomus pacificus* s. lato.

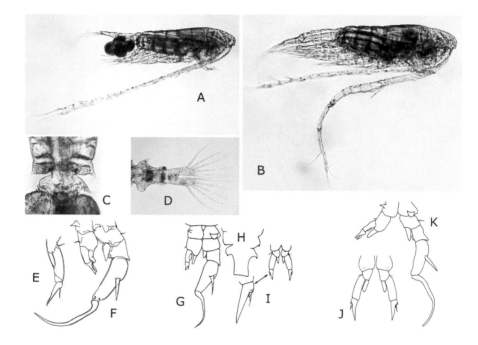

Figure 9.9   *Neutrodiaptomus ostroumovi*. Female: A – habitus lateral, C – 5th pediger and
genital double-somite dorsal, D and H – urosome dorsal, E, I, and J – P5; Male:
B – habitus lateral, F, G, and K – P5. A–D – Vavaiskie lake system, original; J and
K – Zavarzin 2007; E and F – Ueno 1936, G, H, and I – Stepanova 1981.

Stepanova's (Fig. 9.9G and I) for *N. ostoumovi* than to *N. amurensis*. Unfortunately,
Ueno, Borutskii, and Bogoslovskii did not provide drawings for the Vavaiskie lake
system material. However, Kikuchi (1936) noted that the Japanese form has some
differences, one of which is the presence of a hyaline plate on the right male P5
exopod, article I on the medial margin, which is typical for *N. ostroumovi*. We did
not take samples from the Okhotskie freshwater lake group; therefore, the species
affiliation of *Neutodiaptomus* from these reservoirs remains unsolved.

Distribution: Currently known from the lakes of Kamchatka and southern
Sakhalin.

Sakhalin records: Vavaiskie and Chibisanskie lakes, tundra lakes in the basin
of Lake Nevskoe, probably freshwater reservoirs of the Okhotskie lake group
(Fig. 9.10).

Ecology: In freshwater lakes of the Vavaiskie system; one of the dominant species
(Zavarzin 2007). Labay et al. (2010) reports two abundance peaks in the Vavaiskie
Lakes –June–July and October–November. In these lakes, *N. ostoumovi* avoids small
bays with swampy, peat margins, and a pH of 6.5 or lower.

Neutrodiaptomus pachypoditus s. lato (Rylov 1925)

Taxonomic notes: The Sakhalin form (Fig. 9.11A–D and I) is closer to that
described by Streletskaya (1975): she highlights some differences from the specimens

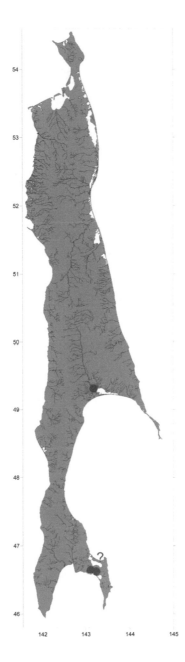

Figure 9.10   Sakhalin records of *Neutrodiaptomus ostroumovi.*

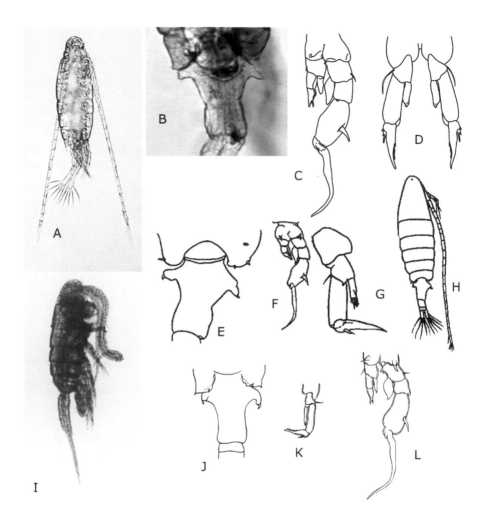

Figure 9.11   *Neutrodiaptomus pachypoditus* s. lato. Female: A and H – habitus dorsal; B, E, and J – 5th pediger and genital double-somite dorsal; D, G, and K – P5. Male: I – habitus lateral; C, F, and L – P5. A, B, C, D, and I – Lake Sladkoe, original; E, F, G, and H – Markevich 1985; J, K, and L – Rylov 1925.

described by Rylov (1925) (Fig. 9.11J–L). Markevich (1985) described a similar form from the Zeya River basin reservoir as a separate species, *N. sklyarovae* Markevich, 1985 (Fig. 9.11 E–H), stating that Streletskaya's form is the same species. The main difference from *N. pachypoditus* s. str. is the expansion of the male P5 exopod in the central part (not in the posterior), the first antennae significantly exceeding body length, plus the form of the caudal rami setae. Males from China have a similar P5 structure (Shen and Song 1979). Apparently, it is a species complex.

Distribution: Eastern Siberia and the Far East of Russia, northeast China (Borutskii et al. 1991), and Japan (Mizuno 2000).

Sakhalin records: Northwest coast freshwater lakes (lakes Sladkoe: Zavarzin 2011a, Zavarzin and Safronov 2001), Shumnoye, and oxbow lakes of the Tym River in the island center (Fig. 9.12).

Ecology: Freshwater species. Borutskii et al. (1991) stated that this species avoids acidic waters, but we have found it in reservoirs where the pH is neutral to slightly acidic (Lake Sladkoe – pH 6.0). In Lake Sladkoe, it is uncommon. We found this species in both the pelagic zone and among macrophytes. In Lake Shumnoye, it was the dominant zooplankton species.

Nordodiaptomus (Wilson, 1951)

Nordodiaptomus aff. *alaskaensis* (Wilson 1951)

Taxonomic notes: Moriya (1979) reported *Nordodiaptomus alaskaensis* from temporary spring ponds in sand dunes on the Ishikari coast (Hokkaido), but did not provide drawings. Specimens from Sakhalin samples fit the diagnosis of *N. alaskaensis* (Wilson 1951, Borutskii et al. 1991) (Fig. 9.13), except for the structure of the female exopod P5 third article (Fig. 9.13C, D and G). Wilson (1951) noted that "a more important difference lies in the third exopod segment, which, is as wide as it is long in *siberiensis*; its width in *alaskaensis* is about twice its length" (SIC). On this basis, individuals from Sakhalin are closer to *N. rylovi* (Borutskii et al. 1991, as *N. siberiensis*). Perhaps *Nordodiaptomus* from Sakhalin and Hokkaido are a new species.

Sakhalin records: The species was found in temporary spring water bodies in southern Sakhalin (Fig. 9.14).

Ecology: Inhabitant of fresh temporary waters.

Temoridae (Giesbrecht 1893)

Heterocope Sars G.O., 1863

Two species occur in Sakhalin, but not together. *Heterocope appendiculata* is found in reservoirs with fish, while *H. borealis* is in reservoirs without fish. Possibly, the smaller *H. appendiculata* does not compete well with the larger *H. borealis.*

Heterocope appendiculata Sars G.O., 1863

Taxonomic notes: Specimens from Sakhalin had no noticeable differences from the typical form.

Distribution: Widespread in the Palearctic boreal and tundra zones (Borutskii et al. 1991, Dussart and Defaye 2002, Vinarski et al. 2015). Glacial relict.

Sakhalin records: Reported in northeast marsh tundra lakes – lakes on the spit of Pil'tun Bay, the lake in the Evay River basin (new record), lakes of the Lake Nevskoe basin (Ueno 1936), and in the northwest coastal lakes to north of Baikal Bay (Fig. 9.15).

Ecology: A typical representative of the temperate freshwater lake pelagic fauna. In the northern part of the range toward the coast, it occurs among macrophytes. The summer form is dicyclic in the south (Lazareva 2003), while it is apparently monocyclic in the north (Borutskii et al. 1991), forming resting eggs, which overwinter. The summer generation develops twice as fast as the spring generation (Fefilova 2015).

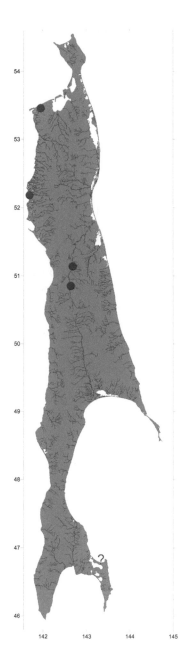

Figure 9.12   Sakhalin records of *Neutrodiaptomus pachypoditus* s. lato.

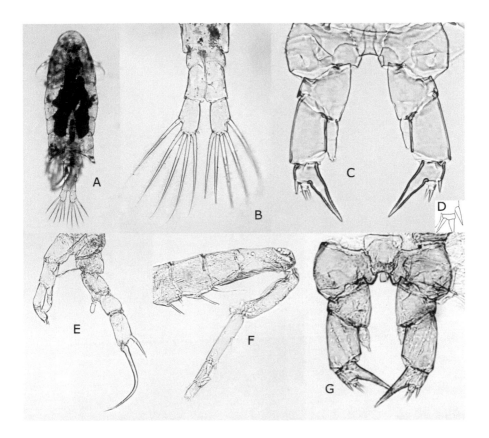

Figure 9.13 *Nordodiaptomus* aff. *alaskaensis* from South Sakhalin (temporary reservoir). Female: A – habitus dorsal, B – urosome, dorsal, C – p5, G – P5 endopod variant, D –exopod P5 third article *N. alaskaensis* s. str.; Male: E – P5, F – right A1, distal articles. A, B, C, E, F, and G – original, D – Wilson 1951.

Nauplii and copepodite stages I–III predominantly feed on detritus and algae; older individuals are obligate predators consuming small crustaceans (Lazareva 2003).

Heterocope borealis (Fischer 1851)

Taxonomic notes: Our specimens had no noticeable differences from the typical form.

Distribution: The main range extends to north of the Arctic Circle throughout the Palaearctic (Borutskii et al. 1991); in Asia, it penetrates slightly to the south, to 51–52° nl (Rylov 1932b).

Sakhalin records: Small tundra lakes: lakes near Baikal Bay (Rylov 1932b) and Lake Svetloe (northwest), lakes Mivka, Sredneye, Golovka, and a number of other nameless lakes around Chaivo and Pil'tun bays (northeast of the island). The maximum abundances occur fishless lakes (Fig. 9.16).

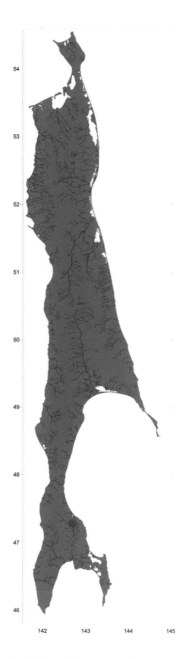

Figure 9.14   Sakhalin records of *Nordodiaptomus* aff. *alaskaensis*.

Figure 9.15   Sakhalin records of *Heterocope appendiculata.*

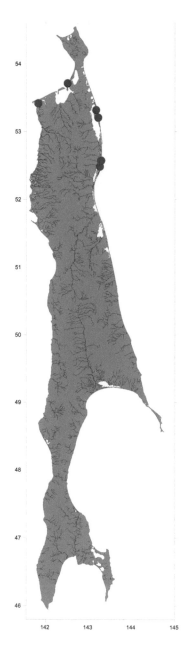

Figure 9.16   Sakhalin records of *Heterocope borealis*.

Ecology: A predatory freshwater species in small fishless water bodies (Pezheva et al. 2016). Like the previous species, it oviposits without forming egg sacs, and over-winters as diapausing eggs. In Europe it has two generations per year (Vekhov 1980).

Eurytemora (Giesbrecht 1881)

Eurytemora are confined mainly to brackish waters (marine lagoons, estuaries, etc.), but often live in both completely freshwater and marine water (Kos 2016) and above (0–40‰) – this is the widest salinity range among all known copepods (Dodson et al. 2010). Six species have been recorded from Sakhalin and the adjacent coastal areas, only two live in the salinity range considered here.

Eurytemora aff. *affinis* (Poppe 1880)

Taxonomic notes: *Eurytemora affinis* was previously considered as a widespread Holarctic species (Borutskii et al. 1991). However, recent molecular and hybridization studies (Alekseev and Souissi 2011) indicate that its scattered populations are a species complex, divided about a million years ago (Knowlton 1993, Lee 1999, 2000, Lee and Frost 2002). Specimens from Sakhalin (Fig. 9.17) are close to those reported for Japan (Chihara and Murano 1997), and apparently belong to an undescribed species (Sukhikh and Alekseev 2013).

Distribution: The form is known from freshwater and brackish water of north ern Japan (Hokkaido, Honshu Island – Mizuno and Miura 1984, as *E. affinis*) and Sakhalin Island.

Sakhalin records: Northeast lagoons (Kafanov et al. 2003, Labay et al. 2016 –as *E. affinis*), Lake Sladkoe (Zavarzin and Safronov 2001, Zavarzin 2011a – as

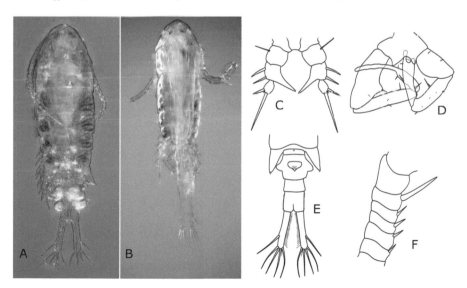

Figure 9.17 *Eurytemora* aff. *affinis* from Sakhalin (A and B – Lake Panitu, C–F – Lake Tunaicha). Female: A – habitus dorsal, C – P5, E – fifth pediger and urosome, dorsal; Male: B – habitus dorsal, D – P5, F – right A1 fragment (original).

E. affinis), lakes Ainskoe (Ueno 1935a,b, Borutskii and Bogoslovskii, 1964) and Tunaicha (Usova et al. 1980 – as *E. affinis*, Samatov et al. 2002, Zavarzin 2003, 2004, 2005 – as *Eurytemora* sp., Labay et al. 2013, 2016 – as *E. affinis*). We have additional records from Lake Panitu and tundra lakes having a connection with the sea, in the desalinated zones of bays and oxbows in river mouths from the northwestern coast and in the central lakes Nevskoe and Protochnoe (Fig. 9.18).

Ecology: On Sakhalin, this species is common in oligohaline and freshwater bodies having a connection with the sea. It dwells in both the pelagial and the littoral, where it is an epibenthic grazer. The species is euryhaline, able to live in 0 to greater than 30‰ salinity with an optimum between 5 and 15‰ (Ishikawa et al. 1999), but we did not find it at a salinity greater than 8‰ (the outer boundary of the α-horohalinicum – Kinne 1971, Khlebovich 1989). Probably, biotic factors (food competition, predation) limit its spread to more saline waters (Ishikawa et al. 1999).

Seasonal dynamics have been studied for Lake Tunaicha. The first nauplii appeared in mid-April, adult females carrying eggs occurred in early May, maximum abundance was observed in September, and in February–March the species was absent (Zavarzin 2005). It apparently produces diapausing eggs. No winter occurrence was reported in Hokkaido (Ban and Minoda 1989). Diapause egg production in Lake Ohnuma was observed in spring and late autumn (Ban 1992b). According to Ban (1992a), diapause egg production is related to short day length, low temperature, and high population density. Eggs produced in spring can remain in the sediment for at least a year and the accumulation of the egg bank serves as an important source of recruitment (Ban 1992b). Ban and Minoda (1989) with reference to Matoda (1950) noted that before the 1950s the occurrence of *E. affinis* in Japan was probably restricted to some brackish waters, and it is suggested that an artificial introduction of smelt *Hypomesus nipponensis* McAllister, 1963 from brackish lakes played an important role in its invasion of freshwaters. If the Japan invasion is related to human activity, then the observations from northern Sakhalin are apparently a natural migration process from the estuaries because there were no artificial introductions of fish or other aquatic organisms.

Eurytemora composita (Keifer 1929)

Taxonomic notes: Male from Sakhalin is similar to the material from the lake Issyk-Kul (Kos 2016) (Fig. 9.19).

Distribution: Alaska, Bering Sea (at St. Lawrence Island), near Paramushir Island (northern Kuril Islands), Korea, Lake Issyk-Kul (Kyrgyzstan) (Chang 2014, Kos 2016).

Sakhalin records: In a freshwater oxbow lake in the estuarine part of the Lyutoga River (one mature male) (Fig. 9.20).

Ecology: Brackish water species of lakes and desalinated coastal areas of seas; occurs at a salinity of 5.7‰ (Heron 1964, Kos 2016).

Superorder: Podoplea (Giesbrecht 1882)
    Order: Cyclopoida (Burmeister 1834)

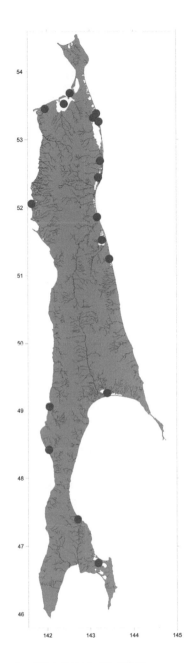

Figure 9.18   Sakhalin records of *Eurytemora* aff. *Affinis*.

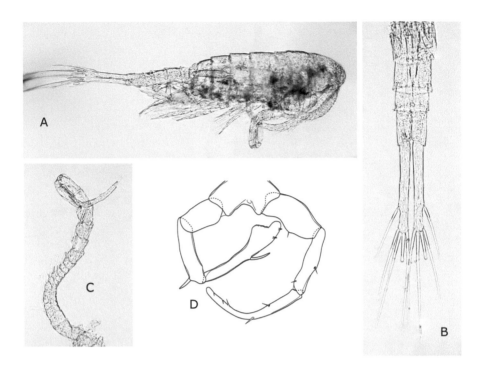

Figure 9.19 *Eurytemora composita* from Sakhalin (oxbow lakes of Lyutoga River). Male: A – habitus lateral, B – urosome, dorsal, C – right A1, D – P5 (original).

Suborder: Cyclopida (Khodami et al. 2019)

The free-living Cyclopoida of Sakhalin, including planktonic forms, have been poorly studied. Widespread and cosmopolitan species are reported: *Mesocyclops leuckarty* (Claus 1857), *Cyclops strenuus* (Fischer 1851), *Thermocyclops oithonoides* (Sars 1863). However, studies have shown that the taxonomy is far more complex. Currently, Cyclopidae of Sakhalin must be treated as "s. lato"; therefore, distribution outside Sakhalin is not considered.

Cyclopidae (Rafinesque 1815)

Cyclops (Müller 1785)

Cyclops strenuus s. lato (Fischer 1851)

Taxonomic notes: Einsle (1996) hypothesized that *C. strenuus* occurs only in Europe. The form on Sakhalin requires verification.

Sakhalin records: We found this form in the lakes of Mount Spamberg above 800 m elevation (Zavarzin 2011a), and in lakes Russkoe and Khvalisekoe (Okhotskie freshwater lake group) (SakhNIRO unpublished data) (Fig. 9.21).

Ecology: In various fresh and oligohaline water bodies, including mountain lakes. Pelagic in small reservoirs, littoral in large reservoirs (Rylov 1948, Monchenko 1974). Only found in freshwater on Sakhalin.

Figure 9.20   Sakhalin records of *Eurytemora composita*.

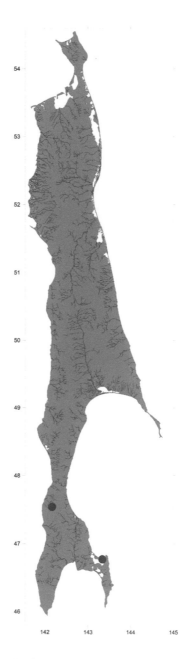

Figure 9.21   Sakhalin records of *Cyclops strenuus* s. lato.

Thermocyclops (Kiefer 1927)

Thermocyclops oithonoides s. lato (Sars 1863)

Taxonomic notes: *Thermocyclops oithonoides* was previously reported from China and Japan (Shen and Tai 1962a, Kikuchi 1930), but later studies demonstrated that it had been replaced by congeners (Guo 1999, Mirabdullayev and Ueda 2001). The form on Sakhalin needs to be clarified.

Sakhalin records: Chernysheva and Sabitov (1980) report this species from freshwater lakes of the Okhotskie lake group (Fig. 9.22).

Ecology: Various fresh and oligohaline water bodies. Most common in the pelagial zone (Rylov 1948, Monchenko 1974). On Sakhalin in small freshwater lakes.

Mesocyclops (Sars 1914)

Mesocyclops leuckarti s. lato (Claus 1857)

Taxonomic notes: *Mesocyclops leuckarti* in Japan has been reexamined and described as different species (Defaye and Kawabata 1993, Ueda et al. 1997, Ishida 1999, Ohtsuka and Ueda 1999). The Sakhalin populations need to be reexamined.

Sakhalin records: Lake Ainskoe (Ueno 1935a,b, Borutskii and Bogoslovskii 1964), Lake Tunaicha (Usova et al. 1980, Labay et al. 2016), Lake Vyselkovoe (Vavaiskie lake system) (Borutskii and Bogoslovskii 1964), Okhotskie lake group (Lake Russkoye, Lake Khvalisekoe (Borutskii and Bogoslovskii 1964, Chernysheva and Sabitov 1980), Lake Khazarskoe, Lake Svobodinskoye (Chernysheva and Sabitov 1980)); Vavaiskie lake system (Chibisanskie Lakes – Ueno 1935a,b, Borutskii and Bogoslovskii 1964, Zavarzin 2007, Labay et al. 2010 – and Vavaiskie Lakes – Zavarzin 2007, Labay et al. 2010). We collected it in Lake Svetloe and in the Nikolskii Pass mountain lake (600 m above sea level). Juvenile *Mesocyclops* were noted in tundra lakes near Moskal'vo village (Rylov 1932b) and Tym' River oxbow lakes (unpublished data of SakhNIRO) (Fig. 9.23).

Ecology: Found in a wide variety of fresh and oligohaline water bodies, from shallow pools to large lakes and rivers. In lakes it is common in the pelagic plankton, but also dwells in the littoral.

Halicyclops (Norman 1903)

Completely unexamined in Sakhalin. All references are only to genus. In the considered salinity range, there are apparently two species. Urosome fragments with P5 for the two forms are depicted (Fig. 9.24).

Sakhalin records: Oligohaline Lake Ainskoe in the southwest, Lake Tunaicha (Zavarzin 2003, 2004, Labay et al. 2016) and Lyutoga River estuary in the south, to lagoon bays in the northeast (Fig. 9.25).

Ecology: Brackish water in lagoons and estuaries.

Suborder: Ergasilida (Khodami et al. 2019)

Ergasilidae (Burmeister 1835)

Ergasilus (von Nordmann 1832)

Fish ectoparasites localized on gills, fins, and body surface. Only adult females are parasitic on fishes, and juveniles and adult males are free living (planktonic) (Zmierzlaya 1972, Kim 2014). They are a common zooplankton in Sakhalin fresh

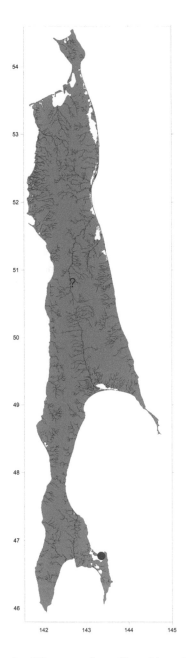

Figure 9.22   Sakhalin records of *Thermocyclops oithonoides* s. lato.

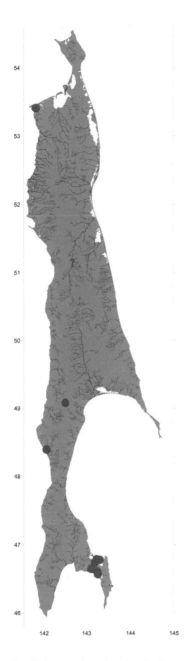

Figure 9.23 Sakhalin records of *Mesocyclops leuckarti* s. lato.

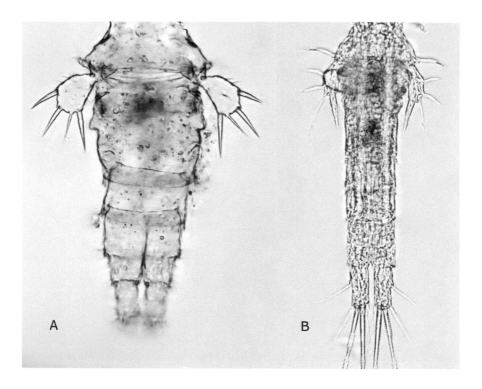

Figure 9.24 *Halicyclops* spp. from Sakhalin (Lyutoga River estuary). A – *Halicyclops* sp.1 urosome, ventral, B – *Halicyclops* sp. 2 urosome, ventral (original).

and brackish waters. In planktonic samples, identification is made only to genus. Vinogradov (2011) and Sokolov et al. (2012) identified adult females to species level from host fish.

Planktonic stages have been found in Sakhalin across the entire island. Freshwaters: Lake Sladkoe in the northwest (Zavarzin and Safronov 2001, Zavarzin 2011a), Lake Panitu and freshwater tundra lakes around Pil'tun Bay (Lake Golovka, Lake Lebyazh'e, etc.) in the northeast, the oxbow lakes of the central Tym' River, Chibisanskie Lakes (Zavarzin 2007, Labay et al. 2010), Okhotskie freshwater lake group (Lake Russkoe, Lake Khvalisekoe, Lake Khazarskoe, Lake Svobodinskoe) (Chernysheva and Sabitov 1981), and the oxbow lakes at the mouth of the Lyutoga River in the south. Oligohaline water bodies: Lake Ainskoe (Ueno 1935a, b) and Lake Protochnoe in the southwest, Lake Tunaicha (Samatov et al. 2002, Zavarzin 2003, 2004, 2005, Labay et al. 2016) in the south, Lake Lebyazh'e (Borutskii and Bogoslovskii 1964) in the southeast, Lake Slonovoe at the base of the Terpeniya peninsula, and Pil'tun Bay in the northeast (our data) (Fig. 9.26).

Ergasilus anchoratus (Markevich 1946)
    Distribution: Amur River basin, the waters of China and Sakhalin Island (Markevich 1946, Gusev 1987, Sokolov et al. 2012).

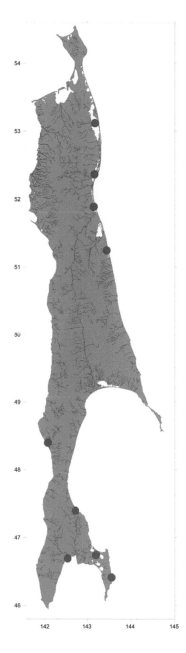

Figure 9.25 Sakhalin records of *Halicyclops* spp.

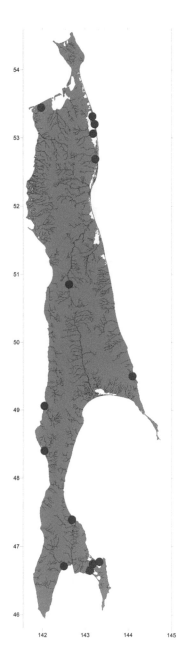

Figure 9.26   Sakhalin records of *Ergasilus* spp. planktonic stages.

Sakhalin records: Lake Sladkoe (northwest of the island) (Sokolov et al. 2012) (Fig. 9.27).

Ecology: Freshwater ectoparasite of catfish, loaches, and carps (Markevich 1946, Gusev 1987, Sokolov et al. 2012).

Ergasilus auritus (Markevich 1940)

Distribution: River basins of the Pacific and Atlantic coasts of North America, river basins of the Sea of Japan and the Sea of Okhotsk (Markevich 1956, Vinogradov 2011).

Sakhalin records: Lake Krestonozhka and adjacent freshwater areas of oligo-haline Lake Tunaicha, Lake Russkoe and Vavaiskie Lakes, and the Lyutoga River (Vinogradov 2011, Vinogradov and Zavarzin 2013). Sokolov et al. (2012) noted *Ergasilus* cf. *auritus* in Lake Sladkoye in the northwest (Fig. 9.28).

Ecology: Ectoparasite of a large number of freshwater fish species (Vinogradov 2011).

Ergasilus briani (Markevich 1932)

Distribution: Widely distributed in Eurasia (Gusev 1987, Sokolov et al. 2012).

Sakhalin records: Lake Sladkoe (northwest) (Sokolov et al. 2012) (see Fig. 9.27).

Ecology: A parasite of carp and loaches (Gusev 1987, Sokolov et al. 2012).

Ergasilus hypomesi (Yamaguti 1936)

Distribution: Sea of Japan basin, reservoirs in Russia, Korea, China, and Japan (Hokkaido) (Markevich 1956, Gusev 1987, Li 2004, Ohtsuka et al. 2004b, Vinogradov 2011).

Sakhalin records: Lakes Tunaicha and Ainskoe (Vinogradov 2011, Vinogradov and Zavarzin 2013) (Fig. 9.29).

Ecology: Ectoparasite of a large number of fish species (Vinogradov 2011). On Sakhalin, only in oligohaline lagoon lakes; juveniles and males keep mainly in the littoral zone among macrophytes.

Ergasilus wilsoni (Markevich 1933)

Distribution: Coastal areas of the Far Eastern seas, and Sea of Japan basin fresh-water bodies (Markevich 1956, Gusev 1987, Vinogradov 2014).

Sakhalin records: Southern marine lagoons: Lake Izmenchivoe and Lake Ptich'e. Oligohaline lagoons: Lake Tunaicha, Lake Ainskoe. Freshwater: Lake Krestonozhka (Lake Tunaicha basin), Lake Bol'shoye Vavaiskoe (Vavaiskie lake system), oxbow lakes of Lyutoga River and the Firsovka River, as well as in the sea coast of the Tatar Strait (Vinogradov 2011, Vinogradov and Zavarzin 2013, Vinogradov 2014) (Fig. 9.30).

Ecology: Brackish limnophilic species, found in desalinated coastal marine areas, brackish water bodies, as well as in river mouths. Ectoparasites of a wide range of fish hosts (Vinogradov 2014). The planktonic stage seasonal dynamics have been studied for Lake Tunaicha. Among many hosts, *E. wilsoni* was acutely domi-nant, suggesting that the vast majority of free-swimming individuals belonged to the same species. It was noted from the third week of June in the plankton, when the water temperature reached 13–14°C. Copepodites continued to be observed until the

Figure 9.27   Sakhalin records of *Ergasilus anchoratus* and *Ergasilus briani*.

Figure 9.28 Sakhalin records of *Ergasilus auritus*.

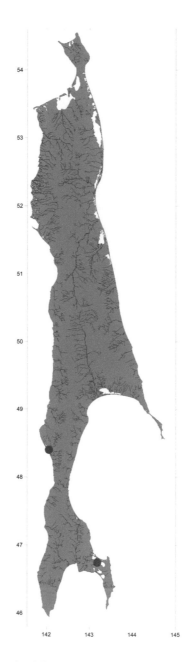

Figure 9.29 Sakhalin records of *Ergasilus hypomesi*.

Figure 9.30   Sakhalin records of *Ergasilus wilsoni*.

first week of November, when the temperature dropped to 6–7°C. The maximum lake plankton abundance was observed during the summer maximum temperatures from mid-August to mid-September.

Thersitina (Norman 1905)

Like *Ergasilus*, this genus has planktonic stages, but they were not found, and were only identified by adult females from the hosts.

Thersitina gasterostei (Pagenstecher 1861)

Distribution: In the coastal zone of the northern Atlantic and Pacific Oceans (Markevich 1956, Gusev 1987).

Sakhalin records: Lake Tunaicha (Vinogradov 2011, Vinogradov and Zavarzin 2013) (Fig. 9.31).

Ecology: Brackish, ectoparasite of a large number of fish taxa (Markevich 1956, Gusev 1987). On Sakhalin encountered only on three-spined stickleback (Gusev 1951, Vinogradov 2011).

9.4 ZOOGEOGRAPHY

Since the Paleozoic, freshwaters have repeatedly been colonized by copepods (Boxshall and Jaume 2000). The formation of the Sakhalin modern fauna is closely related to marine transgressions and regressions. In the Miocene–Pleistocene, the southern parts of Sakhalin were periodically connected by a land bridge to Hokkaido, and during marine transgressions this region was separated from the northern part of the island, which in turn was connected to the mainland. The mouth of the paleo-Amur River passed through Sakhalin (Lindberg 1972, Bezverkhniy et al. 2002, Matjushkov et al. 2014), which created the possibility of freshwater faunal contacts of different regions. The modern freshwater ecosystems were formed mainly in the late Würm regression and subsequent Holocene transgression. At the Pleistocene–Holocene border, in the late Würm climatic minimum, the sea level fell by 130 m, causing Sakhalin, Hokkaido, and the South Kuril Islands to be a single island system connected with Primorsky Krai (mainland Russia). It was probably at this time that paleo-Amur species penetrated into north Sakhalin. The separation from Hokkaido occurred 12,000 to 11,000 years ago, and from the mainland about 7,000 years ago (Bezverkhniy et al. 2002, Bogatov et al. 2007). In the interglacial Nome Pleistocene phase, the Sea of Japan and the Sea of Okhotsk were brackish (Lindberg 1972, Galerkin et al. 1982), which allowed colonization by brackish water fauna.

Zoogeographic zoning of brackish water and freshwater biota on Sakhalin was examined by many researchers (Taranets 1938, Berg 1962, Kruglov and Starobogatov 1993, Safronov and Nikiforov 1995, Starobogatov 1995, Chereshnev 1998, Kafanov and Kudryashov 2000, Nikiforov 2001, Prozorova 2001, Labay 2005, Bogatov et al. 2007). However, current freshwater and brackish water copepod distribution data do not correspond to any of the proposed schemes, due to high zooplankton vagility and to the small number of studies, small number of areas studied, and incomplete taxonomic identification.

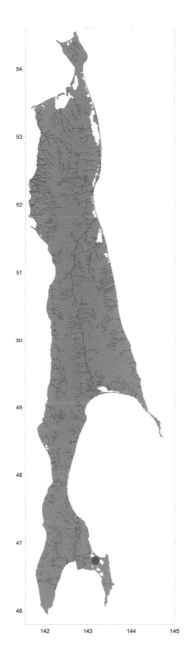

Figure 9.31 Sakhalin records of *Thersitina gasterostei*.

The basis of the Sakhalin brackish water planktonic copepod fauna is the taxocene of *S. tenellus*, *E.* aff. *affinis*, and *P. japonicus*. This group is common in waters of coastal habitats throughout the island. The maximum species abundance of the group is in oligohaline waters, where *Sinocalanus* and *Eurytemora* are often the dominant taxa (lakes Tunaicha, Nevskoe, and Ainskoe). *Eurytemora* aff. *affinis* was also found in absolutely freshwater bodies on the coasts, and we did not find it at salinities higher than α-horohalinicum; the two other species, however, were not detected in freshwater, but found in the coastal areas.

Halicyclops and planktonic stages of *Ergasilus* cooccur with Calanoida in many water bodies (in the south it is *E. wilsoni*, less often *E. hypomesi*). Therefore, the taxocene consists entirely of forms of maritime origin; Pacific near-Asian lower boreal (including lower boreal-subtropical) species. The unity of the group from reservoirs around the island is apparently due to the high euryhalinity of the taxa; nearly all are able to withstand marine salinity. As noted by Ueda et al. (2004) and Sakaguchi et al. (2011), when flooding occurs, they are flushed into the sea with a brackish water mass and some are transported to neighboring estuaries by tidal currents. Thus, brackish water copepods can expand their distribution along the coasts. Apparently, all species of the group (*Ergasilus* and *Halicyclops* are questionable) are also common to Hokkaido.

In oligohaline waters, *Mesocyclops leuckarti* s. lato is also a component of predominantly freshwater continental fauna. The amphiboreal meroplankton ectoparasite species *T. gasterostei* was reported from Lake Tunaicha (only from host fish, was not found in plankton).

Heterocope borealis, *H. appendiculata*, and *A. pacificus* are common in small freshwater tundra lakes. *Heterocope appendiculata* is found in both the northern and southern half of the island, while *H. borealis* and *A. pacificus* are found only in tundra lakes in the north. As noted above, *Heterocope* species are not found sympatrically, while either species can coexist with *A. pacificus* s. lato. *Heterocope* are glacial relicts and their primary distribution is the northern Palaearctic. *Heterocope borealis* is common mainly in the Arctic, while *A. pacificus* complex is apparently endemic to eastern Asia. Of the plankton cyclops in these lakes, *M. leuckarti* s. lato and free-swimming *Ergasilus* are common in water bodies with fish.

Acanthodiaptomus pacificus s. lato is also one of the main zooplankton components in mountain lakes in the south and is often the dominant species (Zavarzin 2004). This is the only Calanoida representative in Sakhalin mountain lakes. Cyclopoida were also found in mountain lakes: *C. strenuus* s. lato in Spamberg Mountain and *M. leuckarti* s. lato in Kamyshovyi Pass.

Representatives of *Neutrodiaptomus* inhabit freshwater oxbows and large freshwater lakes: *N. pachypoditus* in the north and *N. ostroumovi* in the south (*Neutrodiaptomus* of the Okhotskie Lakes remained unidentified). *Neutrodiaptomus pachypoditus* and *N. ostroumovi* belong to the fauna of northeastern Asia. It is interesting that *N. pachypoditus* is reported for Hokkaido (Makino et al. 2018), while we found this species only in northern Sakhalin, with *N. ostroumovi* in the south. *Ergasilus auritus* are common in lakes of these types, apparently both in the south and in the north, with *E. anchoratus* and *E. briani* only in the north. *Ergasilus auritus* and *E. briani* are apparently widespread, *E. anchoratus* is the Amur species and on

Sakhalin is apparently associated with the paleo-Amur fish. Here *M. leuckarti* s. lato can also be found. Representatives of *Thermocyclops* were found only in oxbows.

Mesocyclops leuckarti s. lato is found in oligohaline and freshwater bodies of all types. However, it may be represented by different species in different habitats.

The temporary reservoirs remain almost unexplored. We only found *N.* aff. *alaskiensis* in these habitats, the taxonomic and zoogeographic status of which has not been clarified. Also, the only record of *E. composita* was from an oxbow lake in the estuarine zone of the Lyutoga River, which represents a very disjunct record.

The invasive species *B. triarticulata,* was reported once in 1976 from Lake Russkoe (Okhotskie freshwater lake group); further studies were not conducted in this reservoir.

The native fauna of freshwater and brackish water pelagic copepods apparently does not so far include Sakhalin endemics (*N.* aff. *alaskiensis* remains in question), but contains both northern holarctic glacial relics and components of the oriental fauna typical of the northeast Asia. Although taxonomic identification is not complete, it is clear that the plankton copepod fauna of Sakhalin is similar to the fauna of Hokkaido (Ohtsuka et al. 2004a, Makino et al. 2018). Nine species of Calanoida have been found on Sakhalin. Six are common to Hokkaido: brackish *S. tenellus*, *P. japonicus*, and *E.* aff. *affinis*, freshwater *A. pacificus*, *N. pachypoditus*, and *H. appendiculata*, and *N.* aff. *alaskaensis* remains in question. Of Ergasilidae, *E. auritus*, *E. hypomesi*, *E. wilsoni*, *E. briani*, and *T. gasterostei* are common for both islands. Comparison of Sakhalin and Hokkaido Cyclopidae is impossible due to the lack of reliable species identifications. However, it can be assumed that at least some of Sakhalin species will also be common to Hokkaido.

ACKNOWLEDGMENTS

The author is deeply grateful to Dr. Vyacheslav Labay ("SakhNIRO", Russia) for the idea of this study, to Dr. Tadashi Kawai (Hokkaido Research Organization, Japan) for the possibility of conducting it, to Dr. Hiroshi Ueda (Usa Marine Biological Institute, Kochi University, Japan) for valuable advice, to Dr. D. Christopher Rogers (University of Kansas, USA) for revision and linguistic correction of the text, to Dr. Anna Neretina (A.N. Severtsov Institute of Ecology and Evolution) for the reviewing, as well as to all colleagues who participated in collecting material.

REFERENCES

Aladin, N.V. and Plotnikov, I.S. 2013. The concept of relativity and plurality of barrier salinity zones and forms of existence of the hydrosphere. *Proceedings of the Zoological institute of the Russian Academy of Science* 317:7–21 (in Russian).

Alekseev, V.R. and Souissi, A. 2011. A new species within *the Eurytemora affinis* complex (Copepoda: Calanoida) from the Atlantic Coast of USA, with observations on eight morphologically different European populations. *Zootaxa* 2767:41–56.

Alfonso, G. and Belmonte, G. 2008. Expanding distribution of *Boeckella triarticulata* (Thomson, 1883) (Copepoda: Calanoida: Centropagidae) in Southern Italy. *Aquatic Invasions* 3:247–51.

Asami, H. 2004. Early life ecology of Japanese smelt (*Hypomesus nipponensis*) in Lake Abashiri, a brackish water, eastern Hokkaido. *Scientific Report of Hokkaido Fisheries Experimental Station* 67:1–79.

Asami, H. and Ito, Y. 2003. Population dynamic and production of *Sinocalanus tenellus* (Kikuchi) (Copepoda, Calanoida) in Lake Abashiri, a brackish water, eastern Hokkaido, Japan. *Bulletin of Plankton Society of Japan* 50:67–78.

Ban, S. 1992a. Effects of photoperiod, temperature, and population density on induction of diapause egg production in *Eurytemora affinis* (Copepoda: Calanoida) in Lake Ohnuma, Hokkaido, Japan. *Journal of Crustacean Biology* 12:361–67.

Ban, S. 1992b. Seasonal distribution, abundance and variability of diapause eggs of *Eurytemora affinis* (Copepoda; Calanoida) in the sediment of Lake Ohnuma. Hokkaido. *Bulletin of Plankton Society of Japan* 39:41–48.

Ban, S. and Minoda, T. 1989. Seasonal distribution of *Eurytemora affinis* (Poppe, 1880) (Copepoda; Calanoida) in freshwater Lake Ohnuma. Hokkaido. *Bulletin of Faculty of Fisheries, Hokkaido University* 40:147–53.

Bayly, I.A.E. 1992. The non-marine Centropagidae (Copepoda: Calanoida) of the World. In *Guides to the Identification of the Microinvertebrates of the Continental Waters of the World*, 21–30. The Hague: SPB Academic Publishing.

Berg, L.S. 1962. Razdelenie territorii Palearktiki I Amurskoi oblasti na zoogeograficheskie oblasti na osnovanii rasprostraneniya presnovodnych ryb (The division of the territory of the Palearctic and the Amur region into zoogeographic regions based on the distribution of freshwater fish). In *Selected Proceedings*. Vol. 5, 320–60. Moskow-Leningrad: USSR Academy of Science (in Russian).

Bezverkhniy, V.L., Pletnev, S.P. and Nabiullin, A.A. 2002. Outline of geological structure and development of the Kuril Island System and adjacent regions. In *Flora and Fauna of the Kuril Islands (Materials of International Kuril Island Project)*, 9–22. Vladivostok: Dal'nauka (in Russian).

Bogatov, V.V., Storozhenko, S.U., Barkalov, V.U., et al. 2007. Biogeography of Sakhalin Island by the example of the distribution of terrestrial and freshwater biota. Theoretical and practical problems of studying of invertebrates associations. In *Memory Ya.I. Starobogatov*, 193–224. Moscow: KMK Scientific Press Ltd. (in Russian).

Borutskii, E.V. and Bogoslovskii, A.S. 1964. Zooplankton ozer yuzhnogo Sakhalina (Zooplankton of southern Sakhalin lakes). In *Ozera yuzhnogo Sakhalina i ikh ichtyofauna (Lakes of southern Sakhalin and Their Ichthyofauna)*, 97–140. Moscow: Moscow State University (in Russian).

Borutskii, E.V., Stepanova, L.A. and Kos, M.S. 1991. *Opredelitel' Calanoida presnykh vod SSSR (Identification Guide to Calanoida from Fresh Waters of USSR)*. Leningrad: Nauka (in Russian).

Boxshall, G.A. and Jaume, D. 2000. Making waves: The repeated colonization of fresh water by copepod crustaceans. *Advances in Ecological Research* 31:61–79.

Burckhardt, G. 1913. Zooplankton aus ost- und süd-asiatischen Binnengewässern. *Zoologische Jahrbücher, Abteilung für Systematik, Ökologie und Geographie der Tiere* 34:341–471. (in German).

Chang, C.Y. 2009. *Inland-water Copepoda. Illustrated Encyclopedia of Fauna and Glora of Korea*. Vol. 42. Seoul: Ministry of Education.

Chang, C.Y. 2014. *Continental Calanoida. Invertebrate Fauna of Korea*. Vol. 21. Incheon: National Institute of Biological Resources.

Chereshnev, I.A. 1998. *Biogeography of Freshwater Fish Fauna in the Russian Far East.* Vladivostok: Dal'nauka (in Russian).

Chernysheva, E.R. 1974. O nakhozhdenii novogo dlya Sakhalina vida kopepody v ozere Russkoe (On finding a new copepods species for Sakhalin in the lake Russkoe). *Izvestiya TINRO* 93:112 (in Russian).

Chernysheva, E.R. and Sabitov, E.K. 1980. Zooplankton ozer Okhotskoi gruppy Sakhalinskoi oblasti. Raspredelenie I (Zooplankton of the lakes of the Okhotsk group of the Sakhalin region) *ratsional'noye ispol'zovanie vodnykh bioresursov Sakhalina Kuril'skikh ostrovov. Distribution and Rational Use Water Zooresources of Sakhalin and Kuril Islands),* 17–21. Vladivostok: Far Eastern Scientific Center, USSR Academy of Sciences (in Russian).

Chernysheva, E.R. and Sabitov, E.K. 1981. K nakhozhdeniyu v ozerakh Okhotskoi gruppy (Yuzhnyi Sakhalin) planktonnykh rachkov–parazitov ryb (To finding in the lakes of the Okhotsk group (South Sakhalin) planktonic crustaceans–fish parasites). In *Itogi issledovanii po voprosam ratsional'nogo ispol'zovaniya i okhrany biologicheskikh resursov Sakhalina i Kuril'skikh ostrovov: tez. dokl. nauch.-prakt. konf. (Results of Studies on the Rational Use and Protection of the Biological Resources of Sakhalin and the Kuril Islands,* 43–5. Yuzhno-Sakhalinsk: Thesis of Scientific-Practical Conference (in Russian).

Chihara, M. and Murano, M. 1997. *An Illustrated Guide to Marine Plankton in Japan.* Tokyo: Tokai University Press (in Japanese).

Defaye, D. and Kawabata, K. 1993. *Mesocyclops dissimilis* n. sp. from Lake Biwa, Japan (Copepoda, Cyclopoida). *Hydrobiologia* 257:121 26.

Dodson, S., Skelly, D.A, and Lee, C. 2010. Out of Alaska: Morphological diversity within the genus *Eurytemora* from its ancestral Alaskan range (Crustacea, Copepoda). *Hydrobiologia* 653:131–48.

Dussart, B. and Defaye, D. 2002. *World Directory of Crustacea Copepoda of Inland Waters. I-Calaniformes.* Leiden: Backhuys.

Einsle, U. 1996. *Copepoda: Cyclopoida. Genera Cyclops,* Megacyclops, Acanthocyclops. *Guides to the Identification of the Microinvertebrates of the Continental Waters of the World.* Vol. 10. Amsterdam: SPB Academic.

Fefilova, E.B. 2015. *Copepods (Copepoda). Fauna of the European North-East of Russia.* Vol. 12. Moscow: KMK Scientific Press (in Russian).

Ferrari, I. and Rossetti, G. 2006. New records of the Centropagid Boeckella *triarticulata* (Thomson, 1883) (Copepoda: Calanoida) in Northern Italy: Evidence of a successful invasion? *Aquatic Invasions* 1:219–22.

Ferrari, I., Farabegoli, A., Pugnetti, A. and Stella, E. 1991. The occurrence of a calanoid Australasian species, *Boeckella triarticulata* (Thomson), in fish ponds of Northern Italy. *Verhandlungen Internationale Vereinigung Limnologie* 24:2822–27.

Galerkin, L.I., Barash, M.B., Sapozhnikov, V.V. and Pasternak, F.A. 1982. *Tihij okean* (Pacific Ocean). 1982. Moscow: Mysl'(relational database management system) (in Russian).

Guo, X. 1999. The genus *Thermocyclops* Kiefer, 1927 (Copepoda: Cyclopidae) in China. *Hydrobiologia* 403:87–95.

Gusev, A.V. 1951. Parasiticheskie Copepoda s nekotorykh morskikh ryb (Parasitic Copepoda from some sea fishes). Parasit. *Digest of the Zoological Institute of the USSR Academy of Science* 13:394–463 (in Russian).

Gusev, A.V. 1987. Sem. Ergasilidae Edwards, 1940, Lernaeopodidae Edwards, 1940, Argulidae Müller, 1785. In *Opredelitel' parazitov presnovodnykh ryb fauny SSSR. T. 3 Paraziticheskie mnogokletochnye* (Family Ergasilidae Edwards, 1940, Lernaeopodidae

Edwards, 1940, Argulidae Müller, 1785. In *Identification Guide to Parasites of Freshwater Fishes Fauna of the USSR. 3. Parasitic Multicellular*), 386–432, 473–508, 515–520. Leningrad: Nauka (in Russian).

Heron, G.A. 1964. Seven species of *Eurytemora* (Copepoda) from north-western North America. *Crustaceana* 7:199–211.

Ishida, T. 1999. *Mesocyclops yesoensis* sp. nov., *M. leuckarti* and *M. pehpeiensis* (Crustacea: Copepoda: Cyclopoida) from Hokkaido, northern Japan. *Biogeography* 1:81–5.

Ishikawa, D., Ban, S. and Shiga, N. 1999. Effects of salinity on survival, and embryonic and postembryonic development of *Eurytemora affinis* from a freshwater lake. *Plankton Biology and Ecology* 46:113–9.

Jeffries, H.P. 1967. Saturation of estuarine zooplankton by congenetic associates. *Estuaries* 83:500–8.

Kafanov, A.I. and Kudryashov, V.A. 2000. *Marine Biogeography: A Text-Book*. Moscow: Nauka (in Russian).

Kafanov, A.I., Labay, V.S. and Pecheneva, N.A. 2003. *Biota and Macrobenthic Communities of the Northeast Sakhalin Lagoons*. Yuzhno-Sakhalinsk: Sakhalin Research Institute of Fisheries and Oceanography (in Russian).

Khlebovich, V.V. 1974. *Kriticheskaya solenost' biologicheskikh protsessov (Critical Salinity of Biological Processes)*. Leningrad: Nauka (in Russian).

Khlebovich, V.V. 1989. Critical salinity and chorohalinicum: An update of concepts. *Proceedings of the Zoological Institute of the USSR Academy of Science* 196:5–10 (in Russian).

Kikuchi, K. 1930. A comparison of the diurnal migration of plankton in eight Japanese lakes. *Memoirs of the College of Science, Kyoto Imperial University* 5:27–74.

Kikuchi, K. 1936. Fresh-water and brackish-water calanoid copepods of Japan with notes on their geographical distribution. *Bulletin of the Biogeographical Society of Japan* 6:275–84.

Kim, I.H. 2014. Invertebrate Fauna of Korea. Arthropoda: Maxillopoda: Copepoda: Cyclopoida, Fish-Parasitic Cyclopoid Copepods. *Flora and Fauna of Korea* 21:1–226.

Kinne, O. 1971. Salinity—Invertebrates. *Marine Ecology* I:821–74.

Knowlton, N. 1993. Sibling species in the sea. *Annual Revue Evolution and Systematics* 24:189–216.

Kos, M.S. 1985. K faune Calanoida (Copepoda) pribrezhnykh raionov Yuzhnogo Sakhalina (To fauna Calanoida (Copepoda) of coastal southern Sakhalin regions). *Researches Fauna of Seas* 30:225–58 (in Russian).

Kos, M.S. 2016. *Calanoid Copepods of the Families Stephidae and Temoridae of the Seas of Russia and Adjacent Waters*. Saint-Petersburg: Zoological Institute RAS (in Russian).

Kruglov, N.D. and Starobogatov, Ya.I. 1993. Guide to recent molluscs of northern Eurasia. 3. Annotated and illustrated catalogue of species of the family Lymnaeidae (Gastropoda, Pulmonata, Lymnaeiformes) of Palaearctic and adjacent river drainage areas. Part 1. *Ruthenica* 3:65–92 (in Russian).

Labay, V.S. 2005. Fauna of the Malacostraca (Crustacea) from the fresh and brackish water of Sakhalin Island. In *Flora and Fauna of Sakhalin Island (Materials of International Sakhalin Island Project)*. Part 2, 64–87. Vladivostok: Dal'nauka (in Russian).

Labay, V.S. 2011. Zoogeographical essay of Malacostraca (Crustacea) fauna from fresh and brackish waters of Sakhalin Island. *Transactions of Sakhalin Scientific Research Institute of Fisheries and Oceanography* 12:131–51 (in Russian).

Labay, V.S., Atamanova, I.A., Zavarzin, D.S., et al. 2014. *Reservoirs of Sakhalin Island: From Lagoons to Lakes*. Yuzhno-Sakhalinsk: Sakhalin Regional Museum (in Russian).

Labay, V.S., Dairova, D.S., Zavarzin, D.S., et al. 2012. Freshwater ecosystem researches of the Laboratory of Hydrobiology, SakhNIRO. *Transactions of the Sakhalin Research Institute of Fisheries and Oceanography* 13:152–59 (in Russian).

Labay, V.S., Pecheneva, N.V., Zavarzin, D.S. and Bragina, I.Ju., 2002. Kormovaya baza molodi kety v zalivakh Pil'tun I Nyiskii (Severo-vostochnyi Sakhalin (Feeding base of young keta in Pil'tun and Nyiskii Bays (northeast Sakhalin). In Coastal Fisheries, the 21st Century. International Scientific-Practical Conference 19–21 Sept. 2001. 67–70. Yuzhno-Sakhalinsk: Sakh. knizh. izd. (in Russian).

Labay, V.S., Zavarzin, D.S., Konovalova, N.V., et al. 2013. Results of complex plankton and benthos researches in lagoons of southern Sakhalin. *Transactions of Sakhalin Scientific Research Institute of Fisheries and Oceanography* 14:153–79 (in Russian).

Labay, V.S., Zavarzin, D.S., Konovalova, N.V., et al. 2016. *Water Biota of Tunaicha Lake (Southern Sakhalin) and Conditions of It Dwelling.* Yuzhno-Sakhalinsk: Sakhalin Scientific Research Institute of Fisheries and Oceanography (in Russian).

Labay, V.S., Zavarzin, D.S., Moukhametova, O.N., et al. 2010. *Plankton and Benthos of Vavajskaya Lakes System (Southern Sakhalin) and Conditions of Their Dwelling.* Yuzhno-Sakhalinsk: Sakhalin Scientific Research Institute of Fisheries and Oceanography (in Russian).

Labay, V.S., Zhivogljadova, L.A., Polteva, A.V., et al. 2015. *Watercourses of Sakhalin Island: Life in the Running Water.* Yuzhno-Sakhalinsk: Sakhalin Regional Museum (in Russian).

Lazareva, V.I. 2003. Kontrol' chislennosti rakoobraznykh planktonnym hish'nikom *Heterocope appendiculata* Sars (Copepoda) (Control of the number of crustaceans by plankton predator *Heterocope appendiculata* Sars (Copepoda)). In Troficheskie svyazi v vodnykh soobshchestvakh i ekosistemakh: Materialy Mezhdunarodnoi konferentsii 28–31 oktyabrya 2003 g. (Trophic relationships in aquatic communities and ecosystems: Proceedings of the International Conference October 28–31, 2003). Borok: Papanin Institute for Biology of Inland Waters Russian Academy of Sciences. 72 (in Russian).

Lee, C.E. 1999. Rapid and repeated invasions of fresh water by the copepod *Eurytemora affinis. Evolution* 53:1423–34.

Lee, C.E. 2000. Global phylogeography of a cryptic copepod species complex and reproductive isolation between genetically proximate "populations". *Evolution* 54:2014–27.

Lee, C.E. and Frost, B.W. 2002. Morphological stasis in the *Eurytemora affinis* species complex (Copepoda: Temoridae). *Hydrobiologia* 480:111–28.

Li, H.K. 2004. Copepodid stages of *Ergasilus hypomesi* Yamaguchi (Copepoda, Poecilostomatoida, Ergasilidae) from a brackish lake in Korea, *Korean Journal of Biological Sciences,*8:1–12.

Lindberg, G.U. 1972. *Krupnye kolebaniya urovnya okeana v chetvertichnyi period (Large Fluctuations in Sea Level in the Quaternary).* Leningrad: Nauka (in Russian).

Makino, W. and Tanabe, A. 2009. Extreme population genetic differentiation and secondary contact in the freshwater copepod *Acanthodiaptomus pacificus* in the Japanese Archipelago. *Molecular Ecology* 18:3699–713.

Makino, W., Tanabe, A. and Urabe, J. 2018. Distribution of freshwater calanoid copepods in Japan in the early decades of the 21st Century: Implications for the assessment and conservation of biodiversity. *Limnology and Oceanography* 63:758–72.

Maly, E.J. 1984. Dispersal ability and relative abundance of *Boeckella* and *Calamoecia* (Copepoda: Calanoida) in Australian and New Zealand waters. *Oecologia* 62:173–81.

Maly, E.J. 1991. Dispersal ability and its relation to incidence and geographic distribution of Australian centropagid copepods. *Verhandlungen Internationale Vereinigung Limnologie* 24:2828–32.

Markevich, A.P. 1946. Prazitichni Copepoda rib z bassejnu r. Amura (Paprazitic copepods of fish in the Amur River basin). *Naukovi zapiski Kiivskogo derzhavnogo universiteta* 5:225–45 (in Ukranian).

Markevich, A.P. 1956. *Paraziticheskie veslonogie ryb SSSR (Parasitic Copepods of Fishes of USSR)*. Kiev: USSR Academy of Science (in Russian).

Markevich, G.I. 1985. *Neuthrodiaptomus sklyarovae* sp. n. (Copepoda, Diaptomidae) iz basseina reki Zei (*Neuthrodiaptomus sklyarovae* sp. n. (Copepoda, Diaptomidae) from the Zeya River basin). *Zoological Journal* 64:1098–9 (in Russian).

Matjushkov, G.V., Solovjev, A.V. and Melnikov, O.A. 2014. *Sakhalin Island Geological Past*. Yuzhno-Sakhalinsk: Sakhalin Regional Museum (in Russian).

Matoda, S. 1950. Hokkaido koshoshi (Lakes in Hokkaido). *Scientific Reports of the Hokkaido Fish Hatchery* 5:1–96 (in Japanese).

Mirabdullayev, I.M. and Ueda, H. 2001. A redescription of *Thermocyclops uenoi* (Crustacea, Copepoda). *Vestnik Zoologii* 35:17–22.

Mizuno, T. 2000. Order Calanoida. In *An Illustrated Guide to Freshwater Zooplankton in Japan*, eds. T. Mizuno and E. Takahashi, 2–16. Tokyo: Tokai University Press (in Japanese).

Mizuno, T. and Miura, Y. 1984. Freshwater Copepoda in Japan. In Chinese/*Japanese Freshwater Copepoda*, eds. C.J. Shen and T. Mizuno, 471–646. Yonago: Tatara-shobo (in Japanese).

Monchenko V.I. 1974. *Shchlepnoroti tsiklopopodyni, tsiklopi (Cyclopidae) (Maxillopodans Cyclopoids, Cyclopidae)*. Fauna of Ukraine 27(3)). Kiiv: Naukova Dumka (in Ukrainian).

Moriya, H. 1979. Zooplanktons in temporary pools fed by snow-melt water on the Ishikari coastal area, Hokkaido. 北海道大学研究紀要 (No English title, downloadable PDF from http://hdl.handle.net/2115/37092) 2:23–38. (in Japanese).

Nemchinova, I.A. 2006. Species composition and structure of summer zooplankton in the lagoon-type Lake Izmenchivoye. *Transactions of Sakhalin Scientific Research Institute of Fisheries and Oceanography* 8:89–106 (in Russian).

Nemchinova, I.A. 2011. *Structure and Quantitative Description of Zooplankton in Baikal Bay off North Western Sakhalin in July 2009*. In Proceedings of the 26th International Symposium on Okhotsk Sea and Sea Ice. Mombetsu, Hokkaido, Japan, 20–25 February 2011, 49–54. Mombetsu.

Nielsen, D.L., Brock, M.A., Rees, G.N. and Baldwin, D.S. 2003. Effects of increasing salinity on freshwater ecosystems in Australia. *Australian Journal of Botany* 51:655–65.

Nikiforov, S.N. 2001. Ikhtiofauna presnykh vod Sakhalina i ee formirovanie (The ichthyofauna of fresh waters of Sakhalin Island and its formation). Avtoref. dis. kand. boil. nauk (extended abstract of Ph.D. thesis). Vladivostok: Institute of Marine biology, Far Eastern Branch of Russian Academy of Science:1–25 (in Russian).

Ohtsuka, S. and Nishida, S. 2017. Copepod Biodiversity in Japan: Recent Advances in Japanese Copepodology. In *Species Diversity of Animals in Japan*, eds. M. Motokawa and H. Kajihara, 565–602. Switzerland: Springer.

Ohtsuka, S. and Ueda, H. 1999. Zoogeography of pelagic copepods in Japan and its adjacent waters. *Bulletin of the Plankton Society of Japan* 46:1–20 (in Japanese).

Ohtsuka, S., Ho, J.-S. and Nagasawa, K. 2004a. Ergasilid copepods (Poecilostomatoida) in plankton samples from Hokkaido, Japan, with reconsideration of the taxonomic status of *Limnoncaea* Kokubo, 1914. *Journal of Natural History* 38:471–98.

Ohtsuka, S., Ho, J.-S., Nagasawa, K., et al. 2004b. The identity of *Limnoncaea diuncata* Kokubo, 1914 (Copepoda:Poecilostomatoida) from Hokkaido, Japan, with the relegation of Diergasilus Do, 1981 to a junior synonym of Thersitina Norman, 1905. *Systematic Parasitology* 57:35–44.

Onishchenko, N.I. 1987. *Vodnye resursy Sakhalina i ikh izmeneniya pod vliyaniem khozyaistvennoi deyatel'nosti.* (Water Resources of Sakhalin and Their Changes Caused by Economic Activities). Vladivostok: Dal'nevost. Otd., Akad. Nauk SSSR 1 (in Russian).

Pezheva, M.K., Kazanchev, C.S. and Avalishvili, E.T. 2016. Typological classification of mountainous karst lakes of kabardino-balkarian republic based on zooplankton data. *The Bulletin of KrasGAU* 6:50–56 (in Russian).

Prozorova, L.A. 2001. Features of the distribution of freshwater mollusk fauna in the Far East of Russia and its biogeographical zoning. *Vladimir Ya. Levanidov's Biennial Memorial Meetings* 1:112–25 (in Russian).

Rylov, V.M. 1925. Zur Copepoden Fauna des äussersten Ostens. *Zoologischer Anzeiger* 63:307–18 (in Germany).

Rylov, V.M. 1932a. Nauchnye rezultaty dal'nevostochnoi gydrofaunisticheskoi expeditsii Zoologicheskogo muzeya v 1927 g. 4. Presnovodnye Eucopepoda (Crustacea) Ussuriiskogo kraya (Scientific results of Far East Hydrofaunistic expedition of Zoological Museum 1927. 4. Freshwater Eucopepoda (Crustacea) of Ussuri region). *Proceedings of the Zoological Institute of the USSR Academy of Science* 1:243–80 (in Russian).

Rylov, V.M. 1932b. Zur Kenntnis der Copepoden und Cladocerenfauna der Insel Sachalin. *Zoologischer Anzeiger* 99:101–08. (in Germany).

Rylov, V.M. 1948. *Rakoobraznye. Cyclopoida presnyh vod. (Crustaceans. Cyclopoida of Fresh Waters).* Fauna of the USSR. III (3). Moscow-Leningrad: Nauka (in Russian).

Sabitov, E.K. and Chernysheva, E.R. 1976. K kharakteristike zooplanktona ozer Russkoye i Khvalisekoye (yugo-vostochnyi Sakhalin) (To the characteristics of zooplankton of the lakes Russkoye and Khvalisekoe (southeast Sakhalin). In *Prirodnye usloviya Sakhalina. Sakh. Otdel. GO SSSR. (Natural Conditions of Sakhalin)* Leningrad: Department of Geographical Society of the USSR. 23–28 (in Russian).

Safronov, S.N. and Nikiforov, S.N. 1995. Species composition and distribution of ichthyofauna of fresh and brackish waters of Sakhalin (Report). In Materials of the XXX Scientific and Methodical Conference of Teachers of Yuzhno-Sakhalinsk State Pedagogical Institute (April 1995). Part 2, 112–124. Yuzhno-Sakhalinsk: Yuzhno-Sakhalinsk State Pedagogical Institute (in Russian).

Sakaguchi, S.O. and Ueda, H. 2010. A new species of *Pseudodiaptomus* (Copepoda: Calanoida) from Japan, with notes on the closely related *P. inopinus* Burckhardt, 1913 from Kyushu Island. *Zootaxa* 2623:52–68.

Sakaguchi, S.O. and Ueda, H. 2011. Morphological divergence of *Pseudodiaptomus inopinus* Burckhardt, 1913 (Copepoda: Calanoida) between the Japan Sea and Pacific coasts of western Japan. *Plankton and Benthos Research* 6:124–28.

Sakaguchi, S.O. and Ueda, H. 2018. Genetic analysis on *Pseudodiaptomus inopinus* (Copepoda, Calanoida) species complex in Japan: revival of the species name of *P. japonicus* Kikuchi, 1928. *Plankton and Benthos Research* 13:173–79.

Sakaguchi, S.O., Ueda, H., Ohtsuka, S., et al. 2011. Zoogeography of planktonic brackish-water calanoid copepods in western Japan with comparison with neighboring Korean fauna. *Plankton and Benthos Research* 6:18–25.

Sakhalin Region. 1994. *Geographical Overview.* Yuzhno-Sakhalinsk: Sakh. knizh. izd. (in Russian).

Samatov, A.D., Labay, V.S., Motylkova, I.V., et al. 2002. Short characteristic of water biota of Tunaicha Lake (Southern Sakhalin) in summer period. *Transactions of Sakhalin Scientific Research Institute of Fisheries and Oceanography* 4:258–69 (in Russian).

Shen, C.J. and Song, D.X. 1979. Calanoida, Sars, 1903. In *Fauna Sinica. Crustacea. Freshwater Copepoda*, 53–163, Beijing: Science Press (in Chinese).

Shen, C.J. and Tai, A.Y. 1962a. The Copepoda of the Wu-Li Lake, Wu-Sih, Kiangsu Province. I. Calanoida. *Acta Zoologica Sinica* 14:99–118 (in Chinese).

Shen, C.J. and Tai, A.Y., 1962b. The Copepoda of the Wu-Li Lake, Wu-Sih, Kiangsu Province. II. Cyclopoida. *Acta Zoologica Sinica* 14:225–48 (in Chinese).

Smirnov, S.S. 1929. Beiträge zur Copepodenfauna Ostasiens. *Zoologischer Anzeiger* 81:317–29 (in Germany).

Soh, H.Y., Kwon, S.W Lee, W. and Yoon, Y.H. 2012. A new *Pseudodiaptomus* (Copepoda, Calanoida) from Korea supported by molecular data. *Zootaxa* 3368:229–44.

Sokolov, S.G., Shedko, M.B., Protasov, E.N., and Frolov, E.V. 2012. Parasites of the inland water fishes of Sakhalin Island. In *Flora and Fauna of North-West Pacific Islands (Materials of International Kuril Island and International Sakhalin Island Projects)*, 179–216. Vladivostok: Dal'nauka.

Starobogatov, Ya.I. 1995. Amphipoda, Isopoda. In *Key to Freshwater Invertebrates of Russia and Adjacent Lands. 2. Crustacea*, ed. S.J. Tsalolikhin, 167–173, 184–206. St. Petersburg: Zoological Institute of Russia, Russian Academy of Science. (in Russian).

Stepanova, L.A. 1981. Novyi vid *Ligulodiaptomus* (Copepoda. Calanoida) iz ozera Malogo (Kamchatka) (A new spesies of *Ligulodiaptomus* (Copepoda. Calanoida) from Lake Maloye (Kamchatka). *Zoological Journal* 60:309–12 (in Russian).

Streletskaya, E.A. 1975. K voprosu o sistematicheskom polozhenii nekotorykh presnovodnykh rakoobraznykh (Cladocera, Copepoda) basseina r. Kolymy (To the issues of the systematic position of some freshwater crustaceans (Cladocera, Copepoda) of Kolyma river basin). In *Gidrobiologicheskie issledovaniya vnutrennikh vodoemov Severo-Vostoka SSSR. (Hydrobiological studies of inland waters of the North-East of the USSR)*, 60–138. Vladivostok: Far Eastern Scientific Center, USSR Academy of Sciences.

Sukhikh, N.M. and Alekseev, V.R. 2013. *Eurytemora caspica* sp nov from the Caspian Sea – One more new species within the E-affinis complex (Copepoda: Calanoida: Temoridae). *Proceedings of the Zoological Institute of the Russian Academy of Science* 317:85–100.

Taranetz, A.Y. 1938. K zoogeographii Amurskoy perekhodnoy oblasti na osnove izuchenia presnovodnoy ichtyofauny (To the zoogeography of the Amur transition region based on the study of freshwater ichthyofauna). *Bulletin of the Far Eastern Branch of the USSR Academy of Science* 28:99–115 (in Russian).

Ueda, H. and Sakaguchi, S.O. 2019. *Pseudodiaptomus yamato* n. sp. (Copepoda, Calanoida) endemic to Japan, with redescriptions of the two closely related species *P. inopinus* Burckhardt and *P. japonicus* Kikuchi. *Plankton and Benthos Research* 14:29–38.

Ueda, H., Terao, A., Tanaka, M., et al. 2004. How can river-estuarine planktonic copepods survive river floods? *Ecological Research* 19:625–32.

Ueda, H., Ishida, T. Imai, J.-I. 1997. Planktonic cyclopoid copepods from small ponds in Kyushu, Japan. II. Subfamily Cyclopinae. *Hydrobiologia* 356:61–71.

Ueno, M. 1935a. Crustacea, collected in the lakes of Southern Sakhalin. *Annotationes Zoologicae Japonenses* 15:88–94.

Ueno, M. 1935b. Limnological Reconaissance of Southern Sakhalin. II. Zooplankton. *Bulletin of the Japanese Society of Scientific Fisheries* 4:190–94 (in Japanese).

Ueno, M. 1936. Zooplankton of Lake Taraika and its neighbouring water, Sakhalin. *Transactions of the Sapporo Natural History Society* 14:173–78.

Usova, N.P., Filatova, V.I. and Chernysheva, E.R. 1980. O gidrobiologicheskom sostoyanii ozera Tunaicha. Raspredelenie I ratsional'noye ispol'zovanie vodnykh bioresursov Sakhalina I Kuril'skikh ostrovov (About hydrobiological status Lake Tunaicha. Distribution and rational use water zooresources of Sakhalin and Kuril Islands), 8–17. Vladivostok: Far Eastern Scientific Center, USSR Academy of Sciences. (in Russian).

Vekhov, N.V. 1980. Biologiya veslonogikh rakoobraznykh tundrovykh vodoemov. II. Vremennye vodoemy. Copepods biology in tundra water bodies. II. Temporary water bodies. *Biological Science* 2:44–50 (in Russian).

Vinarski, M.V., Palatov, D.M. and Novichkova, A.A. 2015. The first freshwater molluscs from Wrangel Island, Arctic Russia. *Polar Research* 34:1–4.

Vinogradov, S.A. 2011. Parasitic copepods of Ergasilidae family from fishes of South Sakhalin. *Izvestiya TINRO* 166:208–18. (in Russian).

Vinogradov, S.A. 2014. The first data on the parasitic copepods from fishes in Ptich'e Lake (south-eastern Sakhalin). *The Bulletin of Irkutsk State University. Series Biology. Ecology.* 10:108–16 (in Russian).

Vinogradov, S.A. and Zavarzin, D.S. 2013. Ecology and epizootic significance of parasitic copepods from genus *Ergasilus* in Lake Tunaycha (southern Sakhalin). *Izvestiya TINRO* 174:247–56 (in Russian).

Wilson, M.S. 1951. A new subgenus of *Diaptomus* (Copepoda: Calanoida), including an Asiatic species and a new species from Alaska. *Journal of the Washington Academy of Science* 41:168–79.

Zavarzin, D.S. 2003. Seasonal dynamics of zooplankton from the Tunaycha Lake (southern Sakhalin). *Transactions of Sakhalin Scientific Research Institute of Fisheries and Oceanography* 5:106–12 (in Russian).

Zavarzin, D.S. 2004. Composition and spatial distribution of zooplankton communities from the Tunaicha Lake (southern Sakhalin) by the 2001 summer survey data. *Transactions of Sakhalin Scientific Research Institute of Fisheries and Oceanography* 6:331–38 (in Russian).

Zavarzin, D.S. 2005. Some aspects of seasonal dynamic of zooplankton from Tunaycha Lake (Southern Sakhalin) at the contemporary stage. *Vladimir Ya. Levanidov's Biennial Memorial Meetings* 3:95–105 (in Russian).

Zavarzin, D.S. 2007. Lake zooplankton of the Vavay system (southern Sakhalin) from surveys conducted in July 2004 and 2005. *Transactions of Sakhalin Scientific Research Institute of Fisheries and Oceanography* 9:152–65 (in Russian).

Zavarzin, D.S. 2011a. Zooplankton of Sladkoe Lake (north-west of Sakhalin Island) by results of research in July 2009. *Vladimir Ya. Levanidov's Biennial Memorial Meetings* 5:173–81 (in Russian).

Zavarzin, D.S. 2011b. Zooplankton of the Spamberg Mountain lakes (southwestern Sakhalin). *Transactions of Sakhalin Scientific Research Institute of Fisheries and Oceanography* 12:94–109 (in Russian).

Zavarzin, D.S. 2020. *Pseudodiaptomus japonicus* Kikuchi, 1928 (Copepoda, Pseudodiaptomidae), a brackish-water copepod formerly known as *P. inopinus* Burckhardt, 1913, on Sakhalin Island (Russian Far East). *Crustaceana* 93:541–7.

Zavarzin, D.S. and Atamanova, I.A. 2014. Zooplankton seasonal dynamics in Ptichye Lake and adjoining sea coastal waters of Southern Sakhalin. *Vladimir Ya. Levanidov's Biennial Memorial Meetings* 6:239–49 (in Russian).

Zavarzin, D.S. and Safronov, S.N. 2001. Zooplankton ozera Sladkoe (severo-zapadnyi Sakhalin) Zooplankton of Lake Sladkoye (northern-west Sakhalin). *Vladimir Ya. Levanidov's Biennial Memorial Meetings* 1:187–94 (in Russian).

Zmerzlaya, E.I. 1972. *Ergasilus sieboldi* Nordmann, 1832, ego razvitie, biologiya i epizooto-logicheskoe znachenie (*Ergasilus sieboldi* Nordmann 1832, its development, biology and epizootic significance). *Proceedings of* GosNIORKh 80:132–77 (in Russian).

CHAPTER 10

Conservation of Continental Mysida and Stygiomysida

Mikhail E. Daneliya and Karl J. Wittmann

CONTENTS

10.1 INTRODUCTION

Effective species conservation implies protection at all levels from individuals to ecosystems. However, monitoring and protection of single specimens is usually not practicable with small-sized invertebrates. Unlike vertebrates and large-sized invertebrates, conservation should start with the protection of populations. Most small, nonnoxious aquatic invertebrates are not perceived as being of direct human interest and are rarely used as commercial resources, and their disappearance often not being immediately noticed. Thus, decline of populations is mostly seen in the context of environmental degradation rather than single species protection.

Aquatic invertebrate conservation is poorly developed. Small crustaceans, like mysids, have gained little attention from conservationists. Not surprisingly, their taxonomy, biogeography, and ecology are rather poorly developed: the diversity is still to be fully discovered, many species are known only from original descriptions, and ecological data are fragmentary. Two Mysida species from Bermudan marine caves were included on IUCN Red List (Iliffe 1996a, b). The most comprehensive national-level status assessment was made in Norway (Oug et al. 2015), where two freshwater species were included on the Red List (Oug et al. 2015, Spikkeland et al. 2016). Another five species are present on national lists: three in the Red Data Book of Ukraine (Dovgal 2009a, 33b, Samchishina 2009), one in the Red Book of the Republic of Moldova (Toderaş et al. 2015) and in the Red Data Book of the Pridnestrovian Moldavian Republic (Filipenko 2009), and one in the Red Data Book of Lithuania (Vaitonis 2007). These same species were included in the regional list of Leningrad Oblast of Russia (Alekseev 2002). Stepanyants et al. (2015) remarked that the conservation status for most species has not yet been assessed, only *Diamysis pusilla* G.O. Sars, 1907, as critically endangered. Thus, they recommended the creation of a protected area in this species range.

Marine caves of Bermuda are not covered here; nonetheless, it appears important to stress that the local government established a concise management plan (Glasspool 2003) for the endangered cave fauna, providing effective measures for the protection of mysids at population, species, and community levels. A number of protected areas were established at various levels across the distribution range of mysids listed as endangered (Dovgal 2009a, b, Samchishina 2009) in the Ukraine. This provides partial protection; however, current status estimates are still lacking.

Freshwater mysids of Fennoscandia, represented by three species of *Mysis* and by *Neomysis integer* (Leach, 1814), were well studied; nonetheless, updates are so far available only for the Norwegian Red List (Spikkeland et al. 2016). The status of least concern was attributed to *Mysis relicta* Lovén, 1862, an endemic of the Baltic province within the Palaearctic. *Mysis salemaai* Audzijonytė et Väinöla, 2005, a species with generally more ample distribution in the North Palaearctic, received the status of vulnerable after more data became available compared to a previous assessment, this species being now present in only one lake of Norway, indicative of some risk of environmental degradation. Due to potential range fragmentation, the status of being near threatened in Norway was assigned to *Mysis segerstralei* Audzijonytė et Väinöla, 2005, a circum-Arctic species, known in Fennoscandia only from one lake on the border between Finland and Norway and from a single estuary at the coast of the Barents Sea. *Mysis relicta* is indicated as vulnerable in the Red Data Book of Lithuania (Vaitonis 2007), though it has been recorded there from 15 lakes; and in the Red Data Book of the Nature of Leningrad Oblast in Russia (Alekseev 2002), this assessment was made prior to the taxonomic revision of the *M. relicta* complex, whereby *M. salemaai* was also recorded from this region (Audzijonytė and Väinöla 2005).

Though being capable of occasional predation and cannibalism, freshwater mysids mostly occupy the necto-benthic detritophage and filter-feeding niches, transforming debris into body mass, in turn being important food for fishes. Mysid number declines may cause habitat eutrophication with serious depletions of fish populations. In the mid-20th century, a number of mysid species were introduced in waters beyond their native range in order to increase fish productivity, often resulting in problematic issues of invasive species.

Here, we propose a first assessment of the conservation status for all mysid species inhabiting freshwaters and related continental water bodies, based on literature data and our own unpublished observations. In addition to main biogeographic regions, biogeographic provinces and additional distribution data are indicated below. The conservation status of each species is assessed employing the criteria of the International Union for Conservation of Nature (IUCN 2012).

10.2 WORLD FAUNA

10.2.1 List of Taxa

Mees and Meland (2020) listed 1,181 species of Mysida and 16 species of Stygiomysida. The freshwater fauna contributes about 7% to total numbers according to Porter et al. (2008). Mysids, in a wide sense, are primarily marine crustaceans. A number of euryhaline species penetrate estuaries, river deltas, and coastal lakes; only a small fraction of the world fauna is endemic to fresh and other continental waters. Freshwater lists were more recently updated for the Nearctic (Price 2015a, b), Palaearctic, and Ponto-Caspian regions (Daneliya et al. 2012). Recently, a number of new freshwater mysids have been discovered in the West Palaearctic (Wittmann and Ariani 2012; Wittmann et al. 2016) and Neotropics (Wittmann 2017, 2018); consequently, an updated list is presented here. All Stygiomysida are included, because this group shows a clearly continental distribution, even though some species are found in brackish and marine habitats.

The taxonomic status of most species is rather well established. The subspecies *Paramysis (Serrapalpisis) lacustris turcica* Băcescu, 1948, from Lake Beyşehir in Turkey (Anatolian province of the Ponto-Caspian region), has no type material remaining, and has never been studied after the first description by Băcescu (1948). A potentially new species of *Parvimysis* sp. from Amazonia (Neotropical region) still requires clarification from better preserved material (Wittmann 2018). A presumably undescribed *Stygiomysis* sp. was recorded by Pesce (1976) from groundwater of the Salento Peninsula in Italy (Palaearctic region).

10.2.1.1 Mysida

Within order Mysida, only the Mysidae include representatives from freshwater. Among the ten subfamilies, three have species in continental waters: Mysinae (67), Heteromysinae (1), and Leptomysinae (2). Thus, the total number of continental Mysida is 70 species (Table 10.1). Among representatives of the Mysinae, certain groups have an almost entirely continental distribution. This includes the tribe Diamysini, with eight out of nine genera found largely in freshwaters. Significant inland evolution also occurred in the tribe Paramysini, where four out of five extant genera comprise continental species, with *Paramysis* as the most species-rich.

10.2.1.2 Stygiomysida

This order includes 16 species of continental subterranean mysid-like crustaceans. Two families are recognized: Lepidomysidae (one genus, nine species) and Stygiomysidae (one genus, seven species). Although certain species are found in euhaline conditions, these clearly show a relict continental distribution in tropical and subtropical hypogean environments, and consequently are listed in Table 10.1.

10.3 DISTRIBUTION AND BIOGEOGRAPHIC DIVISION

Concepts of biogeographic division provide important bases for the establishment of protected areas, as particularly implemented in marine conservation (connection between biogeographic divisions and conservation are reviewed in Gubbay 2014). Mysids are traditionally considered in marine biogeography; nonetheless, certain marine and brackish water species gained the potential of penetrating continental waters. Such species still have been characterized in the frame of marine biogeography (here using the terminology of Petryashev 2005, 2009, Daneliya and Petryashev 2011a). In the course of long-term freshening of large isolated water bodies, a number of species became eulimnic, while some of their congeners may have maintained marine connections. This holds true for certain species of *Mysis, Paramysis, Diamysis, Neomysis,* and others (Table 10.1). In other words, part of the species is attributed to marine types of ranges, part continental, and the remaining species to both types.

Many freshwater species and subspecies were just recently discovered in waters of the Adriatic basin (Wittmann and Ariani 2012, Wittmann et al. 2016) and especially from the Neotropics (Wittmann 2017, 2018). Freshwater mysids are now known from all major continental biogeographic regions with exception of the Antarctic. Among other regions, Afrotropical (Ethiopian), Australian, Nearctic, Oriental, Palaearctic, Ponto-Caspian, and Neotropical regions contain 2, 2, 5, 3, 13, 26, and 29 native species and subspecies, respectively, found in fresh or brackish continental waters.

Most continental species are confined to a certain biogeographic region, and may be useful as biogeographic indicators. At least 11 species show an interregional distribution: *Diamysis mesohalobia heterandra* Ariani et Wittmann, 2000, *Hyperacanthomysis longirostris* (Ii, 1936), *Mesopodopsis slabberi* (van Beneden, 1861), *Mysis nordenskioldi* Audzijonytė et Väinölä, 2007, *M. segerstralei* Audzijonytė et Väinölä, 2005, *Neomysis awatschensis* (Brandt, 1851), *N. integer* (Leach, 1814), *N.*

Table 10.1 List of Freshwater Species of the Orders Mysida and Stygiomysida, together with Distribution and Proposed Conservation Status

Taxon	Major region	Type of range	Countries (native range)	Distribution, habitat, and population status remarks	Proposed conservation status
Order Mysida					
Mysidae					
Leptomysinae					
Afromysini					
Tenagomysis					
T. chiltoni W.M. Tattersall, 1923	AU (IP)	NZL	New Zealand	Common in estuaries, lakes, and rivers, but distribution incompletely documented (Jocqué and Blom 2009, and unpublished data)	DD (LC or NT)
T. novaezealandiae Thomson, 1900	AU (IP)	NZL	New Zealand	Common in estuaries, lakes, and rivers, but distribution incompletely documented (Jocqué and Blom 2009, and unpublished data)	DD (LC or NT)
Heteromysinae					
Deltamysis					
D. holmquistae Bowman et Orsi, 1992	NP (NA) (possibly nonnative)	ORE	USA	Estuarine species, occasionally found in freshwater. Originally known from Sacramento-San Joaquin Estuary, but presumably also dwelling in other Californian estuaries. Potential alien species of unclear origin (Bowman and Orsi 1992, Fofonoff et al. 2018)	DD

(Continued)

Table 10.1 (Continued)

Taxon	Major region	Type of range	Countries (native range)	Distribution, habitat, and population status remarks	Proposed conservation status
Mysinae					
Mysini					
Mesopodopsis					
M. orientalis (W.M. Tattersall, 1908)	IP (OR)	IND-MAL	Pakistan (?), India, Bangladesh, Thailand, Malaysia, Indonesia	Common and abundant brackish water species, also found in freshwaters up to 290 riverine-km from the sea (Nesemann et al. 2007, Biju and Panampunnayil 2010). Taxonomic status in Pakistan is unclear (Kazmi and Sultana 2015). Consumed by humans	LC (DD on national levels)
M. slabberi (Van Beneden, 1861)	AA, ES, (PA) (PC)	EASLB	Atlantic coast from Norway to Morocco; entire Mediterranean coast, Marmora Sea, Black Sea, Sea of Azov	Common and abundant marine and brackish water species occasionally found in coastal freshwaters of Europe (Wittmann et al. 2016). Population density up to 130 individuals/m^2 (Azov Sea basin). One of the most widespread and abundant species. Possibility of cryptic diversity (Remerie et al. 2006)	LC (DD on national levels, but LC: Russia, Ukraine)

(Continued)

Table 10.1 (Continued)

Taxon	Major region	Type of range	Countries (native range)	Distribution, habitat, and population status remarks	Proposed conservation status
Mysis					
M. amblyops G.O. Sars, 1907	PC	CAD	Russia, Azerbaijan, Iran	Pelagic in depths of 50–750 m, thus limited to the deep-water districts of the Caspian Sea (Derzhavin 1939, Bondarenko 1991, Daneliya and Petryashev 2011a)	NT
M. caspia G.O. Sars, 1895	PC	CAS	Russia, Azerbaijan, Iran, Turkmenistan, Kazakhstan	In depths of 38–390 m, in all districts of the Caspian province (Derzhavin 1939, Bondarenko 1991, Daneliya and Petryashev 2011a)	LC
M. diluviana Audzijonytė and Väinölä, 2005	NA	LAW-MAK (introduced in NA beyond the native range)	USA, Canada	Common species around the Great Lakes area, extending also to the Arctic. Reported from various US states, but the only reliable identifications are given by Audzijonytė and Väinölä (2005)	LC
M. macrolepis G.O. Sars, 1907	PC	CAD	Russia, Azerbaijan, Iran (?), Turkmenistan (?), Kazakhstan	In depths of 150–425 m, mostly 250–300 m, limited to the deep-water districts of the Caspian Sea, and known from small number of localities (Derzhavin 1939, Bondarenko 1991, Daneliya and Petryashev 2011a)	NT (DD on national levels)

(Continued)

Table 10.1 (Continued)

Taxon	Major region	Type of range	Countries (native range)	Distribution, habitat, and population status remarks	Proposed conservation status
M. microphthalma G.O. Sars, 1895	PC	CAD	Russia, Azerbaijan, Iran (?), Turkmenistan (?), Kazakhstan	In depths of 50–927 m, limited to the deep-water districts of the Caspian Sea, comparatively common (Derzhavin 1939, Bondarenko 1991, Daneliya and Petryashev 2011a)	NT
M. nordenskioldi Audzijonytė and Väinölä, 2007	AA, NP, (NA)	ARB	Russia, USA, Canada, Denmark (Greenland), Norway	Common in coastal brackish water, exceptionally in freshwater (lakes in Greenland). High genetic diversity with low geographic structure: isolate in White Sea (Russia) (Audzijonytė and Väinölä 2007)	LC
M. relicta Lovén, 1862	PA (AA)	BAL (introduced in part of range)	Russia, Finland, Sweden, Norway, N Germany, Poland, Lithuania	Rather common species in deeper lakes and occasionally in streams of Fennoscandia, but rare, fragmented, and at risk in the lakes along the southern coast of the Baltic Sea. Taxonomic status recently changed (Audzijonytė and Väinölä 2005). Documented partial disappearance in Poland (Żmudziński 1990). VU in Leningrad Oblast of Russia (Alekseev 2002), but may have been confounded with *M. salemaai*. Sensitive to warming	LC (DD: Russia LC: Sweden, Norway, Finland. VU: Germany, Lithuania, Poland)

(Continued)

Table 10.1 (Continued)

Taxon	Major region	Type of range	Countries (native range)	Distribution, habitat, and population status remarks	Proposed conservation status
M. salemaai Audzijonytė and Väinölä, 2005	PA, (AA)	NPA	Russia, Finland, Sweden, Norway, Denmark, UK, Ireland. Germany, Poland, Lithuania, Latvia, Estonia	Common in freshwater and brackish water from North Europe to the East Siberian Arctic. Genetically diverse, but barely structured geographically (Audzijonytė and Väinölä 2005, 2006). Single locality at risk in Norway (Spikkeland et al. 2017). Only in one lake of the UK (Audzijonytė and Väinölä 2005), and in ten lakes of Ireland and Northern Ireland (Griffiths et al. 2015). Documented decrease in Northern Ireland (Griffiths et al. 2015). Sensitive to warming	LC (LC: Russia, Finland, Sweden. VU: Norway, Denmark, UK, Ireland, Germany, Poland, Lithuania, Latvia, Estonia)

(Continued)

Table 10.1 (Continued)

Taxon	Major region	Type of range	Countries (native range)	Distribution, habitat, and population status remarks	Proposed conservation status
M. segerstralei Audzijonytė and Väinölä, 2005	HO, AA	NOH, ARC	Russia, Finland, Norway, Denmark (Greenland), Canada, USA	Widespread and common circum-Arctic species (Audzijonytė and Väinölä 2005). High genetic diversity with low geographic structure: isolates in western White Sea (Russia) and Bering Strait (USA) (Audzijonytė and Väinölä 2006). In Norway, in two separate localities with NT status (Spikkeland et al. 2017). In Finland, in the small Lake Pulmankijärvi at the border between Finland and Norway, representing a protected area. Sensitive to warming	LC (LC: Russia, Canada, USA. NT: Norway, Finland)
Paramysini					
Caspiomysis					
C. knipowitschi G.O. Sars, 1907	PC	CAW	Russia, Azerbaijan, Iran, Turkmenistan, Kazakhstan	Mostly in the Central Caspian Sea, but found in all districts and occasionally outside of this biogeographic province, at depths of 8–200 m, mostly 11–60 m (Derzhavin 1939, Bondarenko 1991, Daneliya and Petryashev 2011a)	LC

(Continued)

Table 10.1 (Continued)

Taxon	Major region	Type of range	Countries (native range)	Distribution, habitat, and population status remarks	Proposed conservation status
Katamysis					
K. warpachowskyi G.O. Sars, 1893	PC (invasive in PA)	BAC	Romania, Moldova, Ukraine, Russia, Turkmenistan, Kazakhstan	Despite EN status in Ukraine, invasive species in PA (Hanselmann 2010, Daneliya and Petryashev 2011a). Current population status in the Eastern Caspian Sea (Kazakhstan, Turkmenistan) unknown	LC (DD on national levels)
Paramysis					
P. (Longidentia) adriatica Wittmann, Ariani and Daneliya, 2016	PA	ADR	Italy, Croatia, Bosnia and Herzegovina	Recently discovered in lakes and rivers at the North Adriatic coast (Wittmann et al. 2016). Most localities are within protected areas	DD
P. (Mesomysis) intermedia (Czerniavsky, 1882)	PC (introduced in PA)	BAC	Romania, Moldova, Ukraine, Russia, Azerbaijan, Kazakhstan	Widely distributed shallow water species from freshwater and brackish water of the Black-Azov and Caspian Sea basins, introduced beyond its native range (Daneliya and Petryashev 2011b)	LC
P. (Metamysis) grimmi (G.O. Sars, 1895)	PC	CAS	Russia, Azerbaijan, Iran, Turkmenistan, Kazakhstan	Sublittoral (10–185 m, mostly 30–80 m), found in all Caspian districts (Derzhavin 1939, Bondarenko 1991, Daneliya and Petryashev 2011a)	LC

(Continued)

Table 10.1 (Continued)

Taxon	Major region	Type of range	Countries (native range)	Distribution, habitat, and population status remarks	Proposed conservation status
P. (M.) inflata (G.O. Sars, 1907)	PC	CCA	Russia, Azerbaijan, Kazakhstan	Uncommon endemic of the Central Caspian Sea, found at depths of 10–230 m, mostly 10–80 m (Derzhavin 1939, Bondarenko 1991, Daneliya and Petryashev 2011a)	NT
P. (M.) ullskyi Czerniavsky, 1882	PC (introduced in PA)	BAC	Romania, Moldova, Ukraine, Russia, Azerbaijan, Iran, Turkmenistan, Kazakhstan	Among most common shallow water Ponto-Caspian species, introduced beyond the native range (Derzhavin 1939, Bondarenko 1991, Daneliya and Petryashev 2011a). High genetic diversity with two major geographic isolates (Audzijonytė et al. 2006)	LC
P. (Nanoparamysis) loxolepis (G.O. Sars, 1895)	PC	CAW	Russia, Azerbaijan, Iran, Turkmenistan, Kazakhstan	Among the most abundant Caspian mysids, found at depths from 5 to 900 m, in all biogeographic districts, as well as penetrating a neighboring province (Derzhavin 1939, Bondarenko 1991, Daneliya and Petryashev 2011a).	LC

(Continued)

Table 10.1 (Continued)

Taxon	Major region	Type of range	Countries (native range)	Distribution, habitat, and population status remarks	Proposed conservation status
P. (Paramysis) baeri Czerniavsky, 1882	PC	CAW	Russia, Azerbaijan, Kazakhstan	Taxonomic status recently revised (Daneliya et al. 2007). Distribution, habitat and population status in the Caspian Sea appear unknown. So far documented in the Northern basin and in coastal areas of Central and Southern basins.	DD
P. (P.) bakuensis G.O. Sars, 1895	PC (introduced in PA)	BAC	Romania, Moldova, Ukraine, Russia, Azerbaijan, Kazakhstan	Widespread shallow water species in freshwater and brackish water of the Black-Azov and Caspian Sea basins, introduced beyond the native range (Daneliya et al. 2007), but distribution and habitat in the Caspian Sea are poorly known. Probably EN in Moldova and Pridnestrovie (as the junior synonym *P. baeri bispinosa*) (Filipenko 2009)	LC (DD on national levels)
P. (P.) eurylepis G.O. Sars, 1907	PC	CCA	Russia, Azerbaijan, Kazakhstan	Uncommon endemic of the Central Caspian Sea. Sublittoral, from 10 to 114 m (Derzhavin 1939, Daneliya and Petryashev 2011a). Population status unclear	NT (DD on national levels)

(Continued)

Table 10.1 (Continued)

Taxon	Major region	Type of range	Countries (native range)	Distribution, habitat, and population status remarks	Proposed conservation status
P. (P.) kessleri kessleri (G.O. Sars, 1895)	PC	CAW	Russia, Azerbaijan, Iran, Turkmenistan, Kazakhstan	One of the most common Caspian mysids, found at 0–114 m depth (Derzhavin 1939, Bondarenko 1991, Daneliya and Petryashev 2011a)	LC
P. (P.) k. sarsi (Derzhavin, 1925)	PC	ABL	Romania, Moldova, Ukraine	Endemic of the NW–Black Sea, in freshwater and brackish water. Shallow water subspecies (Daneliya and Petryashev 2011b). Common	LC
P. (Serrapalpisis) incerta (G.O. Sars, 1895)	PC	CAW	Russia, Azerbaijan, Iran, Turkmenistan, Kazakhstan	Depths of 5–94 m, mostly 11–40 m, in all Caspian biogeographic districts (Derzhavin 1939, Bondarenko 1991, Daneliya and Petryshev 2011a)	LC
P. (S.) kosswigi Băcescu, 1948	PC	ANT	Turkey	Stenoendemic, found in Turkish lake Işıklı, including springs and its effluent Büyük Menderes Nehri, and in Küçük Menderes Nehri (Wittmann et al. 2016)	VU

(Continued)

Table 10.1 (Continued)

Taxon	Major region	Type of range	Countries (native range)	Distribution, habitat, and population status remarks	Proposed conservation status
P. (S.) lacustris lacustris (Czerniavsky, 1882)	PC (invasive in PA)	BAC-ANT	Georgia, Turkey, Romania, Moldova, Ukraine, Russia, Azerbaijan, Iran, Turkmenistan, Kazakhstan	Belonging to the most common shallow water Ponto-Caspian mysids, introduced beyond the native range, with high genetic diversity between and within populations. Presence of geographic isolates (Audzijonytė et al. 2006, 2015)	LC (DD: Georgia, Turkey)
P. (S.) I. turcica (Băcescu, 1948)	PC	ANT	Turkey: Anatolia	Stenoendemic of Lake Beyşehir. Taxonomic and population status unclear. According to Kocataş et al. (2003), recorded in 1999 still in masses; however, not found upon a most recent survey	VU
P. (S.) sowinskii Daneliya, 2002	PC	BAC	Russia	Abundant shallow water species in freshwater and brackish water of the Azov Sea basin, with population density up to 50 individuals/m^2, and seemingly high genetic diversity (Audzijonytė et al. 2006). Its population status in the Caspian Sea is unknown	DD

(Continued)

Table 10.1 (Continued)

Taxon	Major region	Type of range	Countries (native range)	Distribution, habitat, and population status remarks	Proposed conservation status
Schistomysis					
S. elegans G.O. Sars, 1907	PC	ECA	Turkmenistan, Kazakhstan	Endemic of the Eastern Caspian Sea, mostly found in the central basin, at depths of 8–100 m, mostly 11–40 m (Derzhavin 1939, Bondarenko 1991, Daneliya and Petryashev 2011a). Current population status unclear	NT (DD on national levels)
Hemimysini					
Hemimysis					
H. anomala G.O. Sars, 1907	PC (invasive in PA, NA)	BAC	Bulgaria, Romania, Moldova, Ukraine, Russia, Turkmenistan, Kazakhstan	Despite EN status in Ukraine, an invasive species in PA and NA. Showing negative phototaxis and staying in deeper parts of rivers. Single native localities in Russia: in Don River and Volga River. In the Caspian Sea, confined mostly to the east coast (Kazakhstan and Turkmenistan), but not recorded there since the 1930s (Derzhavin 1939, Bondarenko 1991, Daneliya and Petryashev, 2011a). Low genetic diversity (Audzijonytė et al. 2008)	LC (CR: Kazakhstan, Turkmenistan. VU: Russia)

(Continued)

Table 10.1 (Continued)

Taxon	Major region	Type of range	Countries (native range)	Distribution, habitat, and population status remarks	Proposed conservation status
Diamysini					
Antromysis					
A. anophelinae W.M. Tattersall, 1951	NT	PAN	Costa Rica	From crab burrows near the mouth of Aranjuez River (Tattersall 1951). Distribution and current status unknown	DD
A. cenotensis Creaser, 1936	NT	YUK	Mexico	Numerous caves and wells in Quintana Roo and Yucatán (Reddell 1981)	DD
A. cubanica Băcescu and Orghidan, 1971	NT	ANL	Cuba	Brackish water lake in Juanello Piedra Cave (Petrescu and Wittmann 2009)	DD
A. juberthiei Băcescu and Orghidan, 1977	NT	ANL	Cuba	Cave on Pine Island (Petrescu and Wittmann 2009)	DD
A. peckorum Bowman, 1977	NT	ANL	Jamaica	Cavernicolous (Bowman 1977).	DD
A. reddelli Bowman, 1977	NT	YUK	Mexico	Cave in the Acatlán of Oaxaca (Bowman 1977)	DD

(Continued)

Table 10.1 (Continued)

Taxon	Major region	Type of range	Countries (native range)	Distribution, habitat, and population status remarks	Proposed conservation status
Diamysis					
D. camassai Ariani and Wittmann, 2002	PA	SAL	Italy: Apulia	Mesohaline karstic subterranean waters (dolinas) near Gulf of Tarent on the west coast of Salento Peninsula (Ariani and Wittmann 2002). Mysids not found upon inspections in 2012. Possibly extinct due to strong pollution by waste deposition	CR
D. fluviatilis Wittmann and Ariani, 2012	PA (ES)	ADR	Italy	Various freshwater habitats of the North Adriatic Sea coast, occasionally in estuaries. Rather narrow range in highly urbanized area (Wittmann and Ariani 2012)	NT
D. hebraica Almeida Prado-Pro, 1981	PA	SYR	Israel	Stenoendemic, only known from three coastal streams (Wittmann et al. 2016)	VU
D. lacustris Băcescu, 1940	PA	DIN: Lake Scutari	Montenegro, Albania	Stenoendemic of Lake Scutari (Wittmann et al. 2016). Most of the lake is protected as a national park.	EN

(Continued)

Table 10.1 (Continued)

Taxon	Major region	Type of range	Countries (native range)	Distribution, habitat, and population status remarks	Proposed conservation status
D. mesohalobia heterandra Ariani and Wittmann, 2000	PA, ES	EMT	Italy: Apulia, Croatia, Bosnia, Greece: Corfu, Turkey	Widespread in an- to euhaline waters of Adriatic and Marmora Sea basins, Lake Deran (Wittmann et al. 2016). Potentially endangered by fishery activities	LC (DD in Greece and Turkey)
D. pengoi (Czerniavsky, 1882)	PC	ABL-ANT	Romania, Moldova, Ukraine, Russia, Georgia, Turkey	Phytophilic species from freshwaters of the Azov and Black Sea basins. Distantly isolated populations. The largest population, in Ukraine, is LC. Range regression in Don River basin (Russia), however populations in the Middle Don reach densities of 100 individuals/m³. Also in one small lake of Georgia (not assessed for a century) and in Turkey	LC (DD: Moldova, Georgia. LC: Romania, Ukraine. NT: Russia, Turkey)
D. pusilla G. O. Sars, 1907	PC	ECA	Kazakhstan, Turkmenistan	Endemic of the eastern part of the Central Caspian Sea, found on a hard bottom at depths from 10–31 m. Not recorded since the 1930s (Derzhavin 1939, Bondarenko 1991). Considered CR (Stepanyants et al. 2015) or even EX (Bondarenko 1991), but still without official status	CR

(Continued)

Table 10.1 (Continued)

Taxon	Major region	Type of range	Countries (native range)	Distribution, habitat, and population status remarks	Proposed conservation status
Gangemysis					
G. assimilis Derzhavin, 1924	IP (OR)	GAN-MAL	India, Bangladesh, Nepal, Malaysia	Estuarine and freshwaters of Indian northeast coast, rivers of Bangladesh and Nepal, up to 1,000 riverine-km from the sea (Nesemann et al. 2007). Disjunct population in the Strait of Malacca (Hanamura et al. 2008), with a protection area in the range. Consumed by humans. Probably common	LC
Limnomysis					
L. benedeni Czerniavsky, 1882	PC (invasive in PA)	PC	Georgia, Turkey, Bulgaria, Romania, Moldova, Ukraine, Russia, Azerbaijan, Iran, Kazakhstan, Turkmenistan	Phytophylic species. One of the most common mysids in freshwaters of Europe, introduced beyond the native range (Wittmann et al. 2016). High genetic diversity with geographic structure (Audzijonytė et al. 2006). Isolated small, native population in Lake Paliastomi, Georgia, where it needs to be protected, but current status unknown	LC (LC: Turkey, Bulgaria, Romania, Moldova, Ukraine, Russia, Azerbaijan, Iran, Kazakhstan, Turkmenistan. DD: Georgia)

(Continued)

(Continued)

Table 10.1 (Continued)

Taxon	Major region	Type of range	Countries (native range)	Distribution, habitat, and population status remarks	Proposed conservation status
Parvimysis					
P. almyra Brattegard, 1977	NT	GVN	Suriname	Freshwater and brackish water (Brattegard 1977). Population status unknown	DD
P. amazonica Wittmann, 2018	NT	AMZ	Brazil	Phytophilic. So far only known from Tupé Lake and its effluent into Rio Negro, and Arara Lake (Wittmann 2018). Located in a protected area	DD
P. fittkaui Wittmann, 2018	NT	AMZ	Brazil	Phytophilic, among leaf litter in lakes, creeks, and rivers. So far known only from tributaries of Rio Negro (Wittmann 2018). Distribution and population status data insufficient	DD
P. fluviatilis Wittmann, 2018	NT	AMZ	Brazil	Phytophilic. So far only known from Urucú River, a tributary of Solimões (Wittmann 2018). Distribution and population status data insufficient	DD
P. lacustris Wittmann, 2018	NT	AMZ	Brazil	Phytophilic. So far only known from Lake Arara drainage into the Solimões, and mouth of the Rio Negro-Solimões confluence (Wittmann 2018). Distribution and population status data insufficient	DD

Table 10.1 (Continued)

Taxon	Major region	Type of range	Countries (native range)	Distribution, habitat, and population status remarks	Proposed conservation status
P. macrops Wittmann, 2018	NT	AMZ	Brazil	Phytophilic. So far only known from Urucú River, a tributary to the Solimões (Wittmann 2018). Distribution and population status data insufficient	DD
P. pisciscibus Henderson et Bamber, 1983	NT	AMZ	Brazil	Among leaf litter. Phytophilic in blackwater lakes, creeks, and rivers. So far known only from tributaries of Rio Negro and Solimões (Wittmann 2018). Distribution and population status data insufficient	DD
P. tridens Wittmann, 2018	NT	AMZ	Brazil	Phytophilic in leaf litter and freshwater creek Igarapé da Cachoeirinha, a tributary to the lower Rio Negro (Wittmann 2018). Distribution and population status data insufficient	DD
Surinamysis					
S. americana (W.M. Tattersall, 1951)	NT	GVN	Suriname	Freshwaters along the coast (Tattersall 1951)	DD
S. merista (Bowman, 1980)	NT	GVN: Orinoco basin	Venezuela	Only known from a freshwater gully into Orinoco River (Bowman 1980)	DD

(Continued)

Table 10.1 (Continued)

Taxon	Major region	Type of range	Countries (native range)	Distribution, habitat, and population status remarks	Proposed conservation status
S. rionegrensis Wittmann, 2017	NT	AMZ	Brazil	So far only known from single locality of Rio Negro, 80 km upstream of Manaus (Wittmann 2017)	DD
S. robertsonae Bamber et Henderson, 1990	NT	AMZ	Brazil	Phytophilic, among leaf litter in lakes, rivers, and creeks, in Rio Negro basin upstream of the Rio Negro-Solimões confluence (Wittmann 2017). Apparently common species, total distribution range unknown	DD
Taphromysis					
T. louisianae Banner, 1953	NA	MIS	USA: Florida, Louisiana, Mississippi, Texas; Mexico	Coastal lakes, marches and ditches, as well as tributaries of upper Mississippi River basin (Brooks et al. 1998)	LC
Troglomysis					
T. vjetrenicensis Stammer, 1933	PA	DIN	Bosnia and Herzegovina	Stenoendemic of Vjetrenica Cave system (Wittmann et al. 2016). The cave is a candidate for UNESCO Heritage Site	EN

(Continued)

Table 10.1 (Continued)

Taxon	Major region	Type of range	Countries (native range)	Distribution, habitat, and population status remarks	Proposed conservation status
Neomysini					
Hyperacanthomysis					
H. longirostris (Ii, 1936)	NP, IP (OR) (PA)	WPSLB	China, Korea, Japan	Common species in shallow brackish waters, occasionally found in freshwater. Introduced to Sacramento-San Joaquin delta (ORE) (Modlin and Orsi 1997, Fukuoka and Murano 2000)	LC
Neomysis					
N. awatschensis (Brandt, 1851)	NP, HO (introduced in PA)	WPWB, ESB-YUN	Korea, Japan, Russia, USA, Canada	Common species in shallow brackish water and freshwater from Japan Sea coast to Alaska Peninsula and Mackenzie River delta. Introduced in freshwaters of West Siberia and Northwest Russia (Petryashev and Daneliya 2014)	LC
N. integer (Leach, 1814)	AA, ES (PA)	EASBL	From Russia to Spain in the West Arctic and Atlantic; Baltic Sea states; Mediterranean coast of France	Common species in shallow brackish waters of East Atlantic subtropical to boreal waters from Mediterranean to the Baltic and Barents Seas, rarely found in freshwaters of Europe (Wittmann et al. 2016). High genetic diversity with geographic structure (Remerie et al. 2009)	LC

(Continued)

Table 10.1 (Continued)

Taxon	Major region	Type of range	Countries (native range)	Distribution, habitat, and population status remarks	Proposed conservation status
N. japonica Nakazawa, 1910	NP, IP (PA) (introduced to NA and AU)	WPSLB	China, Korea, Japan	Common species in shallow brackish waters, occasionally found in freshwaters. Introduced to Sacramento-San Joaquin delta (CAL) and New South Wales of Australia (Hutchings 1983, Fofonoff et al. 2018)	LC (DD: China, Korea)
N. mercedis Holmes, 1896	NP, NA	EPWB, YUN-ORE	Canada, USA	Common species in estuaries and lakes along the Pacific coast (Holmquist 1973).	LC
N. nigra Nakazawa, 1910	NP, IP, PA, OR	WPTS, SIN-JAP	China, Korea, Japan, Russia: Primorye Region	Common species in coastal brackish water and freshwater of Japan (Petryashev and Daneliya 2014), but its status in other parts of the range is unclear, particularly because it has been previously confused with *N. awatschensis*	LC (DD on national levels)
Orientomysis					
O. aspera (Ii, 1936)	NP, IP (OR) (PA)	WPSLB	China, Korea, Japan	Common species in shallow brackish waters, occasionally found in freshwaters. Introduced to Sacramento-San Joaquin delta (ORE) (Modlin and Orsi 1997, Fukuoka and Murano 2005)	LC

(Continued)

Table 10.1 (Continued)

Taxon	Major region	Type of range	Countries (native range)	Distribution, habitat, and population status remarks	Proposed conservation status
Stygiomysida					
Lepidomysidae					
Spelaeomysis					
S. bottazzii Caroli, 1924	PA	SAL	Italy: Apulia	Stenoendemic of subterranean waters and dolinas in several localities of Apulia (SE Italy) (Ariani 1982, Ariani et al. 2000). Endangered by habitat disturbance, and hydrological effects of climate change; urgent need for protection	EN
S. cardisomae Bowman, 1973	NT	PAN-PER	Colombia, Peru	Brackish water habitats (crab burrows) of two distant localities, Archipelago of San Andrés and Boca del Río (Bowman 1973)	DD
S. cochinensis Panampunnayil et Viswakumar, 1991	OR	IND	India: Kerala	Single locality, brackish water prawn pond (Panampunnayil and Viswakumar 1991)	DD
S. longipes (Pillai et Mariamma, 1964)	OR	IND	India: Kerala	Single well (Pillai and Mariamma 1964)	DD

(Continued)

Table 10.1 (Continued)

Taxon	Major region	Type of range	Countries (native range)	Distribution, habitat, and population status remarks	Proposed conservation status
S. nuniezi Băcescu et Orghidan, 1971	NT	ANL	Cuba	Brackish water lake in Juanello Piedra Cave (Petrescu and Wittmann 2009)	DD
S. olivae Bowman, 1973	NT	OAX	Mexico	Single cave (Cueva del Nacimiento del Rio San Antonio) in Oaxaca (Bowman 1973)	DD
S. quinterensis (Villalobos, 1951)	NT	EMX	Mexico	From four caves in Tamaulipas and San Luis Potosi (Reddell 1981).	DD
S. servata (Fage, 1924)	AF	EAF	Kenya, Tanzania: Zanzibar, Seychelles	Cave lakes at the African coast and coral reef pool in Aldabra (Tetè 1983)	DD
S. villalobosi García-Garza, Rodríguez-Almaraz et Bowman, 1996	NT	EMX	Mexico	Two caves and spring in Nuevo León (Garzía-Garza et al. 1996)	DD
Stygiomysidae					
Stygiomysis					
S. aemete Wagner, 1992	NT	ANL	Dominican Republic	Subterranean waters (Wagner 1992)	DD
S. clarkei Bowman, Iliffe et Yager, 1984	NT	ANL	Turks and Caicos Islands	Subterranean waters of Middle Caicos Island and Providenciales Island (Bowman et al. 1984)	DD

(Continued)

Table 10.1 (Continued)

Taxon	Major region	Type of range	Countries (native range)	Distribution, habitat, and population status remarks	Proposed conservation status
S. cokei Kallmeyer et Carpenter, 1996	NT	YUK	Mexico	Several caves near Tulum of Quintana Roo (Kallmeyer and Carpenter 1996)	DD
S. holthuisi (Gordon, 1958)	NT	ANL	St. Maarten, Puerto Rico, Anguilla and Bahamas: Grand Bahama	Subterranean waters (Bowman 1976, Bowman et al. 1984)	DD
S. hydruntina Caroli, 1937	PA	SAL	Italy: Apulia	Stenoendemic of subterranean waters of the Salento Peninsula (Pesce 1976); only very few records after 1937. Endangered by habitat fragmentation and hydrological effects of climate change	CR
S. ibarrae Ortiz, Lalana and Perez, 1996	NT	ANL	Cuba	Single cave (Ortiz et al. 1996)	DD
S. major Bowman, 1976	NT	ANL	Jamaica	Single cave (Bowman 1976).	DD

Major biogeographic regions/realms: AA, Arctatlantic; AF, Afrotropical; AU, Australian; ES, East Atlantic Subtropical; HO, Holarctic; IP, Indo-West-Pacific; NA, Nearctic; NP, North Pacific Boreal; NT, Neotropical; OR, Oriental; PA, Palaearctic; PC, Ponto—Caspian. Parentheses indicate less common occurrence in regions or environments, respectively.

Species ranges: ABL, Azov-Black Sea (freshwater); ABL-ANT, Black Sea-Anatolian; ADR, Adriatic (freshwater); AMZ, Amazonian; ANL, Antillean; ANT, Anatolian; ARB, Arctic-Boreal; ARC, Arctic; BAC, Black Sea—Caspian (freshwater); BAL, Baltic (freshwater); CAD, Deep-Sea Caspian; CAS, Caspian; CAW, Caspian widespread (occasionally in neighboring districts); CBR, Central Brazilian; CCA, Central Caspian; DIN, Dinaric; EAF, East African; EASLB, East Atlantic Subtropical-Low Boreal (marine); EASP, East Atlantic Subtropical-Boreal (marine); ECA, Eastern Caspian; EMT, Eastern Mediterranean (freshwater); EMX, East Mexican; EPWB, East Pacific widespread Boreal (Oregonic-Sitkanic); ESB-YUN, East-Siberian-Yukonic; GAN-MAL, Gangean-Malayan; GVN, Gvianic; IND, Indian; IND-MAL, Indo-Malayan; LAW-MAK, Lawrencian-Mackenzian; LMP, Lower Mississippian; MIS, Mississippian; NPA, North Palaearctic (East Arctic); NOH, North Holarctic; NZL, New Zealandic; OAX, Oaxacan; ORE, Oregonic; PAC, Pacific (freshwater); PAN, Panaman; PAN-PER, Panaman-Peruvian; SAL, Salentian; SIN-JAP, Sino-Japanic; SYR, Syrian; WPSLB, West Pacific Subtropical-Low Boreal; WPTS, West Pacific Tropical-Subtropical; WPWB, West Pacific widespread Boreal; YUK, Yukatanic; YUN-ORE, Yukonic-Oregonian.

Conservation status (IUCN): CR, critically endangered; EN, endangered; VU, vulnerable; NT, nearly threatened; LC, least concern; DD, data deficient.

japonica Nakazawa, 1910, *N. mercedis* Holmes, 1896, *N. nigra* Nakazawa, 1910, and *Orientomysis aspera* (Ii, 1936). These species, except for the first one, are among the most widespread and common mysids; among these a total of four are considered as currently invasive.

Continental biogeography of mysids, especially in tropical and subtropical regions, is in early stages of exploration. Reliable biogeographic division is not yet reasonable at global scales. Three provinces are attributed to the comparatively well-studied Ponto-Caspian region: Anatolian, Black Sea-Caspian, and Caspian (Daneliya and Petryashev 2011a), inhabited by a total of 27 endemic species and subspecies, respectively. Only *Limnomysis benedeni* Czerniavsky, 1882, has been recorded from all three Ponto-Caspian provinces, and due to introductions beyond its native range, it has become one of the most widespread and common continental mysids. Distribution ranges of quite a large number of species covers two provinces: *Diamysis pengoi* (Czerniavsky, 1882), *Paramysis lacustris lacustris* (Czerniavsky, 1882) (in Black Sea-Caspian and Anatolian provinces), *Hemimysis anomala* G.O. Sars, 1907, *Katamysis warpachowskyi* G.O. Sars, 1893, *P. intermedia* (Czerniavsky, 1882), *P. ullskyi* Czerniavsky, 1882 (predominantly in the Black Sea-Caspian, but partly also in the Caspian), *Caspiomysis knipowitschi* G.O. Sars, 1907, *P. baeri* Czerniavsky, 1882, *P. incerta* G.O. Sars, 1895, *P. kessleri kessleri* G.O. Sars, 1895, and *P. loxolepis* (G.O. Sars, 1895) (predominantly in the Caspian, partly in the Black Sea-Caspian). Among these, *P. l. lacustris, H. anomala, K. warpachowskyi, P. intermedia*, and *P. ullskyi* are invasive species.

The Anatolian province is the least studied area of the Ponto-Caspian region. A total of six species and subspecies, respectively, are recorded in this province. The endemic *P. kosswigi* Băcescu, 1948, and *P. lacustris turcica* have rather limited ranges, each confined to a single lake and nearby springs. The essentially marine *M. slabberi* was occasionally recorded in oligohaline coastal lakes (Wittmann et al. 2016). Within the Asia Minor Peninsula, *Paramysis lacustris lacustris* and *D. pengoi* are so far known only from isolated single localities. Nonetheless, the former species is the most common in the Ponto-Caspian region, while the latter is also found in the Black Sea-Caspian province. The largest range is shown by *L. benedeni*, also found in numerous localities of the northern part of the Asia Minor Peninsula (Wittmann et al. 2016).

The Black Sea-Caspian province contains 11 species and subspecies, respectively, only three considered strictly confined to this province: *P. bakuensis, P. k. sarsi*, and *P. sowinskii*. Another four species (*H. anomala, K. warpachowskyi, P. intermedia*, and *P. ullskyi*) are also found in the Eastern Caspian Middle Depth District or occasionally in other parts of the Caspian province. Four species and subspecies, respectively, are shared also with the Anatolian province. This last includes *M. slabberi, D. pengoi, L. benedeni*, and *P. l. lacustris*. This province is divided into five districts: Riverine Azov-Black Sea, Estuarine Azov-Black Sea, Caucasian, Northern Caspian, and Coastal Caspian (Daneliya and Petryashev 2011a). The majority of taxa, namely, the six species, *K. warpachowskyi, P. bakuensis, P. intermedia, P. sowinskii, P. ullskyi*, and *H. anomala*, are distributed among four districts (excluding the Caucasian). *Limnomysis benedeni* and *P. lacustris* are shared by all five districts. *Diamysis pengoi* is found in two districts, Riverine Azov-Black Sea and Caucasian. Finally, *P. k. sarsi* is also found in two districts, Riverine and Estuarine Azov-Black Sea.

Among 13 species of the Caspian province, seven are stenoendemic: *M. amblyops* G.O. Sars, 1907, *M. caspia* G.O. Sars, 1895, *M. macrolepis* G.O. Sars, 1907, *M. microphthalma* G.O. Sars, 1895, *P. eurylepis* G.O. Sars, 1907, *P. grimmi* G.O. Sars, 1895, and *P. inflata* G.O. Sars, 1907. Two additional species, *D. pusilla* G.O. Sars, 1907 and *Schistomysis elegans* G.O. Sars, 1907, occasionally penetrate the adjacent Coastal Caspian District of the Black Sea-Caspian province. Four species or subspecies, respectively, *C. knipowitschi*, *P. incerta*, *P. k. kessleri*, and *P. loxolepis* are often found in the Coastal Caspian District. The Caspian province does not include the Northern Caspian Sea wherefrom all above-mentioned species have not been recorded yet. Among the Caspian Sea endemics, only *P. baeri* is also found in the Northern Caspian and Coastal Caspian districts, but the distribution of this species is rather poorly known since its taxonomic revision (Daneliya et al. 2007). The Caspian province also contains five districts: Northwestern Middle Depth, Eastern Caspian Middle Depth, Southern Caspian Middle Depth, Central Caspian Deep Sea, and Southern Caspian Deep Sea. Four species of this province, *C. knipowitschi*, *P. incerta*, *P. k. kessleri*, and *P. loxolepis*, are widely distributed Caspian mysids found in all five districts, also penetrating the Coastal Caspian District of the Black Sea-Caspian province. The endemics *P. grimmi* and *M. caspia* are found in all five districts as well. The two Central Caspian endemics, *P. eurylepis* and *P. inflata*, are found in three districts, but only in certain parts: Northwestern Middle Depth, Eastern Caspian Middle Depth, and Central Caspian Deep Sea. The three deep-water *Mysis* species, namely, *M. amblyops*, *M. macrolepis*, and *M. microphthlma*, are shared by both deep-water districts of the Caspian Sea. Finally, the Eastern Caspian endemics *D. pusilla* and *S. elegans* are found almost exclusively in the Eastern Caspian Middle Depth District, though occasionally recorded also in the adjacent coastal area, while *S. elegans* occurs in the Central Caspian Deep Sea as well. The latter two species are the rarest Ponto-Caspian mysids.

10.4 GENETIC DIVERSITY

The study of genetic diversity of species appears helpful to reveal population structure, distribution, and potential exchange of genetic material between populations. Interchange of genes between populations allows genes of extinct populations to have been conserved in surviving ones. Genetic studies can also help to reveal potential cryptic diversity within species, as was the case with *Paramysis baeri* and *P. bakuensis* (see Daneliya et al. 2007), and within the *Mysis relicta* species complex (see Audzijonytė and Väinölä 2005).

So far only a small fraction of mysid species has been analyzed genetically. Mitochondrial DNA diversity was recently studied for several Ponto-Caspian

(Audzijonytė and Väinölä 2006, 2009, 2015, Daneliya et al. 2007, Audzijonytė et al. 2008) and Northern Holarctic (glacial relict) mysids (Audzijonytė and Väinölä 2005, 2006, 2007). Some preliminary studies were also made for *Mesopodopsis slabberi* (see Remerie et al. 2006) and *Neomysis integer* (see Remerie et al. 2009), in each case indicating high levels of genetic diversity. *Mesopodopsis slabberi* shows very high intrapopulation diversity, possibly indicative of cryptic allopatric taxa, one in the Atlantic and the other in the Mediterranean. Furthermore, several isolated populations of *M. slabberi* exist in the Mediterranean Sea. The Baltic and North Sea basins are probably inhabited by a single, recently expanded population of *N. integer*, while a number of distinct, long isolated populations are found along the Atlantic coast.

Based on available genetic data, the following three species groups can be considered:

1. *Species with high geographic divergence.* Such species have deeply divergent mtDNA lineages, separated from each other geographically, indicating isolation between populations or even presence of cryptic taxa. In the case that a certain population becomes threatened or extinct, there would be limited or no possibility for natural restoration. *Limnomysis benedeni*, *M. slabberi*, *M. segerstralei*, *N. integer*, *P. lacustris*, and *P. ullskyi* belong to this group.

2. *Species with high genetic diversity, but with low geographic divergence.* This group contains long-term stable species and free genetic exchange across its range. To our current knowledge, the following species are considered here: *M. salemaai*, *P. intermedia*, and *P. sowinskii*.

3. *Species with low genetic diversity.* Species with low diversity went through recent population bottlenecks, the current diversity restored from a limited number of individuals. The only species so far known is *Hemimysis anomala*, an invasive Ponto-Caspian species. However, there are few data from its native range.

10.5 FACTORS OF POPULATION DECREASE

Various factors may pose risks on mysid and stygiomysid populations (Table 1): hydrological and direct temperature effects due to climate change, habitat degradation by pollution, eutrophication, waste disposal, sediment intrusion, watershed deforestation, damage to bottom sediments, and, to a minor extent, illegal commerce.

Climate change: Temperature changes may affect mysids directly by surpassing optimal temperatures for development or indirectly (and probably in complex ways) by changing dissolved oxygen levels, evaporation rates, water level changes, and other hydrological consequences, as well as through creating favorable conditions for invasive species, and by enhancing competition and restructuring of local communities.

Unlike marine mysids from the same latitudes, glacial relict *Mysis* species populations from the Northern Palaearctic and Nearctic, dwelling in lakes and brackish estuaries, are isolated in their habitats and are potentially affected by water temperature changes, putting them at risk of extinction.

Spelaeomysis bottazzii and *Stygiomysis hydruntina* from groundwater of Apulia (southern Italy) are endangered by prolonged periods of drought, enhancing seawater intrusion into the karstic underground, and thus reducing the exchange between shallow and deep groundwater. The resulting hydrological stagnancy is expected to render part of the deep waters uninhabitable, blocking migration pathways. Risks from lowered water tables may concern other subterranean mysids as well.

Pollution appears to be among the major negative factors for mysids. *Diamysis pengoi* was formerly distributed along the entire course of the Don River in Russia (Daneliya 2003). According to our surveys in 1998–2003, it has been extinct in the upper and lower reaches due to the effects of the highly populated and industrialized cities Voronezh and Rostov-on-Don. Its congener *D. camassai*, described from semi-subterranean environments of the Salento Peninsula (SE Italy) in 2002, may already be extinct due to waste disposal into the known localities. *Paramysis (S.) lacustris turcica* was still present in Lake Beyşehir, in Turkey, in surveys of 1999 (Kocataş et al. 2003), but was not recorded in more recent samplings. This may also be due to high environmental degradation, from pollution and other factors, despite establishment of protected areas (Işildar 2010, and subsequent publications). Near extinction of mysids (*D. pusilla*, *H. anomala*, *K. warpachowskyi*, and *S. elegans*) on the East Caspian Sea coast, as well as general risk for all other Caspian Sea species, may largely be due to oil and gas production, developed during 20th century and recently intensified.

10.6 CONSERVATION STATUS

Our knowledge about taxonomy and distribution of mysids is still in a very initial stage, and the majority of species are categorized as Data Deficient (DD) (Table 1). Most studied continental mysids are widely distributed, genetically diverse, and thus are treated as Least Concern (LC).

Based on currently available knowledge about occurrence of continental mysids, we propose that three species should be considered as critically endangered (CR): *Diamysis pusilla*, *D. camassai*, and *Stygiomysis hydruntina*. The first, *D. pusilla*, is endemic of the Eastern Caspian Sea, recorded only from Kazakhstan and Turkmenistan (G.O. Sars 1907, Derzhavin 1939), some 300 km along the coast line. It had been found mostly in the Eastern Caspian Middle Depth District, and also occasionally penetrating adjacent waters of the Coastal Caspian District (Daneliya and Petryashev 2011a), but has not been recorded since 1930s and possibly has gone extinct (Bondarenko 1991). Because of the absence of recent comprehensive sampling in Kazakhstan and Turkmenistan, it was suggested to consider it CR (Stepanyants et al. 2015). Unfortunately, no protected areas exist in the species range. The second mysid, *D. camassai* from brackish semi-subterranean habitats (dolinas), near

Apulian-Ionian coasts of SE Italy, was reported as late as 2002 and is now absent from the type locality as of 2012. This species may possibly become extinct due to strong pollution by waste disposal in the area. Finally, the third species, subterranean *Stygiomysis hydruntina*, also stenoendemic of the Salento Peninsula in SE Italy, has almost disappeared from the groundwater.

Two stenoendemics with stable populations, *Troglomysis vjetrenicensis* from Vjetrenica cave in Bosnia and Herzegovina (ca 7 km long), and *D. lacustris* from Lake Scutari (370 km²), on the border between Montenegro and Albania, are endangered by habitat disturbance. Another stenoendemic *Spelaeomysis bottazzii* from the Salento Peninsula in SE Italy is also strongly affected by habitat disturbance and hydrological effects of climate change, thus all these species are considered endangered (EN). Environments of the first two species are at least partly covered by conservation areas, thus reducing the risk that the species may become CR.

Three stenoendemic species are considered vulnerable (VU): *Paramysis kosswigi* from Lake Işikli in Turkey and several springs; *Paramysis lacustris turcica* endemic of Lake Beyşehir in Turkey, and *D. hebraica*, found only in three coastal streams of Israel. The distribution status of *P. kosswigi* was recently updated by Wittmann et al. (2016), and it may be absent from the Black Sea basin, now appearing to be restricted to southwest Anatolia. The species was sampled repeatedly from Işikli springs during 1990s and early 2000s, also found in the Büyük Menderes Nehri, effluent of Lake Işikli, and more distantly, in the Küçük Menderes Nehri (Wittmann et al. 2016). *Paramysis lacustris turcica* was found only in Lake Beyşehir. Despite the establishment of the Lake Beyşehir National Park in 1993, its environment is under heavy anthropogenic pressure (Işildar 2010). This subspecies was found still in high numbers during a 1999 survey of the lake by Kocataş et al. (2003), whereas recent surveys failed to find any specimens. Considering its narrow range and strong environmental pressures, it is treated here as VU.

Seven species are treated as nearly threatened (NT): Ponto-Caspian *Mysis amblyops, M. macrolepis, M. microphthalma, Paramysis (Metamysis) inflata, P. (P.) eurylepis, Schistomysis elegans*, and Palaearctic *D. fluviatilis*. All these species were present in recent samplings, but need some protection due to narrow distributions in combination with high environmental pressure.

REFERENCES

Alekseev, V.R. 2002. Reliktovaya mizida. *Mysis relicta* Lov. In *Krasnaya Kniga Prirody Leningradskoi Oblasti, T,* ed. G.A. Noskov 3. Sankt-Peterburg: Mir i Semya (Relict mysid. In: Red Data Book of Nature of the Leningrad Oblast) (in Russian).

Ariani, A.P. 1980. *Spelaeomysis bottazzii* Caroli (Crustacea, Mysidacea) nella falda freatica del litorale brindisino. *Annuario dell'Istituto e Museo di Zoologia dell'Università di Napoli* 23(1979–1980):157–66.

Ariani, A.P. 1982. Osservazioni e ricerche su *Typhlocaris salentina* (Crustacea, Decapoda) e *Spelaeomysis bottazzii* (Crustacea, Mysidacea). Approccio idrogeologico e biologico sperimentale allo studio del popolamento acquatico ipogeo della Puglia. *Annuario dell'Istituto e Museo di Zoologia dell'Università di Napoli* 25:201–326.

Ariani, A.P. and Wittmann, K.J. 2000. Interbreeding versus morphological and ecological differentiation in Mediterranean *Diamysis* (Crustacea, Mysidacea), with description of four new taxa. *Hydrobiologia* 441:185–236.

Ariani, A.P. and Wittmann, K.J. 2002. The transition from an epigean to a hypogean mode of life: Morphological and bionomical characteristics of *Diamysis camassai* sp. nov. (Mysidacea, Mysidae) from brackish-water dolinas in Apulia, SE-Italy. *Crustaceana* 74:1241–65.

Ariani, A.P. and Wittmann, K.J. 2004. Mysidacea (Crustacea) as ecological and biogeographical markers in Mediterranean brackish environments. In Rapport du 37ème Congrès de la CIESM. 37th CIESM Congress Proceedings, 7–11 June 2004, Barcelona. Rapports de la Commission Internationale pour l'Exploration Scientifique de la Mer Méditerranée 37:479.

Ariani, A.P., Camassa, M.M. and Wittmann, K.J. 2000. The dolinas of Torre Castiglione (Gulf of Tarent, Italy): Environmental and faunistic aspects of a semi-hypogean water system. *Mémoires de Biospéologie* 27:1–14.

Audzijonytė, A. and Väinölä, R. 2005. Diversity and distributions of circumpolar fresh- and brackish-water *Mysis* (Crustacea: Mysida): Descriptions of *M. relicta* Lovén, 1862, *M. salemaai* n. sp., *M. segerstralei* n. sp. and *M. diluviana* n. sp., based on molecular and morphological characters. *Hydrobiologia* 544:89–141.

Audzijonytė, A. and Väinölä, R. 2006. Phylogeographic analyses of a circumarctic coastal and a boreal lacustrine mysid crustacean, and evidence of fast postglacial mtDNA rates. *Molecular Ecology* 15:3287–301.

Audzijonytė, A. and Väinölä, R. 2007. *Mysis nordenskioldi* n. sp. (Crustacea, Mysida), a circumpolar coastal mysid separated from the NE Pacific *M. litoralis* (Banner, 1948). *Polar Biology* 30:1137–57.

Audzijonytė, A., Baltrūnaitė, L., Väinölä, R. and Arbačiauskas, K. 2015. Migration and isolation during the turbulent Ponto-Caspian Pleistocene create high diversity in the crustacean *Paramysis lacustris*. *Molecular Ecology* 24:4537–55.

Audzijonytė, A., Daneliya, M.E. and Väinölä, R. 2006. Comparative phylogeography of Ponto-Caspian mysid crustaceans: Isolation and exchange among dynamic inland sea basins. *Molecular Ecology* 15:2969–84.

Audzijonytė, A., Wittmann, K.J., Ovcarenko, I. and Väinölä, R. 2009. Invasion phylogeography of the Ponto-Caspian crustacean *Limnomysis benedeni* dispersing across Europe. *Diversity and Distributions* 15:346–55.

Audzijonytė, A., Wittmann, K.J. and Väinölä, R. 2008. Tracing recent invasions of the Ponto-Caspian mysid shrimp *Hemimysis anomala* across Europe and to North America with mitochondrial DNA. *Diversity and Distributions* 14:179–86.

Băcescu, M. 1948. Myside (Racusori evoluti) pontocaspice in apele Anatoliei Sud-Vestice. *Revista Stiinţifică "V. Adamachi"* 34:1–2.

Biju, A. and Panampunnayil, S.U. 2010. Mysids (Crustacea) from the salt pans of Mumbai, India with the description of a new species. *Marine Biology Research* 6:556–69.

Bondarenko, M.V. 1991. Mizidy Kaspiya i ikh rol' v ekosistemakh morya (Mysids of the Caspian Sea and their role in the ecosystems of the sea). Abstract of Ph.D. thesis. Moscow, 24 p. (in Russian).

Bowman, T.E. 1973. Two new American species of *Spelaeomysis* (Crustacea: Mysidacea) from a Mexican cave and land crab burrows. *Bulletin of the Association for Mexican Cave Studies* 5:13–20.

Bowman, T.E. 1976. *Stygiomysis major*, a new troglobitic mysid from Jamaica, and extension of the range of *S. holthuisi* to Puerto Rico (Crustacea: Mysidacea: Stygiomysidae). *International Journal of Speleology* 8:365–73.

Bowman, T.E. 1977. A review of the genus *Antromysis* (Crustacea: Mysidacea), including new species from Jamaica and Oaxaca, Mexico, and a redescription and new records for *A. cenotensis*. Studies on the Caves and Cave Fauna of the Yucatan Peninsula. *Bulletin of the Association for Mexican Cave Studies* 6:27–38.

Bowman, T.E. 1980. *Antromysis* (*Surinamysis*) *merista*, a new freshwater mysid from Venezuela (Crustacea, Mysidacea). *Proceedings of the Biological Society of Washington* 93:208–15.

Bowman, T.E. and Orsi, J.J. 1992. *Deltamysis holmquistae*, a new genus and species of Mysidacea from the Sacramento-San Joaquin estuary of California (Mysidae: Mysinae: Heteromysini). *Proceedings of the Biological Society of Washington* 105:733–42.

Bowman, T.E., Iliffe, T.M. and Yager, J. 1984. New records of the troglobitic mysid genus *Stygiomysis*: *S. clarkei*, new species, from the Caicos Islands, and *S. holthuisi* (Gordon) from Grand Bahama Island (Crustacea: Mysidacea). *Proceedings of the Biological Society of Washington* 97:637–44.

Brattegard, T. 1977. Three species of Mysidacea (Crustacea) from Surinam. *Zoologische Mededelingen* 50:282–93.

Brooks, C.P., Dreves, D.P. and White, D.S. 1998. New records of *Taphromysis louisianae* Banner, 1953 (Mysidae) with notes on its ecology. *Crustaceana* 71:955–60.

Daneliya, M.E. 2003. Mizidy (Crustacea: Mysidacea) basseina Azovskogo morya [Mysids (Crustacea: Mysidacea) of the Sea of Azov Basin]. Abstract of Ph.D. thesis. St. Petersburg, 23 p. (in Russian).

Daneliya, M.E. and Petryashev, V.V. 2011a. Biogeographic zonation of the Black Sea and Caspian Sea basin based on mysid fauna (Crustacea: Mysidacea). *Russian Journal of Marine Biology* 37:85–97.

Daneliya, M.E. and Petryashev, V.V. 2011b. Redescription of three species and a subspecies of the mysid genus *Paramysis* (Mysida, Mysidae) from Ponto-Caspian basin. *Crustaceana* 84:797–829.

Daneliya, M.E., Audzijonyte, A. and Väinölä, R. 2007. Diversity within the Ponto-Caspian *Paramysis baeri* Czerniavsky sensu lato revisited: *P. bakuensis* G.O. Sars restored (Crustacea: Mysida: Mysidae). *Zootaxa* 1632:21–36.

Daneliya, M.E., Petryashev, V.V. and Väinölä, R. 2012. Continental mysid crustaceans of Northern Eurasia. In: *Aktualnye problemy izucheniya rakoobraznykh kontinental-nykh vod. Sbornik lektsii i dokladov mezhdunarodnoi shkoly-konferentsii.* Institut biologii vnutrennikh vod im. I.D. Papanina RAN, Borok, 5–9 noyabrya 2012 (Modern problems of the continental crustacean research. In Proceedings of the Lectures and Presentations of the International School-Conference. I.D. Papanin's Institute of Biology of Continental Waters RAS, Borok, November 5–9, 2012), eds. N.M. Korovchinsky, S.M. Zhdanova, and A.V. Krylov, 21–30. Kostroma: Kostromskoi pechatnyi dom.

Derzhavin, A.N. 1939. *Mizidy Kaspiya (Mysids of the Caspian Sea).* Baku: Izdatelstvo AzFAN (in Russian).

Dovgal, I.V. 2009a. Mizida Varpakhovskogo. *Katamysis warpachowskyi* Sars, 1893. In *Chervona kniga Ukrainy: tvarynnyi svit* (Warpachowsky's mysid. *In: Red Data Book of Ukraine: Animal World*), ed. I.A. Akimov 34. Globalkonsalting, Kiiv. (in Ukrainian).

Dovgal, I.V. 2009b. Mizida zubchasta. *Hemimysis serrata* Bacescu, 1938. In *Chervona kniga Ukrainy: tvarynnyi svit* (Serrated mysid. *In: Red Data Book of Ukraine: Animal World*), ed. I.A. Akimov, 33. Globalkonsalting, Kiiv. (in Ukrainian).

Filipenko, S.I. 2009. Semeistvo Mizidy—Mysidae. In *Krasnaya Kniga Prindestrovskoi Moldavskoi Respubliki*. (Family mysids—Mysidae. In: *Red Data Book of Pridnestrovian Moldavian Republic*), ed. O.A.Kalyakin. B.I., Tiraspol, 141–142 (in Russian).

Fofonoff, P.W., Ruiz, G.M., Steves, B., Simkanin, C. and Carlton, J.T. 2018. National exotic marine and estuarine species information system. Accessed at http://invasions.si.edu/nemesis/ on 21-Aug-2018.

Fukuoka, K. and Murano, M. 2000. *Hyperacanthomysis*, a new genus for *Acanthomysis longirostris* Ii, 1936, and *A. brevirostris* Wang and Liu, 1997 (Crustacea: Mysidacea: Mysidae). *Plankton Biology and Ecology* 47:122–8.

Fukuoka, K. and Murano, M. 2005. A revision of East Asian *Acanthomysis* (Crustacea: Mysida: Mysidae) and redefinition of *Orientomysis*, with description of a new species. *Journal of Natural History* 39:657–708.

García-Garza, M.E., Rodríguez-Almaraz, G.A. and Bowman, T.E. 1996. *Spelaeomysis villalobosi*, a new species of mysidacean from northeastern México (Crustacea: Mysidacea). *Proceedings of the Biological Society of Washington* 109:97–102.

Glasspool, A. 2003. *Management Plan for Bermuda's Critically Endangered Cave Fauna*. Bermuda: Government of Bermuda.

Griffiths, D., Macintosh, K.A., Forasacco, E., Rippey, B., Vaughan, L., McElarney, Y.R. and Gallagher, K. 2015. *Mysis salemaai* in Ireland: New occurrences and existing population declines. *Biology and Environment Proceedings of the Royal Irish Academy* 115B:1–7.

Gubbay, S. 2014. *A Review of the Use of Biogeography and Different Biogeographic Scales in MPA Network Assessment*. JNCC Report No 496. Peterborough: Joint Nature Conservation Committee.

Hanamura, Y., Fukuoka, K., Siow, R. and Chee, Ph.E. 2008. Re-description of a little-known Asian estuarine mysid *Gangemysis assimilis* (Tattersall, 1908) (Peracarida, Mysida) with a range extension to the Malay Peninsula. *Crustacean Research* 37:35–42.

Hanselmann, A.J. 2010. *Katamysis warpachowskyi* Sars, 1877 (Crustacea, Mysida) invaded Lake Constance. *Aquatic Invasions* 5(Supplement 1):31–4.

Holmquist, C. 1973. Taxonomy, distribution and ecology of the three species *Neomysis intermedia* (Czerniavsky), *N. awatschensis* (Brandt) and *N. mercedis* Holmes (Crustacea, Mysidacea). *Zoologische Jahrbücher. Abteilung für Systematik, Ökologie und Geographie der Tiere* 100:197–222.

Hutchings, P.A. 1983. The wetlands of Fullerton Cove, Hunter River, N.S.W. *Coast and Wetlands* 3:12–21.

Iliffe, T.M. 1996a. *Bermudamysis speluncola*. The IUCN Red List of Threatened Species 1996: e.T2766A9478950. Accessed at http://dx.doi.org/10.2305/IUCN.UK.1996.RLTS.T2766A9478950.en on 26-Oct-2018.

Iliffe, T.M. 1996b. *Platyops sterreri*. The IUCN Red List of Threatened Species 1996: e.T17563A7140340. Accessed at http://dx.doi.org/10.2305/IUCN.UK.1996.RLTS.T17563A7140340.en on 26-Oct-2018.

Işıldar, G.Y. 2010. Anthropogenic impacts on Beyşehir Lake National Park: Infrastructure problems and management issues. *Gazi University Journal of Science* 23:271–80.

International Union for Conservation of Nature. 2012. *IUCN Red List categories and criteria.* Version 3.1. 2nd edition. Gland, Switzerland and Cambridge, UK: IUCN, iv+32 p.

Jocqué, M. and Blom, W. 2009. Mysidae (Mysida) of New Zealand; a checklist, identification key to species and an overview of material in New Zealand collections. *Zootaxa* 2304:1–20.

Kallmeyer, D.E. and Carpenter, J.H. 1996. *Stygiomysis cokei*, new species, a troglobitic mysid from Quintana Roo, Mexico (Mysidacea: Stygiomysidae). *Journal of Crustacean Biology* 16:418–27.

Kazmi, Q.B. and Sultana, R. 2015. A new record of genus *Mesopodopsis* (Crustacea: Mysida) from Northern Arabia sea, Pakistan. *International Journal of Fauna and Biological Studies* 2:1–4.

Kocataş, A., Özbek, M., Ustaoğlu, M.R. and Balik, S. 2003. Contribution to the knowledge of the distribution of mysid (Mysidacea, Crustacea) species in Turkish inland waters. In International Symposium of Fisheries and Zoology, 23–26 Sept. 2003, ed. I.K. Oray, M.S. Çelikkale, and G. Özdemir, 269–277. Istanbul.

Mees, J. and Meland, K. 2020. World list of Mysida, Lophogastrida and Stygiomysida. Accessed at http://www.marinespecies.org/mysidacea on 2020-10-18.

Modlin, R.F. and Orsi, J.J. 1997. *Acanthomysis bowmani*, a new species, and *A. aspera* Ii, Mysidacea newly reported from the Sacramento-San Joaquin Estuary, California (Crustacea: Mysidae). *Proceedings of the Biological Society of Washington* 110:439–46.

Nesemann, H. Sharma, S. Sharma, G. Khanal, S.N. Pradhan, B. Shah, D.N. and Tachamo, R.D. 2007. Mollusca, Annelida, Crustacea (in part). In *Aquatic Invertebrates of the Ganga River System.* 1st edition, ed. H. Nesemann, 1–263. Kathmandu: Sunil Uprety, Chandi Media Pvt. Ltd.

Ortiz, M., Lalana, R. and Perez, A. 1996. El primer registro del genero *Stygiomysis* (Crustacea, Mysidacea) en la Isla de Cuba y description de una especie nueva (First record of the genus *Stygiomysis* (Crustacea, Mysidacea) in Cuban Island and a description of a new species). *Revista de Investigaciones Marinas* 17:107–15.

Oug, F., Brattegard, T., Walseng, B. and Djursvoll, P. 2015. Krepsdyr (Crustacea). In *Norsk rødliste for arter 2015*, eds. S. Henriksen and O. Hilmo, Artsdatabanken. Accessed at http://www.artsdatabanken.no/ Rodliste/Artsgruppene/Krepsdyr on 2020-10-18.

Panampunnayil, S.U. and Viswakumar, M. 1991. *Spelaeomysis cochinensis*, a new mysid (Crustacea, Mysidacea) from a prawn culture field in Cochin, India. *Hydrobiologia* 209:71–78.

Pesce, G.L. 1976. On a *Stygiomysis* (Crustacea, Mysidacea) from southern Italy. *Bollettino del Museo civico di Storia naturale di Verona* 2:439–43.

Petrescu, I. and Wittmann, K.J. 2009. Catalogue of the Mysida type collection (Crustacea: Peracarida) from the "Grigore Antipa" National Museum of Natural History (Bucureşti). *Travaux du Muséum National d' Histoire Naturelle "Grigore Antipa"* 52:53–72.

Petryashev, V.V. 2005. Biogeographical division of the North Pacific sublittoral and upper bathyal zones by the fauna of Mysidacea and Anomura (Crustacea). *Russian Journal of Marine Biology* 31 (Supplement 1):9–26.

Petryashev, V.V. 2009. The biogeographical division of the Arctic and North Atlantic by the mysid (Crustacea: Mysidacea) fauna. *Russian Journal of Marine Biology* 35:97–116.

Petryashev, V.V. and Daneliya, M.E. 2014. Taxonomic status of the Western-Pacific mysid species of the *Neomysis awatschensis* (Brandt, 1851) group. *Russian Journal of Marine Biology* 40:165–76.

Pillai, N.K. and Mariamma, T. 1964. On a new lepidomysid from India. *Crustaceana* 7:113–24.

Porter, M.L., Meland K. and Price W. 2008. Global diversity of mysids (Crustacea-Mysida) in freshwater. *Hydrobiologia* 595:213–18.

Price, W.W. 2015a. Order Mysida. In *Keys to Nearctic Fauna. Thorp and Covich's freshwater invertebrates*—Volume II, 4th edition, eds. J.H. Thorpe and D.C. Rogers, 702–10. Cambridge, MA: Academic Press.

Price, W.W. 2015b. Order Stygiomysida. In *Keys to Nearctic Fauna. Thorp and Covich's freshwater invertebrates*—Volume II, 4th edition, eds. J.H. Thorpe and D.C. Rogers, 710–1. Cambridge, MA: Academic Press.

Reddell, J.R. 1981. A review of the cavernicole fauna of Mexico, Guatemala, and Belize. *Bulletin of the Texas Memorial Museum* 27:1–327.

Remerie, T., Bourgois, T., Peelaers, D., Vierstraete, A., Vanfleteren, J. and Vanreusel, A. 2006. Phylogeographic patterns of the mysid *Mesopodopsis slabberi* (Crustacea, Mysida) in Western Europe: Evidence for high molecular diversity and cryptic speciation. *Marine Biology* 149:465–81.

Remerie, T., Vierstraete, A., Weekers, P.H.H., Vanfleteren, J.R. and Vanreusel, A. 2009. Phylogeography of an estuarine mysid, *Neomysis integer* (Crustacea, Mysida), along the north-east Atlantic coasts. *Journal of Biogeography* 36:39–54.

Samchishina, L.V. 2009. Mizida anomalna. *Hemimysis anomala* Sars, 1907. In *Chervona kniga Ukrainy: tvarynnyi svit.* (*Anomalous mysid. In Red Data Book of Ukraine: Animal World*), ed. I.A. Akimov 32. Kiiv: Globalkonsalting (in Ukrainian).

Sars, G.O. 1907. Mysidae. In *Trudy Kaspiiskoi Ekspeditsii* 1:243–313, pls. I–XII.

Spikkeland, I., Kinsten, B., Kjellberg, G., Nilssen, J.P. and Väinölä R. 2016. The aquatic glacial relict fauna of Norway - an update of distribution and conservation status. *Fauna Norvegica* 36:51–65.

Stepanyants, S.D., Khlebovitch, V.V., Alekseev, V.R., Daneliya, M.E. and Petryashev, V.V. 2015. *Opredelitel ryb i bespozvonochnykh Kaspiiskogo morya.* Tom 2. Strekajushchie, grebneviki, mnogoshchetinkovye chervi, veslonogie rakoobraznye i mizidy. St. Petersburg/Moscow: KMK Scientific Press Ltd., 244 p. (Identification keys for fishes and invertebrates of the Caspian Sea. V. 2. Cnidarians, ctenophorans, polychaets, copepods and mysids).

Tattersall, W.M. 1951. A review of the Mysidacea of the United States National Museum. *Bulletin of the United States National Museum* 201:1–292.

Teté, P. 1983. Sur la variabilité de *Spelaeomysis servatus* (Fage) (Mysidacea). Recherches en Afrique de l´Institut de Zoologie de l´Aquila (Italie). VIII. *Crustaceana* 44:216–21.

Toderaş, I., Dediu, I. and Munjiu, O. 2015. *Paramysis baeri bispinosa* (Martynov, 1924). Paramizis ber bispinos. In *Cartea Roşie A Republicii Moldova.* ed. G. Duca. Ştiinţa, Chişinău, 464.

Vaitonis, G. 2007. Reliktinė mizidė. *Mysis relicta* Loven, 1862. In *Lietuvos Raudonoji Knyga*, 1–37. Rašomavičius: V Lututė.

Wagner, H.P. 1992. *Stygiomysis aemete* n.sp., a new subterranean mysid (Crustacea, Mysidacea, Stygiomysidae) from the Dominican Republic, Hispaniola. *Bijdragen tot de Dierkunde* 62:71–79.

Wittmann, K.J. 2017. The genus *Surinamysis* (Mysida, Mysidae, Diamysini) from Amazonia and the coast of Brazil, with descriptions of two new species. *Crustaceana* 90:359–80.

Wittmann, K.J. 2018. Six new freshwater species of *Parvimysis*, with notes on breeding biology, statolith composition, and a key to the Mysidae (Mysida) of Amazonia. *Crustaceana* 91:537–76.

Wittmann, K.J. and Ariani, A.P. 2012. The species complex of *Diamysis* Czerniavsky, 1882, in fresh waters of the Adriatic basin (NE Mediterranean), with descriptions of *D. lacustris* Băcescu, 1940, new rank, and *D. fluviatilis* sp. nov. (Mysida, Mysidae). *Crustaceana* 85:1745–79.

Wittmann, K.J., Ariani, A.P. and Daneliya, M. 2016. The Mysidae (Crustacea: Peracarida: Mysida) in fresh and oligohaline waters of the Mediterranean. Taxonomy, biogeography, and bioinvasion. *Zootaxa* 4142:1–70.

Żmudziński, L. 1990. Past and recent occurrence of Malacostraca glacial relicts in Polish lakes. *Annales Zoologici Fennici* 27:227–30.

CHAPTER 11

Reassessing the Current Conservation Status of the Freshwater Aeglid, *Aegla jaragua* Moraes, Tavares and Bueno, 2016 (Decapoda: Anomura: Aeglidae) as Critically Endangered A Ten-Year Case Study

Sérgio L. S. Bueno, Milena R. Wolf, and Roberto M. Shimizu

CONTENTS

11.1 INTRODUCTION

For over ten years, the freshwater anomuran *Aegla jaragua* Moraes, Tavares, and Bueno, 2016, has been the subject of several studies that resulted in the description of its life cycle (Cohen et al. 2011), assessment of population size (Cohen et al. 2011), and taxonomic and conservation issues (Moraes et al. 2016). In addition, the fieldwork of a biennial course on decapod crustacean populations, offered to undergraduate students under the supervision of two of us (SLSB and RMS) from 2011 through 2019, provided the opportunity to follow up on the abundance and the population size and structure of *A. jaragua*. The population-size estimates of *A. jaragua* for ten years (2009–2019) is a remarkable fieldwork achievement in the scientific literature, without parallel for any other freshwater decapod, let alone a threatened one. We discuss this information, and present an updated taxonomic and ecological assessment of *A. jaragua* and the associated conservation implications.

11.2 INTRODUCTION TO THE AEGLIDS: FROM LONG-GONE MARINE ANCESTORS TO PRESENT-DAY FRESHWATER SURVIVORS

11.2.1 Aeglidae Dana, 1852

The origin of Aeglidae is marine as indicated by fossils from the Cretaceous (Feldmann 1984, Feldmann et al. 1998). Fossil specimens of the two aeglid monotypic genera, *Haumuriaegla glaessneri* (Feldmann 1984) and *Protaegla miniscula* (Feldmann et al. 1998), were collected from marine deposits in present-day New Zealand and Mexico, respectively, suggesting the early Mesozoic Pacific region as the family area of origin (Feldmann 1986, Pérez-Losada et al. 2004).

A third taxon, *Aegla* Leach, 1820, contains the only living representatives. All 87 (as of 2019) nominal species and subspecies of *Aegla* are entirely adapted to freshwater environments. *Aegla* is endemic to temperate and subtropical South America (Schmitt 1942, Bond-Buckup and Buckup 1994, Bueno et al. 2016a). Their distribution includes several major hydrographic systems in the South America southern cone, between latitudes 20° (southeastern Brazil) and 50° (Chile) (Bueno et al. 2007, Oyanedel et al. 2011).

11.2.2 Origin and Dispersal of *Aegla* in Continental South America

Recent investigations combining geological evidence and phylogenetic analysis provide a consistent hypothesis regarding the origin, evolutionary history, and dispersal of *Aegla* (Perez-Losada et al. 2004, Bartholomei-Santos et al. 2020). These studies strongly suggest that the geological events (orogenic uplifting and paleobasin formation) that shaped continental South America during late Mesozoic and early Cenozoic (Potter 1997, Lundberg et al. 1998, Ribeiro 2006) determined the current distribution pattern of present-day aeglids.

It is now well accepted that the Pacific side of South America most probably represented the "port of entry" of ancestral marine aeglids by means of marine transgressions during the early uplifting of the Andes Cordillera in the late Cretaceous–early Tertiary, about 90–60 million years ago (mya) (Pérez-Losada et al. 2004). The earliest *Aegla* representatives that were fully adapted to freshwater environment are estimated to have occurred at least 60 mya (Pérez-Losada et al. 2004). The physiological mechanisms that contributed to freshwater adaptation (McNamara and Farias 2020) were accompanied by adaptive life strategies, including direct postembryonic development (hatching of benthic juveniles from large eggs) and parental care (Bueno et al. 2016a, 2020). These adaptive traits developed independently in freshwater brachyurans and astacideans (Vogt 2016).

The continuing uplifting of the Andes Cordillera contributed to the geographic segregation of *Aegla* populations on the west (Chilean species) and east sides of the mountain range (Pérez-Losada et al. 2004). This also caused a major shift in the flow direction of Pacific watersheds, which began to flow eastward toward the Atlantic (Potter 1997, Lundberg et al. 1998, Almeida et al. 2000). The eastward shift in flow created a gateway interconnection between western watersheds and eastern paleobasins of the Paraná and Uruguay Rivers in the late Eocene–early Oligocene (43–30 mya), opening eastward and northward dispersal routes for aeglids within both paleobasins (Pérez-Losada et al. 2004).

Five monophyletic clades (A to E; sensu Pérez-Losada et al. 2004) (Fig. 11.1) of living freshwater aeglids have been recognized and the phylogenetic relationships are reflected by each distinct dispersal route as well as by each respective present-day distribution pattern along the major hydrographic systems (Pérez-Losada et al. 2004). Clades A and B form the oldest assemblages (~42 and ~41 mya, respectively) and are comprised of *Aegla* species distributed in watersheds on the western (Chilean) and eastern (Argentinean) sides of the Andes Cordillera. Clades C (~33 mya), D (~25 mya), and E (~24 mya) are later clades and include all remaining *Aegla* species distributed across central (Clade D) and eastern regions (Clades C and E) (Fig. 11.1).

11.2.3 Diversity and Geographic Distribution of *Aegla* Species from Clade C

The freshwater ecoregion concept used here follows Abell et al. (2008) and is defined as a large area, which may include one or more watersheds with distinct assemblage of natural freshwater communities and species. Ecoregions are of interest for conservation planning actions aimed at biodiversity conservation (Groves et al. 2002) and is important for conservation planning for aeglids (Pérez-Losada et al. 2009, Gonçalves et al. 2018, Tumini et al. 2019).

Species from Clade C, which includes *A. jaragua*, are distributed in six freshwater ecoregions: Paraiba do Sul, Ribeira de Iguape, Southeastern Mata Atlantica, Upper Uruguay, Upper Parana, and Iguassu, indicating that an ancient dispersal route toward Brazil's southern and southeastern regions was strongly associated with the formation of the Paraná paleobasin (Pérez-Losada et al. 2004).

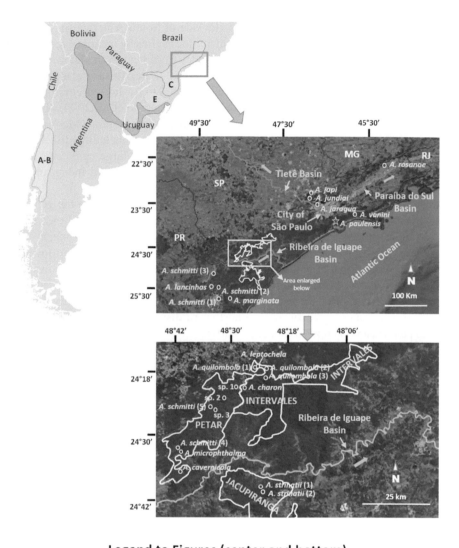

Legend to Figures (center and bottom)

★ *Aegla paulensis* s. str.

● Formerly *Aegla paulensis*

● Other aeglid species from the Atlantic Forest Biome

⟶ Watershed flow direction

Deforested areas (light green)
Atlantic Forest (dark green)

Watershed

State border

Figure 11.1   *Top:* Southern cone of South America depicting approximate geographical extension of clades A through E (modified from Pérez-Losada et al. 2004). *Center:* Aeglid distribution in Atlantic Forest remnants between 22° and 26° Latitude. *Bottom:* Detail of enlarged area indicated in central figure showing sites of occurrence of stygobitic and stygophilic aeglids from the Alto Ribeira karst area. Numerical sequencing after the name of species indicates different sites of occurrence for the taxon. *Abbreviations:* PR (state of Paraná), SP (state of São Paulo), MG (state of Minas Gerais), RJ (state of Rio de Janeiro). PETAR, Intervales and Jacupiranga are State Parks (state of São Paulo). Satellite background maps acquired from GoogleEarth.

Clade C is the most speciose one with 31 nominal species (Table 11.1), representing 35.6% of the 87 recognized species. Thirty species (96.8%) of this clade are endemic to Brazil, the only exception being *A. parana* Schmitt, 1942, which is from Brazil and Argentina (Bond-Buckup and Buckup 1994). Another species from Clade C, *A. franca* Schmitt, 1942, from Claraval County, southeastern Brazil, represents the septentrional distribution limit (latitude 20°S) for the genus (Bueno et al. 2007). Clade C also includes all stygobitic and stygophilic species (Türkay 1972, Bond-Buckup and Buckup 1994, Fernandes et al. 2013, Maia et al. 2013, Bueno et al. 2017).

Many Clade C species are found in the Atlantic Forest Biome remnants in subtropical Brazil between 22° and 25° south latitude (Fig. 11.1). These remnants are circumscribed in the Serra do Mar biogeographic subregion, one of five Atlantic Forest biogeographic subregions that are recognized as areas of endemism and undescribed biodiversity (Silva and Casteleti 2003). This makes this area one of the most important and most threatened biodiversity hotspots in the world (Câmara 2003, Galindo-Leal and Câmara 2003).

The Serra do Mar biogeographic subregion represents contiguous areas of preserved evergreen forests oriented east–west along the Serra do Mar and Serra da Mantiqueira coastal ridges, which run parallel to the continental coast of southeastern Brazil. It also defines an important biodiversity corridor (the Serra do Mar Corridor) for the coastal montane landscape, covered by dense tracts of predominantly ombrophilous forests (Galindo-Leal and Câmara 2003, Silva and Casteleti 2003).

The Paraíba do Sul and the Ribeira de Iguape freshwater ecoregions are restricted to the Serra do Mar biogeographic subregion (Fig. 11.1). These ecoregions are coastal hydrographic basins that ultimately flow and drain directly into the Atlantic Ocean. The Tietê River, however, does not drain into the sea and is one of the main tributaries that contribute to the formation of the Upper Paraná basin. The Upper Tietê sub-basin is limited to the Serra do Mar biogeographic subregion and flows through mosaic patches of preserved Atlantic Forest. No significant contiguous remnants of Atlantic Forest exist along the remaining Tietê River before it drains into the Paraná River (Fig. 11.1).

11.2.4 The Conservation Status of *Aegla* Species from Clade C

The International Union for Conservation of Nature (IUCN) categories and criteria (IUCN 2013) have been used for assessing aeglid species and subspecies conservation status (Bueno et al. 2016a,b, Jara et al. 2006, Santos et al. 2017, Tumini et al. 2019, Boos et al. 2020). *Aegla* is the most threatened taxon among Neotropic freshwater decapods (Bueno et al. 2016a).

Over 70% (22 out of 31) of Clade C species have been assessed as Critically Endangered (CR; 11 species; 35.5%), Endangered (EN; 6 species; 19.4%), or Vulnerable (VU; 5 species; 16.1%) (Table 11.1). The remaining species have been assessed as Least Concern (LC; 19.35%), and the conservation status of *A. marginata* s. str. Bond-Buckup and Buckup, 1994 and two recently described species,

Table 11.1 **Diversity, Distribution According to Ecoregion (sensu Abell et al. 2008), Habitat, and Conservation Status of Aeglids of Clade C**

Ecoregion	Species	Habitat	Conservation status and criteria	References Taxonomy (I); Conservation (II)
Ribeira de Iguape (330)	Aegla cavernicola Türkay, 1972	Stygobite, lotic	CR – B2ab(iii, v)	Türkay (1972) (I) Maia et al. (2013) (II)
	Aegla charon Bueno and Moraes, 2017	Stygobite, lentic	CR – B2ab(iii)	Bueno et al. (2017) (I) (II)
	Aegla lancinhas Bond-Buckup and Buckup, 2015	Epigean, lotic	EN – B1B2ab(iii)	Bond-Buckup and Buckup (1994) (as A. paulensis) (I) Santos et al. (2015) (I) (II)
	Aegla leptochela Bond-Buckup and Buckup, 1994	Stygobite, lotic	CR – B2ab(iii,v)	Bond-Buckup and Buckup (1994) (I) Maia et al. (2013) (II)
	Aegla microphthalma Bond-Buckup and Buckup, 1994	Stygobite, lotic	CR – A4ae+B2ab(iii, v)	Bond-Buckup and Buckup (1994) (I) Maia et al. (2013) (II)
	Aegla quilombola Moraes, Tavares and Bueno, 2017	Stygophile lotic	NE	Bond-Buckup and Buckup (1994) (as A. marginata) (I) Moraes et al. (2017) (I)
	Aegla strinatii Türkay, 1972	Stygophile, lotic	EN – B2ab(iii)	Türkay (1972) (I) Bueno et al. (2016a) (II)
Upper Parana (344)	Aegla castro Schmitt, 1942	Epigean, lotic	LC	Schmitt (1942) (I) Pérez Losada et al. (2009) (II)
	Aegla franca Schmitt, 1942	Epigean, lotic	CR – B2ab(iii)	Schmitt (1942) (I) Bueno et al. (2016a) (II)
	Aegla japi Moraes, Tavares and Bueno, 2016	Epigean, lotic	VU – B2aD2	Moraes et al. (2016) (I) (II)
	Aegla jaragua Moraes, Tavares and Bueno, 2016	Epigean, lotic	CR – A4eB2a	Bond-Buckup and Buckup (1994) (as A. paulensis) (I) Moraes et al. (2016) (I) (II)
	Aegla jundiai Moraes, Tavares and Bueno, 2016	Epigean, lotic	VU – B2aD2	Pérez-Losada et al. (2004) (as A. paulensis) (I) Moraes et al. (2016) (I) (II)

(Continued)

Table 11.1 (Continued) **Diversity, Distribution According to Ecoregion (sensu Abell et al. 2008), Habitat, and Conservation Status of Aeglids of Clade C**

Ecoregion	Species	Habitat	Conservation status and criteria	References Taxonomy (I); Conservation (II)
	Aegla lata Bond-Buckup and Buckup, 1994	Epigean, lotic	CR – B1ab(i, iii, iv)	Bond-Buckup and Buckup (1994) (I) Bueno et al. (2016a) (II)
	Aegla loyolai Bond-Buckup and Santos, 2015	Epigean, lotic	EN – B1+B2ab (iii)	Santos et al. (2015) (I) (II)
	Aegla paulensis Schmitt, 1942	Epigean, lotic	VU – B2aD2	Schmitt (1942) (as *A. odebrechtii paulensis*) (I) Bond-Buckup and Buckup (1994) (as *A. paulensis*) (I) Moraes et al. (2016) (as *A. paulensis* s. str.) (I) (II)
	Aegla perobae Hebling and Rodrigues, 1977	Epigean, lotic	CR – B2ab(III) c(iv)	Hebling and Rodrigues (1977) (I) Bueno et al. (2016a) (II)
	Aegla schmitti Hobbs III, 1978	Epigean and stygophile, lotic	LC	Hobbs III (1978) (I) Pérez-Losada et al. (2009) (II)
	Aegla vanini Moraes, Tavares and Bueno, 2016	Epigean, lotic	VU – B2aD2	Bond-Buckup and Buckup (1994) (as *A. paulensis*) (I) Moraes et al. (2016) (I) (II)
Paraiba do Sul (329)	*Aegla rosanae* Campos Jr., 1998	Epigean, lotic	CR – B2ab(iii)	Campos Jr. (1998) (I) Bond-Buckup and Buckup (2000) (as *A. paulensis*) (I) Moraes et al. (2016) (as *A. rosanae*, revalidated) (I) (II)
Southeastern Mata Atlantica (331)	*Aegla marginata* Bond-Buckup and Buckup, 1994	Epigean, lotic	NE as *A. marginata* s. str.	Bond-Buckup and Buckup (1994) (I) Moraes et al. (2017) (as *A. marginata* s. str.) (I) Pérez-Losada et al. (2009) (II)
	Aegla odebrechtii Müller, 1876	Epigean, lotic	LC	Müller (1876) (I) Pérez-Losada et al. (2009) (II)

(Continued)

Table 11.1 (Continued) **Diversity, Distribution According to Ecoregion (sensu Abell et al. 2008), Habitat, and Conservation Status of Aeglids of Clade C**

Ecoregion	Species	Habitat	Conservation status and criteria	References Taxonomy (I); Conservation (II)
	Aegla parva Bond-Buckup and Buckup, 1994	Epigean, lotic	LC	Bond-Buckup and Buckup (1994) (I) Pérez-Losada et al. (2009) (II)
	Aegla pomerana Bond-Buckup and Buckup, 2010	Epigean, lotic	EN – B1ab(iii)	Bond-Buckup et al. (2010) (I) Bueno et al. (2016a) (II)
Iguassu (346)	*Aegla meloi* Bond-Buckup and Santos, 2015	Epigean, lotic	CR – B2ab(iii)	Santos et al. (2015) (I) (II)
	Aegla okora Páez and Teixeira, 2018	Epigean, lotic	NE	Páez et al. (2018) (I)
	Aegla parana Schmitt, 1942	Epigean, lotic	LC	Schmitt (1942) (I) Pérez-Losada et al. (2009) (II)
Upper Uruguay (333)	*Aegla brevipalma* Bond-Buckup and Santos, 2012	Epigean, lotic	CR – B2ab (iii)	Santos et al. (2012) (I) (II)
	Aegla camargoi Buckup and Rossi, 1977	Epigean, lotic	EN – B2ab(iii)	Buckup and Rossi (1977) (I) Bueno et al. (2016a) (II)
	Aegla jarai Bond-Buckup and Buckup, 1994	Epigean, lotic	LC	Bond-Buckup and Buckup (1994) (I) Pérez-Losada et al. (2009) (II)
	Aegla oblata Bond-Buckup and Santos, 2012	Epigean, lotic	EN – B1ab(iii)	Santos et al. (2012) (I) Bueno et al. (2016a) (II)
	Aegla spinosa Bond-Buckup and Buckup, 1994	Epigean, lotic	VU – B1ab(iii)	Bond-Buckup and Buckup (1994) (I) Bueno et al. (2016a) (II)

For species with occurrence in more than one ecoregion (i.e., *Aegla schmitti* and *A. parva*), only the ecoregion that includes the type locality has been indicated to avoid repetition. IUCN categories: CR (Critically Endangered), EN (Endangered), VU (Vulnerable), LC (Least Concern), NE (Not Evaluated). See IUCN (2013) for details on criteria.

A. quilombola Moraes, Tavares and Bueno, 2017 and *A. okora* Páez and Teixeira, 2018 have yet to be assessed (Moraes et al. 2017, Páez et al. 2018).

Some Clade C species assessed as Least Concern are still of conservation concern (IUCN 2013). *Aegla schmitti* Hobbs III, 1979, for instance, has been assessed as LC (Bueno et al. 2016b) because its geographic distribution includes large areas that are restricted within three freshwater ecoregions (Ribeira de Iguape, Iguassu, and Southeastern Mata Atlantica) (Trevisan and Masunari 2010). Results of morphometric analyses, however, suggest that several populations of *A. schmitti* may actually represent cryptic species (Trevisan et al. 2014). Should this be correct, these taxa will have to have their conservation status reevaluated. This was the case for *A. paulensis* Schmitt, 1942 after the taxon was recently reviewed (discussed below).

11.3 *AEGLA PAULENSIS* SCHMITT, 1942 S. LAT: FROM LEAST CONCERN TO SEVERAL ENDANGERED SPECIES

11.3.1 Taxonomic Remarks

Schmitt (1942: 490–493) described *Aegla odebrechtii paulensis* from southeastern Brazil based on specimens from Alto da Serra de Cubatão (currently Alto da Serra de Paranapiacaba Biological Reserve), located in the Upper Tietê sub-basin (Upper Parana ecoregion) in São Paulo state. In the description, he mentioned that the morphology of the specimens very much resembled that of *A. odebrechtii* Müller, 1876, from Itajaí basin, in the state of Santa Catarina (Southeastern Mata Atlantica ecoregion). Schmitt also recognized, however, that those specimens he had at hand were not fully developed, so he treated the material from Alto da Serra as a subspecies.

Half a century later, Bond-Buckup and Buckup (1994) raised *A. paulensis* to full specific status. The authors also reported several new localities in the Upper Tietê sub-basin, and also added records from a different and disjunctive watershed, the Ribeira de Iguape Basin. Later, a third disjunctive watershed yielded *A. rosanae* Campos Jr. 1998, from a small stream in the Paraíba do Sul Basin, also in the state of São Paulo (Campos Jr., 1998), and was deemed as synonym of *A. paulensis* (Bond-Buckup and Buckup, 2000).

More recently, Santos et al. (2015) demonstrated the population of *A. paulensis* from the Ribeira de Iguape Basin as a new species: *A. lancinhas* Bond-Buckup and Buckup, 2015. Based on morphological and molecular analyses, Moraes et al. (2016) argued that *A. paulensis* actually comprised an assemblage of cryptic species, which led to the recognition and description of four new species from the Upper Tietê sub-basin: *A. vanini* Moraes, Tavares and Bueno, 2016, *A. japi* Moraes, Tavares, and Bueno, 2016, *A. jundiai* Moraes, Tavares and Bueno, 2016, and *A. jaragua* Moraes, Tavares and Bueno, 2016. Moraes et al. (2016) also provided a redescription of *A. paulensis* s. str. from the type locality (Upper Tietê sub-basin) and revalidated *A. rosanae*, so far the only known aeglid from the Paraíba do Sul Basin (Campos Jr. 1998, Moraes et al. 2016).

11.3.2 Distribution and Habitat Characterization

All taxa mentioned in the previous section exhibit a similar distribution pattern in the Atlantic Forest Biome. They show high local endemism, each limited to a single locality (except *A. lancinhas*, for which more than one location has been

Figure 11.2   (A) The Benfica stream, city of Piquete, Paraíba do Sul Basin, type locality of
Aegla rosanae. (B) The Pai Zé stream, Jaraguá State Park, Upper Tietê sub-
basin, type locality of *Aegla jaragua.*

reported – Santos et al. 2015) with no congeners present. These localities share several ecological similarities, allowing some habitat characterization generalizations
(Fig. 11.2). The following generalized description of such habitat fits perfectly with
the ecological characterization of the headwater zone according to the river continuum concept as proposed by Vannote et al. (1980), and of Atlantic Forest freshwater
streams as put forward by Castro (1999):

• The localities are at high altitudes (between 600 and 900 m above sea level – a.s.l.)
  in low-order streams of foothill forests.
• The headwater zone of low-order streams are typically flanked by dense and high
  riparian vegetation, which partially blocks direct sunlight, producing a characteristic diffuse weak ambient light, contributing to cool water temperatures with little
  diel variation.
• Aquatic gross primary production is reduced or negligible. The headwater streams
  are greatly dependent on allochthonous debris and soluble organic compounds
  from the adjacent terrestrial environment to maintain the nutrient cycle.
• They are pristine, clear, fast-flowing water over bedrock. Shallow depths and the
  uneven rocky substrates produce myriad waterfalls of various sizes, which provide
  a well-mixed and well-oxygenated water column.

Aquatic macroinvertebrate feeding strategies in the headwater zone is characterized by the relative dominance (from high to low) of collectors (= fine and ultrafine
particulate filter feeders), shredders, predators, and grazers (Vanotte et al. 1980).
Aeglids are omnivorous (Rodrigues and Hebling 1978, Bueno and Bond-Buckup
2004, Castro-Souza and Bond-Buckup 2004, Santos et al. 2008, Colpo et al. 2012)
and mostly belong to both shredder and predator functional feeding groups. As
shredders, aeglids are important in the processing and consumption of leaf litter
and its microbial community (Cogo and Santos 2013). As predators, aeglids actively
feed on small aquatic insect larvae (Magni and Py-Daniel 1989, Castro-Souza and
Bond-Buckup 2004).

11.4 *AEGLA JARAGUA* MORAES, TAVARES AND BUENO, 2016: A TEN-YEAR CASE STUDY

11.4.1 Type Locality and Distribution

Aegla jaragua is one of the recently described aeglid species following the taxonomic revision of *A. paulensis* s. lat. (Moraes et al. 2016). The species is restricted to its type locality, the Pai Zé stream (23°27′27.9″S; 46°45′32.3″W), a first-order stream with the source located within the boundaries of the Jaraguá State Park (approx. 800 m a.s.l) in Jaraguá District, near the city of São Paulo, Brazil.

Jaraguá State Park (JSP) is a 492-hectare conservation unit and one of the few municipal areas where Atlantic Forest Biome remnants are preserved. It contains a tourist attraction, the "Pico do Jaraguá", the highest mountain peak (altitude 1,135 m), near the city of São Paulo. The JSP is visited daily by the local population for recreation and sport activities or simply for a nice view of the São Paulo megalopolis.

However, the JSP has been under chronic anthropogenic pressure even before the park was created in 1961. The pressure comes from surrounding ongoing and uncontrolled urbanization. Presently, the JSP is completely surrounded by urban areas (Fig. 11.3). It has become a tiny island of preserved forest within a vast, continuous degraded land extending in all directions.

Figure 11.3   Aerial view of Jaraguá State Park completely surrounded by permanent human occupation in the Jaraguá district. The blue dot indicates the type locality of *Aegla jaragua*. The red dot indicates the lake that contains an established population of the exotic red crawfish *Procambarus clarkii*. Background satellite photo acquired from GoogleEarth.

In spite of this, the JSP is a safe haven for *A. jaragua*. The estimated area of occupancy (AOO) of *A. jaragua* is less than 0.001 km² because this species' distribution is restricted to the stretch of the Pai Zé stream (length: approx. 500 m; average width 1.5 m; average depth: 10 cm) within the park's boundaries. There is no chance of survivorship of viable populations of the species outside the protected area of the park.

11.4.2 Biology

The first report of aeglids (as *A. paulensis* s. lat.) in the JSP comes from Bond-Buckup and Buckup (1994). Thirteen years would pass before field studies started in 2007 (Cohen et al. 2011). The following summarized description of the biology of the species is based on field data obtained from sampling specimens with baited traps (Cohen et al. 2011).

Aegla jaragua is a small aeglid. Adult males attain a larger size than adult females (maximum carapace length – rostrum excluded: 18.84 and 17.12 mm, respectively). Males and females reach morphometric maturity (puberty molt) at a similar age (males: 14.1 months; females: 14.5 months).

Temporal variation in sex ratios is associated with the reproductive cycle. Males predominate over females most of the year. The sex ratio reverses at the beginning of the markedly seasonal reproductive period, when females bear fully developed ovaries (indicating impending oviposition) and mating is imminent. The reproductive season is five to six months long, extending from austral autumn to midwinter. This is in accordance with the marked seasonal reproductive pattern observed in other aeglid species from low latitudes (Rodrigues and Hebling 1978, Swiech-Ayoub and Masunari 2001, Bueno and Shimizu 2008, 2009, Rocha et al. 2010, Grabowski et al. 2013, Takano et al. 2016, Marçal et al. 2018).

Eggs are attached to the female pleopods and incubated in the brood chamber formed by the folded abdomen. Postembryonic development is direct and a complete morphological description of the first juvenile instar is provided by Moraes and Bueno (2013). Juveniles hatch during austral midwinter (Cohen et al. 2011) and remain protected in the brood chamber for a few days (Moraes and Bueno 2013). Direct postembyonic development and parental care are exclusive to *Aegla* among anomurans (see Bond-Buckup et al. 1996, 1999, Bueno and Bond-Buckup 1996, López-Greco et al. 2004, Francisco et al. 2007, Teodósio and Masunari 2007, Moraes and Bueno 2015, Silva et al. 2017 for additional examples). Because oviposition is total and the reproductive period is seasonal, there is one recruitment pulse per year. Males grow faster than females, but females live longer (33.9 and 40.2 months, respectively). Both sexes live long enough to reproduce twice during their lifetime.

11.4.3 Population-Size Estimates

11.4.3.1 Concept, Terminology, and Methods

Indices of abundance are excellent indicators to determine population sizes. Sequential data of population size estimated on a regular basis over periods of time becomes invaluable information for detecting temporal population variations and, consequently, to assess conservation status (IUCN 2013).

The methods of Petersen, Schnabel, and Schumacher-Eschmeyer (all methods are described in detail in Krebs 1999) applied to closed populations are preferred for aeglid population-size estimates (Bueno and Bond-Buckup 2000, Bueno et al. 2007, 2014, Cohen et al. 2013, Dalosto et al. 2014). These methods employ mark and recapture techniques and are based on the premise that the proportion of recaptured individuals (captured, marked released back to the environment in a previous sampling event) in relation to the total number of sampled individuals is the same as the proportion of marked individuals that have been released in the environment in relation to the total number of individuals in the population. The concept of a closed population presupposes that the number of individuals will not change during the estimation time period, meaning migration, recruitment, and mortality effects are negligible (Seber 1982, Krebs 1999).

The Petersen method is the simplest of the three, requiring one recapture event only, i.e., animals sampled on day one are marked and released back in the sampling area. After enough time (usually a full day) is given for marked animals to evenly disperse in the population, a second sampling (= recapture) event is carried out. This method was used to estimate the population size of *Aegla platensis* Schmitt, 1942 (Bueno and Bond-Buckup 2000, Dalosto et al. 2014).

The Schnabel and the Schumacher-Eschmeyer methods are extensions of the Petersen method, because both estimators involve multiple and sequential recapture events (Seber 1982, Krebs 1999). With each new sampling event, both unmarked and marked individuals are counted; unmarked individuals are then marked and released back to the sampling site along with the previously marked (= recaptured) ones, thus building up the relative number of marked individuals in relation to that of not yet sampled and unmarked ones in a closed population condition. Both methods show several similarities in how data are collected as well as in the calculation procedures (see Krebs 1999 for details). The Schnabel method was used to estimate the population size of *A. franca* and *A. longirostris* Bond-Buckup and Buckup, 1994 (Bueno et al. 2007, Baumart et al. 2018), and the Schumacher-Eschmeyer method was employed to estimate the population size of *A. franca*, *A. perobae* Hebling and Rodrigues, 1977, *A. longirostris*, and *A. jaragua* (Bueno et al. 2007, 2014, Cohen et al. 2013, Baumart et al. 2018, this chapter).

All three mark and recapture methods mentioned above are based on the following assumptions (Krebs 1999):

- The population is closed.
- Samples are randomly taken.
- Marking does not affect the catchability of marked animals.
- Marking is not lost during the sampling period.

The Schumacher-Eschmeyer method is deemed the most robust of the three methods (Seber 1982, Krebs 1999). The advantage this method has over the other two is that fulfillment of the assumptions can be verified by applying a "through the origin" linear regression (Zar 1996) on the set of [number of marked animals at large, at each sampling event (M_t), proportion of marked animals in each sample (R_i/C_i)] data points (all terms and abbreviations are from Krebs 1999). A significant positive slope means that as the marked individuals accumulate in the environment (M_t gets higher) at each sampling event, R_i/C_i increases proportionally, in accordance to a linear function. This, in turn, indicates that every assumption required to validate the method has been fulfilled.

Recently, Baumart et al. (2018) carried out a comparative study among the Schnabel, the Schumacher-Eschmeyer methods, and the Bayesian model. According to the authors, all three population-size estimators provided similar results, though the Bayesian model was regarded as the one that tends to produce a more conservative approach when conservation issues are concerned.

11.4.3.2 Fieldwork Procedures and Recommendations

To provide a closed population condition, the upstream and downstream limits of a stream stretch are isolated by setting nets across the stream between opposing margins, thus defining a selected working area and to prevent aeglids from moving into or away from it (Figs. 11.4A and B). Since most aeglid populations from the Atlantic Forest are found in high-altitude low-order streams, which are typically narrow and shallow, the procedure of isolating a well-defined working site should not be difficult. Caution should be taken to guarantee that the lower portion of the isolating net be firmly positioned underwater and kept in full contact with the streambed to avoid breach passages for aeglids to move about underneath the net. This was achieved by placing flat rocks on top of the submerged lower section of net (Fig. 11.4A). Also, nets should be checked for integrity and cleaned twice a day (morning and afternoon) to remove accumulated debris.

Aeglids were captured using baited traps. Affordable homemade simple design traps were assembled with low-cost plastic parts (Fig. 11.4C). The traps consisted of a perforated plastic container, except for its entrance, which bears a plastic tube (hair curler) tilted upward inside the container (Fig. 11.4C). A perforated canister containing bait was kept locked in position on the trap lid just before the trap was positioned underwater. These perforations were large enough to allow water to flow through and the attractant in the bait to be continuously released downstream. These perforations, however, were small enough to prevent the aeglids from picking up pieces of food inside and consuming them. This is important because the attractant in the bait should be equally effective to both marked (previously captured) and unmarked individuals in any sampling event. Our experience has shown that dry, fish flavored, cat food makes an excellent bait for sampling aeglids, in addition to being commercially available anywhere, easy to store, and ready to use (Bueno et al. 2007).

Because aeglids are more active at night (Sokolowicz et al. 2007, Ayres-Peres et al. 2011), traps should be set late in the afternoon and checked the following morning (Bueno et al. 2007, 2014, Rocha et al. 2010, Cohen et al. 2013, Takano et al. 2016). Traps should be randomly distributed across the stream bed as demographic categories of interest (adult males, ovigerous or non-ovigerous females) may show

Figure 11.4   (A) Net stretched across the downstream limit of the isolated working area. Notice the line of flat rocks over the submerged lower section of net. (B) Net stretched across the upstream limit of the isolated working area. (C) Model of the plastic trap used for collecting *Aegla jaragua*. A perforated 135 mm film canister attached to the inner side of the lid is used to hold the bait inside. (D) Trap set in position on the streambed. A large flat rock is positioned on top of the trap lid to keep the trap submerged and in place overnight.

exploratory preference for different microhabitats, such as shallow areas near the margins and deeper areas farther away from the margin, and places showing varied hydrodynamic conditions (calmer water in natural pools, or near fast-flowing water turbulence). Regardless of the place where the trap is set underwater, its entrance must always be positioned facing downstream so that aeglids are lured by the bait scent to move upstream and toward the entrance, the only access to enter inside the trap (Fig. 11.4D).

The bait should be replaced after every recapture event. Used bait should not be discarded in the stream to prevent it from affecting the study. Traps are also set outside the upstream and downstream limits of the study area to indicate breaches on the isolation nets should marked individuals be captured by any of these traps. Should this happen, the marked individual is to be released back in the study area (and not counted as a recaptured individual) and immediate care should be taken to rectify any possible isolating problem associated with the net.

The population-size estimation is based on the number of mature individuals (Criteria A and C, see IUCN 2013), so, it is important that the mesh size selected will prevent mature individuals from entering or leaving the study area (= migration).

Mature (= adult) aeglids are those that have gone through the single puberty molt, which marks the transition from the juvenile to the adult phase (morphometric maturity) (Bueno and Shimizu 2009, Takano et al. 2016). Following the puberty molt, significant changes in the allometric rate of some specific body parts can be detected by relative growth analysis (Bueno and Shimizu 2009, Takano et al. 2016, Marçal et al. 2018). Also, secondary sexually dimorphic traits become evident as adult male chelae (first pereopods) are heavier than in adult females, and adult females exhibit a wider pleonal region than adult males (Bueno and Shimizu 2009, Oliveira and Santos 2011, Trevisan and Santos 2012, Copatti et al. 2015, Takano et al. 2016, da Silva et al. 2017, Marçal et al. 2018, Adam et al. 2018).

The study population average size at the onset of morphometric maturity (ASOMM) should be determined in advance, to serve as a reference size when sorting juveniles from adults. All unmarked individuals (Fig. 11.5A) should have their carapace length (CL) measured, but only adults (CL ≥ ASOMM) should be marked before they are returned to the study area (Fig. 11.5B). Juveniles (CL < ASOMM) are not marked and are released outside the study area. The data collection of juvenile size is valuable information for investigating population structure in parallel to population size (see Section 4.3.3).

In our studies, animals were marked with a mixture of super glue (cyanoacrylate) and fine powdered commercial dye, which is prepared at the time of use. A thin coat is spread over a specific area on the carapace (Fig. 11.5C). This coat dries within seconds, is water resistant, and is not removed by friction (Cohen et al. 2013). In closed population working conditions, all individuals receive similar markings so that marked individuals can be readily distinguished from unmarked ones in subsequent sampling events (Krebs 1999). In the event that a series of estimations of population size in the same population should be performed a few months apart from one another (see Bueno et al. 2014 for the species *A. perobae*), then a different marking pattern is adopted at each period (Fig. 11.5D) with respect to the period underway.

The marking technique does not affect the catchability of marked animals. Both marked and unmarked individuals are sampled randomly (Fig. 11.5E) and the proportion rate between marked and unmarked individuals increases linearly as a result of cumulative increase in relative number of marked individuals (Figs. 11.5F and 11.6).

11.4.3.3 *Aegla jaragua*: Population-Size Estimates (2009–2019)

The *A. jaragua* population size was estimated by Cohen et al. (2013) in September 2009 (late austral winter/dry season) and March 2010 (late austral summer/rainy season). The numbers were estimated as 943 (density = 6.7 individuals/m²) and 1,434 (density = 11.5 individuals/m²), respectively. From 2011 through 2019, the population size was estimated biennially as part of the activities of the undergraduate course "Studies on populations of decapod crustaceans: an analytical approach". Samples for population size and structure were taken in July 2011, 2013, 2015, 2017, and 2019. An additional estimate was performed in August 2018 to follow up on the population at an annual interval, after a marked decrease in size was detected in 2017 (see below).

Figure 11.5   (A) Collection of unmarked specimens of *Aegla jaragua* captured in the first
sampling event. (B) Collection of adult specimens of *A. jaragua* after they were
marked. Floating leaves were used to provide shelter and to reduce stress due
to agonistic confrontations among individuals. (C) Mark on the cardiac area of
two adult males of *A. jaragua*. (D) Adult female of *A. perobae* marking types
from different sampling periods. (E) Collection of unmarked and marked specimens of *A. jaragua* inside a trap in a recapture event. (F) Five marked adult
specimens of *A. jaragua* (yellow arrowheads) that had just been released back
into the sampling site.

Population-size estimates may vary with season (Bueno et al. 2007, Cohen et
al. 2013, Dalosto et al. 2014, Baumart et al. 2018). These fluctuations may indicate
differential intensity of locomotory behavior between sexes during reproductive and
nonreproductive periods (Shimizu and Bueno 2020), and this may explain temporal
sex ratio variation patterns observed in *A. jaragua* (Cohen et al. 2011). Therefore,

the description and analysis of the ten-year period of population-size estimation of *A. jaragua* that follows is restricted to the winter season data, which includes all estimates obtained in the 2011–2019 period and one in September 2009 by Cohen et al. (2013).

The same stream reach (approx. 70 m long) was isolated by two fishing nets (4 mm mesh) during sampling (Figs. 11.4A and B). The isolated reach was estimated as the sum of areas of successive trapezoids, calculated with width values (lengths of parallel bases) measured at 5 m intervals (height) along the reach.

The capturing, measuring, marking, and returning procedures followed Cohen et al. (2013; see Section 4.3.2), except for the number of traps (15 instead of 20) and number of days in each sampling period (four instead of seven). All population-size estimation calculations in 2009 were redone restricting the data to those collected in the first four days to standardize the results. Gross data from each estimating period is shown in Table 11.2.

The population size for each period was estimated according to the Schumacher-Eschmeyer method (Krebs 1999), and size structure was described by frequency distributions in 1 mm interval classes of carapace length (see Cohen et al. 2013). Each estimated number of individuals and the respective confidence limits were converted to density according to the respective value of the area of the isolated stretch in order to provide proper temporal comparisons.

All datasets provided a significant "through the origin" regression of the proportion of marked individuals in samples (R_i/C_i) vs. marked individuals at large in the stream (M_t) relationship (Fig. 11.6), indicating that the assumptions of the Schumacher-Eschmeyer method were fulfilled (Krebs 1999).

Our results strongly indicate that the impact of markings lost with ecdysis was negligible. First and foremost, the high significance ($p < 0.05$) of the linear equation obtained from all estimates performed (Fig. 11.6) is the best evidence for validating the Schumacher-Eschmeyer regression method (Seber 1982). This validation strongly indicates that none of the working assumptions intrinsically associated with the method were violated, including loss of markings. Second, the high precision of the estimate can be attested by the narrow range of the 95% confidence intervals around most estimates (Fig. 11.7). Third, based on the above, the short time period of four sampling events adopted had no negative effect on the high precision of the Schumacher-Eschmeyer method. Other studies on aeglid population-size estimations (Bueno et al. 2007, Cohen et al. 2013, Bueno et al. 2014, Baumart et al. 2018) performed seven consecutive sampling events before they were concluded, and the estimate results were equally significant. By decreasing the number of sampling events to four, however, the chance of ecdysis occurring (if at all) in some marked individuals is lessened. Seber (1982) recommended that fieldwork be performed over a short period of time if the assumptions of a closed population are to be maintained. Moreover, periodic ecdysis in adults occurs less often than in juveniles, strengthening the argument that multiple recapture events should consider only adults and be performed within the shortest period of time.

Adult individual density values estimated in the 2011–2015 period are nearly equivalent to the 2009 estimates by Cohen et al. (2013) in spite of the disproportionate

Table 11.2 *Aegla jaragua*: Gross Field Data from Population-Size Estimations during the Winter Season of 2009, 2011, 2013, 2015, 2017, 2018, and 2019

Month/year	Sampling events	C_t	R_t	C_t-R_t	M_t	R_t/C_t
September 2009	Day 1	223	0	223	0	0.000
	Day 2	147	32	115	223	0.218
	Day 3	117	62	55	338	0.530
	Day 4	119	56	–	393	0.471
July 2011	Day 1	110	0	110	0	0.000
	Day 2	117	21	96	110	0.180
	Day 3	109	20	89	206	0.184
	Day 4	100	23	–	295	0.230
July 2013	Day 1	141	0	141	0	0.000
	Day 2	135	21	114	141	0.156
	Day 3	194	40	154	255	0.206
	Day 4	154	45	–	409	0.292
July 2015	Day 1	168	0	168	0	0.000
	Day 2	111	23	88	168	0.207
	Day 3	127	41	86	256	0.323
	Day 4	144	52	–	342	0.361
July 2017	Day 1	94	0	94	0	0.000
	Day 2	86	36	50	94	0.419
	Day 3	103	46	57	144	0.447
	Day 4	101	69	–	201	0.683
August 2018	Day 1	79	0	79	0	0.000
	Day 2	57	19	38	79	0.333
	Day 3	59	25	34	117	0.424
	Day 4	60	25	–	151	0.417
July 2019	Day 1	115	0	115	0	0.000
	Day 2	111	29	82	115	0.261
	Day 3	104	44	60	197 (–1 dead)	0.423
	Day 4	118	63	–	256	0.534

Data from 2009 are from the first four sampling events of Cohen et al. (2013). (C_t): total number of individuals sampled; (R_t): number of marked individuals in sample; (C_t-R_t): number of newly marked individuals; (M_t): cumulative number of individuals previously marked and released.

95% confidence interval width observed in the 2011 estimate due to the comparatively low determination of the R_t/C_t vs. M_t regression. The 2017 adult density decreased to 22–30% of the estimated values of the previous period and remained low in the two years that followed (Fig. 11.7).

The population structure also differed between 2011–2015 and 2017–2019 (Fig. 11.8). The proportion of juveniles was markedly higher in the latter period (61.0%) than in the previous period (26.3%). This difference was also apparent in each sampling event: the 2017–2019 proportion of juveniles (26.9–48.5%) was higher than

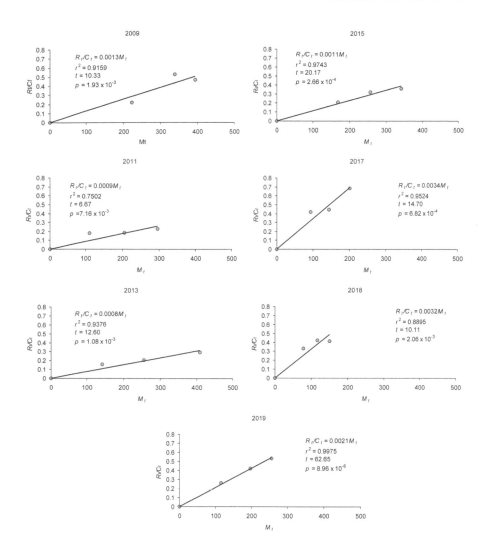

Figure 11.6 *Aegla jaragua.* Linear regression of the proportion of marked adult aeglids in the sample (R_t/C_t) against the number of marked aeglids at large (M_t), with the coefficient of determination (r^2) and the results of the Student's t-test for significance of regression included. The 2009 regression was recalculated using the data from the first four samples obtained by Cohen et al. (2013) to standardize the result in relation to those of the 2011–2019 period. Gross field data for each estimation period are shown in Table 11.2.

those of the previous period (15.4–22.7%), particularly in July 2019, when juveniles comprised almost half of captured individuals (Fig. 11.9).

The distinction between the periods of high and low adult densities becomes less clear when the juvenile numbers (results expressed in number of individuals/trap/ day) are considered. Only the July 2019 value is noticeably higher than those of the 2011–2018, nearly twice the 2013 value, and the second highest

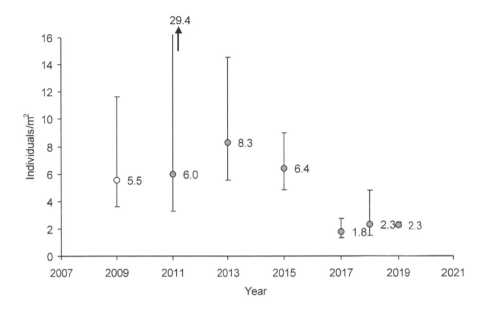

Figure 11.7   *Aegla jaragua*. Estimated population densities and respective 95% confidence intervals during the 2009–2019 sampling period. The 2009 value (empty point) was recalculated using the data from the first four samples obtained by Cohen et al. (2013) to standardize the result in relation to those of the 2011–2019 period.

number of juveniles in this period (Fig. 11.10). This strongly suggests that the increase in the proportion of juveniles in the 2017 and 2018 samples (Fig. 11.9) was largely due to the decrease in the number of adults, rather than an actual increase in abundance of juveniles. Values fluctuated around a mean of 1.6 individuals/trap/day from 2011 to 2018, showing no definite temporal trend (Fig. 11.10). These high values in 2019 indicate that abundance and proportion did actually increase that year.

These results indicate that a persistent restriction in reproductive and/or recruitment intensity occurred in the latter period and it is very likely that it led to low adult densities from 2017 onward. The 2019 increase of proportion and number of juveniles suggests a possible recovery of reproduction and/or recruitment levels.

Patterns of interannual fluctuations in abundance have been attributed to internal population regulation mechanisms or a long-term influence of climatic variables (Lipcius and Van Engel 1990). In our study, both high and low juvenile numbers occurred when adult density was higher (2009–2015). This does not clearly support a density-dependent influence on reproduction and/or recruitment, although a more detailed study of population density and size structure would be required for a more conclusive interpretation.

Possible climatic condition influences were examined through temperature and rainfall records of the 2009–2019 period and the previous decade (1999–2008), compiled from the city of São Paulo climatic station website (www.estacao.iag.usp .br/boletim.php). The mean monthly temperature was compared between the two

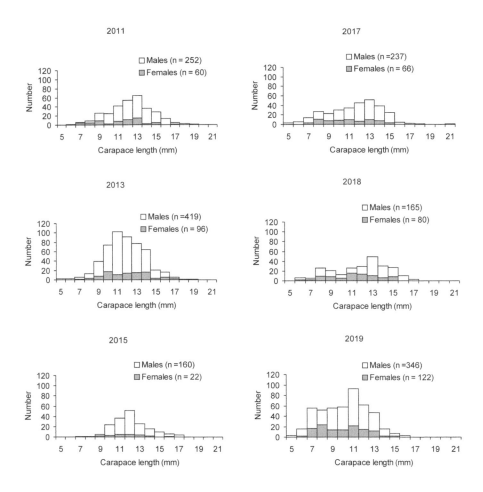

Figure 11.8 *Aegla jaragua*. Frequency distribution in classes of carapace length in the month of July (August in 2018) in the 2011–2019 sampling period. For frequency distribution obtained in July 2008 and 2009, see Cohen et al. (2011).

periods, separately for the warm and cold seasons (months in which values were higher and lower than 20°C, respectively). The total monthly precipitation was compared separately for the rainy and dry seasons (months in which values were higher and lower than 100 mm, respectively). No significant difference of temperature (Mann-Whitney test) or of its variance (F test) was found in either season, or for dry season precipitation. Rainy season precipitation variance of the 2009–2019 decade was significantly higher than that of the previous one (Table 11.3). This was largely due to the uncommonly high values (> 400 mm) recorded in January 2010, January 2011, and March 2019 and low peaks (near 200 mm) in the rainy season of 2014 and 2018 (Fig. 11.11). Similar extreme values were not observed in the 1999–2008 period.

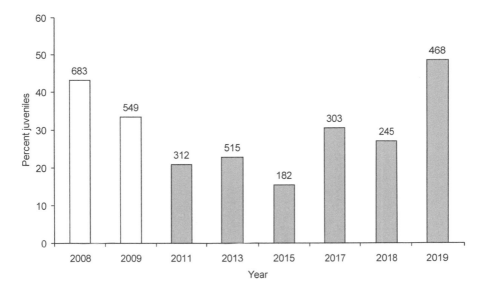

Figure 11.9 *Aegla jaragua*. Proportion of juveniles in the samples of the month of July (August in 2018) in the 2009–2019 sampling period (gray bars). White bars depict the values of July 2008 and June 2009 from Cohen et al. (2011), which were included to complement the temporal sequence. The total number of individuals sampled on each occasion is indicated above the corresponding bar.

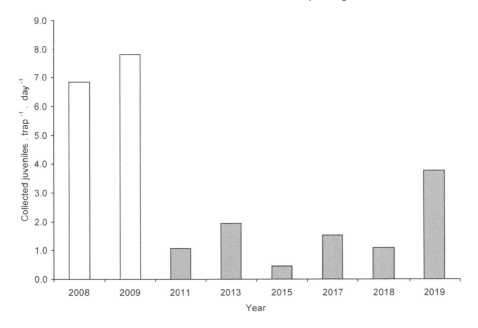

Figure 11.10 *Aegla jaragua*. Number of collected juveniles in the month of July (August in 2018) in the 2009–2019 sampling period (gray bars). White bars depict the values of July 2008 and June 2009 from Cohen et al. (2011), which were included to complement the temporal sequence.

Table 11.3 Mean and Variance of Mean Monthly Temperature in the Warmer and Colder Seasons and Total Monthly Precipitation in the Rainy and Dry Seasons of the 1999–2008 and 2009–2019 Periods in the City of São Paulo, Brazil

| | Mean monthly temperature (°C) | | | | Total monthly precipitation (mm) | | | |
| | Warmer season (previous year November–April) | | Colder season (May–October) | | Rainy season (previous year October–March) | | Dry season (April–September) | |
	1999–2008	2009–2019	1999–2008	2009–2019	1999–2008	2009–2019	1999–2008	2009–2019
n	60	66	60	62	55	66	55	63
Mean	21.5	21.8	17.462	17.627	179.6	207.0	59.5	73.6
Variance	1.5	2.0	2.2048	2.2155	7071.2	12016.0	2700.8	3347.8

Data source: www.estacao.iag.usp.br/boletim.php. Only the variance of precipitation in the rainy season differed between the two periods ($F = 1.699$; $p = 0.046$).

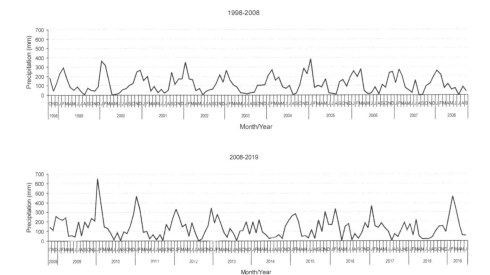

Figure 11.11   Monthly variation of precipitation in the rainy and dry seasons of the 1999–2008 and 2009–2019 sampling periods in the city of São Paulo, Brazil (data source. http://www.estacao.iag.usp.br/boletim.php).

Precipitation variations cause alterations in flow, habitat availability, and physical and chemical conditions of stream environments, and consequently affect macroinvertebrates (Resh et al. 2013). The impact of rainfall irregularity on aeglid population size has been documented for *A. perobae* (Bueno et al. 2014): an abnormally high precipitation in July 2007, comparable to rainy season levels, coincided with juvenile hatching. This event impaired successful recruitment and consequently influenced population size, size structure, and reproduction in the following years. Reduced abundance persisted for over three years after this climatic irregularity. In the case of the *A. jaragua* population, the density and size structure alterations coincided with a period of high interannual precipitation variability in the rainy seasons, although juveniles hatched in the previous dry season attained sizes less susceptible to negative effects of increased flow (5–6 mm of carapace length; Cohen et al. 2011). Conversely, ovary development occurs during this season, which might be affected by rainfall irregularities. This suggests that if rainfall actually played a major part in the changes observed, its effects should have been chronic, due to fluctuations across years.

The direct influence of human activity as a casual factor of alterations in the population is highly improbable since the entire population of *A. jaragua* is located inside the limits of an area of permanent protection (JSP). Thus, we assume that the observed fluctuations of density and changes in the proportions of the demographic groups are natural.

11.5 REMARKS ON THE CONSERVATION STATUS
OF *AEGLA JARAGUA* AND OF OTHER SPECIES
FROM THE ATLANTIC FOREST BIOME

The *A. paulensis* species complex demonstrates a highly endemic distribution pattern, with an extent of occurrence restricted to each species type locality (Moraes et al. 2016). What once was a taxon assessed as LC, now is recognized as a set of several threatened taxa (Table 11.1).

It is worth mentioning that each species from the Upper Tietê area has its known extent of occurrence entirely circumscribed within the boundaries of conservation units: the Alto da Serra de Paranapiacaba Biological Reserve (*A. paulensis* s. str.), the Boracéia Biological Station (*A. vanini*), the Jaragua State Park (*A. jaragua*), the Serra do Japi Biological Reserve (*A. japi*), and Serra do Japi-Guaxinduva-Jaguacoara (*A. jundiai*). Moraes et al. (2016) suggested that all these species (except *A. jaragua*) be assessed as Vulnerable (VU) under criteria B2aD2, since each species range is limited to areas of permanent protection, even though each of these species is known from one location each, and the area of occurrence for each is estimated at less than 10 km².

Moraes et al. (2016) recommended that *A. jaragua* be assessed as Critically Endangered (CR) instead, under criteria A4eB2a, based on the estimated low AOO < 10 km² (criterion "B2"), limited to one single location (subcriterion "a") and the inferred potential decrease in population size (criterion A4, subcriterion e) caused by the introduction of exotic species. The argument behind this recommendation is that there is a permanent lake harboring a well-established population of the exotic red crawfish, *Procambarus clarkii* (Girard, 1852) within the park boundaries, not too far away from the species type locality (see Loureiro et al. 2015, 2018). This invasive species represents a continuous and potential threat to the only known population of *A. jaragua*.

The alien *P. clarkii* is a very robust and aggressive freshwater decapod, which is capable of colonizing different types of freshwater habitats and withstanding harsh environmental conditions (Gherardi 2007). Introductions throughout the world tend to yield serious impacts to endemic biota, most notably biodiversity loss and modification of food web dynamics (Renai and Gherardi 2004, Rodríguez et al. 2005, Correia and Anastácio 2008, Cruz et al. 2008). Established *P. clarkii* populations are usually associated with lentic habitats in the city of São Paulo (Silva and Bueno 2005, Loureiro et al. 2015). Once a population is established, the drainage systems that connect lentic water bodies provide a network that facilitates the dispersal of the alien crawfish (Loureiro et al. 2015).

No occurrence of *P. clarkii* has been reported from high-altitude first-order stream in an Atlantic Forest remnant, such as the type locality of *A. jaragua*. In the JSP, the mountainous terrain prevents connections between the Pai Zé stream and the lake infested with *P. clarkii*, although these two water bodies are only a few hundred meters away from each other (Fig. 11.3). However, the mouth of the Pai Zé stream drains into a large artificial pond, which was built for recreational purposes for local visitors. So far, all surveys for *P. clarkii* in this artificial pond were fruitless

(SLS Bueno, unpublished data). The possibility of *P. clarkii* being accidentally introduced into this pond is potentially high, considering that this alien crawfish has been established in several locations in the city of São Paulo and outskirt areas (Loureiro et al. 2015). Since the size of the *A. jaragua* population shows decreases under current conditions within a 10-year period (see Section 4.3.2), we recommend that criteria A4e be maintained for assessing the conservation status of this species, given the potential threat that *P. clarkii* might present to the *A. jaragua* population.

Boos et al. (2020) suggested that *A. jaragua* should be assessed under criteria A4e only and that criteria B2a should be omitted, since it would not be needed to categorize the species as CR. We recommend, however, that the B2a criteria be maintained to properly assess the actual conservation status of the species, because criterion "B2" (AOO < 10 km²) used jointly with subcriterion "a" (known distribution limited to one location) clearly demonstrates the actual restricted geographical distribution (assessing the species as CR), while criteria A4e is an inferred or suspected potential impact caused by the invasive *P. clarkii*.

Neither *A. lancinhas* nor *A. rosanae* occur in protected areas and continuing habitat decline in their respective ranges has been reported as a result of human occupation or recreation (Santos et al. 2015, Moraes et al. 2016). The AOO of each species is estimated at less than 10 km². *Aegla lancinhas*, for which two locations have been reported, has been assessed as Endangered under the criteria B1–B2ab(iii) (Santos et al. 2015), whereas *A. rosanae* has been assessed as Critically Endangered, under criterion B2ab(iii), with one location reported (Moraes et al. 2016).

The known AOO of all critically endangered and highly endemic stygobitic species from the Ribeira de Iguape ecoregion are entirely within the boundaries of State Parks (Fig. 11.1) in the Atlantic Forest Biome. *Aegla leptochela* Bond-Buckup and Buckup, 1994, and *A. charon* Bueno and Moraes, 2017, are found at Intervales State Park, and *A. microphthalma* Bond-Buckup and Buckup, 1994 and *A. cavernicola* Türkay, 1972 are found at PETAR (Parque Estadual Turístico do Alto Ribeira). Fernandes et al. (2013) reported the occurrence of three new stygobitic aeglid species from PETAR that are still awaiting descriptions (Fig. 11.1).

It is clear that the survivorship of highly endemic species from the Atlantic Forest Biome is ultimately linked to preservation of headwaters biotic and abiotic components in their entirety. The Serra do Mar subregion is still under anthropogenic pressure, as human occupation and land exploration in the Atlantic Forest continues. With approximately 30% of its original extent remaining (Silva and Casteleti 2003), one can but guess how many endemic aeglid species have probably been lost from the Serra do Mar biogeographic subregion.

ACKNOWLEDGMENTS

We would like to express our gratitude to COTEC (Comisão Técnico-Científica), scientific committee of the Secretaria do Meio Ambiente, Instituto Florestal, for providing the license approval to our research team and students to work within the Jaraguá State Park conservation area. We are also very grateful to Angela Christine

Charity, Ana Teresa de Siqueira Bueno, and D. Christopher Rogers for critically reviewing the English text. Special thanks to all the undergraduate students for their help during the population-size estimations at Jaragua State Park (2011–2019), as part of the field work of the biennial course "Studies on populations of decapod crustaceans: an analytical approach": Adriana da Silva, Ana Carolina Buratto, Ana Terra Maraschin Irala, Carolina Rossi, Gisele Salgado Heckler, Larissa Caroline Meneghin Vieira, Naomi Nakao, Pedro Ivo Chiquetto Machado, Thiago da Silva, Ana Paula Brandão de Oliveira, Caio Costa, Camila Garcia Fernandes, Carolina Perozzi Guedes de Azevedo, Juliano Franco de Moraes, Erick Santos, Diego Hernandes, Kleber Mathubara Leite, Priscila Mourão de Almeida, Milena R Wolf, Victor Igor Aguiar Alarcon, Carlos Diego Neves Ananias, Elton Popp Antunes, Flávia Belarmino da Silva, Luiz Pimentel Mattos Neto, Carolina Moraes Martins de Barros, Jeniffer Kim, Daiane Marin de Souza, and Klaus Becker. We would like to thank FAPESP (Fundação de Amparo à Pesquisa do Estado de São Paulo) for providing financial support (Temático BIOTA – INTERCRUSTA Proc. 2018/13685-5) that will guarantee a follow-up field study on population size of critically endangered aeglids (including *Aegla jaragua*) for the next five years (2020–2024). This article conforms with the "Plano de Ação Nacional para a Conservação de Espécies de Peixes e Eglas Ameaçados de Extinção da Mata Atlântica (PAN-PEMA), issued by the Ministério do Meio Ambiente/Instituto Chico Mendes de Conservação da Biodiversidade (Portaria No. 370, issued 1 August 2019).

REFERENCES

Abell, R., Thieme, M.L., Revenga, C., et al. 2008. Freshwater ecoregions of the world: A new map of biogeographic units for freshwater biodiversity conservation. *BioScience* 58:403–14.

Adam, C.L., Marochi, M.A. and Masunari, S. 2018.Ontogenetic shape changes and sexual dimorphism in *Aegla marginata* Bond-Buckup and Buckup, 1994. *Anais da Academia Brasileira de Ciências* 90:1521–32.

Almeida, F.F.M., Neves, B.B.B. and Carneiro, C.D.R. 2000. The origin and evolution of the South American Platform. *Earth-Science Reviews* 50:77–111.

Ayres-Peres, L., Coutinho, C., Baumart, J.S., Gonçalves, A.S., Araújo, P.B. and Santos, S. 2011. Radio-telemetry techniques in the study of displacement of freshwater anomurans. *Nauplius* 19:41–54.

Bartholomei-Santos, M.L., Santos, S., Zimmermann, B.L., Pérez-Losada, M. and Crandall, K.A. 2020. Evolutionary history and phylogenetic relationships of Aeglidae. In *Aeglidae: Life History and Conservation Status of Unique Freshwater Anomuran Decapods*, eds. S. Santos, and S.L.S. Bueno, 1–27. Boca Raton, FL: CRC Press/Taylor & Francis Group.

Baumart, J.S., Cogo, G.B., Morales, F.E.C. and Santos, S. 2018. Population size of *Aegla longirostri* Bond-Buckup and Buckup, 1994 (Crustacea, Decapoda, Anomura): Comparison of methods with the mark-recapture technique in closed population. *Nauplius* 26:e2018016.

Bond-Buckup, G. and Buckup, L. 1994. A família Aeglidae (Crustacea, Decapoda, Anomura). *Arquivos de Zoologia* 32:159–346.

Bond-Buckup, G. and Buckup, L. 2000. *Aegla rosanae* Campos Jr., um novo sinônimo de *Aegla paulensis* Schmitt (Crustacea, Aeglidae). *Revista brasileira de Zoologia* 17:385–86.

Bond-Buckup, G., Bueno, A.P. and Keunecke, K.A. 1996. Primeiro estágio juvenil de *Aegla prado* Schmitt (Crustacea, Decapoda, Anomura, Aeglidae). *Revista brasileira de Zoologia* 13:1049–61.

Bond-Buckup, G., Bueno, A.P. and Keunecke, K.A. 1999. Morphological characteristics of juvenile specimens of *Aegla* (Decapoda, Anomura, Aeglidae). In *Crustacean and the Biodiversity Crisis: Proceedings of the Fourth International Crustacean Congress*, eds. F.R. Schram, and J.C. von Vaupel Klein, 371–81. Amsterdam: Koninklijke Brill NV.

Bond-Buckup, G., Jara, C.G., Buckup, L., Pérez-Losada, M., Bueno, A.A.P., Crandall, K. and Santos, S. 2010. New species and new records of endemic freshwater crabs from the Atlantic Forest in Southern Brazil (Anomura: Aeglidae). *Journal of Crustacean Biology* 30:495–502.

Boos, H., Salge, P.G. and Pinheiro, M.A.A. 2020. Conservation status and threats of Aeglidae: Beyond the assessment. In *Aeglidae: Life History and Conservation Status of Unique Freshwater Anomuran Decapods*, eds. S. Santos, and S.L.S. Bueno, 233–55. Boca Raton, FL: CRC Press/Taylor & Francis Group.

Buckup, L. and Rossi, A. 1977. O gênero *Aegla* no Rio Grande do Sul, Brasil (Crustacea, Decapoda, Anomura, Aeglidae). *Revista Brasileira de Biologia* 37:879–92.

Bueno, A.A.P. and Bond-Buckup, G. 1996. Os estágios juvenis iniciais de *Aegla violacea* Bond-Buckup and Buckup (Crustacea, Anomura, Aeglidae). *Nauplius* 4:39–47.

Bueno, A.A.P. and Bond-Buckup, G. 2000. Dinâmica populacional de *Aegla platensis* Schmitt (Crustacea, Decapoda, Aeglidae). *Revista Brasileira de Zoologia* 17:43–9.

Bueno, A.A.P. and Bond-Buckup, G. 2004. Natural diet of *Aegla platensis* Schmitt and *Aegla ligulata* Bond-Buckup and Buckup (Crustacea, Decapoda, Aeglidae) from Brazil. *Acta Limnologica Brasiliensia* 16:115–27.

Bueno, S.L.S, Camargo, A.L. and Moraes, J.C.B. 2017. A new species of stygobitic aeglid from lentic subterranean waters in southeastern Brazil, with an unusual morphological trait: Short pleopods in adult males. *Nauplius* 25:e201700021.

Bueno, S.L.S., Santos, S., Rocha, S.S., Gomes, K.M., Mossolin, E.C. and Mantelatto, F.L. 2016b. Avaliação dos Eglídeos (Decapoda: Aeglidae). In *Livro Vermelho dos crustáceos do Brasil: Avaliação 2010–2014*, eds. M. Pinheiro, and H. Boos, 35–63. Porto Alegre: Sociedade Brasileira de Carcinologia—SBC.

Bueno, S.L.S. and Shimizu, R.M. 2008. Reproductive biology and functional maturity in females of *Aegla franca* (Decapoda: Anomura: Aeglidae). *Journal of Crustacean Biology* 28:652–62.

Bueno, S.L.S. and Shimizu, R.M. 2009. Allometric growth, sexual maturity and adult chelae dimorphism in *Aegla franca* (Decapoda: Anomura: Aeglidae). *Journal of Crustacean Biology* 29:317–28.

Bueno, S.L.S., Shimizu, R.M. and Moraes, J.C.B. 2016a. A remarkable anomuran: The taxon *Aegla* Leach, 1820. Taxonomic remarks, distribution, biology, diversity and conservation. In *A Global Overview of the Conservation of Freshwater Decapod Crustaceans*, eds. T. Kawai, and N. Cumberlidge, 23–64. Switzerland: Springer.

Bueno, S.L.S., Shimizu, R.M. and Moraes, J.C.B. 2020. Postembryonic development, parental care, and recruitment. In *Aeglidae: Life History and Conservation Status of Unique Freshwater Anomuran Decapods*, eds. S. Santos, and S.L.S. Bueno, 155–179. Boca Raton, FL: CRC Press/Taylor & Francis Group.

Bueno, S.L.S., Shimizu, R.M. and Rocha, S.S. 2007. Estimating the population size of *Aegla franca* (Decapoda: Anomura: Aeglidae) by mark-recapture technique from an isolated section of Barro Preto stream, County of Claraval, State of Minas Gerais, Southeastern Brazil. *Journal of Crustacean Biology* 27:553–59.

Bueno, S.L.S., Takano, B.F., Cohen, F.P.A, Moraes, J.C.B., Chiquetto-Machado, P.I., Vieira, L.C.M. and Shimizu, R.M. 2014. Fluctuations in the population size of the highly endemic *Aegla perobae* (Decapoda, Anomura, Aeglidae) caused by a disturbance event. *Journal of Crustacean Biology* 34:165–73.

Câmara, I.G. 2003. Brief history of conservation in the Atlantic Forest. In *The Atlantic Forest of South America: Biodiversity Status, Threats, and Outlook*, eds. C. Galindo-Leal, and I.G. Câmara, 31–42. Washington, DC: Island Press.

Campos, Jr, O. 1998. Nova espécie do gênero *Aegla* da bacia do Rio Paraíba, Brasil (Anomura, Aeglidae). *Iheringia, Série Zoologia* 85:137–40.

Castro, R.M.C. 1999. Evolução da ictiofauna de riachos sul-americanos: padrões gerais e possíveis processos causais. In *Ecologia de riachos*, eds. E.P. Caramaschi, R. Mazzoni, R., and P.R. Peres-Neto, 139–55. Rio de Janeiro: Série Oecologia Brasiliensis.

Castro-Souza, T. and Bond-Buckup, G. 2004. O nicho trófico de duas espécies simpátricas de *Aegla* Leach (Crustacea, Aeglidae) no tributário da bacia hidrográfica do Rio Pelotas, Rio Grande do Sul, Brasil. *Revista Brasileira de Zoologia* 21:805–13.

Cogo, G.B. and S. Santos. 2013. The role of aeglids in shredding organic matter in neotropical streams. *Journal of Crustacean Biology* 33:519–26.

Cohen, F.P.A, Takano, B.F., Shimizu, R.M. and Bueno, S.L.S. 2011. Life cycle and population structure of *Aegla paulensis* (Decapoda: Anomura: Aeglidae). *Journal of Crustacean Biology* 31:389–95.

Cohen, F.P.A, Takano, B.F., Shimizu, R.M. and Bueno, S.L.S. 2013. Population size of *Aegla paulensis* (Decapoda: Anomura: Aeglidae). *Latin American Journal of Aquatic Research* 41:746–52.

Colpo, K.D., Ribeiro, L.C., Wesz, B. and Ribeiro, L.O. 2012. Feeding preference of the South American endemic anomuran *Aegla platensis* (Decapoda, Anomura, Aeglidae). *Naturwissenschaften* 99:333–36.

Copatti, C.E., Machado, J.V.V. and Trevisan, A. 2015. Morphological variation in the sexual maturity of three sympatric aeglids in a river in Southern Brazil. *Journal of Crustacean Biology* 35:59–67.

Correia, A.M. and Anastácio, P.M. 2008. Shifts in aquatic macroinvertebrate biodiversity associated with the presence and size of an alien crayfish. *Ecological Research* 23:729–34.

Cruz, M.J., Segurado, P., Sousa, M. and Rebelo, R. 2008. Collapse of the amphibian community of the Paul do Boquilobo Natural Reserve (central Portugal) after the arrival of the exotic American crayfish *Procambarus clarkii*. *Herpetological Journal* 18:197–204.

Da Silva, A.R., Paciencia, G.P., Bispo, P.C. and Castilho, A.L. 2017. Allometry and sexual dimorphism of the Neotropical freshwater anomuran *Aegla marginata* Bond-Buckup and Buckup, 1994 (Crustacea, Anomura, Aeglidae). *Nauplius* 25:e2017016.

Dalosto, M.M., Palaoro, A.V., Oliveira, D., Samuelsson, E. and Santos, S. 2014. Population biology of *Aegla platensis* (Decapoda: Anomura: Aeglidae) in a tributary of the Uruguay River, state of Rio Grande do Sul, Brazil. *Zoologia* 31:215–22.

Feldmann, R.M. 1984. *Haumuriaegla glaessneri* n. gen. and sp. (Decapoda; Anomura; Aeglidae) from Haumurian (late Cretaceous) rocks near Cheviot, New Zealand. *New Zealand Journal of Geology and Geophysics* 27:379–85.

Feldmann, R.M. 1986. Paleogeography of two decapods taxa in the Southern Hemisphere; global conclusions with sparse data. In *Crustacean Biogeography, Crustacean Issues*, eds. R.H. Gore, and K.L. Keck, 5–19. Rotterdam: A.A. Balkema.

Feldmann, R.M., Vega, F.J., Applegate, S.P. and Bishop, G.A. 1998. Early Cretaceous arthropods from the Tlayúa formation at Tepexi de Rodríguez, Puebla, México. *Journal of Paleontology* 72:79–90.

Fernandes, C.S., Bueno, S.L.S. and Bichuette, M.E. 2013. Distribution of cave-dwelling *Aegla* spp. (Decapoda: Anomura: Aeglidae) from the Alto Ribeira karstic area in southeastern Brazil based on geomorphological evidence. *Journal of Crustacean Biology* 33:567–75.

Francisco, D.A., Bueno, S.L.S. and Kihara, T.C. 2007. Description of the first juvenile of *Aegla franca* Schmitt, 1942 (Crustacea, Decapoda, Aeglidae). *Zootaxa* 1509:17–30.

Galindo-Leal, C. and Câmara, I.G. 2003. Atlantic Forest hotspot status: an overview. In *The Atlantic Forest of South America: Biodiversity Status, Threats, and Outlook*, eds. C. Galindo-Leal, and I.G. Câmara, 3–11. Washington, DC: Island Press.

Gherardi, F. 2007. Understanding the impact of invasive crayfish. In *Biological Invaders in Inland Waters: Profiles, Distribution, and Threats*, ed. F. Gherardi, 507–41. Dordrecht: Springer.

Gonçalves, A.S., Costa, G.C., Bond-Buckup, G., Bartholomei-Santos, M.L. and Santos, S. 2018. Priority areas for conservation within four freshwater ecoregions in South America: A scale perspective based on freshwater crabs (Anomura, Aeglidae). *Aquatic Conservation: Marine and Freshwater Ecosystems* 28:1077–88.

Grabowski, R.C., Santos, S. and Castilho, A.L. 2013. Reproductive ecology and size of sexual maturity in the anomuran crab *Aegla parana* (Decapoda: Aeglidae). *Journal of Crustacean Biology* 33:1–7.

Groves, C.R., Jensen, D.B., Valutis, L.L., et al. 2002. Planning for biodiversity conservation: Putting conservation science into practice. *BioScience* 52:499–512.

Hebling, N.J. and Rodrigues, W. 1977. Sobre uma nova espécie brasileira do gênero *Aegla* Leach, 1820 (Decapoda, Anomura). *Papéis Avulsos de Zoologia* 30:289–94.

Hobbs III, H.H. 1978. A new species of the endemic South American genus *Aegla* from Paraná, Brazil (Crustacea, Anomura, Aeglidae). *Proceedings of the Biological Society of Washington* 91:982–88.

IUCN (International Union for Conservation of Nature). 2013. Guidelines for Using the IUCN Red List Categories and Criteria. Version 10. Prepared by the Standards and Petitions Subcommittee. http://www.iucnredlist.org/documents/RedListGuidelines.pdf.

Jara, C.G., F.H. Rudolph and Gonzáles E.R. 2006. Estado de conocimiento de los malacostraceos dulceacuicolas de Chile. *Gayana* 70:40–9.

Krebs, C.J. 1999. *Ecological Methodology*. 2nd Edition. Menlo Park: Benjamin/ Cummings.

Lipcius, R.N. and Van Engel, W. 1990. Blue crab population dynamics in Chesapeake Bay: Variation in abundance (York River, 1972–1988) and stock-recruit functions. *Bulletin of Marine Science* 46:180–94.

López-Greco, L.S., Viau, V., Lavolpe, M., Bond-Buckup, G. and Rodríguez, E.M. 2004. Juvenile hatching and maternal care in *Aegla uruguayana* (Anomura, Aeglidae). *Journal of Crustacean Biology* 24:309–13.

Loureiro, T.G., Anastácio, P.M.S.G. Bueno, S.L.S. and Araujo, P.B. 2018. Management of invasive populations of the freshwater crayfish *Procambarus clarkii* (Decapoda, Cambaridae): Test of a population-control method and proposal of a standard monitoring approach. *Environmental Monitoring and Assessment* 190:559.

Loureiro, T.G., Anastácio, P.M.S.G., Bueno, S.L.S., Araujo, P.B., Souty-Grosset, C. and Almerão, M.P. 2015. Distribution, introduction pathway, and invasion risk analysis of the North American crayfish *Procambarus clarkii* (Decapoda: Cambaridae) in southeastern Brazil. *Journal of Crustacean Biology* 35:88–96.

Lundberg, J.G., Marshall, L.G., Guerrero, J., Horton, B., Malabarba, M. C.S.L. and Wesselingh, F. 1998. The stage for Neotropical fish diversification: A history of tropical South American rivers. In *Phylogeny and Classification of Neotropical Fishes. Part 1—Fossils and Geological Evidence*, ed. L.R. Malabarba, 13–48. Porto Alegre: Edipucrs.

Magni, S.T. and Py-Daniel, V. 1989. *Aegla platensis* Schmitt, 1942 (Decapoda: Anomura) um predador de imaturos de Simuliidae (Diptera: Culicomorpha). *Revista da Saúde Pública* 23:258–259.

Maia, K.P., Bueno, S.L.S. and Trajano, E. 2013. Ecologia populacional e conservação de eglídeos (Crustacea: Decapoda: Aeglidae) em cavernas da área cárstica do Alto Ribeira, em São Paulo. *Revista da Biologia* 10:40–45.

Marçal, I.C., Ioshimura, L.M., Rosa, J.J.S. and Teixeira, G.M. 2018. Population structure and sexual maturity of *Aegla castro* (Decapoda, Anomura), an endemic freshwater crab from Brazil. *Invertebrate Reproduction and Development* 62:35–42.

McNamara, J.C. and Faria, S.C. 2020. Physiological ecology: Osmoregulation and metabolism of the aeglid anomurans. In *Aeglidae: Life History and Conservation Status of Unique Freshwater Anomuran Decapods*, eds. S. Santos, and S.L.S. Bueno, 203–31. Boca Raton, FL: CRC Press/Taylor & Francis Group.

Moraes, J.C.B. and Bueno, S.L.S. 2013. Description of the newly-hatched juvenile of *Aegla paulensis* (Decapoda, Anomura, Aeglidae). *Zootaxa* 3635:501–19.

Moraes, J.C.B. and Bueno, S.L.S. 2015. Description of the newly-hatched juvenile of *Aegla perobae* (Crustacea: Decapoda: Aeglidae). *Zootaxa* 3973:491–510.

Moraes, J.C.B., Tavares, M. and Bueno, S.L.S. 2017. Taxonomic review of *Aegla marginata* Bond-Buckup and Buckup, 1994 (Decapoda, Anomura, Aeglidae) with description of a new species. *Zootaxa* 4323:519–33.

Moraes, J.C.B., Terossi, M., Buranelli, R.C., Tavares, M., Mantelatto, F.L. and Bueno, S.L.S. 2016. Morphological and molecular data reveal the cryptic diversity among populations of *Aegla paulensis* (Decapoda, Anomura, Aeglidae), with descriptions of four new species and comments on dispersal routes and conservation status. *Zootaxa* 4193:1–48.

Müller, F. 1876. *Aeglea odebrechtii* n. sp. *Jenaische Zeitschrift für Naturwissenschaft* 10:13–24.

Oliveira, D. and Santos, S. 2011. Maturidade sexual morfológica de *Aegla platensis* (Crustacea, Decapoda, Anomura) no Lajeado Bonito, norte do estado do Rio Grande do Sul, Brasil. *Iheringia* 101:127–30.

Oyanedel, A., Valdovinos, C., Sandoval, N., Moya, C., Kiessling, G., Salvo, J. and Olmos, V. 2011. The southernmost freshwater anomurans of the world: Geographic distribution and new records of Patagonian aeglids (Decapoda: Aeglidae). *Journal of Crustacean Biology* 31:396–400.

Páez, F.P., Marçal, I.C., Souza-Shibatta, L., Gregati, A.G., Sofia, S.K. and Teixeira, G.M. 2018. A new species of *Aegla* Leach, 1820 (Crustacea, Anomura) from the Iguaçu River basin, Brazil. *Zootaxa* 4527:335–46.

Pérez-Losada, M., Bond-Buckup, G., Jara, C.G. and Crandall, K.A. 2004. Molecular systematics and biogeography of the southern South American freshwater "crabs" *Aegla* (Decapoda: Anomura: Aeglidae) using multiple heuristic tree search approaches. *Systematic Biology* 53:767–80.

Pérez-Losada, M., Bond-Buckup, G., Jara, C.G. and Crandall, K.A. 2009. Conservation assessment of southern South American freshwater ecoregions on the basis of the distribution and genetic diversity of crabs from the genus *Aegla*. *Conservation Biology* 23:692–702.

Potter, P.E. 1997. The Mesozoic and Cenozoic paleodrainage of South America: A natural history. *Journal of South American Earth Sciences* 10:331–44.

Renai, B. and Gherardi, F. 2004. Predatory efficiency of crayfish: Comparison between indigenous and non-indigenous species. *Biological Invasions* 6:89–99.

Resh, V.H., Bêche, L.A., Lawrence, J.E. et al. 2013. Long-term population and community patterns of benthic macroinvertebrates and fishes in Northern California Mediterranean-climate streams. *Hydrobiologia* 719:93–118.

Ribeiro, A.C. 2006. Tectonic history and the biogeography of the freshwater fishes from the coastal drainages of eastern Brazil: An example of faunal evolution associated with a divergent continental margin. *Neotropical Ichthyology* 4:225–46.

Rocha, S.S., Shimizu, R.M. and Bueno, S.L.S. 2010. Reproductive biology in females of *Aegla strinatii* (Decapoda: Anomura: Aeglidae). *Journal of Crustacean Biology* 30:589–96.

Rodrigues, W. and Hebling, N.J. 1978. Estudos biológicos em *Aegla perobae* Hebling and Rodrigues, 1877 (Decapoda, Anomura). *Revista Brasileira de Biologia* 38:383–90.

Rodríguez, C.F., Bécares, E., Fernandez-Aláez, M. and Fernandez-Aláez, C. 2005. Loss of diversity and degradation of wetlands as a result of introducing exotic crayfish. *Biological Invasions* 7:75–85.

Santos, S., Ayres-Peres, L., Cardoso, R.C.F. and Sokolowicz, C.C. 2008. Natural diet of the freshwater anomuran *Aegla longirostri* (Crustacea, Anomura, Aeglidae). *Journal of National History* 42:1027–37.

Santos, S., Bond-Buckup, G., Buckup, L., Bartholomei-Santos, M.L., Pérez-Losada, M., Jara, C.G. and Crandall, K.A. 2015. Three new species of Aeglidae (*Aegla* Leach, 1820) from Paraná state, Brazil. *Journal of Crustacean Biology* 35:839–49.

Santos, S., Bond-Buckup, G., Gonçalves, A.S., Bartholomei-Santos, M.L., Buckup, L. and Jara, C.G. 2017. Diversity and conservation status of *Aegla* spp. (Anomura, Aeglidae): an update. *Nauplius* 25:e2017011.

Santos, S., Bond-Buckup, G., Buckup, L., Pérez-Losada, M., Finley, M. and Crandall, K.A. 2012. Three new species of *Aegla* (Anomura) freshwater crabs from the upper Uruguay river hydrographic basin in Brazil. *Journal of Crustacean Biology* 32:529–40.

Schmitt, W.L. 1942. The species of *Aegla*, endemic South American fresh-water crustaceans. *Proceedings of the United States National Museum* 91:431–524.

Seber, G.A.F. 1982. *The Estimation of Animal Abundance and Related Parameters*. Second edition. London: Charles Griffin and Co. Ltd.

Shimizu, R.M. and Bueno, S.L.S. 2020. Sampling and data analysis for population studies on the life history of *Aegla* spp. In *Aeglidae: History and Conservation Status of Unique Freshwater Anomuran Decapods*, eds. S. Santos, and Bueno, 257–77. Boca Raton, FL: CRC Press/Taylor & Francis Group.

Silva, H.L.M. and Bueno, S.L.S. 2005. Population size estimation of the exotic crayfish *Procambarus clarkii* (Girard) (Crustacea, Decapoda, Cambaridae) in the Alfredo Volpi City Park, São Paulo, Brazil. *Revista Brasileira de Zoologia* 22:93–98.

Silva, J.M.C. and Casteleti, C.H. 2003. Status of the biodiversity of the Atlantic forest of Brazil. In *The Atlantic Forest of South America: Biodiversity Status, Threats, and outlook*, eds. C. Galindo-Leal, and I.G. Câmara, 43–59. Washington, DC: Island Press.

Silva, L.S.A., Guerrero-Ocampo, C.M., Negreiros-Fransozo, M.L. and Teixeira, G.M. 2017. Description of the newly-hatched juvenile of *Aegla castro* Schmitt, 1942 (Crustacea, Anomura, Aeglidae). *Zootaxa* 4237:167–80.

Sokolowicz, C.C., Ayres-Peres, L. and Santos, S. 2007. Atividade nictimeral e tempo de digestão de *Aegla longirostri* (Crustacea, Decapoda, Anomura). *Iheringia Série Zoologia* 97:235–38.

Swiech-Ayoub, B.P. and Masunari, S. 2001. Biologia reprodutiva de *Aegla castro* Schmitt (Crustacea, Anomura, Aeglidae) no Buraco do Padre, Ponta Grossa, Paraná, Brasil. *Revista Brasileira de Zoologia* 18:1019–30.

Takano, B.F., Cohen, F.P.A., Fransozo, A., Shimizu, R.M. and Bueno, S.L.S. 2016. Allometric growth, sexual maturity and reproductive cycle of *Aegla castro* (Decapoda: Anomura: Aeglidae) from Itatinga, state of São Paulo, southeastern Brazil. *Nauplius* 24:e2016010.

Teodósio, E.A.F.M.O. and S. Masunari. 2007. Description of first two juvenile stages of *Aegla schmitti* Hobbs III, 1979 (Anomura: Aeglidae). *Nauplius* 15:73–80.

Trevisan, A., Marochi, M.Z., Costa, M., Santos, S. and Masunari, S. 2014. Effects of the evolution of the Serra do Mar mountains on the shape of the geographically isolated populations of *Aegla schmitti* Hobbs III, 1979 (Decapoda: Anomura). *Acta Zoologica* 97:34–41.

Trevisan, A. and Masunari, S. 2010. Geographical distribution of *Aegla schmitti* Hobbs III, 1979 (Decapoda Anomura Aeglidae) and morphometric variations in male populations from Paraná State, Brazil. *Nauplius* 18:45–55.

Trevisan, A. and Santos, S. 2012. Morphological, sexual maturity, sexual dimorphism and heterochely in *Aegla manuinflata* (Anomura). *Journal of Crustacean Biology* 32:519–27.

Tumini, G., Giri, F., Williner, V., Collins, P. and Morrone, J.J. 2019. Selecting and ranking areas for conservation of *Aegla* (Crustacea: Decapoda: Anomura) in southern South America integrating biogeography, phylogeny and assessments of extinction risk. *Aquatic Conservation: Marine and Freshwater Ecosystems* 29:693–705.

Türkay, M. 1972. Neue Höhlendekapoden aus Brasilien (Crustacea). *Revue Suisse de Zoologie* 79:415–18.

Vannote, R.L., Minshall, G.W., Cummins, K.W., Sedell, J.R. and Cushing, C.E. 1980. The river continuum concept. *Canadian Journal of Fisheries and Aquatic Sciences* 37:130–37.

Vogt, G. 2016. Direct development and posthatching brood care as key features of the evolution of freshwater Decapoda and challenges for conservation. In *A Global Overview of the Conservation of Freshwater Decapod Crustaceans*, eds. T. Kawai, and N. Cumberlidge, 169–98. Switzerland: Springer.

Zar, J.H. 1996. *Biostatistical Analysis*. 3rd Edition. Upper Saddle River: Prentice Hall.

CHAPTER **12**

New Insights on Biodiversity and Conservation of Amphidromous Shrimps of the Indo-Pacific islands (Decapoda: Atyidae: *Caridina*)

V. de Mazancourt, W. Klotz, G. Marquet, B. Mos, D.C. Rogers, and P. Keith

CONTENTS

12.1 INTRODUCTION

Atyid shrimps (Crustacea: Decapoda) are essential components of tropical freshwater ecosystems, playing a role of cleaner by shredding fallen leaves or filtering organic particular matter and being preys to a number of organisms (Covich et al. 1999, Crowl et al. 2001). Among them, the genus *Caridina* H. Milne Edwards, 1837, is particularly diversified, comprising more than 300 described species, making it the most diversified of the infra-order Caridea (De Grave et al. 2015). In Indo-Pacific Islands, the majority of *Caridina* species have an original diadromous lifestyle, with a planktonic marine larval phase and a benthic freshwater adult phase (McDowall 2007, Bauer 2013). This strategy allows the species to colonize isolated habitats, such as the rivers of volcanic islands, and to survive the instability of these environments (floods, droughts, volcanism, etc.) by keeping a stock of larvae in the ocean, ready to recolonize depopulated rivers (Keith et al. 2010). Amphidromy, by isolating populations, has also contributed to establishing the high diversity that exists among caridean shrimps. This diversity, however, is the source of an extreme taxonomic confusion that impedes the monitoring of the species and the establishment of appropriate conservation programs (Klotz and Rintelen 2014). With the development of new molecular biology techniques, integrative taxonomy is increasingly used by systematists to achieve more precise species delineations by combining morphological data with genetic, ecological, and/or biogeographical information (Mazancourt et al. 2017).

12.2 BIODIVERSITY

12.2.1 Taxonomy

There is substantial confusion surrounding the taxonomy of many *Caridina* species resulting in the creation of several species complexes (we consider a species complex a monophyletic species group that share a common taxonomic history in having been synonymized at one point, and/or that can be identified by morphological characters). Indo-Pacific island amphidromous shrimps have been grouped into seven main complexes: (1) *Caridina nilotica* (P. Roux, 1833) complex (Johnson 1963, Holthuis 1978, Jalihal et al. 1984, Choy 1991, Richard and Clark 2005, Karge and Klotz 2007), (2) *C. weberi* De Man, 1892 complex (Richard and Chandran 1994, Cai and Shokita 2006a), (3) *C. gracilirostris* De Man, 1892 complex (Cai and Ng 2001, 2007), (4) *C. typus* H. Milne Edwards, 1837 complex (Karge and Klotz 2007, Bernardes et al. 2017), (5) *C. serratirostris* De Man, 1892 complex (Cai and Shokita 2006b), (6) *C. brevicarpalis* De Man, 1892 complex (Short 2009), and (7) *C. propinqua* De Man, 1908 complex (W. Klotz, unpublished). Little work has been done to differentiate among the species clustered within these complexes, possibly due to the difficulty and costs associated with genetic analyses and the highly similar morphology of shrimps from different populations.

Egg size is often a good indicator of amphidromous species. Small eggs indicate that the species has an indirect development with several planktonic stages that are often marine. Conversely, species with large eggs are most likely landlocked, with direct development limiting their dispersal abilities, meaning they are often endemic. However, some species can have small eggs and still be endemic – such as *C. longicarpus* and *C. meridionalis*, both endemic to New Caledonia (Marquet et al. 2003, Mazancourt et al. 2018) or *C. futunensis*, endemic to Futuna Island (Mazancourt et al. 2019a). This can be due to particular conditions that limit the dispersal of marine larvae. For example, oceanic currents or physical barriers (e.g., closed coral lagoons) may prevent dispersal to nearby islands even if larvae are capable of surviving in oceanic conditions. It is not unusual therefore for different types of larval development to occur within one species group. Moreover, in some cases, like in *C. meridionalis* (see Mazancourt et al. 2018) or *C. gracilipes* (W. Klotz, unpublished), egg size varies within a single species, depending on the habitat, with generally larger eggs found in upstream localities and smaller eggs near the estuary, suggesting the possibility of facultative amphidromy.

12.2.2 Phylogeny

Mazancourt et al. (2019b) produced a molecular phylogeny of two species complexes from the Indo-Pacific islands: the *C. nilotica* and *C. weberi* complexes. This demonstrated the importance of an integrative taxonomic approach in understanding *Caridina* diversity. Nonintegrated approaches are likely to underestimate the number of species for a given locality, with many cryptic/pseudocryptic species confused under a single name, or more rarely, morphotypes of a single variable species described as separate species. As a result, geographical ranges and ecological data concerning poorly delineated species are often misleading, impeding the establishment of programs that provide sufficient protection and management. Mazancourt et al. (2019b) also found that habitat was often the most important factor leading to the separation of complexes. Species belonging to different complexes live in different habitat types. Identification to species level (or at least species complex level) and monitoring of *Caridina* diversity over time may therefore provide an early warning system that provides information about changes in their environment.

12.2.3 Habitat

Habitat is a critical factor influencing *Caridina* distribution and it can often be used as a species identification tool (Mazancourt et al. 2019b). *Caridina* inhabit different types of freshwater ecosystems, divided between lentic environments, such as lakes (No. 7 in Figs. 12.1 and, 12.2L) or swamps (No. 8 in Figs. 12.1 and 12.2K), and lotic (flowing) environments in surface as well as subterranean waters (No. 9 in Fig. 12.1). Amphidromous shrimps are mainly found in rivers or lentic environments connected to the sea.

Tropical mountainous islands typical of the Indo-Pacific are subject to important natural disturbances like droughts, floods, volcanic eruptions, or earthquakes. River flow can therefore be subject to great temporal variations (strong during floods or zero flow in temporary waterways) as well as spatial variation (changes in slope, substrate, or plant cover). River morphology, for example, is important in determining environmental condition variability. This variability is often unique to a specific watershed, forming isolated, discontinuous units by the sea and geological formations, such as ridgelines. River length depends on island morphology, ranging from short streams a few meters in length to large rivers of several kilometers. Examples of the great river length diversity in the Indo-Pacific include the Diahot River in New Caledonia and the Sigatoka River in Fiji.

Keith et al. (2010) proposed an altitudinal zonation scheme for lotic habitats, divided into five zones with flexible boundaries, depending on the island studied to better characterize variable environmental conditions experienced by shrimp. The five zones are the spring zone (No. 1 in Figs. 12.1 and 12.2I), the higher course (No.

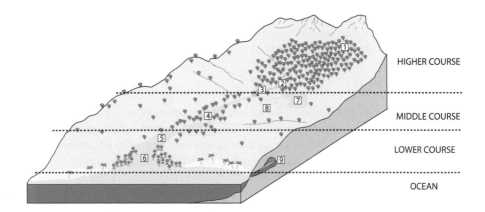

Figure 12.1   Schematic representation of different freshwater habitats. 1: spring zone; 2: higher course; 3: waterfall; 4: middle course; 5: lower course; 6: estuary; 7: lake; 8: swamp; 9: subterranean waters.

2 in Figs. 12.1 and 12.2A and B), the middle course (No. 4 in Fig. 12.1 and 12.2C and D), the low part of the stream (No. 5 in Figs. 12.1 and 12.2E and F), and the lower course (No. 6 in Figs. 12.1 and 12.2G and H). An additional sub-zonation exists, depending on current velocity, distinguishing calm zones (Figs. 12.2C and E), or pools (Fig. 12.2B) from riffles (Figs. 12.2A, D, and F). Usually the altitudinal zonation is further simplified into three functional levels (Fig. 12.1):

- The higher course, characterized by a steep slope (generally >10%), a fast current, and a boulder substrate.
- The middle course, characterized by an intermediate slope (<10%) and a substrate of pebbles and rocks or sand in slow flow areas. This level is typically separated from the higher course by a topographical feature, like a cascade (No. 3 in Figs. 12.1 and 12.2J).
- The lower course, characterized by a very low or zero slope, with a slow current and a pebble substrate with a progressive size decrease when approaching the estuary. The estuary is the lowest end of the river where environmental conditions are influenced by the tidal flux, as distinguished from the purely limnic area. This chemical boundary between seawater and freshwater is an essential habitat for some amphidromous species. The estuary size can vary, from very broad and long for large rivers (like the Jordan River in Santo, Vanuatu) to absent in some small streams flowing directly from the mountains to the sea.

Species living in rivers are subject to a range of biotic and abiotic conditions, depending on the river zone where they live. Differences in biotic and abiotic conditions often form gradients that apply over the entire length of rivers (Table 12.1):

- A current velocity gradient mainly linked to elevation. Water currents are typically fastest at high altitudes and slowest in the estuary, although local variations at a given altitude are also important, depending on the configuration of the river (calm zones vs. riffles).

Figure 12.2   Examples of habitat types. (A) Higher course riffle, lotic mode (River Poitete, Kolombangara Island, Solomon Islands, credit PK). (B) Higher course pool, lentic mode (Ciu Waterfall, New Caledonia, credit VM). (C) Middle course, lentic mode (River Wénou, New Caledonia, credit VM). (D) Middle course riffle, lotic mode (River Nekouri, New Caledonia, credit VM). (E) Lower course, lentic mode (River Hienghène, New Caledonia, credit VM). (F) Lower course riffle, lotic mode (River Negropo, New Caledonia, credit VM). (G) Mangrove estuary (River Tanghène, New Caledonia, credit VM). (H) Estuary, low tide (Nera estuary, New Caledonia, credit VM). (I) Mountain stream (Unnamed tributary of the river Ciit, New Caledonia, credit VM). (J) Waterfall (Colnett waterfall, New Caledonia, credit VM). (K) Swamp (Plaine des Lacs, New Caledonia, credit VM). (L) Lake (Lac en Huit, New Caledonia, credit VM).

Figure 12.2   Continued.

- A temperature and oxygen gradient, with cool and oxygen-rich waters typical of the higher course and warm and low-oxygen waters typical of the lower course.
- A width and depth gradient, from small rivulets in the spring zone to large rivers near the mouth.
- A salinity gradient that is typically limited to the lower course. Salinity tends to decline rapidly outside of the areas influenced by tidal flux.
- A sediment-size gradient, from bedrock in the spring zone to boulders and rocks in the higher course, pebbles in the middle course, and sand and silt in the lower course.
- A particulate organic matter gradient, with clear, oligotrophic waters in the spring area to turbid, eutrophic waters in the estuary.
- A riparian vegetation gradient, with riverbanks consisting of naked rocks in the higher course switching to dense riparian vegetation in the lower course.
- A predation gradient, with the greatest density and diversity of predatory species (birds, fish) present in the lower course. Predators of shrimps in the higher course are mostly limited to eels (*Anguilla* spp.) or insects (dragonfly larvae, etc.).

Gradients in abiotic and biotic factors overlap in many combinations to create a considerable variety of microhabitats (Gehrke et al. 2011) to which *Caridina* species have adapted. We can observe a species vertical zonation, depending on the variation of these factors linked to elevation, as well as horizontal variables linked to the river configuration, according to lotic or lentic facies in the area or microhabitats (sunken

Table 12.1 Variation of Environmental Gradients (Abiotic and Biotic) along the River Course

Variable	Low altitude	High altitude
Current		
Temperature		
Oxygen		
Width and depth		
Salinity		
Sediment size		
Organic matter		
Vegetation		
Predation		

wood, roots, aquatic macrophytes, rocks, etc.). Indeed, different species assemblages are observed at different levels of the rivers and in different flowing facies. The lower course supports species of the *C. nilotica* species complex (Mazancourt et al. 2018), the *C. gracilirostris* species complex, or species allied to *C. serratirostris* and *C. brevicarpalis*. Higher in the river, two species groups are found: *C. weberi* complex in lotic areas and *C. nilotica* complex in lentic areas. Species of the *C. typus* group seem to have a broader range of habitats, present virtually the entire river length (de Mazancourt pers. obs.).

As observed for fish in tropical rivers by Pouilly et al. (2006) and Lorion et al. (2011) in continental systems and by Keith et al. (2015) in insular systems, *Caridina* demonstrates an altitudinal diversity gradient with maximum diversity in the lower course, decreasing with elevation. This diversity gradient is explained by the greater concentration of organic particulate matter in the lower reaches, allowing establishment of higher biomass (Angermeier and Karr 1983). Alternatively, diversification may be promoted by higher habitat heterogeneity (Gorman and Karr 1978) and stronger predator pressure in the lower course (Deacon et al. 2018). The lower course proximity to the ocean also presents the advantage of reducing the distance that dispersing larvae and returning juveniles have to travel. In contrast, resources are more limited but predator pressure is weaker at higher altitudes, which often leads to low biodiversity but high abundance (Leberer and Nelson 2001). Some species living in these environments have adopted a life cycle completed entirely in freshwater, with reduced larval stages or even direct development, avoiding migration hazards (Hancock 1998).

It is important to note that diversity patterns across zones can sometimes be more nuanced than the zonation framework may suggest. Diversity patterns can be dominated by a single parameter, such as the nature of the substrate. In New Caledonia, ultramafic rocks of the south of the island leach high concentrations of metals (nickel, chromium, cobalt) into the water of the rivers. These metals are toxic for most species, but some species have evolved to tolerate these high levels of dissolved metals (Marquet et al. 2003). These rivers thus exhibit a different fauna than rivers on sedimentary substrate, with different altitudinal zonations.

12.2.4 Biogeography

Caridina can be found in virtually all tropical Indo-Pacific and African rivers, ranging from West Africa (Richard and Clark 2009), Egypt (Hussein and Obuid-Allah 1990, Richard and Clark 2005, Mazancourt et al. 2018), and South Africa (Mirimin et al. 2015) to mainland Japan (Saito et al. 2012), Polynesia (Keith et al. 2013), and South Australia (Davie 2002), extending to the Middle East (Christodoulou et al. 2016) (Fig. 12.3).

Regional differences in *Caridina* diversity exist in the Indo-Pacific as expected for such a wide-ranging genus. Focusing on endemic *Caridina* species (Fig. 12.4A), the maximum diversity occurs around the Coral Triangle (Veron et al. 2009), between the Philippines, Indonesia, and Papua-New Guinea. The endemic species number tends to be substantially lower in rivers further away from this area, although large continental islands such as Madagascar or Sri Lanka, and archipelagos such as the Solomon Islands and Sulawesi, constitute local endemism hotspots (De Grave et al. 2008, 2015, Klotz and Rintelen 2013, Cumberlidge et al. 2017). Islands of relatively old geological origin like New Caledonia, Fiji, or Mauritius also appear to harbor a high diversity of endemic species. In contrast, younger and often isolated islands like those of Polynesia or Micronesia harbor fewer endemic species. Localities at the northern and southern limits of the distribution area (Japan and South Australia, for example) show low endemism.

Regional patterns in Indo-Pacific *Caridina* diversity are explained by island age, size, and latitude. Diversity tends to be high on large continental islands close to the equator and low on small young volcanic islands as well as in areas away from the equator (Paulay 1994, Willig et al. 2003). Geologically old islands have had more opportunity to be colonized and had time for colonizers to adapt and specialize. Chen and He (2009) showed that the number of island speciations, and thus their biodiversity, increases with time, whereas the colonization rate decreases. According

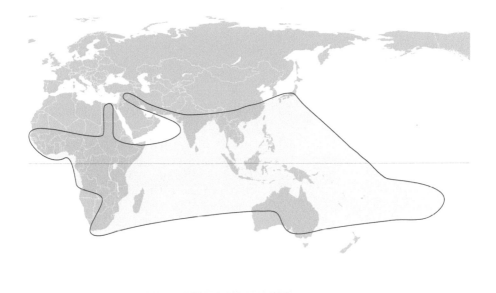

Figure 12.3   Distribution area of *Caridina*.

to the same study, the number of endemic species correlates to the size of the island, due to greater habitat diversity (geological or climatic differences, for example), which agrees with our observations for large islands such as Madagascar, Sri Lanka, and New Caledonia.

Patterns in endemic diversity (Fig. 12.4C) vary considerably from the patterns in amphidromous species diversity (Fig. 12.4A). The greatest amphidromous species diversity is in the Solomon Islands and Vanuatu. This may also include the Coral Triangle, as *Caridina* diversity from the area is poorly known. The high *Caridina* diversity in this region is explained by the complex geological history of the area, which lies at the convergence of three tectonic plates. The region contains islands of continental and volcanic origin, as well as historical connections to Papua in the north.

The lowest amphidromous species diversity is found in Polynesia and at the borders of the *Caridina* distribution area in general (Fig. 12.4C). Important areas of endemism, such as Madagascar, are not diversity hotspots for amphidromous species. Localities with low endemism, like the Polynesian or Micronesian islands, often have a higher proportion of amphidromous species.

Amphidromous species diversity patterns may be explained by the recent colonization of the islands by taxa with great dispersal abilities, allowing them to colonize without becoming reproductively isolated. Small islands often exhibit unstable conditions in which endemic landlocked populations could not maintain themselves. In contrast, amphidromous species can repopulate islands following disturbance as

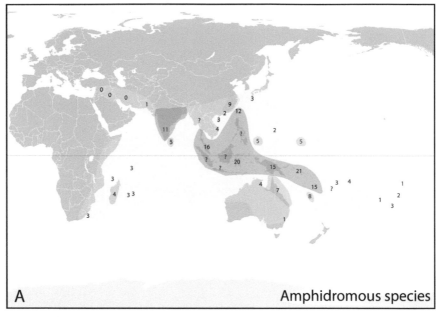

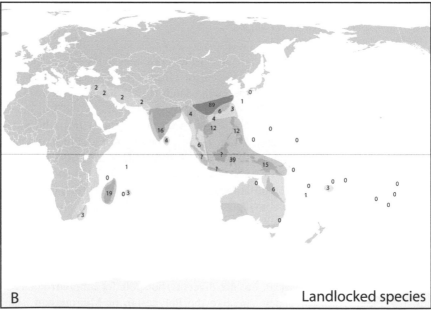

Figure 12.4 Heat maps representing the species richness by locality. (A) Numbers of
amphidromous species (eggs < 0.8 mm). (B) Numbers of landlocked species
(eggs > 0.8 mm).

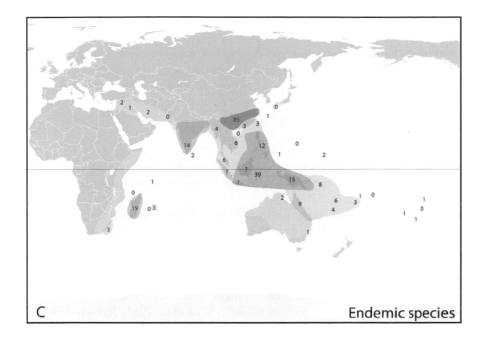

Figure 12.4   (C) Numbers of endemic species (known from a single locality/area).

they maintain a stock of larvae in the ocean, ready to recolonize depopulated habitats (McDowall 2007).

The landlocked or amphidromous lifestyle of a species is not always linked to it being endemic or widespread; amphidromous species are known from a single island. The diversity patterns of landlocked species are shown in Fig. 12.4B. The highest diversity of landlocked species globally is found in the Coral Triangle with a diminution by distance, until they disappear at the distribution area borders. However, the difference between landlocked and endemic or amphidromous species is that landlocked species are largely absent from volcanic islands such as Polynesia, Micronesia, Solomon Islands, or Vanuatu, where the majority of species are amphidromous. In contrast, landlocked species are often the majority of species in continents (India, China, Australia, South Africa) or large continental islands (Madagascar). This suggests that habitats of landlocked species have been available for sufficient time for new species to evolve a landlocked lifestyle from amphidromous ancestors, or for species to diverge from landlocked ancestors that were already present.

12.3 CONSERVATION

Taxonomic confusion has been a substantial obstacle for the establishment of programs aimed at *Caridina* conservation. Some ill-delineated species like *C. longirostris* or *C. weberi* were thought to be widespread. However, studies using an integrative taxonomic approach (Mazancourt et al. 2018) showed that these "species" are actually comprised of several species with restricted distributions (some being endemic). Furthermore, a better knowledge of systematics allows greater understanding of these species habitat and ecology, since each species is adapted to a different environment. These new data will impact the way in which the conservation status of these species will be assessed. Species with small distributions and specific habitat needs are more vulnerable to extinction than widespread and tolerant species (Purvis et al. 2000).

12.3.1 Threats to Amphidromous *Caridina*

The main threats faced by *Caridina* almost always originate from anthropogenic factors (Table 12.2).

Habitat degradation. Many anthropic activities induce habitat degradation. The direct impacts of habitat degradation occur through pollution of the rivers by pesticides (Kumar et al. 2010a,b), industrial accidents (Dominique 2014), illegal poisoning (Greathouse et al. 2005), use of the rivers as dumps (Keith et al. 2006), eutrophication by fertilizers, poisoning and causing the proliferation of algae clogging rivers, or increased sedimentation from mining or construction (dams, roads, etc.), and increasing water turbidity (Boseto et al. 2007, Keith et al. 2013). Habitat destruction also occurs through the water extraction for agriculture, industry, or tourism, causing physical changes (e.g., lower flow rates) and the disappearance of lotic zones (Keith et al. 2013) that are the exclusive habitat of some species (e.g., *Caridina weberi* species complex). The cause and effects of habitat destruction can also be associated with human activities that do not impact rivers directly but have flow effects that impact shrimp habitats and threaten the viability of populations (i.e., indirect effects). For example, the disappearance of perennial rivers and associated loss of habitat for shrimp due to deforestation in Rodrigues Island (Keith et al. 2006) in Madagascar for rosewood, in Anjouan (Mirhani et al. 2014) for firewood, and Ylang-ylang plantations in the Marquesas islands and Rapa for goats and livestock, or in New Caledonia due to bushfires. As some species of amphidromous shrimps live in mangroves (like those of the *C. gracilirostris* complex), they are threatened during both their larval and adult stages by the degradation of their habitat by aquaculture. Together, the impacts of aquatic and terrestrial habitat destruction affect the physiology, reproduction, and migration of caridean shrimps and remain one of the most important threats for carideans and other freshwater species (De Grave et al. 2015).

Reduced connectivity. The construction of instream barriers suppresses habitat connectivity (Dudgeon 2000, March et al. 2003). The construction of dams, such as the Yaté hydroelectric plant in New Caledonia, are instream barriers that result in the rapid disappearance of amphidromous species across the upstream watershed. Indeed, shrimps need to be able to migrate along the river course, depending on water level in order to not get trapped during the dry season. For amphidromous species, it is critical to have a continuum between the adult habitat and the sea in order to maintain populations.

Introduced species. The introduction of exotic species can lead to local extinctions of indigenous species due to predation, competition, habitat alteration, or

Table 12.2   Potential Direct and Indirect Effects of Climate Change on Amphidromous and Endemic (Freshwater) Caridean Shrimp in the Indo-Pacific

	Amphidromous species	Endemic species (limited to freshwater)	Status of knowledge	References
Direct effects				
Reduction in pH of water due to increased CO_2 dissolution from atmosphere (ocean acidification)	• Unknown	• Unknown (likely species-specific)	Crustaceans appear to be highly robust to increased CO_2 levels and/or decreased pH	Wittmann and Portner (2013)Kroeker et al. (2013)
Increased mean water temperature	• Effects likely to be species-specific • Small changes are likely to increase growth rates, reduce time to reproductive maturity, and decrease overall size • Large changes may exceed thermal tolerances, with consequences for the maintenance of populations • Changes in species distributions as species move to new habitat • Changes in phenology (timing of natural events, such as reproduction, migration, etc.)	• Effects likely to be species specific • Small changes are likely to increase growth rates, reduce time to reproductive maturity, and decrease overall size • Large changes are likely to exceed thermal tolerances, with the potential to reduce population size due to limited options for migration to new habitat • Changes in phenology (timing of natural events, such as reproduction, migration, etc.)	Little research on effects for estuarine or freshwater crustaceans. Effects on marine crustaceans are well studied	Kroeker et al. (2013)
Lower dissolved oxygen levels due to increased temperatures	• Unlikely to be important given strong turbulence promoting dissolution of oxygen in upper catchment, except following weather events that cause high levels of organic matter to be deposited in waterways.	• Unlikely to be important given strong turbulence promoting dissolution of oxygen in upper catchment, except following weather events that cause high levels of organic matter to be deposited in waterways.	Little research on effects for estuarine or freshwater crustaceans.	Ficke et al. (2007)

(Continued)

Table 12.2 (Continued)   Potential Direct and Indirect Effects of Climate Change on Amphidromous and Endemic (Freshwater) Caridean Shrimp in the Indo-Pacific

	Amphidromous species	Endemic species (limited to freshwater)	Status of knowledge	References
Hypoxic zones	• May be encountered in estuaries or in oceans during the larval stage • Adults and juveniles only likely to encounter hypoxic zones following weather events that cause high levels of organic matter to be deposited in waterways	• Unlikely to be encountered except following weather events that cause high levels of organic matter to be deposited in waterways	Research on the effects for crustaceans focused in marine systems	Vaquer-Sunyer and Duarte (2008)Breitburg et al. (2018)
Stronger rainfall events, increased likelihood of stronger and more frequent storms and cyclones	• Induce modifications in habitats (e.g., more depth, more flow) • Exacerbate existing problems (e.g., land clearing, turbidity). • Increased likelihood of pollution events (e.g., sewage overflow) • Changes in phenology (timing of natural events, such as reproduction, migration, etc.)	• Induce modifications in habitats (e.g., more depth, more flow) • Exacerbate existing problems (e.g., land clearing, turbidity) • Increased likelihood of pollution events (e.g., sewage overflow) • Changes in phenology (timing of natural events, such as reproduction, migration, etc.)	Substantial number of studies modeling changes in the probability of adverse weather events	Gehrke et al. (2011)Lough et al. (2011)
Changes in ocean currents	• Changes in dispersal of larvae • Interruptions to population connectivity • Potential to isolate populations	• No direct effect	Substantial number of studies on modeling changes in larval dispersal for marine invertebrates, no studies for *Caridina*	Keith et al. (2015)

(Continued)

Table 12.2 (Continued) Potential Direct and Indirect Effects of Climate Change on Amphidromous and Endemic (Freshwater) Caridean Shrimp in the Indo-Pacific

	Amphidromous species	Endemic species (limited to freshwater)	Status of knowledge	References
Changes in multi-year weather patterns (e.g., El Niño)	• Likely to exacerbate the effects of adverse weather events (see above) • Limited rainfall causing reduction in available habitat	• Likely to exacerbate effects of adverse weather events (see above) • Limited rainfall causing reduction in available habitat	Climate change is closely linked with changes in multi-year weather patterns	Cai et al. (2018)
Rising sea levels	• Likely to increase the amount of estuarine habitat available • Influx of seawater into freshwater systems, reducing available habitat. Most severe for small islands • Increased barriers to migration due to engineered structures designed to prevent storm surges and combat sea level rise	• Reduction in potential habitat due to influx of seawater into freshwater habitat • Potential disappearance of freshwater habitat on islands inundated by sea level rise	Many studies modeling projected seawater inundation, but it is not clear the extent to which this will reduce the availability of freshwater habitats and/or increase the availability of estuarine habitats	Mcleod et al. (2010)
Effects of climate and nonclimate stressors combined	• Additive, synergistic, and antagonistic effects possible, dependent on the stressors combined	• Additive, synergistic, and antagonistic effects possible, dependent on the stressors combined	The effects of combined stressors in freshwater and estuarine systems remain understudied	Staudt et al. (2013)Jackson et al. (2016)
Indirect effects of climate change				
Movement of species	• New ecological interactions • Exposure to new pathogens and parasites	• New ecological interactions • Exposure to new pathogens and parasites	Movement of species into naive freshwater habitats has been documented. Little known about the ecological effects	Comte and Grenouillet (2013)Pecl et al. (2017)Mos et al. (2017)

(Continued)

Table 12.2 (Continued) Potential Direct and Indirect Effects of Climate Change on Amphidromous and Endemic (Freshwater) Caridean Shrimp in the Indo-Pacific

	Amphidromous species	Endemic species (limited to freshwater)	Status of knowledge	References
Changes in the abundance of ocean plankton (linked to changes in ocean currents and stronger rainfalls)	• Changes in food availability for larvae • Potential for changes in the timing of reproduction to match phytoplankton production • Changes in the success of larval migration to freshwater habitat due to reduced larval energy reserves • Potential for the larval stage to be a bottleneck, limiting recruitment to some populations	• No direct effects	No studies available for amphidromous species. Limited research for marine larvae	Hays et al. (2005)Beaugrand and Kirby (2018)
Changes in the diversity of ocean plankton (linked to changes in ocean currents and stronger rainfalls)	• Changes in the quality of food for larvae • Changes in the success of larval migration to freshwater habitat due to reduced larval energy reserves • Potential for the larval stage to be a bottleneck, limiting recruitment to some populations	• No direct effects	No studies available	Rosenblatt and Schmitz (2016)
Changes in carbon (C) uptake by algae and plants	• Alter C to N ratios in food • Potential changes in diet to seek out N-rich foods	• Alter C to N ratios in food • Potential changes in diet to seek out N-rich foods	The importance of C to N ratios is well documented across a variety of taxa	Rastetter et al. (1992)Sterner and Elser (2002)

disease transmission (Keith 2002a, b, De Grave et al. 2015). Another important but little studied threat associated with introduced species is the potential for hybridization with nonnative caridean species. The ease with which caridean shrimp hybridize is well noted in the hobby aquarium literature. There have been no reports of hybridization of introduced and native carideans in the wild, but the extinction of local populations due to hybridization following translocation within a catchment has occurred in other atyid genera (Hughes et al. 2003, Fawcett et al. 2010).

Overharvesting. Shrimps are harvested for human consumption. In the Philippines, Madagascar, Indonesia, India, and China, caridean shrimp are caught in large quantities using nets to be eaten fresh, dried, or salted, and to feed animals or used as fertilizer (Holthuis 1980), or in Reunion Island where they are called "chevaquines" (GM, pers. comm.). Another threat for *Caridina* spp. is their harvest for the aquarium trade, particularly for the colorful species living in Sulawesi (De Grave et al. 2015) or "Bee shrimps" from mainland China, some species being known only from a single stream of about 150 m in length (WK pers. comm.). While most species impacted are lacustrine landlocked, in the future growing global demand could see the expansion of the collection of amphidromous species. Currently, only a few amphidromous species appear in the trade, but whether these are collected legally, poached, or bred in captivity is difficult to determine due to long and complex supply chains. It is unclear how many shrimps are traded globally, but there is substantial anecdotal evidence suggesting local populations experience severe declines due to overcollection, particularly where a local color or pattern variant becomes highly sought-after in the trade. For example, Klotz and Lukhaup (2014) reported the disappearance of *Caridina trifascata* Yan and Cai, 2003, from locales near Zuhai, China, over a period of less than 12 months, likely associated with overharvesting for the aquarium trade.

Climate change. Climate change is likely already impacting and will continue to impact caridean shrimp through a variety of direct and indirect mechanisms (Table 12.1). Unfortunately, while crustaceans generally appear to be robust to some changes in their environment associated with climate change (e.g., ocean acidification), little research has been done to address the gaps in our knowledge about indirect effects of climate stressors, and the effects of interactions of climate stressors (e.g., temperature) with nonclimate stressors (e.g., pesticide pollution). This is concerning given studies that have examined the indirect and interactive effects of climate stressors on marine taxa have highlighted how these effects may be more important than the direct effects of climate stressors (e.g., Kroeker et al. 2012, Boyd and Brown 2015, Rosenblatt and Schmitz 2016, Kamya et al. 2017).

12.3.2 Conclusions and Recommendations

Integrative study of *Caridina* led to the discovery of unsuspected diversity in Indo-Pacific Islands, with recognized species complexes that present differences in habitat use. Therefore, when making river freshwater fauna inventories, if it is not possible to identify specimens to species level, the species complex can be sufficient to provide useful data. Keys and diagnoses of the different species complexes are detailed in another publication (Mazancourt *et al.* 2020). Now that species are better delineated through integrative taxonomy, their distribution (including micro- and macrohabitats) can be more informative and their specific conservation status (threats, IUCN status, etc.) needs to be reassessed.

Our recommendations to protect these amphidromous *Caridina* species revolve around three main points:

1. Taking Biological and Ecological Specificities into Consideration

Arguably, the best way to enable conservation is by changing people's mindset directly through education and example. Managing caridean species requires resource managers to have a minimum knowledge of their biology (i.e., amphidromous life cycle) and to take these specificities into consideration when making decisions. Understanding amphidromous species life cycles highlights the importance of maintaining the natural flow of waterways. Maintaining population connectivity through natural water flows supports a range of life stages. This limits local or regional extinction potential as breeding populations depend on the larval pool for recruitment (and conversely, the larval pool depends on the breeding populations). Successful implementation of conservation measures designed to maintain and enhance natural water flows may be best achieved through the inclusion of all project stakeholders (e.g., farmers, fishers, water management authorities, local government, community conservation groups, international conservation organizations, etc.).

2. Limiting Anthropic Impacts on Amphidromous Species

The most efficient way to protect amphidromous species is to limit anthropic impacts. This requires an integrated management approach, which accounts for both terrestrial and aquatic impacts. For example, water catchments should always have a minimum flow in line with their natural seasonal variations to support the upstream juvenile migration, which is often triggered by greater freshwater pulses to the sea during floods. Therefore, management decisions effects, such as not altering water flow, designing roads and dams that maintain connectivity between sea and mountains, maintaining healthy riparian corridors, and preserving estuaries need to be considered in an integrated manner.

3. Establishing priorities for species conservation

The priority must be to protect *Caridina* habitats, as conservation status is not known for many species. Identification of those habitats most in need of special protection, with management and implementation of the recommendations from the previous points would be the best starting point. Caridean shrimps seem to be good indicators of water and habitat quality. They may be a very useful tool for resource managers as indicators of impacts in tropical aquatic systems. Preventing exotic species introductions is another important recommendation. Finally, supporting research on the biology and ecology of these species would help to improve our knowledge and assess the threats they face.

REFERENCES

Angermeier, P.L. and Karr, J.R. 1983. Fish communities along environmental gradients in a system of tropical streams. *Environmental Biology of Fishes* 9:117–35.

Bauer, R.T. 2013. Amphidromy in shrimps: A life cycle between rivers and the sea. *Latin American Journal of Aquatic Research* 41:633–50.

Beaugrand, G. and Kirby, R.R. 2018. How do marine pelagic species respond to climate change? Theories and observations. *Annual Review of Marine Science* 10:169–97.

Bernardes, S.C., Pepato, A.R., Rintelen (von), T., Rintelen (von), K., Page, T.J., Freitag, H. and de Bruyn, M. 2017. The complex evolutionary history and phylogeography of *Caridina typus* (Crustacea: Decapoda): Long-distance dispersal and cryptic allopatric species. *Scientific Reports* 7:9044.

Boseto, D., Morrison, C., Pikacha, P. and Pitakia, T. 2007. Biodiversity and conservation of freshwater fishes in selected rivers on Choiseul Island, Solomon Islands. *The South Pacific Journal of Natural Science* 3:16–21.

Boyd, P.W. and Brown, C.J. 2015. Modes of interactions between environmental drivers and marine biota. *Frontiers in Marine Sciences* 2:1–7.

Breitburg, D., Levin, L.A., Oschlies, A., Grégoire, M., Chavez, F.P., Conley, D.J., et al. 2018. Declining oxygen in the global ocean and coastal waters. *Science* 359:eaam7240.

Cai, W., Wang, G., Dewitte, B., Wu, L., Santoso, A., Takahashi, K., Yang, Y., Carréric, A. and McPhaden, M.J. 2018. Increased variability of eastern Pacific El Niño under greenhouse warming. *Nature* 564:201–06.

Cai, Y. and Ng, P.K.L. 2001. The freshwater decapod crustaceans of Halmahera, Indonesia. *Journal of Crustacean Biology* 21:665–95.

Cai, Y. and Ng, P K.L. 2007. A revision of the *Caridina gracilirostris* De Man, 1892, species group, with descriptions of two new taxa (Decapoda; Caridea; Atyidae). *Journal of Natural History* 41:1585–602.

Cai, Y. and Shokita, S. 2006a. Atyid shrimps (Crustacea: Decapoda: Caridea) of the Ryukyu Islands, southern Japan, with descriptions of two new species. *Journal of Natural History* 40:2123–72.

Cai, Y. and Shokita, S. 2006b. Report on a collection of freshwater shrimps (Crustacea: Decapoda: Caridea) from the Philippines, with descriptions of four new species. *The Raffles Bulletin of Zoology* 54:245–70.

Chen, X.-Y. and He, F. 2009. Speciation and endemism under the model of island biogeography. *Ecology* 90:39–45.

Choy, S.C. 1991. The atyid shrimps of Fiji with description of a new species. *Zoologische Mededelingen* 65:343–62.

Christodoulou, M., Anastasiadou, C., Jugovic, J. and Tzomos, T. 2016. Freshwater shrimps (Atyidae, Palaemonidae, Typhlocarididae) in the broader Mediterranean region: Distribution, life strategies, threats, conservation challenges and taxonomic sssues. In *A Global Overview of the Conservation of Freshwater Decapod Crustaceans*, eds. T. Kawai, and N. Cumberlidge, 199–236. Switzerland: Springer.

Comte, L. and Grenouillet, G. 2013. Do stream fish track climate change? Assessing distribution shifts in recent decades. *Ecography* 36:1236–46.

Covich, A.P., Palmer, M.A. and Crowl, T.A. 1999. The role of benthic invertebrate species in freshwater ecosystems. *BioScience* 49:119–28.

Crowl, T.A., McDowell, W.H., Covich, A.P. and Johnson, S.L. 2001. Freshwater shrimp effects on detrital processing and nutrients in a tropical headwater stream. *Ecology* 82:775–83.

Cumberlidge, N., Rasamy Razanabolana, J., Ranaivoson, C.H., Randrianasolo, H.H. and Sayer, M. 2017. Updated extinction risk assessments of Madagascar's freshwater decapod crustaceans reveal fewer threatened species but more Data Deficient species. *Malagasy Nature* 12:32–41.

Davie, P.J.F. 2002. Crustacea: Malacostraca: Phyllocarida, Hoplocarida, Eucarida (Part 1). In *Zoological Catalogue of Australia*, eds. A. Wells, and W.W.K. Houston, 1–551. Australia, Melbourne: CSIRO Publishing.

De Grave, S., Cai, Y. and Anker, A. 2008. Global diversity of shrimps (Crustacea: Decapoda: Caridea) in freshwater. *Freshwater Animal Diversity Assessment* 595:287–93.

De Grave, S., Smith, K.G., Adeler, N.A., Allen, D.J., Alvarez, F., Anker, A., et al. 2015. Dead Shrimp Blues: A global assessment of extinction risk in freshwater shrimps (Crustacea: Decapoda: Caridea). *PLos One* 10:e0120198.

de Mazancourt, V., Boseto, D., Marquet, G., & Keith, P. (2020). Solomon's Gold Mine: Description or redescription of 24 species of Caridina (Crustacea: Decapoda: Atyidae) freshwater shrimps from the Solomon Islands, including 11 new species. *European Journal of Taxonomy* (696): 1–86. https://doi.org/10.5852/ejt.2020.696

Deacon, A.E., Jones, F.A.M. and Magurran, A.E. 2018. Gradients in predation risk in a tropical river system. *Current Zoology* 64:213–21.

Dominique, Y. 2014. *Constat de pollution suite à un déversement d'effluent industriel au sein du creek de la Baie Nord*. Nouméa, New Caledonia: Oeil.

Dudgeon, D. 2000. Large-scale hydrological changes in tropical Asia: Prospects for riverine biodiversity: The construction of large dams will have an impact on the biodiversity of tropical Asian rivers and their associated wetlands. *BioScience* 50:793–806.

Fawcett, J.H., Hurwood, D.A. and Hughes, J.M. 2010. Consequences of a translocation between two divergent lineages of the *Paratya australiensis* (Decapoda: Atyidae) complex: Reproductive success and relative fitness. *Journal of the North American Benthological Society* 29:1170–80.

Ficke, A.D., Myrick, C.A. and Hansen, L.J. 2007. Potential impacts of global climate change on freshwater fisheries. *Reviews in Fish Biology and Fisheries* 17:581–613.

Gehrke, P.C., Sheaves, M.J., Boseto, D.T., Figa, B.S. and Wani, J. 2011. Chapter 7: Vulnerability of freshwater and estuarine fish habitats in the tropical Pacific to climate change. In *Vulnerability of Tropical Pacific Fisheries and Aquaculture to Climate Change*, eds. J.D. Bell, J.E. Johnson, and A.J. Hobday, 369–431. Auckland, New Zealand: Secretariat of the Pacific Community.

Gorman, O.T. and Karr, J.R. 1978. Habitat structure and stream fish communities. *Ecology* 59:507–15.

Greathouse, E.A., March, J.G. and Pringle, C.M. 2005. Recovery of a tropical stream after a harvest-related chlorine poisoning event. *Freshwater Biology* 50:603–15.

Hancock, M.A. 1998. The relationship between egg size and embryonic and larval development in the freshwater shrimp *Paratya australiensis* Kemp (Decapoda: Atyidae). *Freshwater Biology* 39:715–23.

Hays, G.C., Richardson, A.J. and Robinson, C. 2005. Climate change and marine plankton. *Trends in Ecology and Evolution* 20:337–44.

Holthuis, L.B. 1978. A collection of Decapod Crustacea from Sumba, Lesser Sunda islands, Indonesia. *Zoologische Verhandelingen* 162:1–55.

Holthuis, L.B. 1980. *FAO Species Catalogue. Volume 1-Shrimps and Prawns of the World (An Annotated Catalogue of Species of Interest to Fisheries Vol. 1)*. Rome, Italy: Food and Agriculture Organization of the United Nations.

Hughes, J.M., Goudkamp, K., Hurwood, D.A., Hancock, M.A. and Bunn, S. 2003. Translocation causes extinction of a local population of the freshwater shrimp *Paratya australiensis*. *Conservation Biology* 17:1007–12.

Hussein, M.A. and Obuid-Allah, A.H. 1990. External morphology of the freshwater prawn *Caridina africana* collected from Egypt. *Zoology in the Middle East* 4:71–84.

Jackson, M.C., Loewen, C.J., Vinebrooke, R.D. and Chimimba, C.T. 2016. Net effects of multiple stressors in freshwater ecosystems: A meta-analysis. *Global Change Biology* 22:180–89.

Jalihal, D.R., Shenoy, S. and Sankolli, K.N. 1984. Five new species of freshwater atyid shrimps of the genus *Caridina* H. Milne Edwards from Dharwar area (Karnataka State, India). In *Records of the Zoological Survey of India*. Miscellaneous Publication, Occasional Paper 69:1–40. West Bengal, India: Zoological Survey of India.

Johnson, D.S. 1963. Distributional and other notes on some freshwater prawns (Atyidae and Palaemonidae) mainly from the Indo-West Pacific region. *Bulletin of the National Museum of Singapore* 32:5–30.

Kamya, P.Z., Byrne, M., Mos, B., Hall, L. and Dworjanyn, S.A. 2017. Indirect effects of ocean acidification drive feeding and growth of juvenile crown-of-thorns starfish *Acanthaster planci*. *Proceedings of the Royal Society of London Series B: Biological Sciences* 284: 20170778.

Karge, A. and Klotz, W. 2007. *Süßwassergarnelen aus aller Welt*. Ettlingen, Germany: Dähne Verlag.

Keith, P. 2002a. Freshwater fish and decapod crustacean populations on Réunion Island, with an assessment of species introductions. *Bulletin Français de la Pêche et de la Pisciculture* 364:97–107.

Keith, P. 2002b. Introduction of freshwater fishes and decapod crustaceans in French Polynesia, a review. *Bulletin Français de la Pêche et de la Pisciculture* 364:147–60.

Keith, P., Lord, C. and Maeda, K. 2015. *Indo-Pacific Sicydiine Gobies: Biodiversity, Life Traits and Conservation*. Paris: Société Française d'Ichtyologie.

Keith, P., Marquet, G., Gerbeaux, P., Vigneux, E. and Lord, C. 2013. *Poissons et crustacés d'eau douce de Polynésie*. Paris: Société Française d'Ichtyologie.

Keith, P., Marquet, G., Lord, C., Kalfatak, D. and Vigneux, E. 2010. *Vanuatu Freshwater Fish and Crustaceans*. Paris: Société Française d'Ichtyologie.

Keith, P., Marquet, G., Valade, P., Bosc, P. and Vigneux, E. 2006. *Atlas des poissons et des crustacés d'eau douce des Comores, Mascareignes et Seychelles (Collection Patrimoines naturels Vol. 65)*. Paris, France: Muséum national d'Histoire naturelle.

Klotz, W. and Lukhaup, C. 2014. *Breeders'n'Keepers Wildshrimp China Special*. Vinningen, Germany: Dennerle GmbH.

Klotz, W. and Rintelen (von), K. 2013. Three new species of *Caridina* (Decapoda: Atyidae) from Central Sulawesi and Buton Island, Indonesia, and a checklist of the islands' endemic species. *Zootaxa* 3664:554–70.

Klotz, W. and Rintelen (von), T. 2014. To "bee" or not to be - on some ornamental shrimp from Guangdong Province, Southern China and Hong Kong SAR, with descriptions of three new species. *Zootaxa* 3889:151–84.

Kroeker, K.J., Kordas, R.L., Crim, R., Hendriks, I.E., Ramajo, L., Singh, G.S., Duarte, C.M. and Gattuso, J.-P. 2013. Impacts of ocean acidification on marine organisms: Quantifying sensitivities and interaction with warming. *Global Change Biology* 19:1884–96.

Kroeker, K.J., Micheli, F. and Gambi, M.C. 2012. Ocean acidification causes ecosystem shifts via altered competitive interactions. *Nature Climate Change* 3:156–59.

Kumar, A., Correll, R., C., Grocke, S. and Bajet, C. 2010a. Toxicity of selected pesticides to freshwater shrimp, *Paratya australiensis* (Decapoda: Atyidae): Use of time series acute toxicity data to predict chronic lethality. *Ecotoxicology and Environmental Safety* 73:360–69.

Kumar, A., Doan, H., Barnes, M., Chapman, J.C. and Kookana, R.S. 2010b. Response and recovery of acetylcholinesterase activity in freshwater shrimp, *Paratya australiensis* (Decapoda: Atyidae) exposed to selected anti-cholinesterase insecticides. *Ecotoxicology and Environmental Safety* 73:1503–10.

Leberer, T. and Nelson, S.G. 2001. Factors affecting the distribution of atyid shrimps in two tropical insular rivers. *Pacific Science* 55:389–98.

Lorion, C.M., Kennedy, B.P. and Braatne, J.H. 2011. Altitudinal gradients in stream fish diversity and the prevalence of diadromy in the Sixaola River basin, Costa Rica. *Environmental Biology of Fishes* 91:487–99.

Lough, J.M., Meehl, G.A. and Salinger, M.J. 2011. Chapter 2: Observed and projected changes in surface climate of the tropical Pacific. In *Vulnerability of Tropical Pacific Fisheries and Aquaculture to Climate Change*, eds. J.D. Bell, J.E. Johnson, and A.J. Hobday, 49–99. Auckland, New Zealand: Secretariat of the Pacific Community.

March, J.G., Benstead, J.P., Pringle, C.M. and Scatena, F.N. 2003. Damming tropical islands streams: Problems, solutions and alternatives. *BioScience* 53:1069–78.

Marquet, G., Keith, P. and Vigneux, E. 2003. *Atlas des poissons et des crustacés d'eau douce de Nouvelle-Calédonie (Collection Patrimoines naturels Vol. 58)*. Paris, France: Muséum national d'Histoire naturelle.

Mazancourt (de), V., Klotz, W., Marquet, G. and Keith, P. 2018. Integrative taxonomy helps separate four species of freshwater shrimps commonly overlooked as *Caridina longirostris* (Crustacea: Decapoda: Atyidae) in Indo-West Pacific islands. *Invertebrate Systematics* 32:1422–47.

Mazancourt (de), V., Klotz, W., Marquet, G. and Keith, P. 2019a. Revision of freshwater shrimps belonging to *Caridina weberi* complex (Crustacea: Decapoda: Atyidae) from Polynesia with discussion on their biogeography. *Journal of Natural History* 53:815–847.

Mazancourt (de), V., Klotz, W., Marquet, G., Mos, B., Rogers, D.C. and Keith, P. 2019b. The complex study of complexes: The first well-supported phylogeny of two species complexes within genus *Caridina* (Decapoda: Caridea: Atyidae) sheds light on evolution, biogeography, and habitat. *Molecular Phylogenetics and Evolution* 131:164–80.

Mazancourt (de), V., Marquet, G., Klotz, W., Keith, P. and Castelin, M. 2017. When morphology and molecules work together: Lines of evidence for the validity of *Caridina buehleri* Roux, 1934 (Crustacea: Decapoda: Atyidae) and *Caridina gueryi* Marquet, Keith and Kalfatak, 2009 as its junior synonym. *Invertebrate Systematics* 31:220–30.

McDowall, R. 2007. On amphidromy, a distinct form of diadromy in aquatic organisms. *Fish and Fisheries* 8:1–13.

Mcleod, E., Poulter, B., Hinkel, J., Reyes, E. and Salm, R. 2010. Sea-level rise impact models and environmental conservation: A review of models and their applications. *Ocean and Coastal Management* 53:507–17.

Mirhani, N., Taïbi, A.N., Ballouche, A. and Razakamanana, T. 2014. De la problématique de l'eau au modèle numérique d'aménagement en milieu tropical humide insulaire : le bassin versant d'Ouzini-Ajaho (Anjouan - Comores). In *Eau, milieux et aménagement. Une recherche au service des territoires*, ed. A. Ballouche, 117–134. France: Presses Universitaires d'Angers.

Mirimin, L., Kitchin, N., Impson, D.N., Clark, P.F., Richard, J., Daniels, S.R. and Roodt-Wilding, R. 2015. Genetic and morphological characterization of freshwater shrimps (*Caridina africana* Kingsley, 1882) reveals the presence of alien shrimps in the Cape Floristic Region, South Africa. *Journal of Heredity* 106:711–18.

Mos, B., Ahyong, S.T., Burnes, C.N., Davie, P.J.F. and McCormack, R.B. 2017. Range extension of a euryhaline crab, *Varuna litterata* (Fabricius, 1798) (Brachyura: Varunidae), in a climate change hot-spot. *Journal of Crustacean Biology* 37:258–62.

Paulay, G. 1994. Biodiversity on oceanic islands: Its origin and extinction. *American Zoologist* 34:134–44.

Pecl, G. T., Araújo, M.B., Bell, J.D., Blanchard, J., Bonebrake, T.C., Chen, I.C., et al. 2017. Biodiversity redistribution under climate change: Impacts on ecosystems and human well-being. *Science* 355: eaai9214.

Pouilly, M., Barrera, S. and Rosales, C. 2006. Changes of taxonomic structure of fish assemblages along an environmental gradient in the Upper Beni watershed (Bolivia). *Journal of Fish Biology* 68:137–56.

Purvis, A., Gittleman, J.L., Cowlishaw, G. and Mace, G.M. 2000. Predicting extinction risk in declining species. *Proceedings of the Royal Society of London Series B: Biological Sciences* 267:1947–52.

Rastetter, E.B., McKane, R.B., Shaver, G.R. and Melillo, J.M. 1992. Changes in C storage by terrestrial ecosystems: How CN interactions restrict responses to CO_2 and temperature. In *Natural sinks of CO_2*, eds. J. Wisniewski, and A.E. Lugo, 327–44. Palmas Del Mar, Puerto Rico: Springer.

Richard, J. and Chandran, M.R. 1994. A systematic report on the fresh water prawns of the atyid genus *Caridina* H. Milne Edwards 1837, from Madras (Tamilnadu: India). *Journal of the Bombay Natural History Society* 91:241–59.

Richard, J. and Clark, P.F. 2005. *Caridina nilotica* (P. Roux, 1833) (Crustacea: Decapoda: Caridea: Atyidae) from East Africa, with descriptions of four new species. *Proceedings of the Biological Society of Washington* 118:706–30.

Richard, J. and Clark, P.F. 2009. African *Caridina* (Crustacea: Decapoda: Caridea: Atyidae): redescriptions of *C. africana* Kingsley, 1882, *C. togoensis* Hilgendorf, 1893, *C. natalensis* Bouvier, 1925 and *C. roubaudi* Bouvier, 1925 with descriptions of 14 new species. *Zootaxa* 1995:1–75.

Rosenblatt, A.E. and Schmitz, O.J. 2016. Climate change, nutrition and bottom-up and top-down food web processes. *Trends in Ecology and Evolution* 31:965–75.

Saito, M., Yamashiro, T., Hamano, T. and Nakata, K. 2012. Factors affecting distribution of freshwater shrimps and prawns in the Hiwasa River, southern Japan. *Crustacean Research* 41:27–46.

Short, J.W. 2009. *Freshwater Crustacea of the Mimika Region, New Guinea*. Kuala Kencana, Timika: PT Freeport Indonesia.

Staudt, A., Leidner, A.K., Howard, J., Brauman, K.A., Dukes, J.S., Hansen, L.J., Paukert, C., Sabo, J. and Solorzano, L.A. 2013. The added complications of climate change: Understanding and managing biodiversity and ecosystems. *Frontiers in Ecology and the Environment* 11:494–501.

Sterner, R.S. and Elser, J.J. 2002. *Ecological Stoichiometry: The Biology of Elements from Molecules to the Biosphere*. Princeton, NJ, USA: Princeton University Press.

Vaquer-Sunyer, R. and Duarte, C.M. 2008. Thresholds of hypoxia for marine biodiversity. *Proceeding of the National Academy of Sciences* 105:15452–57.

Veron, J.E.N., Devantier, L.M., Turak, E., Green, A.L., Kininmonth, S., Stafford-Smith, M. and Peterson, N. 2009. Delineating the Coral Triangle. *Galaxea, Journal of Coral Reef Studies* 11:91–100.

Willig, M.R., Kaufman, D.M. and Stevens, R.D. 2003. Latitudinal gradients of biodiversity: Pattern, process, scale, and synthesis. *Annual Review of Ecology, Evolution, and Systematics* 34:273–309.

Wittmann, A.C. and Pörtner, H.O. 2013. Sensitivities of extant animal taxa to ocean acidification. *Nature Climate Change* 3:995.

CHAPTER 13

Updated Extinction Risk Assessment of the Colombian Freshwater Crabs (Brachyura: Pseudothelphusidae, Trichodactylidae) Reveals an Increased Number of Threatened Species

Ada Acevedo-Alonso and Neil Cumberlidge

CONTENTS

13.1 INTRODUCTION

Of the five families of primary freshwater crabs found in tropical and subtropical freshwater habitats, two (Pseudothelphusidae and Trichodactylidae) occur in the Neotropical region. Colombia has the highest freshwater crab biodiversity in the Neotropics (106 species, 24 genera, 2 families), and is the second most species-rich country in the world for this group (Yeo et al. 2008, Cumberlidge et al. 2009, Campos and Campos 2017). The high freshwater crab diversity in Colombia is the result of large areas of suitable freshwater crab habitat that includes five major river basins (the Magdalena-Cauca, Caribe, Pacific, Orinoco, and Amazon) and three major mountain ranges with their associated valleys, streams, and rivers (Fig. 13.1). The Magdalena and Cauca River drainages include high-altitude Andean valleys where the freshwater systems offer a range of habitats with different temperatures and precipitation rates. The Caribe Region includes the Sierra Nevada de Santa Marta and its associated mountain streams and rivers and a major complex of wetlands and marshes at lower altitudes. The Pacific Region includes the Andean piedmont mountain range with its moist tropical forests. The Orinoco River basin has aquatic systems that experience a drier climate, while the Amazon River basin flows through tropical forests with the highest rainfall in the country. These five main watersheds coincide for the most part with the freshwater ecoregions found in Colombia (Thieme et al. 2005, Abell et al. 2008), most of which are named for the principal river basin (except for the Andean ecoregion, which corresponds to the basins of the Magdalena–Cauca Rivers). A sixth freshwater ecoregion includes the Caribbean and the Pacific offshore islands, but in terms of hydrographic distribution these are considered part of the Caribbean and Pacific River basins.

Freshwater crabs are found throughout Colombia's freshwater systems from mountain streams to rivers, wetlands, and caves, and the majority of the species (81%) are country endemics (Cumberlidge et al. 2009). Pseudothelphusid species (15 genera, 91 species) are predominantly semiterrestrial and (with a few exceptions) prefer high elevation habitats (between 400 and 3,000 m asl), while most trichodactylid species (9 genera, 15 species) are fully aquatic and prefer lower elevations (at and below 100 m asl) (Campos 2005, 2014, Cumberlidge et al. 2014).

Freshwater crabs are important components of Colombia's freshwater ecosystems because these macroinvertebrates comprise a high percentage of the biomass, are significant consumers involved in nutrient cycling, and are major food web components (Cumberlidge et al. 2009). Although wild-caught crustaceans provide only a small part of the protein needs for Colombians (just 1%) (Bello et al. 2014), freshwater crabs are nevertheless an important subsistence food for rural communities, particularly in remote areas of the Amazon, Orinoco, and Pacific Slope basins (Campos and Lasso 2015). Freshwater crabs are also medically important, with 22 pseudothelphusid species recognized as the second intermediate host of the trematode lung fluke *Paragonimus* spp., which causes paragonimiasis in humans (Rodríguez and Magalhães 2005). Additionally, freshwater crabs are water quality indicators because adult crabs (and especially juvenile crabs) are sensitive to temperature extremes and chemical pollutants (Campos 2014).

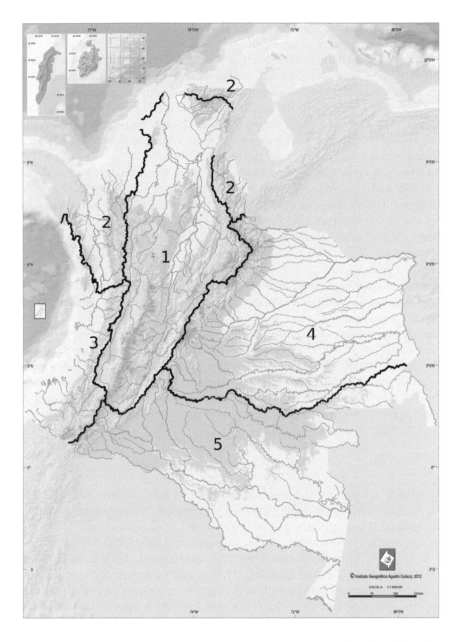

Figure 13.1   The five major river basins in Colombia considered in this work: 1, Magdalena–
Cauca; 2, Caribe; 3, Pacific; 4, Orinoco; and 5, Amazon. Modified from an open-
access map supplied by the Instituto Geográfico Agustín Codazzi.

Recent studies in Colombia indicate that many freshwater fisheries are rapidly declining due to climate change and environmental neglect, and may even be in danger of collapse (Bello et al. 2014). The principal threats to Colombian freshwater habitats include urban development, pollution, and deforestation, and these threats have already caused an 85% decline in the fisheries in the Magdalena-Cauca River basin over the past 20 years (Bello et al. 2014). The preference of pseudothelphusids for small high-altitude streams makes them particularly vulnerable to environmental disturbance caused by anthropogenic threats, while the trichodactylids are less affected because they prefer major rivers.

Colombian freshwater crab conservation status has been assessed twice using the IUCN Red List protocols: first in 2008 as part of the global freshwater crab assessment (Cumberlidge et al. 2009, 2014), and then in 2015 as part of a country-level assessment (Campos and Lasso 2015). The IUCN Red List global assessment (Cumberlidge et al. 2009) found 13.9% of Colombia's freshwater crab species to be threatened, while the 2015 country-level assessment assigned 24.3% of Colombia's freshwater crab species to a threatened category (Campos and Lasso 2015). The assessment by Campos and Lasso (2015) made three main changes to the conservation status of Colombia's freshwater crabs: the number of EN species increased, while the number of VU and DD species decreased. This worsening situation was underlined by the increase in EN category species numbers, which grew from none (Cumberlidge et al. 2009) to 25 species (Campos and Lasso 2015). This chapter addresses the results of a third conservation assessment of all the 106 species of freshwater crabs currently known from Colombia using the IUCN Red List protocols. Despite Colombia being the focus of this chapter on conservation, this assessment is mostly a global one, given the high number of species endemic to this country.

13.2 METHODS

The updated extinction risk reassessments reported here are based on all available data for each species (including ecology, distribution, threats, population trends, and protection) and used the IUCN (2001) Red List Categories and Criteria (version 3.1). Species were assigned to one of eight categories: Extinct (EX), Extinct in the Wild (EW), Critically Endangered (CR), Endangered (EN), Vulnerable (VU), Near Threatened (NT), Data Deficient (DD), or Least Concern (LC). A species was assessed as threatened with extinction if it met the criteria for one of the threatened categories (CR, EN, and VU). The IUCN Red List protocols evaluate the extinction risk of a species according to five criteria: (A) population size reductions, (B) geographic range changes, (C) small population sizes that are declining, (D) very small or restricted populations, and (E) quantitative analyses of extinction risk. In general, the majority of assessment data available on Colombian freshwater crabs includes information on distribution, habitats, and threats (criterion B, the geographic range, either B1 – extent of occurrence, EOO – or B2 – area of occupancy, AOO – or both). Unfortunately, there are very few data available on population size and trends for most Colombian freshwater crab species, making it difficult to employ criteria

A, C, and D. In a few cases, criterion D2 VU was applied for species that either had a restricted area of occupancy or number of locations together with a plausible threat that could drive the taxon to CR or EX in a very short time. Geographical and ecological data were compiled from literature, museum records, and specialist knowledge; both EOO and AOO were estimated from point locality data through an open-source browser, Geospatial Conservation Assessment Tool (GeoCAT) (Bachman et al. 2011).

13.3 RESULTS

13.3.1 IUCN Red List Categorization

Forty-five out of the 106 species assessed here (42.5%) were assigned to a threatened category (either EN or VU), 14.2% were assessed as NT, 22.6% as LC, and 20.8% as DD (Table 13.1).

The majority of the VU and EN species were assessed under criterion B1, and had an extent of occurrence below one of the thresholds, a low number of locations, and a declining area and/or quality of habitat. Two of the threatened species were assessed as VU D2, based on a restricted range (EOO), a low area of occupancy (AOO), a small number of locations, and a plausible threat.

13.3.2 Extinction Risk by Family

The Pseudothelphusidae comprises 86% of the Colombian freshwater crab fauna, and is the most threatened family with 43 of the 91 species (47.3%) assessed in a threatened category (Table 13.2), while only two of the 15 species of Trichodactylidae (13.3%) were found to be threatened.

13.3.2.1 Pseudothelphusidae

Eight out of the 15 genera of pseudothelphusids include threatened species, most of which (87%) belong to three genera: *Neostrengeria* (18 species), *Hypolobocera* (11 species), and *Strengeriana* (7 species) (Table 13.3). Other threatened pseudothelphusid species belong to three monospecific genera (*Colombiathelphusa*, *Eudaniela*, and *Orthothelphusa*), while the single species of *Prionothelphusa* is NT, and the

Table 13.1   **Summary of the Red List Assessment Results for 106 Species of Colombian Freshwater Crabs**

IUCN category	CR	EN	VU	NT	LC	DD	Total
Number of Species	0	28	17	15	24	22	106
Percentage (%)	0	26.4	16.0	14.2	22.6	20.8	100

CR = Critically Endangered, EN = Endangered, VU = Vulnerable, NT = Near Threatened, LC = Least Concern, DD = Data Deficient.

Table 13.2 **Summary of the Red List Assessment of Colombian Freshwater Crabs by Family**

IUCN category	CR	EN	VU	NT	LC	DD	Total	Thr.	Thr. %
Pseudothelphusidae	0	27	16	15	11	22	91	43	47.3%
Trichodactylidae	0	1	1	0	13	0	15	3	13.3%

CR = Critically Endangered, EN = Endangered, VU = Vulnerable, NT = Near Threatened, LC = Least Concern, DD = Data Deficient, Total = Total Number of Species, Thr. = Number of Species in a Threatened Category

Table 13.3 **Summary of the Red List Assessments of the Pseudothelphusid Genera in Colombia**

Genus	CR	EN	VU	NT	LC	DD	Total	Thr.	Thr.%
Chaceus	0	2	0	0	3	1	6	2	2.2
Colombiathelphusa	0	1	0	0	0	0	1	1	1.1
Eidocamptophallus	0	0	0	0	0	1	1	0	0.0
Eudaniela	0	0	1	0	0	0	1	1	1.1
Fredius	0	0	0	0	0	1	1	0	0.0
Hypolobocera	0	6	5	2	3	6	22	11	12.1
Lindacatalina	0	1	1	1	0	1	4	2	2.2
Martiana	0	0	0	0	0	1	1	0	0.0
Moritschus	0	0	0	1	0	2	3	0	0.0
Neostrengeria	0	11	7	5	0	4	27	18	19.8
Orthothelphusa	0	1	0	0	0	0	1	1	1.1
Phallangothelphusa	0	0	0	2	1	1	4	0	0.0
Potamocarcinus	0	0	0	0	1	1	2	0	0.0
Prionothelphusa	0	0	0	1	0	0	1	0	0.0
Strengeriana	0	5	2	4	2	3	16	7	7.7
Total	**0**	**27**	**16**	**16**	**10**	**22**	**91**	**43**	**47.3**

CR = Critically Endangered, EN = Endangered, VU = Vulnerable, NT = Near Threatened, LC = Least Concern, DD = Data Deficient, Total = Total Number of Species, Thr. = Number of Species in a Threatened Category

species in the three remaining monospecific genera are DD (Table 13.3). The highest number of DD species (13/22) are pseudothelphusids in three genera: *Neostrengeria* (4 species), *Hypolobocera* (6 species), and *Strengeriana* (3 species). Seven genera (with 13 species) lacked threatened species: 4 species were NT, 2 were LC, and 7 were DD.

13.3.2.2 Trichodactylidae

Two of three Colombian *Botiella* species were assigned to a threatened category, while all the remaining trichodactylid species were judged to be LC (Table 13.4).

Table 13.4 **Summary of the Red List Assessments of the Genera of Colombian Trichodactylidae.**

Genera	CR	EN	VU	NT	LC	DD	Total	Thr.	Thr. %
Botiella	0	1	1	0	1	0	3	2	13.3
Dilocarcinus	0	0	0	0	1	0	1	0	0
Fredilocarcinus	0	0	0	0	1	0	1	0	0
Forsteria	0	0	0	0	1	0	1	0	0
Moreirocarcinus	0	0	0	0	2	0	2	0	0
Poppiana	0	0	0	0	1	0	1	0	0
Sylviocarcinus	0	0	0	0	3	0	3	0	0
Trichodactylus	0	0	0	0	2	0	2	0	0
Valdivia	0	0	0	0	1	0	1	0	0
Total	0	1	1	0	13	0	15	2	13.3

CR = Critically Endangered, EN = Endangered, VU = Vulnerable, NT = Near Threatened, LC = Least Concern, DD = Data Deficient, Total = Total Number of Species, Thr. = Number of Species in a Threatened Category

13.3.3 Numbers of Threatened Species Found in the Different River Basins

The Magdalena–Cauca River basin has the highest number of species (44), followed by the Pacific Slope (22 species), the Caribe River (21 species), the Orinoco River (15 species), and the Amazon River (11 species) basins. Table 13.5 shows the results of the extinction risk assessment with the number of species assigned to each category by river basin. Note that the total number of species is 113 (rather than 106) because 7 species occur in more than one river basin. Most of the threatened species (19/44) occur in the Magdalena–Cauca River basin, followed by the Caribe River basin (11/21), the Pacific Slope basins (9/22), the Orinoco River basin (7/15), and the Amazon River basin (1/11). The highest number of DD species are found in the Magdalena–Cauca River basin (8/44 species) and the Pacific Slope basins (7/22 species).

The number of threatened freshwater crab species was highest in the Magdalena–Cauca River basin, which has the most diverse freshwater crab fauna. The Caribe and Orinoco River basins have the highest proportion of threatened species (52.4% and 46.7%, respectively), while the Magdalena–Cauca, Pacific Slope, and Amazon River basins have the lowest proportions of threatened species (43.2%, 40.9%, and 9.1%, respectively) (Table 13.6).

13.3.3.1 Magdalena–Cauca River Basin

This river basin has 19 threatened species (Table 13.7): 18 threatened species of pseudothelphusids (*Neostrengeria* and *Strengeriana*), and 1 threatened species of trichodactylid (*Botiella medemi* – Smalley & Rodríguez 1972) that is also found in

Table 13.5 **Summary of the IUCN Red List Categories of Species of Colombian Freshwater Crabs by River Basin**

River basin	CR	EN	VU	NT	LC	DD	Total	Thr.
Magdalena–Cauca	0	11	8	10	7	8	44	19
Caribe	0	8	3	1	5	4	21	11
Pacific	0	5	4	4	2	7	22	9
Orinoco	0	5	2	2	6	0	15	7
Amazon	0	0	1	0	7	3	11	1
Total	**0**	**29**	**18**	**17**	**27**	**22**	**113**	**47**

CR = Critically Endangered, EN = Endangered, VU = Vulnerable, NT = Near Threatened, LC = Least Concern, DD = Data Deficient, Total = Total Number of Species, Thr. = Number of Species in a Threatened Category

Table 13.6 **Summary of the IUCN Red List Categories of Species of Freshwater Crabs Showing the Percentage of Threatened Species by River Basin**

River basin	CR	EN	VU	NT	LC	DD	Thr.%
Magdalena	0	25.0	18.2	22.7	15.9	18.2	43.2
Caribe	0	38.1	14.3	4.8	23.8	19.0	52.4
Pacific	0	22.7	18.2	18.2	9.1	31.8	40.9
Orinoco	0	33.3	13.3	13.3	40.0	0.0	46.7
Amazon	0	9.1	0.0	0.0	63.6	27.3	9.1

CR = Critically Endangered, EN = Endangered, VU = Vulnerable, NT = Near Threatened, LC = Least Concern, DD = Data Deficient, Thr. = % of species in a threatened category.

Table 13.7 **Summary of the IUCN Red List Categories of Species of Colombian Freshwater Crabs in Two Families Showing the Number of Threatened Species Found in the Magdalena–Cauca River Basin**

Family	CR	EN	VU	NT	LC	DD	Total	Thr.
Pseudothelphusidae	0	10	8	10	5	8	41	18
Trichodactylidae	0	0	1	0	2	0	3	1

CR = Critically Endangered, EN = Endangered, VU = Vulnerable, NT = Near Threatened, LC = Least Concern, DD = Data Deficient, Thr. = Number of Species in a Threatened Category.

the Caribe River basin. Two of the pseudothelphusids from the Magdalena–Cauca River basin (*Hypolobocera bouvieri* – Rathbun 1898 – and *Lindacatalina latipens* – Pretzmann 1968) also occur in the Caribe and Amazon River basins (respectively).

13.3.3.2 Caribe River Basin

This river basin has 11 threatened species (Table 13.8), of which two are tricho-dactylids (one EN and one VU); one of these (*Botiella medemi*, EN) is also found in the Magdalena–Cauca River basin. Half of the pseudothelphusid species found in the Caribe River basin are threatened (seven EN and two VU), the majority of

Table 13.8 **Summary of the IUCN Red List Categories of Species of Colombian Freshwater Crabs in Two Families Showing the Number of Threatened Species Found in the Caribe River Basin**

Family	CR	EN	VU	NT	LC	DD	Total	Thr.
Pseudothelphusidae	0	7	2	1	4	4	18	9
Trichodactylidae	0	1	1	0	1	0	3	2

CR = Critically Endangered, EN = Endangered, VU = Vulnerable, NT = Near Threatened, LC = Least Concern, DD = Data Deficient, Total = Total Number of Species, Thr. = Number of Species in a Threatened Category.

Table 13.9 **Summary of the IUCN Red List Categories of Species of Freshwater Crabs in Two Families Showing the Number of Threatened Species Found in the Pacific Slope Basins**

Family	CR	EN	VU	NT	LC	DD	Total	Thr.
Pseudothelphusidae	0	5	4	4	2	7	22	9
Trichodactylidae	0	0	0	0	0	0	0	0

CR = Critically Endangered, EN = Endangered, VU = Vulnerable, NT = Near Threatened, LC = Least Concern, DD = Data Deficient, Total = Total Number of Species, Thr. = Number of Species in a Threatened Category

which are species of *Neostrengeria*. Three of the pseudothelphusid species in the Caribe River basin are also found in other river basins: *Hypolobocera bouvieri* in the Magdalena–Cauca River basin and *H. noanamensis* (Rodríguez et al. 2002) and *H. velezi* (Campos 2000) in the Pacific Slope.

13.3.3.3 Pacific Slope Basins

Nine out of the 22 species of pseudothelphusids found in the Pacific Slope basins are threatened (5 EN, 4 VU) (Table 13.9), eight of which belong to the genus *Hypolobocera*. Three species extend their distribution to other river basins: *Hypolobocera noanamenis* and *H. velezi* are also found in the Caribe River basin, and *Lindacatalina orientalis* (Pretzmann 1968) is also found in the Magdalena–Cauca River basin. Trichodactylidae is not found in the Pacific Slope basins.

13.3.3.4 Orinoco River Basin

A majority (7/10 species) of the pseudothelphusids from this river basin is threatened with extinction (five EN, two VU) (Table 13.10), six of which belong to the genus *Neostrengeria*. All the trichodactylids from this river basin were evaluated as LC.

13.3.3.5 Amazon River Basin

The Amazon River basin in Colombia has only one threatened species, the pseudothelphusid *Lindacatalina sumacensis* (Rodríguez & von Sternberg 1998) (EN)

Table 13.10 **Summary of the IUCN Red List Categories of Species of Freshwater Crabs in Two Families Showing the Number of Threatened Species Found in the Orinoco River Basin**

Family	CR	EN	VU	NT	LC	DD	Total	Thr.
Pseudothelphusidae	0	5	2	2	1	0	10	7
Trichodactylidae	0	0	0	0	5	0	5	0

CR = Critically Endangered, EN = Endangered, VU = Vulnerable, NT = Near Threatened, LC = Least Concern, DD = Data Deficient, Total = Total Number of Species, Thr. = Number of Species in a Threatened Category

(Table 13.11). None of the seven species of trichodactylids found in the Amazon River basin are threatened (seven LC). Two species from the Amazon River basin are also found in the Orinoco: *Moreirocarcinus emarginatus* (Milne Edwards 1853) and *Valdivia serrata* (White 1847).

13.3.4 Taxonomic Changes

The 2008 IUCN Red List freshwater crab conservation assessment focused on 87 species of pseudothelphusids and 14 species of trichodactylids from Colombia (Cumberlidge et al. 2009, 2014). Our reassessment of this fauna is based on an updated species list that reflects the removal of nine species of pseudothelphusids from the country list because they are no longer recognized as valid species (they are either junior synonyms or their presence in Colombia cannot be confirmed). For example, *Hypolobocera triangula* (Ramos-Tafur 2006) is now treated as a junior synonym of *H. rotundilobata* (Rodríguez 1994), and *H. solimani* (Ramos-Tafur 2006) is a junior synonym of *H. alata* (Campos 1989) (see Campos and Guerra 2008). Table 13.12 shows the updated distribution of six other species that do not occur in Colombia, despite earlier records of their presence in the country. For example, *Orthothelphusa venezuelensis* (Rathbun 1905) was incorrectly recorded from Colombia (Cumberlidge 2008), but has never been considered part of the fauna.

Table 13.11 **Summary of the IUCN Red List Categories of Species of Freshwater Crabs in Two Families Showing the Number of Threatened Species Found in the Amazon River Basin**

Family	CR	EN	VU	NT	LC	DD	Total	Thr.
Pseudothelphusidae	0	1	0	0	0	3	4	1
Trichodactylidae	0	0	0	0	7	0	7	0

CR = Critically Endangered, EN = Endangered, VU = Vulnerable, NT = Near Threatened, LC = Least Concern, DD = Data Deficient, Total = Total Number of Species, Thr. = Number of Species in a Threatened Category

Table 13.12   **Updated Distribution of Six Species of Pseudothelphusids That Were Previously Assigned to Colombia but Are no Longer Listed**

Species	Distribution
Hypolobocera aequatorialis (Ortmann 1897)	Ecuador: Tributary of Jubones river, west of Santa Isabel
Hypolobocera conradi (Nobili 1897)	Ecuador: Amazon drainage
Hypolobocera esmeraldensis (Rodríguez & von Sternberg 1988)	Ecuador: Esmeraldas Province Manabi (Chone River)
Hypolobocera exuca (Pretzmann 1977)	Ecuador: Esmeraldas River
Hypolobocera ucuyalensis (Rodríguez & Suárez 2004)	Peru: Pucallpa, Ucayali, km. 59 of Basadre highway, Loreto Department
Microthelphusa viloriai (Suárez 2006)	Endemic to the Venezuela Andes, in Trujillo State, only known from the type

Their previous distribution was based on an extrapolation of their range to all parts of a river basin, including the part that extends into Colombia, despite an absence of locality records.

13.4 DISCUSSION

13.4.1 Threats

The World Wildlife Fund Colombia (WWF-Colombia 2017) reported five principal threats to the country: a change in land use, land degradation, biological invasions, pollution, and climate change. The report also identified three major threats to freshwater ecosystems: modification of natural systems, use of biological resources, and pollution. Colombian freshwater crabs assigned to a threatened category have three main threats: deforestation, mining, and urban development. Urban development is a particular problem in the Magdalena–Cauca and Caribe River basins, while the other river basins are more remote and less impacted by threats of this kind. Water pollution and the draining of wetlands are threats that are associated with mining and urban development and both are specific threats to freshwater crabs, particularly pseudothelphusids. Illegal and unregulated mining operations pose a particular threat to Colombian freshwater ecosystems because up to 80% of all mining operations in this country are illegal (CGR 2017) and are increasing as drug trafficking decreases. Threats to freshwater crabs from pollution events associated with illegal mining are greatest in the more rural Pacific Slope and Amazon River basins, while the impact of legal mining is the major threat to freshwater organisms in the Magdalena–Cauca and Caribe River basins. In addition, exploration and extraction of oil resources in the Orinoco and Caribe River basins pose significant threats to freshwater organisms because of the danger of pollution by oil spills (either accidental or sabotage). Moreover, Colombia is one of the main consumers of commercial fertilizers in Latin America, and these chemicals pose a big risk to freshwater environments because the majority of monitored water stations report a nutrient surplus (WWF-Colombia 2017).

Deforestation in Colombia represents the greatest threat to freshwater crabs. Riparian canopy cover loss or reduction increases stream mean temperatures, reduces litter detritus volume in streambeds, and increases periphyton algal biomass on the substrate (Bojsen and Jacobsen 2003). Human population increases and associated changes in land use have affected around 40% of Colombia (Moreno et al. 2016) including significant and ongoing deforestation and forest degradation (IDEAM 2018, Ramírez-Delgado et al. 2018). The Andean region of Colombia has the highest deforestation rate, with 36% of natural forest remaining according to estimates made in 2015 (González et al. 2018), while the Caribe region has the highest percentage of forest loss, followed by the Orinoco region and the Amazon region (WWF-Colombia 2017).

The Andean region has the highest number of threatened freshwater crab species (11 species EN, 8 species VU, 43.2%), and is the region where the greatest destruction of natural forests for agriculture or urban development have contributed most to biodiversity decline and loss (Moreno et al. 2016). The Andean region's great habitat diversity (Magdalena–Cauca River basins) has resulted in a species-rich crab fauna that is dominated by species of *Neostrengeria* and *Strengeriana*, with many endemic species. Despite the high proportion of threatened species in the Magdalena–Cauca region (43.2%), this region ranks only third for threatened species behind the Caribe (52.4%) and Orinoco (46.7%) regions (Table 13.13).

Additionally, the Magdalena–Cauca River basin flow regimes have been heavily modified, the natural resources overexploited, and the basin aquatic resources (including fisheries) have suffered degradation and reduction (WWF-Colombia 2017). Furthermore, the invasive crayfish *Procambarus clarkii* (Girard 1852) is present in two areas of the Magdalena–Cauca River basins (Valle del Cauca Department and the Altiplano Cundiboyacense [Cundinamarca and Boyacá departments]) (Arias-Pineda and Rodríguez 2008) affecting the 16 freshwater crab species in three genera (*Hypolobocera* – five, *Neostrengeria* – ten, and *Phallangothelphusa* – one) (Table 13.14).

The Caribe River basin has the least forest cover remaining (González et al. 2018) and the highest percentage of threatened freshwater crab species (52.4%) (Table 13.13), mostly species of *Neostrengeria* and *Botiella*. The Caribe River basin is assessed by the IUCN Red List of Ecosystems (RLE) as a critically endangered region (CR) that is threatened by mining and oil extraction (Etter et al. 2017). The Orinoco River basin was found to have the second highest percentage of threatened

Table 13.13   **Summary of the Total Number of Threatened Species of Freshwater Crabs in Colombia by River Basin**

River basin	No. species	% species	Thr.	Thr. %
Magdalena–Cauca	44	41.5	19	43.2
Caribe	21	19.8	11	52.4
Pacific	22	20.8	9	40.9
Orinoco	15	14.2	7	46.7
Amazon	11	10.4	1	9.1

Thr. = number of threatened species

Species of Freshwater Crabs Impacted by the Presence of the Crayfish ***Procambarus clarkii* and the Parts of Colombia That Are Most Affected by the Threat**

Species	Department
Hypolobocera bouvieri (Rathbun 1915), *H. beieri* (Pretzmann 1968), *H. buenaventurensis* (Rathbun 1905), *H. cajambrensis* (Von Prahl 1987), and *H. dentata* (Von Prahl 1987)	Valle del Cauca
Neostrengeria alexae (Campos 2010), *N. aspera* (Campos 1992), *N. bataensis* (Campos & Pedraza 2008), *N. botti* (Rodríguez & Turkay 1978), *N. boyacensis* (Rodríguez 1980), *N. gilberti* (Campos 1992), *N. lasallei* (Rodríguez 1980), *N. lemaitrei* (Campos 2004), *N. lindigiana* (Rathbun 1897), and *N. macropa* (Milne Edwards 1853)	Cundinamarca and Boyacá
Phallangothelphusa dispar (Zimmer 1912)	Cundinamarca

freshwater crab species (46.7%, mostly *Neostrengeria*), and has been assessed by the RLE as a vulnerable (VU) region (Table 13.13), threatened by livestock, large-scale agriculture, and oil extraction.

The Pacific Slope basins freshwater ecosystems are significantly affected by illegal mining and deforestation (IDEAM 2018) but are, nevertheless, listed as LC in the IUCN RLE. This region has a high percentage of threatened freshwater crab species (40.9%), all in *Hypolobocera*, which are principally threatened by mining operations, 90% of which are illegal (WWF-Colombia 2017). Similarly, the Amazon River basin freshwater ecosystems are significantly affected by deforestation (IDEAM 2018), but are listed as LC in the IUCN RLE perhaps because this remote region is less susceptible to biodiversity loss (Moreno et al. 2016). Consequently, the Amazon River basin has a low percentage of threatened freshwater crab species (9.1%) with *Lindacatalina sumacensis* as the only threatened species (Table 13.13).

The vast and remote Orinoco and Amazon River basins freshwater crab faunas are dominated by trichodactylids that are found in one of the least threatened regions in Colombia. Perhaps not surprisingly, these river basins have the fewest numbers of threatened freshwater crab species, because these trichodactylids have wide distributions, a low endemism rate (only 1/15 species: *Botiella medemi*), and low threatened species numbers (only two *Botiella* species from the more disturbed Caribe River basin).

Most threatened freshwater crab species (95.6%) are in the Pseudothelphusidae and are found in the mountainous parts of the Andes, Caribe, and Pacific regions where there are high habitat diversity, endemism, and threat levels. More than half (18/27) of *Neostrengeria* species from the Andes, Caribe, and Orinoco are threatened. Monospecific genera are particularly at risk, because in some cases (*Colombiathelphusa, Eudaniela, Orthothelphusa*) all lineages are threatened (Table 13.15), as well as a large proportion of species in small genera *(Lindacatalina, 2/4, 50%, Chaceus, 2/6, 33%)*. It is of great concern that the highest numbers of threatened species are in the most species-rich genera (*Neostrengeria* 18/27, 88.7%, *Hypolobocera* 11/22, 50%, *Strengeriana* 7/16, 43.8%).

Table 13.15 **Summary of the Total Number of Species in Each Colombian Pseudothelpusid Genus, with the Numbers and Proportions of Threatened Species by Genera**

Genus	Total	Thr.	Thr. %
Chaceus	6	2	33.3
Colombiathelphusa	1	1	100.0
Eidocamptophallus	1	0	0.0
Eudaniela	1	1	100.0
Fredius	1	0	0.0
Hypolobocera	22	11	50.0
Lindactalina	4	2	50.0
Martiana	1	0	0.0
Moritschus	3	0	0.0
Neostrengeria	27	18	66.7
Orthothelphusa	1	1	100.0
Phallangothelphusa	4	0	0.0
Potamocarcinus	2	0	0.0
Prionothelphusa	1	0	0.0
Strengeriana	16	7	43.8

Total = Total Number of Species, Thr. = Number of Threatened Species

13.4.2 Conservation

Freshwater crabs are found in 20 of Colombia's 59 protected areas (18 pseudothelphusids and 4 trichodactylids) but 9 of these protected areas are threatened by deforestation and illegal coca farming (UNODC 2017). The Amazon, Orinoco, and Pacific forests of Colombia have been affected by illegal armed forces whose presence has contributed to habitat conservation (by keeping the human population small) and to habitat degradation (from coca farming) until recently. This situation may change following the peace accord signing with the FARC (Revolutionary Armed Forces of Colombia), which is expected to bring a decrease in illegal agriculture but a growth in human population and development that will accelerate deforestation, land use changes, and water pollution rates.

Since the first Colombian freshwater crab assessment in 2008, a number of political changes have been accompanied by an extinction risk increase in these animals (Fig. 13.2). The first Colombian freshwater crab extinction risk assessment (Cumberlidge et al. 2009, 2014) included 101 species of which 13.9% were assigned to a threatened category. Seven years later Campos and Lasso (2015) reassessed the freshwater crab fauna of Colombia, which by then included 104 species. Of these, they reassessed 49 species and assumed that another 46 species had the same extinction risk that was assigned in 2008, and also this study added several newly described species.

The results here show that 53.1% of the 49 species reassessed by Campos and Lasso (2015) were in a threatened category, which indicates a sharp rise in extinction

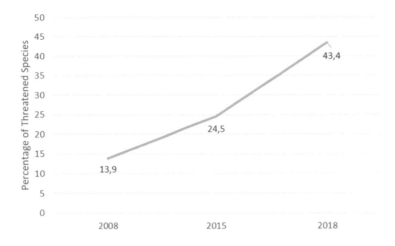

Figure 13.2 Changes in the percentage of threatened freshwater crab species in Colombia between 2008 and 2018.

risks for this fauna. Specifically, the number of endangered species (EN) rose from zero in 2008 to 23.6% in 2015, while the number of DD species fell from 40.6% in 2008 to 18.9% in 2015 (Campos and Lasso 2015). The single CR species reported in 2008 was reassigned to a less threatened category in both the 2015 and the present studies (Table 13.15). Here we can confirm the continuing increase in threatened species reported by those authors (from 24.5% in 2015 to 42.5% VU or EN), but with more near threatened species (from 7.5% in 2015 to 14.2% NT) and fewer least concern species (from 44.3% in 2015 to 22.6% LC) (Table 13.16).

Table 13.16 **Summary of the Red List Assessment Results for Colombian Freshwater Crabs in 2008, 2015, and 2018 (This Study)**

CR	2008	2015	2018
EN	1.0	0.0	0.0
VU	0.0	23.6	26.4
NT	12.9	0.9	16
LC	1.0	7.5	14.2
DD	44.6	44.3	22.6
NE	40.6	18.9	20.8
Threatened	0.0	4.7	0.0
	13.9%	**24.5%**	**42.5%**

CR = Critically Endangered, EN = Endangered, VU = Vulnerable, NT = Near Threatened, LC = Least Concern, DD = Data Deficient.

The percentage of threatened freshwater crab species has increased from 13.9% to 42.5% in the ten years between the first and latest assessments (2008–2018), most of which is due to an increase in the number of EN species (from 0% to 26.4%). The percentage of NT species increased from 1% to 14.2%, while the proportion of LC species decreased from 44.6% to 22.6%, and the proportion of DD species decreased from 40.6% to 20.8%. The differences in the results of these three evaluations are attributed to our increased knowledge of the threats to freshwater crab populations in many parts of Colombia, rather than to an increase in our knowledge of species distributions or population levels. Pseudothelphusidae has the least number of species records, the majority of species (91%) are known from less than six localities, and only 1% of these species are found in more than ten locations (Fig. 13.3). This is why the main focus of the present study is on the threats to species, rather than on specific distributional patterns.

Cumberlidge et al. (2009) pointed out that the high numbers of DD species could lead to underestimations of threatened species in Colombia, and considered the impact of the DD species on extinction risk under three different assumptions. The first assumption considered all DD species to be nonthreatened (the best-case scenario), the second assumption considered that the DD species were threatened in the same proportion as the non-DD species, and the third assumption was that all DD species were threatened (the worst-case scenario) (Table 13.17).

There were no DD trichodactylid species in 2008 and so the threatened species proportion under Assumptions 1, 2, and 3 was 14%; but in 2015 and 2018 there was one more threatened species in each study. The trichodactylid threatened species proportions under Assumptions 1, 2 and 3 were 20%, 21%, and 27%, respectively, for 2015, and 13% for all three assumptions in 2018. There were 41 DD species of pseudothelphusids in 2008 and the proportions of threatened species under

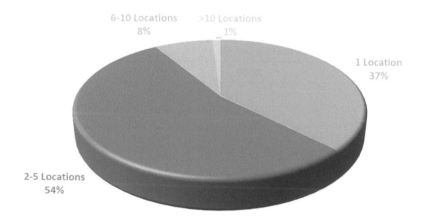

Figure 13.3 Percentage of species of the Pseudothelphusidae of Colombia by number of locations. The ranges match up with the thresholds for IUCN Red List B1 and B2 criteria (1 = CR, 2–5 = EN, 6–10 = VU, >10 = not in a threatened category).

Table 13.17 **Summary of the Red List Assessment Results in 2008, 2015, and 2018 (This Study), and the Proportion of the Colombian Freshwater Crab Fauna Threatened with Extinction under Three Different Assumptions of Risk (See Text for Details) When Today's DD Species Are Assessed in the Future**

	All Colombian FW Crabs			Pseudothelphusidae			Trichodactylidae		
	2008	2015	2018	2008	2015	2018	2008	2015	2018
CR	1	0	0	1	0	0	0	0	0
EN	0	25	28	0	23	27	0	2	1
VU	13	1	17	11	0	16	2	1	1
NT	1	8	15	1	8	15	0	0	0
LC	45	48	24	33	36	11	12	11	13
DD	41	20	22	41	19	22	0	1	0
Total	101	101	106	87	86	91	14	15	15
Assumption 1 (%)	14	26	42	14	27	47	14	20	13
Assumption 2 (%)	23	32	54	26	34	62	14	21	13
Assumption 3 (%)	54	46	63	61	49	71	14	27	13

CR = Critically Endangered, EN = Endangered, VU = Vulnerable, NT = Near Threatened, LC = Least Concern, DD = Data Deficient.

Assumptions 1, 2, and 3 were 14%, 26%, and 61% respectively. The number of DD species of pseudothelphusids decreased in 2015 from 41 to 19, and so the proportions of threatened species under Assumptions 1, 2 and 3 were 27%, 34%, and 49%, respectively; while in 2018 there were 22 DD species, and so the proportions of threatened species under Assumptions 1, 2, and 3 were even higher: 47%, 62%, and 71% respectively. Assumption 2 for all Colombian freshwater crabs (where DD species are eventually assessed at the average rate as the rest of the fauna) indicates a rapid worsening of the extinction risk for this group: from 23% (2008), to 32% (2015), to 54% (2018). Pseudothelphusidae is the most affected, where the extinction risk (Assumption 2) has increased dramatically from 26% in 2008, to 34% in 2015, to 62% in 2018.

Even the best-case scenario (Assumption 1) for all Colombian freshwater crabs (where all DD species are eventually assessed in a nonthreatened category) indicates that there would be a rapid increasing extinction risk from 14% (2008), to 26% (2015), to 42% (2018).

The extinction risk increases for Colombian freshwater crabs over the past decade are being driven by increasing deforestation and human population growth rates. Although Colombian protected areas cover 11.6% of the national territory, the threats to freshwater ecosystems and the organisms that depend on them have worsened. The vast majority of threatened species (VU and EN) are found in the Andes region where many of the aquatic ecosystems are negatively impacted by human activities. This deterioration is due to a combination of social and political difficulties (including illegal armed forces), together with a lack of investment in the sciences, especially biodiversity studies, targeted field research, and active conservation measures. This situation is made more urgent by the fact that a remarkably high percentage of the freshwater crab fauna (80%) are endemic to Colombia. Clearly, Colombia's freshwater crabs and the threatened ecosystems where they are found present great conservation challenges, which in turn calls for a greater investment in field research and active conservation actions to prevent extinctions of this important but understudied group.

ACKNOWLEDGMENTS

We thank Martha R. Campos from the Universidad Nacional de Colombia (UNAL) for her collaboration in the acquisition of data.

REFERENCES

Abell, R., Thieme, M.L., Revenga, C. et al. 2008. Freshwater ecoregions of the World: A new map of biogeographic units for freshwater biodiversity conservation. *BioScience* 58:403–14.

Arias-Pineda, J.Y. and Rodríguez, W.D. 2008. First record of the invasive species *Procambarus* (*Scapulicambarus*) *clarkii* (Girard 1852) (Crustacea, Decapoda, Cambaridae) from the Colombian Eastern Cordillera. *Boletín de la Sociedad Entomológica Aragonesa* 51:313–15.

Bachman, S., Moat, J., Hill, A.W., de Torre, J. and Scott, B. 2011. Supporting red list threat assessments with GeoCAT: Geospatial conservation assessment tool. *ZooKeys* 150:117–26.

Bello, J.C., Báez, M., Gómez, M.F., Orrego, O. and Nägele, L. 2014. *Biodiversidad 2014. Estado y tendencias de la biodiversidad continetal de Colombia*. Bogotá, DC, Colombia: Instituto de Investigación de Recursos Biológicos Alexander Von Humboldt.

Bojsen, B.H. and Jacobsen, D. 2003. Effects of deforestation on macroinvertebrate diversity and assemblage structure in Ecuadorian Amazon streams. *Hydrobiologie* 158:317–42.

Campos, M.R. 2005. *Freshwater Crabs from Colombia. A Taxonomic and Distributional Study*. Bogotá, DC, Colombia: Academia Colombiana de Ciencias Exactas, Físicas y Naturales. Colección Jorge Álvarez Lleras.

Campos, M.R. 2014. *Crustáceos decápodos de agua dulce de Colombia*. Bogotá, DC, Colombia: Universidad Nacional de Colombia (Sede Bogotá), Facultad de Ciencias, Instituto de Ciencias Naturales.

Campos, M.R. and Campos, D. 2017. Species diversity of freshwater decapod crustaceans (crabs and shrimps) from Colombia. *Crustaceana* 90:883–908.

Campos, M.R. and Guerra, L.A. 2008. Propuesta de sinonimia para las especies de cangrejo dulceacuícola *Hypolobocera solimani* e *Hypolobocera triangula* de Colombia. *Revista de Biología Tropical* 56:987–94.

Campos, M.R. and Lasso, C.A. 2015. *Libro rojo de los cangrejos dulceacuícolas de Colombia*. Bogotá, DC, Colombia: Instituto de Investigación de Recursos Biológicos Alexander von Humboldt (IAvH), Instituto de Ciencias Naturales de la Universidad Nacional de Colombia.

CGR (Contraloria General de la Republica). 2017. Minería ilegal sigue arrasando regiones y la debilidad institucional del Estado colombiano se dejó ganar esta batalla, dice el Contralor–Boletines de Prensa–2017–CGR. Available from: https://www.contraloria.g ov.co/contraloria/sala-de-prensa/boletines-de-prensa/boletines-prensa-2017/-/asset_pu blisher/y0hcpbxJNnDG/content/mineria-ilegal-sigue-arrasando-regiones-y-la-debili dad-institucional-del-estado-colombiano-se-dejo-ganar-esta-batalla-dice-el-contralor? inheritRedirect=false (Downloaded on 19th November, 2018).

Cumberlidge, N. 2008. *Orthothelphusa venezuelensis. The IUCN Red List of Threatened Species* 2008:e.T134403A3951218. http://dx.doi.org/10.2305/IUCN.UK.2008.RLTS.T 134403A3951218.en (Downloaded on 20th November 2018).

Cumberlidge, N., Alvarez, F. and Villalobos, J.L. 2014. Results of the global conservation assessment of the freshwater crabs (Brachyura, Pseudothelphusidae and Trichodactylidae): The Neotropical region, with an update on diversity. *ZooKeys* 457:133–57.

Cumberlidge, N., Ng, P.K.L., Yeo, D.C.J. et al. 2009. Freshwater crabs and the biodiversity crisis: Importance, threats, status, and conservation challenges. *Biological Conservation* 142:1665–73.

Etter, A., Andrade, A., Saavedra, K., Amaya, P., Arévalo, P., Cortés, J., Pacheco, C. and Soler, D. 2017. *Lista Roja de ecosistemas de Colombia*. Bogotá, DC, Colombia: Pontificia Universidad Javeriana and Conservación Internacional.

González, J.J., Cubillos, Á., Chadid, M.A., Cubillos, A., Arias, M., Zúñiga, E., Joubert, F., Pérez, I.R. and Berrio, V. 2018. *Caracterización de las principales causas y agentes de la deforestación a nivel nacional período 2005–2015*. Bogotá, DC, Colombia: Instituto de Hidrología, Meteorología y Estudios Ambientales–IDEAM.

IDEAM (Instituto de Hidrología, Meteorología y Estudios Ambientales). 2018. Sistema de Monitoreo de Biomasa y Carbono. Available from: http://smbyc.ideam.gov.co/Moni toreoBC-WEB/reg/indexLogOn.jsp (Downloaded on 19th November 2018).

Moreno, L.A., Andrade, G.I. and Ruíz-Contreras, L.F. 2016. *Biodiversity 2016. Status and Trends of Colombian Continental Biodiversity*. Bogotá, DC, Colombia: Research Institute of Biological Resources Alexander von Humboldt.

Ramírez-Delgado, J.P., Galindo, G.A., Yepes, A.P. and Cabrera, E. 2018. *Estimación de la degradación de bosques de Colombia a través de una análisis de fragmentación*. Bogotá, DC, Colombia: Instituto de Hidrología, Meteorología y Estudios Ambientales–IDEAM.

Rodríguez, G. and Magalhães, C. 2005. Recent advances in the biology of the Neotropical freshwater crab family Pseudothelphusidae (Crustacea, Decapoda, Brachyura). *Revista Brasileira de Zoologia* 22:354–65.

Thieme, M.L., Abell, R., Stiassny, M.L.J., Skelton, P., Lehner, B., Teugels, G.G., Dinerstein, E., Kamdem Toham, A., Burgess, N. and Olson, D. 2005. *Freshwater Ecoregions of Africa and Madagascar: A Conservation Assessment*. Washington, DC: Island Press.

UNODC (Oficina de las Naciones Unidas contra la Droga y el Delito). 2017. *Colombia*. Colombia: Monitoreo de territorios afectados por cultivos ilícitos 2016.

WWF-Colombia. 2017. *Colombia Viva: un país megadiverso de cara al futuro, Informe 2017*. Cali, Colombia: WWF-Colombia.

Yeo, D., Ng, P.K.L., Cumberlidge, N., Magalhães, C., Daniels, S. and Campos, M.R. 2008. Global diversity of crabs (Crustacea: Decapoda: Brachyura) in freshwater. *Hydrobiologia* 595:275–86.

CHAPTER 14

Threats to Endemic Colombian Freshwater Crabs (Decapoda: Pseudothelphusidae, Trichodactylidae) Associated with Climate Change and Human-Mediated Activities

David M. Hudson, Gillian Phillips, Carlos A. Lasso, and Martha R. Campos

CONTENTS

14.1 INTRODUCTION

Challenges abound when approaching conservation in the Neotropics. Colombia and surrounding countries are among the most biodiverse in the world, making this country a biodiversity hotspot (Myers et al. 2000, Hobohm 2003, Mittermeier 2004, Hutter et al. 2013, Collen et al. 2014, Butler 2016). Colombia itself is first in the world by area for aquatic fish diversity (Jiménez-Segura et al. 2016), and its freshwater vertebrates and decapod crustaceans have high endemism (Hutter et al. 2013, Collen et al. 2014). The human population in Colombia also places it among the fastest-growing countries and the high rate of human development is greatly impacting its natural ecosystems (Jiménez-Segura et al. 2016, González-Salazar et al. 2017). Direct anthropogenic alterations of environmental conditions compounded with climate change may be catastrophic to species survival, particularly in the tropical forests (Pimm and Raven 2000, Williams et al. 2003). Since risks to survival, endemism, and species richness do not necessarily overlap (Orme et al. 2005), it can be difficult to prioritize conservation areas spatially due to social challenges. Even though conservation of certain taxa has had some success, invertebrates are rarely included in conservation planning, regardless of their importance to ecosystem function (Collen et al. 2014). A large number of freshwater crustaceans are at risk of extinction both in Colombia and worldwide (Cumberlidge et al. 2009, Campos and Lasso 2015). We therefore provide a glimpse into the conservation challenges for the riverine systems in Colombia through a case study of a broadly distributed, common species of pseudothelphusid freshwater crab.

The abiotic factors associated with Colombia's topography and geological isolation have contributed to the species richness of this country (Myers et al. 2000, Hobohm 2003, Mittermeier 2004, Hutter et al. 2013, Collen et al. 2014, Butler 2016). Colombia's Eastern Cordillera Mountains are the widest in the country and have a higher species richness compared to the other two mountain ranges. The Eastern Cordillera mountains' eastern slope has high humidity due to the northeastern trade winds, and these slopes face the Colombian and Venezuelan savannas in the north and Amazonian lowlands in the south, facilitating biological exchange between these regions (Orme et al. 2005).

In this context, human-mediated threats to freshwater species loom large, with habitat loss/degradation, pollution, and exploitation ranking as the top threats to the survival of aquatic species (Collen et al. 2014), and climate change is affecting certain Andean lakes (Michelutti et al. 2015). Threats include deforestation, mining, agriculture, aquaculture, and development in riparian systems that all reduce habitat quality. Other threats to freshwater species in Colombia include nonindigenous species introductions, with two well-established globally invasive crustaceans now competing with endemic species (Campos 2005, Álvarez-León and Gutiérrez-Bonilla 2007). These multiple threats are of particular concern because they increase the likelihood of species loss.

14.2 FRESHWATER CRAB ECOLOGICAL IMPORTANCE AND CONSERVATION STATUS

Freshwater crabs are a food source for finfish species that provide a source of livelihood for humans in freshwater systems (Blair et al. 2008, Cumberlidge et al. 2009). Freshwater crabs are detritivores (Dobson et al. 2002, 2007a,b, Cumberlidge et al. 2009), and are part of aquatic food webs that provide ecosystem services by recycling carbon. These crustaceans also contribute by serving as hosts to parasites such as trematode lung flukes whose life cycles involve molluscan and mammalian hosts (Phillips et al. 2019). Trematode parasites in freshwater crabs have been found throughout Colombia, and as far inland as the Sabana de Bogotá (Vélez et al. 1995, Vélez et al. 2003, Casas et al. 2008, Uruburu et al. 2008, Phillips et al. 2019). This parasitism rate suggests the importance of crabs as prey for predators.

The extent of freshwater crab roles in aquatic food webs is understudied (Hobohm 2003, Mittermeier 2004, Hutter et al. 2013, Collen et al. 2014, Guarnizo et al. 2015, Butler 2016, Jiménez-Segura et al. 2016). As amphibious animals, freshwater crabs also interact with forest and terrestrial food webs. Currently, only 25% of upland vegetation remains and more is likely to be cut down as societal pressures increase (Myers et al. 2000, Alvarez-Berríos and Mitchell Aide 2015, Baptiste et al. 2017).

Colombia has the highest freshwater crab diversity in South America (Campos and Lasso 2015). Two freshwater crab families occur: Pseudothelphusidae (94 species) and Trichodactylidae (15 species in 9 genera). Trichodactylidae is found up to 900 m, while Pseudothelphusidae occur from sea level to 3,000 m (Campos 2014, Campos and Lasso 2015, Campos 2017, Campos et al. 2019). The International Union for the Conservation of Nature (IUCN) Red List of Threatened Species global conservation assessment (Cumberlidge et al. 2014) and the Libro Rojo de Los Cangrejos Dulceacuícolas de Colombia regional assessment (Campos and Lasso 2015) classify 56% of pseudothelphusids and 17% of trichodactylids as Data Deficient. Fourteen Colombian freshwater crab species are listed as threatened, the majority are narrow-range endemic species impacted by anthropogenic activities. Colombia's complex topography and ecology are favorable to speciation (Cumberlidge et al. 2014).

Most Colombian freshwater crab species (85/105, 81%) are localized endemics (Cumberlidge et al. 2009, Campos 2014, Campos and Lasso 2015), with the number of new species described growing yearly, particularly for cave species and species in isolated habitats (Campos 2017, Campos et al. 2019, Campos and Camacho 2019). The genera are highly endemic also (9/21 genera, 43%), but not at family level. Pseudothelphusids comprise 49 genera and 301 species, and there are 15 genera and 50 species of trichodactylids (Campos 2014). The Colombian endemic species are mainly found in isolated mountain streams and isolated stretches of forested rivers (Campos 2005). In-country protection of these species is progressing, with the passage of a ministry resolution protecting inland fisheries (2017 Colombian Ministry of Environment and Sustainable Development Resolution #1912) that protects 23 pseudothelphusids and 3 trichodactylids. However, another resolution passed in 2019 (Resolución de la Autoridad Nacional Pesquera – AUNAP), classifies the pseudothelphusid *Neostrengeria macropa* (H. Milne-Edwards, 1853) as a fisheries resource, even though the in-country red book (Campos and Lasso 2015) recommended it for endangered listing.

14.3 A CASE STUDY ON THREATS FROM CLIMATE CHANGE

The various threats noted earlier associated with climate change are illustrated by the widely occurring pseudothelphusid crab species, *Hypolobocera bouvieri* (Rathbun,

1898). Climate change will impact tropical mountain systems differently than lowlands through isolation of populations and their potential subsequent extinction (Carr and Tognelli 2016). Movement of species to new areas is determined by each species' physiological optimum (Case and Taper 2000). Some species have more unstable range evolution than others (Roy et al. 2001), especially as mobile ectotherms demonstrate a range of critical thermal maxima through behavioral management of thermal stress (Bates et al. 2014). Depending on a species adaptive capacity, some authors have suggested facilitating fish and invertebrate dispersion past barriers and monitoring species upon which endangered species rely for survival (Carr and Tognelli 2016).

14.3.1 The Model Methods: Maximum Entropy Modeling (MaxEnt)

A maximum entropy (MaxEnt) model was used here to identify the distribution and habitat selection of wildlife using presence-only location data (Baldwin 2009). Even with low numbers of collection localities, MaxEnt allows for predictions to be made that can drive identification of future collection sites. This program estimates the most uniform distribution of sampling points compared with background locations given the constraints from lack of available collection data (Baldwin 2009), with the added advantage of both continuous and categorical variables (Baldwin 2009). We selected MaxEnt for this study because other programs require physiological and behavioral reaction norm information, and this is currently limited for Neotropical freshwater crustaceans. We are working on a mechanistic model for future studies that can provide higher resolution for the prediction of future ranges of species affected by climate change.

Regularization causes the overfitting of results in the predicted distribution and therefore the end prediction to be clustered around collection location points. However, here regularization has been added to estimate distribution, allowing for an average value to be assigned to the approximate empirical average but not to equal it (Baldwin 2009). It is a multivariate, nonparametric approach that handles nonlinearities in the predictor variables well, and is largely unaffected by high cross-correlations and spatial autocorrelation among variables (Phillips et al. 2006). MaxEnt models are useful for identifying areas that have similar environmental conditions to known species, though they have some limitations for determining the true limits of species' ranges (Pearson et al. 2006, Suárez-Seoane et al. 2008). The predicted ranges could either be (1) currently inhabited by this species, or one or more other species, or (2) could be vacant. Since these species have probably arisen due to allopatric speciation, possible endemic ranges are likely to overlap across closely related species.

The jackknife method was used to determine the level of influence of each variable on the model. Here, multiple versions of the model are run, with the exclusion of one variable at a time. This also helps identify highly correlated variables and to evaluate skew (Baldwin 2009). To evaluate different models, the area under the curve (henceforth, AUC) between 0.5 (no better than random) and 1.0 (perfect fit) is input in a receiver operating conditions (henceforth, ROC) plot (Phillips et al. 2006,

Baldwin 2009). If the AUC is greater than 0.75, it is considered to be an acceptable model (Elith et al. 2006, Barnhart and Gillam 2014). One can also have selection thresholds, though this may not be possible in a presence/absence dataset (Baldwin 2009). To train the model, one needs at least 30 collection localities, with increased prediction plateauing around 50 sampling locations. There is a regularization proce-dure that can be used for as low as five sample locations to create useful models, but there could be sampling biases (Baldwin 2009).

WorldClim data were downloaded for altitude and current BioClim datasets (Fick and Hijmans 2017). Climate projections from global climate models GISS-E2-R or NASA data for representative concentration pathways (RCP), RCP 8.5 (RCP85) were downloaded for 2070, derived from the most recent climate change report (IPCC 2014). The data were clipped to the area of interest using ArcGIS (Version10.6.1), then raster data were converted to ASC files that could be used in the MaxEnt program in R statis-tical software. A Maximum Entropy Species Distribution Model (version 3.3.3k) was completed as in Mantilla-Meluk and Muñoz-Garay (2014) and Phillips et al. (2006). The environmental layers included were altitude, annual precipitation, annual mean temperature, mean temperature of the warmest month, and mean temperature of the coldest temperature, taking into account the impact of temperature and precipitation on prediction of extirpation (McCain and Colwell 2011). Response curves and jack-knife products for the relative importance of each variable were examined for model fitness of individual variables. Present time environmental layers were compared with the projection layers for 2070 at the higher RCP levels (8.5) by MaxEnt. The files were then imported into GIS for mapping. Georeferencing was necessary as the clipped data were stripped of the physical coordinates. Using the Geoprocessing functions control points, they were set based on the original clip of the data. The control points were then saved using the Update Georeferencing option. MaxEnt automatically uses a regu-larization of values, which reduces overfitting (Suárez-Seoane et al. 2008). Sampling biases close to cities certainly exist for Colombia, due to a lack of transportation infra-structure to remote areas and social barriers for the past few decades, which may have affected previous models and the one produced here.

ArcGIS analysis methods followed those of Young et al. (2011). Data from WorldClim (2.0 updated 1 June 2016), a monthly average of climate data for mini-mum, mean, and maximum temperate and precipitation for 1970–2000 (http://world-clim.org/version2) with a 30-s resolution (~1 km^2) were downloaded and clipped to relevant extents using ArcGIS v10.6.1. The crustacean collection location data were provided by Prof. Martha H. Rocha De Campos of the Instituto de Ciencias Naturales at the Universidad Nacional de Colombia, and divided into shrimp, trichodactylid, and pseudothelphusid. The data were converted to CSV files with only relevant infor-mation for MaxEnt. ESRI ArcGIS was used to clip BioClim and other variable files to the geographic extents of Colombia, with the creation of a bio_1_env file to use to set the geographic mask for the rest of the variables. This procedure was repeated for all possible variables for MaxEnt, and then converted from raster to ASCII for the model. One cosmopolitan pseudothelphusid species, *Hypolobocera bouvieri*, was chosen for its broad distributional range that includes a number of river sys-tems in Colombia. Maxent 3.4.1 was used to predict the potential geographic extent with all BioClim variables for this species with jackknife, response curves, 25% test data, and 15 replicates, to identify relevant variable for this model. Additionally, to predict climate change scenarios, we used future variables from 2070 (30s resolution as well) under carbon emissions conditions RCP 8.5 from the climate model GISS-E2-R. These were clipped and converted, and the analysis of extent was run as well. The zonal statistics tool under spatial analyst was used within ArcGIS and ArcCatalog to calculate area of the model output with a probability over 0.75.

14.3.2 MaxEnt Model Results

Under current conditions, the model indicates a broad distributional range for *H. bouvieri* throughout Colombia (Fig. 14.1). Some of the original collection locations fell outside of the areas predicted to have a 75% or more probability of occurrence, though these are mainly in areas with a low sampling effort that are either hard to reach or are historically inaccessible for social reasons. However, this indicates a need for more observations for those areas, particularly in Chocó Department in the northwestern portion of the country, and in Córdoba and Sucre Departments in the north. *Hypolobocera bouvieri* occupies a wide distributional range across both the Eastern and Western Cordillera mountain ranges, and future collections may provide evidence that populations are sufficiently different to be considered subspecies or entirely separate species. Indeed, many populations of *H. bouvieri* are treated as subspecies by Campos (2014). As for riverine connectivity (Fig. 14.1), one can predict some connected areas could also result in a more informed model through a watershed analysis within GIS.

The worst-case scenario for carbon dioxide emissions in 2070 (RCP 8.5) using the most effective model predicts an area of 75% probability of the range of *H. bouvieri* increasing by 1,532.5 km^2 (from 64.5 to 1,597 km^2, Fig. 14.2). While this calculation is likely to be an underestimation and skewed based on the conversion of coordinate systems, it does predict a tangible increase in suitable habitat based solely on environmental factors. This can be seen in Fig. 14.2, where the current collection locations fall out of the higher probability areas of the model, and the connectivity of the river systems affects the increased distributional range in 2070.

14.3.3 Discussion of Interactive Risks from the Case Study

The area available for a widely distributed species such as *H. bouvieri* could grow in the future, particularly when the low-temperature limit is released from areas of higher altitude, though further isolation and breakdown of gene flow between metapopulations could result in additional speciation. However, the variability of temperature and desiccation tolerance in the population is currently unknown, and would need to be worked out in order to make a mechanistic model of the animals' finer use of space. In work we previously completed for another species, *Neostrengeria macropa* H. Milne Edwards, 1853 (Hudson et al. 2016), the temperature changes expected to occur in its range would not cause it the same level of risk as this species. The narrow niche width of this range-restricted species may play a key role in understanding

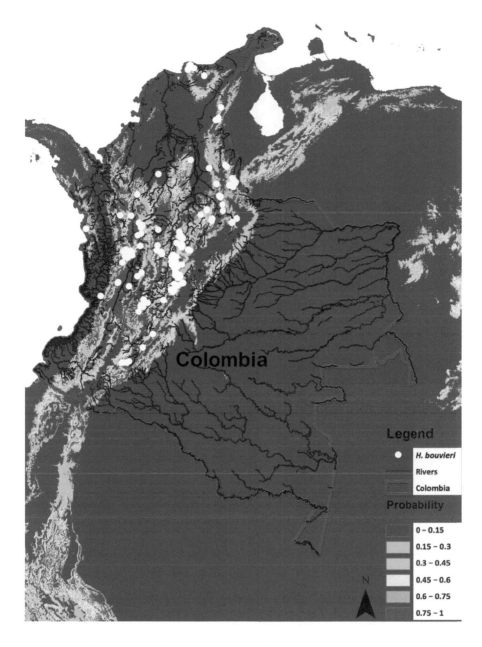

Figure 14.1   MaxEnt probability map for likely locations based on current climatic conditions for *Hypolobocera bouvieri*. Collection locations in white, with rivers as a proxy for conductivity for freshwater crabs.

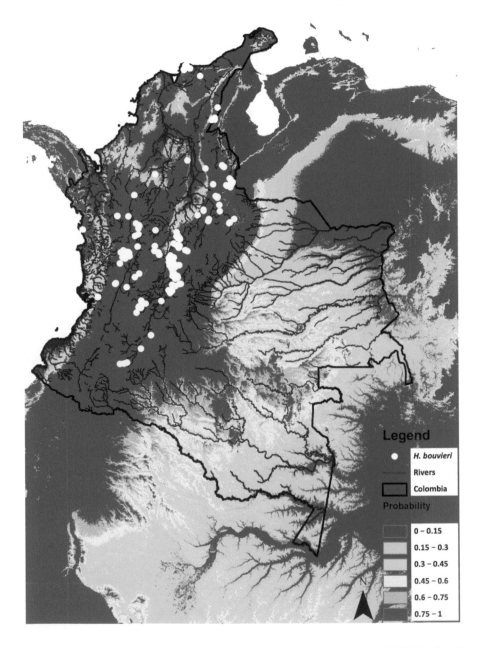

Figure 14.2   MaxEnt probability map for likely locations based on 2070 RCP 8.5 climatic
               condition scenario for *Hypolobocera bouvieri*. Collection locations in white, with
               rivers as a proxy for conductivity for freshwater crabs.

how ranges change over time in other species of the genus. Additionally, it is well-supported that mountaintop, polar, and other range-restricted species are at high risk for extirpation with climate change (Parmesan 2006). The geographic features of the central plain in Colombia meant that there was little change in the current expected extent and no major change in probable distribution for *N. macropa*. However, threats from habitat encroachment by human development, chemical spills, invasive species (crayfish), and other associated changes to freshwater tributaries and marshes still place this species as endangered (Campos and Lasso 2015).

Climate variables such as temperature or evapotranspiration are strong predictors of aquatic biodiversity (Hawkins et al. 2003, Hillebrand 2004, Burgmer et al. 2007). Tropical South America plays a dominant role in atmospheric convection and circulation, and so changes in physical conditions there can influence other systems. Variation in climate in the tropics in limited timescales influences higher latitude climates, and though no simple relationship exists between climate indices and diversity, species composition changes directly with variability in characteristics like temperature (Burgmer et al. 2007). These changes may be influenced by species-specific tolerances to temperature and rainfall, as when the northern equatorial Atlantic Ocean has unusually low sea surface temperatures and then the Amazon has a higher rainfall (Burgmer et al. 2007). The Andes Mountains are affected by climate variability driven by the El Niño Southern Oscillation (ENSO), and are projected to be affected by climate change, particularly the Andean lakes (Michelutti et al. 2015).

In the event of climate zones becoming rearranged by long-term changes, it is expected that species will track their optimal habitat (Peterson et al. 2002, Thomas et al. 2004, Loarie et al. 2008). Increasing global temperatures in lowland areas are predicted to drive lowland species to higher altitudes with cooler temperatures, but local conditions should be taken into account when making predictions. While land area is reduced as one moves up a mountain, effectively reducing habitat area, a temporal factor also exists, as animals are likely to take advantage of short-term variability and microhabitats (Farallo and Miles 2016). However, range-restricted endemic species may not be able to respond successfully when faced with climate-induced alterations (Carr and Tognelli 2016). Reductions in range size will have a drastic impact on endemic species, which are already range-restricted (Manne and Pimm 2001), though increases in range size for broad-tolerance species as we observe with *H. bouvieri* may also be expected. Some cases may see species expand their range when lower level temperature restrictions are lifted (Neumann et al. 2013, Sadowski et al. 2018). With this in mind, it is important to model not only the potential distributions of widely distributed species, but also those with restricted ranges. Habitat loss in a region of such high endemism like Colombia could push species extinctions much higher than the conservative assumptions from climate change alone (Collen et al. 2014, Baptiste et al. 2017). Nevertheless, even under conservative conditions, the Tropical Andes may lose up to 47% of its species if atmospheric carbon dioxide levels double (Malcolm et al. 2006).

Compounding this climate effect is the effect of human activity. Industrial activities, reviewed in the next section, pose a unique threat to freshwater ecosystems and the organisms that depend on them (Fig. 14.3) even over short distances, because the potential for damage can impact the entire range of a species with a restricted

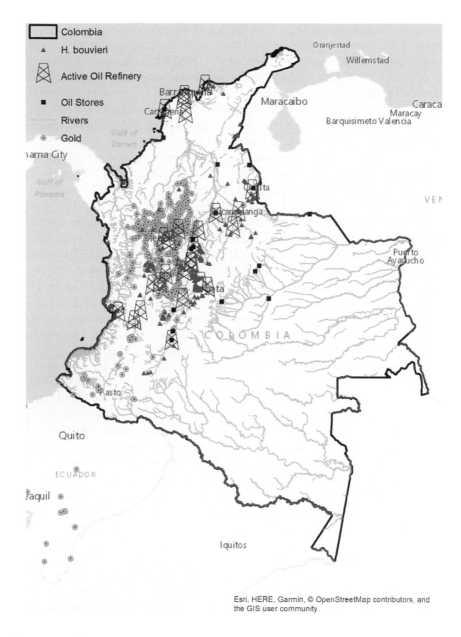

Figure 14.3   Base map with collection location data for *Hypolobocera bouvieri* (red triangles) and industry within Colombia. Types of industry included: petroleum mined as well as stores, mineral mines, and phosphate.

distribution. Colombia's rapid development could cause negative impacts on critically endangered endemic species of freshwater crabs. There is considerable overlap between mining, petroleum extraction, and the modeled locations of this species. Historically, as in other industrializing countries, development has extended from navigable waterways toward the interior. This increases the risk that spills and effluent impose on aquatic species, and the risks for human health in the areas since these are often the main sources of drinking water. These species are at unprecedented risk of declining populations when the threats from global heating and from encroachment by invasive species act together. Ecological changes impact not only the conservation of these crab species, but also the human interests of food procurement and avoidance of disease (Casas et al. 2008, Arias et al. 2011).

14.4 HUMAN-MEDIATED RISKS TO CRUSTACEAN SPECIES

Work to address overfishing, habitat loss, increased human population density, longitudinal habitat fragmentation, introduced species, and pollution are underway (Barletta et al. 2016). The risks for crustaceans, while similar to fishes, probably more closely parallel those of mollusks, but fall into the same categories (Lasso et al. 2016). Climate change is not explored in great detail below, but all risks may be enhanced by climate change, particularly changes to species interactions and microhabitats that restrict many isolated species from migrating (Carr and Tognelli 2016).

14.4.1 Increasing Connectivity and Human Activity

The rate of species introductions has increased worldwide, exacerbating other effects (Pimentel et al. 2005, Kareiva et al. 2007, Bates et al. 2014). Conflict over food security and conservation in global hotspots is a major issue worldwide (Molotoks et al. 2017). Regions suffering food insecurity tend to devalue conservation actions (Baptiste et al. 2017, Suarez et al. 2018). There is concern that food security could produce further pressure on these crustaceans. Increased connectivity and the pressures of human activities also pressure troglobitic species, particularly with pollution, destruction of karst in subterranean habitats, and uncontrolled tourism (Lasso et al. 2018).

14.4.2 Development

Development in riparian and coastal systems causes changes in land use that impact freshwater, estuarine, and marine systems (Collen et al. 2014, Molotoks et al. 2017). Changes in land use due to development are the greatest threat to biodiversity in freshwater ecosystems (Collen et al. 2014). Many species considered for endangered status in the 2014 analysis (Campos and Lasso 2015) were considered high risk specifically due to urban development and increased human population density, for example, pseudothelphusids and trichodactylids living in areas along the Magdalena River Basin and coastal Caribbean. Pseudothelphusids are at risk throughout Andean

montane areas that are desirable for development. Areas of high biodiversity are
also likely to be desirable development locations (e.g., coasts, dunes, proximity to
water resources, etc.), leaving native species vulnerable to habitat loss and extinction
(Baptiste et al. 2017). For example, about 2.6 billion people live within 100 km of a
coast worldwide (Sale et al. 2014), with a projected 938 million in the more at-risk
low-elevation coastal zone by 2030 (Neumann et al. 2015). Humans live dispropor-
tionately near waterways, modifying riparian zones and discharging nutrients, sediments, and contaminants. Streams and rivers are more vulnerable to these effects
than lakes due to watershed area (Sala et al. 2000).

14.4.3 Deforestation

One of the principal risks for freshwater crustaceans at higher altitudes is the loss
of riparian buffers. Colombia has an estimated loss of 1 million hectares of forests
in the last 50 years (Baptiste et al. 2017). Deforestation exacerbates environmental
problems due to runoff and siltation from erosion, impeding riparian buffers and
water quality levels (Dudgeon 1992, 2000, Roulet et al. 2001, Dudgeon et al. 2006,
Strayer 2006, Jiménez-Segura et al. 2016). The physicochemical changes affect food
abundance, create physiological challenges, and potentially eliminate microhabitats
(Roulet et al. 2001, Jiménez-Segura et al. 2016).

Deforestation in the Magdalena River Valley has contributed a large portion of
global deforestation (Alvarez-Berríos and Mitchell Aide 2015), and is also threatened by gold mining, crude oil extraction, and palm oil cultivation. The conversion
of tropical forest to palm oil plantations is driven by global demand for palm oil
(Pardo Vargas et al. 2015) and is slated for significant expansion (Ocampo-Peñuela et
al. 2018). However, spatial data exists, and should be used in development planning
and to minimize conservation risks.

14.4.4 Mining and Petrochemical Extraction

Colombia has major fossil fuel and mineral deposits (Gonzalez-Salazar et al. 2017).
The consequences of exploiting these resources may threaten water resources and
conservation outcomes if not managed properly (Lia and Kjøk 2004). There have
been petroleum spills caused by equipment failure (Katz 2018), recent regional con-
flicts, and terrorism (Brodzinsky 2015). Industrial fracking chemicals, known to
have toxic effects on reproductive success (Webb et al. 2014, Elliott et al. 2016),
are a concern for water quality risk analysis (Entrekin et al. 2015), considering that
oil and gas fracking was recently approved on a pilot basis in Colombia (Cobb and
Acosta 2019).

Mining also poses major runoff and storage issues, and often scar the landscape.
Gold mining regionally produces freshwater mercury pollution (Roulet et al. 1999,
2001), and has resulted in the highest per capita mercury pollution in the world
(Cordy et al. 2011). It is estimated that 9% of total deforestation in Colombia is attributed to gold mining, and there is a great deal of illegal mining (either subsistence
or larger scale) as well (Alvarez-Berríos and Mitchell Aide 2015). Combined with

deforestation, mining can cause mercury content increases in fish populations and pass the contaminant up the food chain to humans and other vertebrates (Olivero et al. 2002, Olivero-Verbel et al. 2011, Alvarez et al. 2012a,b). Salt mining runoff can cause increased salinity of soils and freshwater, and higher exposures to chloride (Trombulak and Frissell 2000, Evans and Frick 2001, Findlay and Kelly 2011, Corsi et al. 2015). Freshwater animals are physiologically selected to deal with low salinities, and are not often equipped to maintain internal osmotic balance if exposed to salinities outside their normal tolerance range.

14.4.5 Agriculture and Aquaculture Activities

Colombia has a diverse and extensive agricultural sector to sustain a growing population of almost 50 million people, including operational inland fish and invertebrate aquaculture. Agriculture has changed the landscape, replacing lowland forests with palm oil plantations, livestock ranching, horticulture farms, coffee, and cacao (Clerici et al. 2019). This does not include the illicit cultivation of coca and opium in areas that were controlled for decades by armed groups, though it is difficult to convince farmers to grow less profitable alternatives due to the unparalleled demand for drugs in developed economies (Baptiste et al. 2017, Suarez et al 2018, Clerici et al. 2019). Though some agricultural activities, particularly coffee and cacao cultivation, can be achieved more sustainably through forest floor cocultivation that leaves a lot of forest intact, the historical effects of clear-cutting have left scars on the landscape (Baptiste et al. 2017). Encouragement of sustainable agricultural practices moving forward, along with the protection of large swaths of habitat, will help enhance conservation priorities in the country.

Some aquaculture farms are coculturing species to help relieve subsistence-fishing pressure. However, fish and invertebrates not native to certain river valleys are being cultivated in those places. This could be a source of in-country invasion across boundaries. Aquaculture and some regional government entities introduced a number of nonnative invertebrates, including at least two crustaceans that are now established: the Louisiana red-clawed crayfish, *Procambarus clarkii* (Girard 1952), and a species of giant river prawn, *Macrobrachium rosenbergii* De Man, 1879 (Campos 2005, Álvarez-León and Gutiérrez-Bonilla 2007). Inland fisheries crashed from 60,000 tons in 1975 to 10,000 tons in 2014 from overfishing, and remain largely artisanal (Jiménez-Segura et al. 2016). Shifting subsistence anglers to more sustainable food production methods and economic value is imperative for recovery. Crabs are often the subjects of artisanal fisheries, which, depending on consumption methods, can provide a parasite vector to humans (Arias et al. 2011).

14.4.6 Transportation Infrastructure Impacts

Increased development of road and rail networks in Colombia, along with hydroelectric dam construction, increases habitat fragmentation and poses a further threat to narrow-range endemic species. Most goods are transported to market via road and ship, with rail playing a minor role. The pending construction of the Puente

Terrestre Inter-Oceánico land bridge (rail, road, canal, and oil pipeline) will likely have a major environmental impact and result in further deforestation. The under-developed wastewater treatment facilities in much of the country means that many areas have major problems with untreated waste effluent. Plastic waste is inundating rivers and streams, and introducing chemicals that affect the physiology of aquatic life (McGinnis et al. 2012). Compounded with the migrant crisis both internally and with neighboring countries, desperate people may cause additional stress to freshwater species (Barletta et al. 2010, Barletta et al. 2016).

14.5 CONCLUSIONS AND RECOMMENDATIONS

With potential increases in area for widely tolerant species as observed in the model here, a reduction in biodiversity from competition could mean that species extirpations could accelerate as range-expanding species act as if they are invasive. This case study highlights the need for additional studies of widely distributed species because there could be regional differences between populations beyond morphology alone. A more dire situation is likely for range-restricted animals that are at risk from stochastic events such as industrial spills.

Integrating anthropogenic influences into conservation plans for freshwater species is critical for saving species that will be affected by climate change. This case study has demonstrated the influence of continued human impacts, including industrial extraction sites for mining and petroleum on the area available for survival of a crustacean species that will be affected by climate change. Species extinctions are likely to occur from the multiple stressors affecting these populations (Collen et al. 2014). There is certainly a need to combine these threats when determining species assessments as part of IUCN Red List assessments to inform future field work. Much has been done to assess the climate risk to fishes, mollusks, and other invertebrates (Carr and Tognelli 2016, Lasso et al. 2016), which will hopefully lead to additional assessments of other taxa. The impact on undersampled areas in Colombia is obvious based on sampling for all taxa in the country, but hopefully this will provide a focus for potential sampling efforts for this and other freshwater species. Cave species are particularly restricted (Lasso et al. 2018) and need additional collection effort to determine their geographic extents.

In light of environmental pressures on populations, efficient infrastructure for the collection and processing of species data is imperative for effective species management in megadiverse developing countries (Paknia et al. 2015). Overcorrections can limit biodiversity research too, so policy that allows for the collection of critical information on species is important, particularly for developing nations (Prathapan et al. 2018). This underlines the need to update the conservation status for all species, especially the data-deficient species. Genetic and morphological information for pseudothelphusids would be helpful to aid in species identification. It is also important to document biogeographic shifts as the climate crisis intensifies and to observe any changes to the distributional ranges of species (especially range-restricted species, see Parmesan (2006)). All of these recommendations fit into providing additional information for conservation and policy actions in the region, without which these aquatic species stand little chance of continuing to coexist in a rapidly developing region.

ACKNOWLEDGMENTS

The authors would like to thank the Instituto de Ciencias Naturales, Universidad Nacional de Colombia, Bogotá, DC, Colombia (ICN) for continued collaboration and support, along with the many student contributors to Professor Martha R. Campos' dataset on *Hypolobocera bouvieri* collection locations. We would also like to thank Rebecca Kulp (the Maritime Aquarium at Norwalk) for her valuable comments on the manuscript. This work was supported through the Maritime Aquarium at Norwalk's conservation ask program and the George S. and Pamela M. Humphrey Fund. Additionally, the authors would like to thank the editors and reviewers for their valuable comments on the manuscript.

REFERENCES

Alvarez, S., Jessick, A.M., Palacio, J.A. and Kolok, A.S. 2012a. Methylmercury concentrations in six fish species from two Colombian Rivers. *Bulletin of Environmental Contamination and Toxicology* 88:65–68.

Alvarez, S., Kolok, A.S., Jimenez, L.F., Granados, C. and Palacio, J.A. 2012b. Mercury concentrations in muscle and liver tissue of fish from marshes along the Magdalena River, Colombia. *Bulletin of Environmental Contamination and Toxicology* 89:836–40.

Alvarez-Berríos, N.L. and Mitchell Aide, T. 2015. Global demand for gold is another threat for tropical forests. *Environmental Research Letters* 10:014006.

Álvarez-León, R. and Gutiérrez-Bonilla, F. de P. 2007. Situación de los invertebrados acuáticos introducidos y transplantados Colombia: antecedentes, efectos y perspectivas. *Revista de la Academia Colombiana de Ciencias Exactas, Físicas y Naturales* 31:557–74.

Arias, S.M., Salazar, L.M., Casas, E., Henao, A. and Velásquez, L.E. 2011. *Paragonimus* sp. en cangrejos y sensibilización de la comunidad educativa hacia los ecosistemas acuáticos de La Miel y La Clara, Caldas, Antioquia. *Biomédica* 31:209.

Baldwin, R.A. 2009. Use of maximum entropy modeling in wildlife research. *Entropy* 11:854–66.

Baptiste, B., Pinedo-Vasquez, M., Gutierrez-Velez, V.H., et al. 2017. Greening peace in Colombia. *Nature Ecology and Evolution* 1:0102.

Barletta, M., Cussac, V.E., Agostinho, A.A., et al. 2016. Fisheries ecology in South American river basins. In *Freshwater Fisheries Ecology*, ed. J.F. Craig, 311–48. Hoboken, NJ: Wiley Blackwell.

Barletta, M., Jaureguizar, A.J., Baigun, C., et al. 2010. Fish and aquatic habitat conservation in South America: A continental overview with emphasis on neotropical systems. *Journal of Fish Biology* 76:2118–76.

Barnhart, P.R. and Gillam, E.H. 2014. The impact of sampling method on maximum entropy species distribution modeling for bats. *Acta Chiropterologica* 16:241–48.

Bates, A.E., Pecl, G.T., Frusher, S., et al. 2014. Defining and observing stages of climate-mediated range shifts in marine systems. *Global Environmental Change* 26:27–38.

Blair, D., Agatsuma, T. and Wang, W. 2008. Paragonimiasis. In *Food-borne Parasitic Zoonoses*, eds. K.D. Murrell, B. Fried, 117–150. New York: Springer.

Brodzinsky, S. 2015. Farc rebels bomb new section of Colombian oil pipeline. The Guardian. Downloaded from https://www.theguardian.com/world/2015/jun/29/colombia-farc-rebels-bomb-oil-pipeline (accessed 29 June 2015).

Burgmer, T., Hillebrand, H. and Pfenninger, M. 2007. Effects of climate-driven temperature changes on the diversity of freshwater macroinvertebrates. *Oecologia* 151:93–103.

Butler, R. 2016. The top 10 most biodiverse countries. Downloaded from https://news. mongabay. com/2016/05/top-10-biodiverse-countries/ (accessed 31 March 2019).

Campos, M.R. 2005. *Procambarus (Scapulicambarus) clarkii* (Girard, 1852), (Crustacea: Decapoda: Cambaridae). Una langostilla no nativa en Colombia. *Revista de la Academia Colombiana de Ciencias* 29:295–302.

Campos, M.R. 2014. *Crustáceos decápodos de agua dulce de Colombia*. Bogotá, DC, Columbia: Universidad Nacional de Colombia.

Campos, M.R. 2017. Two new species of freshwater, cave-dwelling crabs of the genus *Neostrengeria* Pretzmann, 1965, from Colombia (Crustacea: Decapoda: Pseudothelphusidae). *Zootaxa* 4247:157–64.

Campos, M.R. and Camacho, R. 2019. A new species of freshwater crab of the genus *Strengeriana* Pretzmann, 1971, from El Jardín Natural Reserve, Quindío, Colombia (Crustacea: Decapoda: Pseudothelphusidae). *Zootaxa* 4671:595–600.

Campos, M.R. and Lasso CA. 2015. *Libro rojo de los cangrejos dulceacuícolas de Colombia*. Bogotá, DC, Colombia: Instituto de Investigación de Recursos Biológicos Alexander von Humboldt (IAvH), Instituto de Ciencias Naturales de la Universidad Nacional de Colombia.

Campos, M.R., Lasso, C.A. and Arias, M. 2019. A new species of freshwater crab of the genus *Phallangothelphusa* Pretzmann, 1965, from the foothills of the Serranía Yariguíes of Colombia (Crustacea: Decapoda: Pseudothelphusidae). *Zootaxa* 4550:579–84.

Carr, J. and Tognelli, M.F. 2016. Evaluación de la vulnerabilidad al cambio climático de las especies de agua dulce de los Andes Tropicales. In *Estado de conservación y distribución de la biodiversidad de agua dulce en los Andes tropicales*, eds. M.F. Tognelli, C.A. Lasso, C.A. Bota-Sierra, L.F. Jiménez-Segura, and N.A. Cox, 127–55. Gland, Switzerland: IUCN.

Casas, E., Gomez, C., Valencia, E., Salazar, L. and Velásquez, L.E. 2008. Estudio de foco de paragonimosis en Ruente Clara, Robledo, area periurbana de Medellin, Antioquia. *Biomédica* 28:396–403.

Case, T.J. and Taper, M.L. 2000. Interspecific competition, environmental gradients, gene flow, and the coevolution of species' boarders. *American Naturalist* 155:583–605.

Clerici, N., Salazar, C., Pardo-Díaz, C., Jiggins, C.D., Richardson, J.E. and Linares, M. 2019. Peace in Colombia is a critical moment for Neotropical connectivity and conservation: Save the northern Andes-Amazon biodiversity bridge. *Conservation Letters* 12: e12594.

Cobb, J.S. and Acosta, L.J. 2019. Fracking could nearly triple Colombia oil and gas reserves: Minister. Reuters. Downloaded from https://www.reuters.com/article/us-colombia-energy/fracking-could-nearly-triple-colombia-oil-and-gas-reserves-minister-idUSKCN1QP1OO (accessed 11 November 2019).

Collen, B., Whitton, F., Dyer, E.E., et al. 2014. Global patterns of freshwater species diversity, threat and endemism: Global freshwater species congruence. *Global Ecology and Biogeography* 23:40–51.

Cordy, P., Veiga, M.M., Salih, I., et al. 2011. Mercury contamination from artisanal gold mining in Antioquia, Colombia: The world's highest per capita mercury pollution. *Science of The Total Environment* 410–411:154–60.

Corsi, S.R., De Cicco, L.A., Lutz, M.A. and Hirsch, R.M. 2015. River chloride trends in snow-affected urban watersheds: Increasing concentrations outpace urban growth rate and are common among all seasons. *Science of the Total Environment* 508:488–97.

Cumberlidge, N., Alvarez, F. and Villalobos, J.L. 2014. Results of the global conservation assessment of the freshwater crabs (Brachyura, Pseudothelphusidae and Trichodactylidae): The Neotropical region, with an update on diversity. *ZooKeys* 457:133–57.

Cumberlidge, N., Ng, P.K.L., Yeo, D.C.J., et al. 2009. Freshwater crabs and the biodiversity crisis: Importance, threats, status, and conservation challenges. *Biological Conservation* 142:1665–73.

Dobson, M., Magana, A., Mathooko, J.M. and Ndegwa, F.K. 2002. Detritivores in Kenyan highland streams: More evidence for the paucity of shredders in the tropics? *Freshwater Biology* 47:909–19.

Dobson, M., Magana, A.M., Lancaster, J. and Mathooko, J.M. 2007a. Aseasonality in the abundance and life history of an ecologically dominant freshwater crab in the Rift Valley, Kenya. *Freshwater Biology* 52:215–25.

Dobson, M., Magana, A.M., Mathooko, J.M. and Ndegwa, F.K. 2007b. Distribution and abundance of freshwater crabs (*Potamonautes* spp.) in rivers draining Mt Kenya, East Africa. *Fundamental and Applied Limnology/Archiv für Hydrobiologie* 168:271–79.

Dudgeon, D. 1992. Endangered ecosystems: A review of the conservation status of tropical Asian rivers. *Hydrobiologia* 248:167–91.

Dudgeon, D. 2000. The ecology of tropical Asian rivers and streams in relation to biodiversity conservation. *Annual Review of Ecology and Systematics* 31:239–63.

Dudgeon, D., Arthington, A.H., Gessner, M.O., et al. 2006. Freshwater biodiversity: Importance, threats, status and conservation challenges. *Biological Reviews* 81:163–82.

Elith, J., Graham, C.H., Anderson, R.P., et al. 2006. Novel methods improve prediction of species' distributions from occurrence data. *Ecography* 29:129–51.

Elliott, E.G., Ettinger, A.S., Leaderer, B.P., Bracken, M.B. and Deziel, N.C. 2016. A systematic evaluation of chemicals in hydraulic-fracturing fluids and wastewater for reproductive and developmental toxicity. *Journal of Exposure Science and Environmental Epidemiology* 27:90–99.

Entrekin, S.A., Maloney, K.O., Kapo, K.E., Walters, A.W., Evans-White, M.A. and Klemow, K.M. 2015. Stream vulnerability to widespread and emergent stressors: A focus on unconventional oil and gas. *PLoS ONE* 10:e0137416.

Evans, M. and Frick, C. 2001. The effects of road salts on aquatic ecosystems. Environment Canada, Burlington, Ontario, Canada, Report No.: National Water Resources Institute Series No. 02–308: 298. Downloaded from http://scec.ca/pdf/the_effects_road_salts .pdf (accessed 11 November 2019).

Farallo, V.R. and Miles, D.B. 2016. The importance of microhabitat: A comparison of two microendemic species of *Plethodon* to the widespread *P. cinereus*. *Copeia* 104:67–77.

Fick, S.E. and Hijmans, R.J. 2017. WorldClim 2: New 1-km spatial resolution climate surfaces for global land areas. *International Journal of Climatology* 37:4302–15.

Findlay, S.E.G. and Kelly, V.R. 2011. Emerging indirect and long-term road salt effects on ecosystems: Findlay and Kelly. *Annals of the New York Academy of Sciences* 1223:58–68.

Gonzalez-Salazar, M.A., Venturini, M., Poganietz, W-R., Finkenrath, M., L.V. and Leal, M.R. 2017. Combining an accelerated deployment of bioenergy and land use strategies: Review and insights for a post-conflict scenario in Colombia. *Renewable and Sustainable Energy Reviews* 73:159–77.

Guarnizo, C.E., Paz, A., Muñoz-Ortiz, A., Flechas, S.V., Méndez-Narváez, J. and Crawford, A.J. 2015. DNA barcoding survey of Anurans across the Eastern Cordillera of Colombia and the impact of the Andes on cryptic diversity. *PLoS ONE* 10:e0127312.

Hawkins, B.A., Field, R., Cornell, H.V., et al. 2003. Energy, water, and broad-scale geographic patterns of species richness. *Ecology* 84:3105–17.

Hillebrand, H. 2004. On the generality of the latitudinal diversity gradient. *The American Naturalist* 163:192–211.

Hobohm, C. 2003. Characterization and ranking of biodiversity hotspots: Centres of species richness and endemism. *Biodiversity and Conservation* 12:279–87.

Hudson, D.M., Brittain, V. and Phillips, G. 2016. Behavioral response to temperature change by the freshwater crab *Neostrengeria macropa* (H. Milne Edwards, 1853) (Brachyura: Pseudothelphusidae) in Colombia. *Journal of Crustacean Biology* 36:287–94.

Hutter, C.R., Guayasamin, J.M. and Wiens J.J. 2013. Explaining Andean megadiversity: The evolutionary and ecological causes of glass frog elevational richness patterns. *Ecology Letters* 16:1135–44.

IPCC (Intergovernmental Panel on Climate Change). 2014. Climate change 2014: Synthesis Report. In *Contribution of Working Groups I, II and III to the Fifth Assessment Report of the Intergovernmental Panel on Climate Change*, eds. Core Writing Team, R.K. Pachauri, and L.A. Meyer, 1–151. Geneva, Switzerland: IPCC.

Jiménez-Segura, L.F., Galvis-Vergara, G., Cala-Cala, P., et al. 2016. Freshwater fish faunas, habitats and conservation challenges in the Caribbean river basins of north-western South America: Freshwater Fishes of North-West South America. *Journal of Fish Biology* 89:65–101.

Kareiva, P., Watts, S., McDonald, R. and Boucher, T. 2007. Domesticated nature: Shaping landscapes and ecosystems for human welfare. *Science* 316:1866–69.

Katz, B. 2018. Oil spill in Colombia kills 2,400 animals. Smithsonian Magazine. Downloaded from https://www.smithsonianmag.com/smart-news/oil-spill-colombia-has-killed-2400-animals-180968653/ (accessed 11 November 2019).

Lasso, C.A., Campos, M.R. and Fernández Auderset, J. 2018. Cangrejos cavernícolas de Colombia [cave-dwelling crabs of Colombia]. In Conference paper. I Congreso Colombiano de Espeleología y VIII Congreso Espeleológico de América Latina y el Caribe, San Gil (Santander), Colombia, 59–63.

Lasso, C.A. Correoso, M., Lopes-Lima, M., Ramírez, R. and Tognelli, M.F. 2016. Estado de conservación y distribución de los moluscos de agua dulce de los Andes Tropicales. In *Estado de conservación y distribución de la biodiversidad de agua dulce en los Andes Tropicales*, eds. M.F. Tognelli, C.A. Lasso, C.A. Bota-Sierra, L.F. Jiménez-Segura, and N.A. Cox, 57–66. Gland, Switzerland: IUCN.

Lia, B. and Kjøk, Å. 2004. Energy supply as terrorist targets? Patterns of "petroleum terrorism." 1968–99. In *Oil in the Gulf: Obstacles to Democracy and Development*, eds. D. Heradstveit, and H. Hveem, 100–124.Aldershot, Hants, England: Ashgate; Burlington, VT: Routledge.

Loarie, S.R., Carter, B.E., Hayhoe, K., et al. 2008. Climate change and the future of California's endemic flora. *PLoS ONE* 3:e2502.

Malcolm, J.R., Liu, C., Neilson, R.P., Hansen, L. and Hannah, L. 2006. Global warming and extinctions of endemic species from biodiversity hotspots. *Conservation Biology* 20:538–48.

Manne, L.L. and Pimm, S.L. 2001. Beyond eight forms of rarity: which species are threatened and which will be next? *Animal Conservation* 4:221–29.

Mantilla-Meluk, H. and Munoz-Garay, J. 2014. Biogeography and taxonomic status of *Myotis keaysi pilosatibialis* LaVal 1973 (Chiroptera: Vespertilionidae). *Zootaxa* 3793:060–070.

McCain, C.M. and Colwell, R.K. 2011. Assessing the threat to montane biodiversity from dis-cordant shifts in temperature and precipitation in a changing climate: Climate change risk for montane vertebrates. *Ecology Letters* 14:1236–45.

McGinnis, C.L., Encarnacao, P.C. and Crivello, J.F. 2012. Dibutyltin (DBT) an endo-crine disrupter in zebrafish. *Journal of Experimental Marine Biology and Ecology* 430–431:43–47.

Michelutti, N., Wolfe, A.P., Cooke, C.A., Hobbs, W.O., Vuille, M. and Smol, J.P. 2015. Climate change forces new ecological states in tropical Andean lakes. *PLoS ONE* 10:e0115338.

Mittermeier, R.A. (ed.). 2004. *Hotspots Revisited: Earth's Biologically Richest and Most Endangered Terrestrial Ecoregions.* Mexico City: CEMEX.

Molotoks, A., Kuhnert, M., Dawson, T. and Smith, P. 2017. Global hotspots of conflict risk between food security and biodiversity conservation. *Land* 6:67.

Myers, N., Mittermeier, R.A., Mittermeier, C.G., da Fonseca, G.A.B. and Kent, J. 2000. Biodiversity hotspots for conservation priorities. *Nature* 403:853–58.

Neumann, B., Vafeidis, A.T., Zimmermann, J. and Nicholls, R.J. 2015. Future coastal popula-tion growth and exposure to sea-level rise and coastal flooding—A global assessment. *PLoS ONE* 10:e0118571.

Neumann, H., de Boois, I., Kröncke, I. and Reiss, H. 2013. Climate change facilitated range expansion of the non-native angular crab *Goneplax rhomboides* into the North Sea. *Marine Ecology Progress Series* 484:143–53.

Ocampo-Peñuela, N., Garcia-Ulloa, J., Ghazoul, J. and Etter, A. 2018. Quantifying impacts of oil palm expansion on Colombia's threatened biodiversity. *Biological Conservation* 224:117–21.

Olivero, J., Johnson, B. and Arguello, E. 2002. Human exposure to mercury in San Jorge river basin, Colombia (South America). *Science of The Total Environment* 289:41–47.

Olivero-Verbel, J., Caballero-Gallardo, K. and Negrete-Marrugo, J. 2011. Relationship between localization of gold mining areas and hair mercury levels in people from Bolivar, north of Colombia. *Biological Trace Element Research* 144:118–32.

Orme, C.D.L., Davies, R.G., Burgess, M., et al. 2005. Global hotspots of species richness are not congruent with endemism or threat. *Nature* 436:1016–19.

Paknia, O., Rajaei, Sh, H. and Koch, A. 2015. Lack of well-maintained natural history col-lections and taxonomists in megadiverse developing countries hampers global biodi-versity exploration. *Organisms Diversity and Evolution* 15:619–29.

Pardo Vargas, L.E., Laurance, W.F., Clements, G.R. and Edwards, W. 2015. The impacts of oil palm agriculture on Colombia's biodiversity: what we know and still need to know. *Tropical Conservation Science* 8:828–45.

Parmesan, C. 2006. Ecological and evolutionary responses to recent climate change. *Annual Review of Ecology, Evolution, and Systematics* 37:637–99.

Pearson, R.G., Raxworthy, C.J., Nakamura, M. and Townsend Peterson, A. 2006. Predicting species distributions from small numbers of occurrence records: a test case using cryptic geckos in Madagascar: Predicting species distributions with low sample sizes. *Journal of Biogeography* 34:102–17.

Peterson, A.T., Ortega-Huerta, M.A., Bartley, J., Sanchez-Cordero, V., Soberon, J., Buddemeier, R.H. and Stockwell, D.R.B. 2002. Future projections for Mexican faunas under global climate change scenarios. *Nature* 416:626–29.

Phillips, G., Hudson, D.M. and Chaparro-Gutiérrez, J.J. 2019. Presence of *Paragonimus* spe-cies within secondary crustacean hosts in Bogotá, Colombia. *Revista Colombiana de Ciencias Pecuarias* 32:150–57.

Phillips, S.J., Anderson, R.P. and Schapire, R.E. 2006. Maximum entropy modeling of spe-cies geographic distributions. *Ecological Modelling* 190:231–59.

Pimentel, D., Zuniga, R. and Morrison, D. 2005. Update on the environmental and economic costs associated with alien-invasive species in the United States. *Ecological Economics* 52:273–88.

Pimm, S.L. and Raven, P. 2000. Extinction by numbers. *Nature* 403:843–45.

Prathapan, K.D., Pethiyagoda, R., Bawa, K.S., Raven, P.H., Rajan, P.D. and 172 co-signatories from 35 countries. 2018. When the cure kills-CBD limits biodiversity research. *Science* 360:1405–06.

Rathbun, M.J. 1898. A contribution to a knowledge of the fresh-water crabs of America. The Pseudothelphusidae. *Proceedings of the United States National Museum* 21:507–37.

Roulet, M., Lucotte, M., Farella, N., et al. 1999. Effects of recent human colonization on the presence of mercury in Amazonian ecosystems. *Water, Air and Soil Pollution* 112:297–313.

Roulet, M., Lucotte, M., Canuel, R., et al. 2001. Spatio-temporal geochemistry of mercury in waters of the Tapajós and Amazon rivers, Brazil. *Limnology and Oceanography* 46:1141–57.

Roy, K., Jablonski, D. and Valentine, J.W. 2001. Climate change, species range limits and body size in marine bivalves. *Ecology Letters* 4:366–70.

Sadowski, J.S., Gonzalez, J.A., Lonhart, S.I., Jeppesen, R., Grimes, T.M. and Grosholz, E.D. 2018. Temperature-induced range expansion of a subtropical crab along the California coast. *Marine Ecology* 39:e12528.

Sala, O.E., Chapin III, S.F., Armesto, J.J., et al. 2000. Global biodiversity scenarios for the year 2100. *Science* 287:1770–74.

Sale, P.F., Agardy, T., Ainsworth, C.H., et al. 2014. Transforming management of tropical coastal seas to cope with challenges of the 21st century. *Marine Pollution Bulletin* 85:8–23.

Strayer DL. 2006. Challenges for freshwater invertebrate conservation. *Journal of the North American Benthological Society* 25:271–87.

Suarez, A., Árias-Arévalo, P.A. and Martínez-Mera, E. 2018. Environmental sustainability in post-conflict countries: Insights for rural Colombia. *Environment, Development and Sustainability* 20:997–1015.

Suárez-Seoane, S., García de la Morena, E. L., Morales Prieto, M. B., Osborne, P. E. and de Juana, E. 2008. Maximum entropy niche-based modelling of seasonal changes in little bustard (*Tetrax tetrax*) distribution. *Ecological Modelling* 219:17–29.

Thomas, C.D., Cameron, A., Green, R.E., et al. 2004. Extinction risk from climate change. *Nature* 427:145–48.

Trombulak, S.C. and Frissell, C.A. 2000. Review of ecological effects of roads on terrestrial and aquatic communities. *Conservation Biology* 14:18–30.

Uruburu, M., Granada, M. and Velasquez, L.E. 2008. Distribucion parcial de Paragonimus en Antioquia, por presencia de metacercarias en cangrejon dulciacuicolas. *Biomédica* 28:562–68.

Vélez, I., Velásquez, L.E. and Vélez, I.D. 2003. Morphological description and life cycle of *Paragonimus* sp. (Trematoda: Troglotrematidae): Causal agent of human Paragonimiasis in Colombia. *Journal of Parasitology* 89:749–55.

Vélez, I.D., Ortega, J., Hurtado, M. and Salazar, A.L. 1995. La paragonimosis en al comunidad indígena Emberá de Colombia. *Biomédica* 27:51–54.

Webb, E., Bushkin-Bedient, S., Cheng, A., Kassotis, C.D., Balise, V. and Nagel, S.C. 2014. Developmental and reproductive effects of chemicals associated with unconventional oil and natural gas operations. *Reviews on Environmental Health* 29:307–18.

Williams, S.E., Bolitho, E.E. and Fox, S. 2003. Climate change in Australian tropical rainforests: an impending environmental catastrophe. *Proceedings of the Royal Society B: Biological Sciences* 270:1887–92.

Young, N., Carter, L. and Evangelista, P. 2011. *A MaxEnt model v3. 3.3 e tutorial (ArcGIS v10)*. Fort Collins, CO: Natural Resource Ecology Laboratory at Colorado State University and the National Institute of Invasive Species Science.

CHAPTER 15

A New Morphotype of the Crayfish *Cambarus hubrichti* (Decapoda: Cambaridae) from a Karst Spring Cave System, with Comments on Its Ecology

Teresa M. Carroll, D. Christopher Rogers, David B. Stern, and
Keith A. Crandall

CONTENTS

15.1 INTRODUCTION

Karst topography is characterized by sinkholes, losing streams, rapid-flowing drainage patterns, caves, and large springs (Vineyard and Fender 1982, Adamski et al. 1995). These geomorphological features are organized by the process of karstification in the porous, soluble rock (primarily limestone and dolomite), typical of karst terrains. Subterranean karst caves and cave streams (above the water table) in North America harbor what is often termed troglobitic or terrestrial, cave-limited species (cf. Culver and Sket 2000). These generally include oligochaete worms, free-living flatworms (cave planarians), nematodes (roundworms), crustaceans (amphipods, isopods, copepods, and decapods), arachnids (Acari; ticks and mites), centipedes and millipedes, salamanders, and a few insect species (Culver et al. 2000, Culver et al. 2003, Romero 2009). Compared to surface streams, diversity in these caves is low, often due to the fragmented nature of cave habitats and limited dispersal abilities (Skeet 1999).

Fauna associated with phreatic (underwater) limestone caves (phreatic conduits) harbor stygobitic aquatic, cave-limited species (cf. Culver and Sket 2000). These cave communities encompass most of the same taxonomic groups as terrestrial cave habitats (annelids, crustaceans, arachnids, and insects), albeit at markedly lower densities (Coineau 2000). Interestingly, crustaceans comprise 43% of the obligate phreatic-dwelling species (Gibert and Culver 2009). Once considered biologically depauperate ecosystems (Romero 2009), both terrestrial and aquatic caves represent small restricted hotspots of subterranean biodiversity (Culver and Sket 2000).

In this chapter, we describe a new morphotype of the Salem Cave Crayfish, *Cambarus hubrichti* captured after being sighted during several diving expeditions in phreatic karst caves in the Salem Plateau region of Missouri, USA. The North American genus *Cambarus* (family Cambaridae) was recently comprised of approximately ten subgenera (Hobbs 1989), which were clearly nonmonophyletic groups (Breinholt et al. 2012), with *Cambarus hubrichti* placed in the subgenus *Erebicambarus*. This subgenus was comprised of six species, most of which are associated with caves and springs (Hobbs 1989). However, Crandall and De Grave (2017) recently updated the crayfish taxonomy and eliminated the subgeneric designations as there was clear evidence against these taxonomic groups representing phylogenetic groupings. In North America, populations of cave crayfish within *Cambarus* occur in karstic (limestone) areas of West Virginia, the Cumberland Plateau of the Southern Appalachians, and the Ozarks Plateau region of Kansas, Oklahoma, Arkansas, and Missouri (Hobbs and Barr 1960, Reynolds et al. 2013), representing the largest invertebrates of subterranean ecosystems in the Missouri Ozarks (Koppelman and Figg 1995). The existence of these populations in subterranean habitats places them within the 95% of obligate cave-dwelling species considered "imperiled" by the Nature Conservancy (Culver et al. 2000).

Caves have been characterized as evolutionary laboratories, unique for studying natural selection and adaptations in cave fauna (White 1969). Thus, cave crayfish within *Cambarus* exhibit a number of morphological, behavioral, and physiological adaptations that help them survive in these dark habitats (Hobbs et al. 1977).

Adaptations in cave crayfish include loss of pigmentation and eyes, and elongated limbs and antennae, which enhance sensory perceptions (Li and Coper 2002). It is also assumed that olfactory sensory structures may have evolved functionality (Cooper et al. 2001) possibly for finding food and mates. Cave-adapted crayfish also exhibit long life spans, with estimates ranging from 9 to 10 years (Weingartner 1977) to as long as 16 (Streever 1996) or 22 years (Venarsky et al. 2012). In a large cave in Alabama crayfish, age estimates ranged between 37 and 176 years (Cooper 1975).

In this study, we also examined preliminary ecological data and various biological and ecological explanations for microhabitat choice and trophic feeding pathways of this new morphotype to inform subsequent studies. Until 1960, contributions to biospeology by American biologists (beyond taxonomic characterizations) were deficient (Romero 2009). Hence, our understanding of cave biology and ecology is minimal relative to other ecosystems. Detailed biodiversity assessments are limited in these habitats by several factors. Researchers have reported that assessments are difficult because invertebrate densities in most phreatic systems are low (Gibert and Deharveng 2002). Moreover, there are obvious risks associated with scientific study of deep phreatic cave environments. The inaccessibility of submerged subterranean habitats has consequently limited our knowledge of underground fauna to that primarily studied through artificial access points (wells, boreholes), and explains why these caves remain one of the most poorly understood ecosystems (Gracning and Brown 2003).

Karst systems are highly vulnerable to contamination and other anthropogenically induced environmental pressures (i.e., mining, land use), which affect the condition of subsurface ecosystems by impacting the quantity and quality of the groundwater (Danielopol et al. 2003). Consequently, *C. hubrichti* is threatened not only by changes in hydrological regime (due to quarrying and water extraction), but also changes in water quality (due to sewage and pesticides, herbicides, fertilizer, sediment) (Davis and Bell 1998, Lewis 2002, The Nature Conservancy 2003). The impacts of these threats are likely to increase due to climate change and increases in population and urbanization in the area. These threats are compounded by the long life span of *C. hubrichti*, their limited natural range, which makes this species vulnerable to stochastic effects, and the lack of food resources in phreatic cave habitats (Owen et al. 2015). It is therefore critical to understand the genetic structure of a species (Schrimpf et al. 2014), as well as its biological and ecological constraints (Carroll, in prep.) in order to protect and conserve its intraspecific genetic diversity and to appropriately define conservation units (Crandall et al. 2000).

15.2 METHODS

15.2.1 Site Description

Cambarus hubrichti occurs only in the Salem Plateau section of the east central Missouri Ozarks, USA (Hobbs et al. 1977, Fig. 15.1A). Its range extends through Camden, Carter, Crawford, Howell, Oregon, Phelps, Pulaski, Shannon, and Ripley

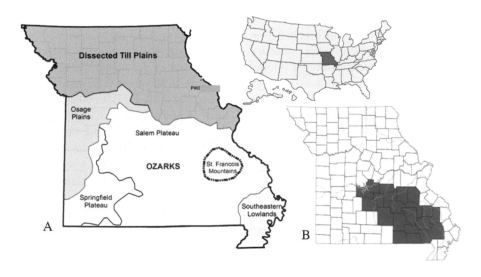

Figure 15.1   (A) Map showing physiographic regions of Missouri. Missouri Department of
Natural Resources, Geological Survey and Resource Assessment Division. (B)
Inset Map of Missouri showing the distribution of *Cambarus hubrichti* (shaded
area). Missouri Department of Conservation (MDC).

counties (Pflieger 1996, Fig. 15.1B). All records are from deep subterranean lakes,
cave streams, and the mouths of large springs at depths of 12–50 m (Hobbs et al.
1977, Pflieger 1996).

Cambarus hubrichti has been listed as a Sensitive Species by the USDA Forest
Service (Lewis 2002), stable by the American Fisheries Society (Taylor et al. 2007)
and given the heritage G4 rank (Apparently Secure) by NatureServe (2015). This
species has also been given a status of "Data Deficient" by the International Union
for the Conservation of Nature (Cordeiro and Thoma 2010). Although a subterranean
obligate in most cases, the species occurs in many Missouri caves on a variety of substrates in both lotic and lentic cave habitats. There is insufficient information on the
effect of pollution on these populations. Fertilizer pollution has been noted to cause
mass mortality at one site, but this phenomenon has not been documented in other
sites. Otherwise, the species appears stable throughout its range. More research is
required to list this species as anything other than Data Deficient by the International
Union for the Conservation of Nature. However, due to its limited range (Fig. 15.1B)
and habitat restrictions, the Missouri Department of Conservation (MDC) has given
it a status of "Species of Conservation Concern" and a state rank of vulnerable (S3,
MDC 2016).

The Salem Plateau is approximately 70,448 km² and is underlain by interbed-
ded Ordovician and Cambrian dolomites (Fig. 15.2), which together form the Ozark
aquifer, used throughout this area as a source of water for public and domestic use.
Dissolution of the dolomites in this aquifer has created intense karst terrain characterized by strongly fluctuating discharge, multiple sinkholes, springs and losing

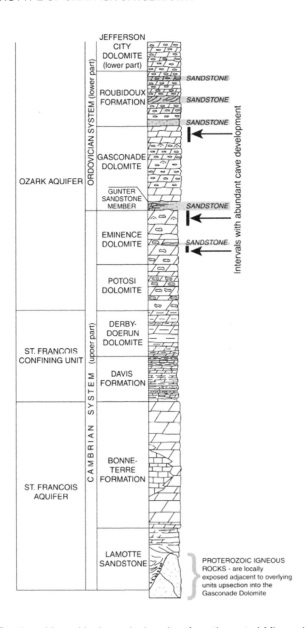

Figure 15.2 Stratigraphic and hydrogeologic units of south-central Missouri showing strati-graphic intervals with abundant cave development and highlighting sandstone horizons. United States Geological Survey (USGS). Cave location noted with *.

streams within a few miles of Roubidoux Spring, allowing for deep groundwater circulation (Vineyard and Fender 1982, Adamski et al.1995).

Roubidoux Spring is situated in the Salem Plateau region of Pulaski County, Missouri (37.8249° N, 92.2015° W). This second magnitude spring (~2 m³/s) lies at the base of a tall dolomite bluff along Roubidoux Creek, a large tributary creek

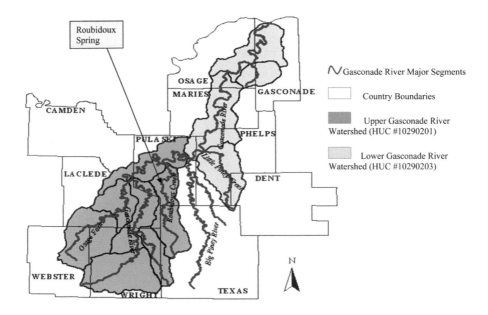

system of the Gasconade River (Fig. 15.3). The Gasconade River Basin lies within
the Upper Gasconade Watershed (Hydrologic Unit Code #10290201; 239 m a.s.l.)
This watershed is rural with low population density but high agricultural use with
33% and 49% cropland and grassland used for livestock production, respectively.

The Roubidoux Spring cave system, like many large underwater karst cave systems in Missouri, consists of one large tunnel with no major branches, compared to
more complicated, multibranching Florida springs. After the first 120 m, the phreatic conduit of the spring drops to depths ranging from 37 m to more than 61 m. The
conduit feeding Roubidoux Spring is unique in that it is uncharacteristically large
and alternates between a narrow river-like channel and large open cavernous rooms.
Water velocity is greatly reduced between these two hydrogeological settings, resulting in large deposits of fine sediment. Crayfish have most often been found on silt
mounds in these cavernous rooms.

15.2.2 Sample Collection

Crayfish were collected by members of the Ozark Cave Diving Alliance (OCDA;
Shannon Wallace, Dirk Bennett, Ben Perkins, Dave Buranu, Steve Gridley) under
the Missouri Department of Conservation Wildlife Collector's Permit 16231. The
collections were made on 14 September 2013. Special permission was not required
to access the sampling location. The OCDA began exploring the Roubidoux submerged cave system in 1998 and has since accessed and surveyed 3,431 m of cave
passage at depths of up to ~80 m. Samples were collected from the phreatic conduit

and the large cavernous room known as "Lithuania" at depths ranging from ~50 to 70 m. Sampling was limited to six individuals during one sampling event in order to protect the welfare of the overall population. Crayfish were collected using hand dip nets and placed in sterile plastic vials (along with spring water), sealed and brought to the surface by the divers (see Crandall 2016 for a general discussion of crayfish collecting, including cave habitats). Once at the surface, each crayfish was placed in individual compartments within a multi-partitioned plastic tackle box, covered with spring water and placed in a cooler with ice for transport back to the laboratory (J.W. Fetzner pers. comm.). Pleurobranch tissue was dissected from each specimen and placed in sterile glass vials containing 95% EtOH and stored for DNA extraction. The remainder of each specimen was held in 95% EtOH for 48 h and then preserved in 75% EtOH for later morphological examination. These whole adult crayfish have been deposited at the National Museum of Natural History, Smithsonian Institution, and serve as voucher specimens for caves to help elucidate relationships between *C. hubrichti* and the other *Erebicambarus* species from Missouri and Tennessee.

Several general water quality parameters were measured *in situ* every 30 s along the Roubidoux conduit and within the cave for 24 h at depths ranging from 13 m just inside the spring mouth to greater than 50 m within the large submerged cave. Dissolved oxygen, pH, water temperature, turbidity, and specific conductivity were measured using a YSI 6920 Environmental Monitoring System Sonde. After taking measures along the conduit, the sonde was deployed in the cavernous room where most of the crayfish were collected for a period of 24 h. Three replicate cave sediment samples were collected using 2" × 4" (circumference × length, respectively) clear plastic PVC pipes with lids at each site where crayfish were collected for later geochemical and microbial studies.

15.2.3 Morphological Analyses

Crayfish specimens were identified using standard dichotomous keys, including Hobbs (1972), Pflieger (1996), and Thoma (2016). Crayfish specimens were observed using an illuminated magnifying glass and a Wild M-8 dissection microscope. Drawings were made freehand (Fig. 15.4A and B).

15.2.4 DNA Extraction, PCR, and Sequencing

Total DNA was extracted with the Qiagen DNeasy DNA extraction kit following manufacturer's instructions. We targeted both mitochondrial genes (16S, 12S, COI) and a nuclear (28S) gene for sequencing. Polymerase chain reaction (PCR) products for three mitochondrial genes – partial 16S rDNA (~460 bp; using the primer 16sf-cray [Buhay and Crandall 2005] and 16s-1472r [Crandall and Fitzpatrick 1996]), partial COI (~659 bp; with primers LCO1-1490 and HCO1-2198 [Folmer et al. 1994], and partial 12S rDNA (~390 bp; using the primers 12sf and 12sr; Mokady et al. 1999) – were amplified using a protocol following Porter et al. (2005) and Crandall and Fitzpatrick (1996). We also PCR-amplified the partial nuclear gene 28S (~800–1000 bp, with primers 28sF-cray and 28sR-cray as described in Breinholt et al. (2012).

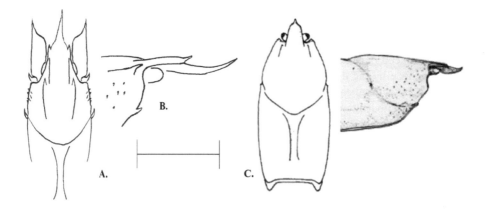

Figure 15.4 *Cambarus hubrichti*, new morphotype. (A) Anterior part of head and carapace,
dorsal view. (B) Anterior most portion of carapace, eyestalk and rostrum, right
lateral view. Scale bar = 2 cm. (C) *Cambarus hubrichti*, Hobbs (1972).

Bidirectional sequences for each gene were generated on an ABI Prism 3730XL capillary sequencer using the ABI Big Dye Ready-Reaction kit following standard cycle
sequencing protocols, with an exception of 1/16th of the standard reaction volume.
The mitochondrial genes 16S, COI, and 12S have differing amounts of variation and
are commonly used for phylogenetic analysis in crayfish (e.g., Fetzner and Crandall
2003, Buhay and Crandall 2005, Breinholt et al. 2009). The region of 28S used tends
to be the most variable region of 28S among Crustacea (Toon et al. 2009) and was
sampled for a subset of taxa to serve as a nuclear genome marker for estimating
phylogenetic relationships among the species of the genus *Cambarus* (Breinholt et
al. 2012).

15.2.5 Phylogenetic Analyses

The rRNA sequences (16S, 12S, 28S) were aligned with MAFFT v7.305b (Katoh et
al. 2013). COI sequences were aligned using TranslatorX (Abascal et al. 2010), which
translates protein-coding sequences to amino acids, aligns protein sequences with
MAFFT v7.305b (Katoh et al. 2013), and then back-translates to nucleotides. This
aids in the identification and removal of sequences with premature stop codons and
indels that are indicative of numts (Song et al. 2008).

Best-fit models of evolution (Posada et al. 2001) and an optimal data-partitioning
scheme were chosen using Partition Finder v1.1.1 (Lanfear et al. 2012) with each
rRNA gene and COI codon position chosen as *a priori* data subsets and using the
Bayesian Information Criterion (BIC) for model selection. A maximum-likelihood
tree was estimated in RAxML 8.2 (Stamatakis 2014) using the partitioning scheme
selected with Partition Finder. One thousand bootstrap replicates were followed by
ten maximum-likelihood tree searches under the GTRCAT model with final optimization under GTRGAMMA.

15.3 RESULTS

Six crayfish specimens were collected: four form II males and two females. Body lengths (rostral apex to telson apex) were 32.2, 41.1, 54.1, and 66.5 mm for males and 55.9 and 56.4 mm for females.

15.3.1 General Water Quality

Measures of the chemical and physical general water quality parameters (mean ± SD) in the phreatic conduit were as follows: water temperature of 13.18°C ± 0.01; pH of 6.95 ± 0.01; turbidity of 0.02 NTU ± 0.05; dissolved oxygen of 5.75 mg/l ± 0.08, and conductivity of 374.5 μS/cm ± 0.98. The cavernous room from which the crayfish were collected had the following mean physical and chemical characteristics: water temperature of 13.19°C ± 0.00; pH of 7.03 ± 0.03; turbidity of 0.03 NTU ± 0.04; dissolved oxygen of 5.62 mg/l ± 0.06, and specific conductivity of 568.3 μS/cm ± 5.26. Other than dissolved oxygen, all parameters complied with water quality standards for the state of Missouri (MDNR 2014). Several measures of dissolved oxygen (DO), however, were slightly below the state of Missouri's minimum water quality standard of 6.0 mg/l for cool water streams/springs (MDNR 2014). The range in DO concentrations between that recorded at the spring mouth (1 m deep) to >50 m deep in the cave was 5.5–6.6 mg/l.

15.3.2 Diagnosis

The crayfish material was determined as *Cambarus* (*Aviticambarus*) *hamulatus* (Cope and Packard 1881) using Hobbs (1972) and Thoma (2016), which cover all crayfish species of North America. Hobbs (1972) alone brought the specimens to *C. tenebrosus* (Hay 1902), and other Ohio River epigean species. The keys in Pflieger (1996) identified our material as *C. hubrichti* (*C. hamulatus* is not known from Missouri, thus was not in this reference). However, the morphology of our material did not match the *C. hubrichti* descriptions in Hobbs (1952), Hobbs et al. (1977), and Pflieger (1996). The salient new morphological characters are as follows: cephalothorax subcylindrical in cross-section; eyes reduced, white with a central vague pigment spot; antennal scale length ~2× width; rostrum length ~2× basal width, acute with one pair of lateral spines (Fig. 15.4); acumen elongate, slightly arcing with apex directed dorsodistally, postorbital spines present and stout in Fig. 15.4; carapace hepatic region with numerous scattered spines, directed anteriorly; one stout branchiostegial spine present (Fig. 15.4); one or two large, lateral cervical spines present; areola seven to eight times as long as wide (Fig. 15.4), with four to five punctae across narrowest portion; chelae and gonopod as for *C. hubrichti* typical form II; annulus ventralis as for *C. hubrichti*.

15.3.3 Sequence Data

We generated 12S and 16S sequences from six animals total, with COI from two and 28S from four, and these new sequences have been deposited in GenBank under accession numbers MK484617-MK484622, MK484692-MK484697, MK484698-MK484701, and MK492377-MK492378. The COI sequences have also been deposited on the BOLD database www.boldsystems.org/ for DNA barcoding reference (numbers provided upon acceptance). We combined these data with data from a variety of other crayfish species, including a previous sample of *Cambarus hubrichti* (see Breinholt et al. 2012). Our phylogeny from the combined dataset clearly shows these new samples (KC 8467-8472), cluster with the previous sample of *C. hubrichti* with 100% bootstrap support (Fig. 15.5). This clade is also very distinct from the clade of other *Cambarus* cave species from the Cumberland Plateau, including *C. hamulatus*.

15.4 DISCUSSION

Our material represents a new morphotype of *C. hubrichti*. In Hobbs' (1972) key, our material easily was identified to *Erebicambarus*; however, because of the presence of marginal rostrum spines, the presence of cervical spines, and the relatively wide areola, the material keyed to *C. tenebrosus* and Ohio River epigean species.

Using the keys in Thoma (2016), our *C. hubrichti* material corresponded to *C. hamulatus,* specifically: eyes not pigmented, chela palm with multiple rows of tubercles, rostrum with well-defined angle at apex and bearing lateral spines, cervical spines present, areola wide, and hepatic region bearing numerous spines. In couplet 6 (Thoma 2016), again it is the presence of the marginal rostrum spines that caused our morphological misidentification.

Furthermore, using the original descriptions and subsequent figures and redescriptions (Hobbs 1952, 1989, Hobbs et al. 1977, Pflieger 1996), secondary characteristics in our material corresponded to *C. hamulatus*. The rostrum length is ~2× basal width with one pair of lateral spines in our material, as opposed to ~1.5× basal width and lacking lateral spines, or at most with lateral tubercles as described for *C. hubrichti*. Strong, stout postorbital spines are present, whereas only postorbital ridges are present in the *C. hubrichti* descriptions. The carapace hepatic region bears numerous scattered spines and a large branchiostegial spine, whereas the descriptions of *C. hubrichti* mention no such spines. These character states agree with the various descriptions and depictions of *C. hamulatus*, but our material differs from that species in the form of the gonopod and the areola, matching instead the descriptions of *C. hubrichti* (Hobbs 1952, 1989, Hobbs et al. 1977, Pflieger 1996). The curved form of the acumen does not correspond to either taxon.

15.4.1 Distribution

So far, this new morphotype of *C. hubrichti* has only been collected from Roubidoux Spring in Pulaski County, Missouri. All six specimens were collected from the phreatic conduit or cave nearly a kilometer back from the spring mouth. The new morphotype described here could become confined to this deep saturated karst habitat. Crayfish dispersal ability is strongly reduced in karstic phreatic systems due to

Figure 15.5 Estimated phylogenetic relationships of samples from Roubidoux Spring with other known crayfish species based partial gene sequence data from the mitochondrial genome (on a combination of 12S, 16S, and COI) and partial 28S data from the nuclear genome. The phylogeny was estimated using maximum-likelihood implemented in RAxML with concatenated sequence data and an optimized model of evolution. Bootstrap values are shown on nodes supported with > 70% bootstrap support based on 1,000 pseudo-replicates.

progressive physical fragmentation (karstification) that occurs in these submerged systems (Lefébure et al. 2007). This results in hydrogeomorphically induced genetic isolation and explains why these fragmented phreatic systems are known to be environments of high endemism (Barr and Holsinger 1985). The distributional patterns of this new morphotype also indicate, in accordance with other authors, that habitat plays a key role in the evolution of this genus (Breinholt et al. 2012).

15.4.2 Life History

Stable or predictable environments, like springs (Death et al. 2004, Glazier 2012), typically favor K-selected species with long life histories, delayed maturity, low reproductive output, and some degree of parental care (Venarsky et al. 2012). Research in systems above the water table (caves and cave streams) has shown that cave crayfish are long-lived with female age-at-first-reproduction of 5–6 years, validating that they are K-selected (Venarsky et al. 2012). While there is a paucity of information about reproduction or life history patterns of stygobiotic species, the information available indicates that they have lengthened developmental cycles (Strayer 1994), lack larval dispersal stages (Holsinger 1993), and produce larger and fewer eggs than their epigean counterparts (Hobbs and Lodge 2010). Poulson and White (1969) hypothesized that these life history patterns are related to reduced genetic variability (evidenced in decreased phenotypic variation in morphological traits) caused by isolation and adaptations to the constancy of underground environments. The discovery of the specimens described here fails to support this theory, but rather suggests, as Buhay and Crandall (2005) found, that long evolutionary history patterns have allowed these stygobites to persist and accumulate genetic variability. Additionally, phreatic caves, like Roubidoux, form below the water table through karstification, and therefore represent expanding groundwater habitats for stygobites (White 1988). Hence, our findings also support the notion that hydrogeomorphically induced expansion, followed by long intervals of isolation, serves as a driving force for increased variation in cave crayfish species.

15.4.3 Ecological Preferences of Stygobionts; Habitat Notes

Dissolved oxygen levels were slightly below that specified as meeting standards for waters, like Roubidoux that are designated for cold-water fishery (MDNR 2014). Crayfish, in general however, have been reported to have a broad tolerance to ranges of temperature and dissolved oxygen (Reynolds and Souty-Grosset 2012, Veselý et al. 2015). Members of the genus *Cambarus* have even been characterized as tolerant of impaired conditions (Peake et al. 2004). While some research has shown that water chemistry was of minor importance in determining stygobiotic occurrences (Hahn 2002, Paran et al. 2005), others have found that oxygen concentration was an important factor associated with crayfish presence (Datry et al. 2005, Haddaway et al. 2015). Geology has been reported to be the key factor regulating stygobiotic species distribution because of its effect on groundwater oxygen supplies (Dole-Oliver et al. 2009). Hence, because karstic systems maintain strong hydrological connectivity with surface environments (Graening and Brown 2003), oxygen to Roubidoux sty-gobionts was likely not a limiting factor.

15.4.4 Crayfish Feeding

Crayfish are polytrophic, functioning simultaneously as herbivores, omnivores, and detritivores (Hobbs 2001, Reynolds et al. 2013). The diets of epigean crayfish consist of algae, macrophytes, invertebrates, fish (and fish eggs), and plant detritus whose nutritional value comes from the epiphytic bacterial and fungal constituents (Hogger 1988, Hobbs and Lodge 2010). The diets of subterranean crayfish are relatively lim-ited because groundwater habitats are characteristically oligotrophic (Strayer 1994). It is well documented that the primary source of energy for invertebrate consumers living in these aphotic systems is allochthonous particulate organic carbon (POC) carried into the caves (Culver 1986, Cumberlidge et al. 2015). These inputs are spatially and temporally variable, related to hydrologic inputs (Simon 2008), and minimal compared to surface systems, making groundwater one of the most severe ecosystems on the planet (Malard et al. 2009). Phreatic karst cave systems like Roubidoux represent even more extreme habitats than their terrestrial subterranean counterparts (caves and cave streams). While both are associated with subterranean environments, and therefore suffer from the absence of autochthonous production, availability of organic carbon sources for troglofauna (in systems above the water table) is considerably greater than for stygofauna due to greater access to terrestrial inputs (i.e., bat guano). Little is known of the feeding patterns of strictly stygobitic crayfish (Pflieger 1996). Nonetheless, given the level of energy limitation in these systems, it has been reported that population densities are low, food chains are short, and consumers must perform as trophic generalists in order to survive (Culver 1986, Gibert et al. 1994a).

15.4.5 Trophic Relationships and Probable Stygobitic Feeding Pathways

Many karst cave food webs are exclusively heterotrophic, while others are fueled by chemolithoautotrophic microorganisms (Bacteria and Archaea; Simon 2008). They, along with small protozoans and fungi, form microcolonies and thin biofilms that attach to fine sediments (Pederson 2001, Griebler et al. 2002, Engel et al. 2004). Distribution of subterranean macrofauna in chemolithoautotrophy-based systems often correlates with these high-protein microbial biofilms (Bärlocher and Murdoch 1989, Pusch 1996). Heterotrophic and chemoautotrophic food webs both are typi-cally comprised of three trophic levels: the base, invertebrate grazers, and predators (Sarbu et al. 1996, Graening and Brown 2003, Simon 2008). These truncated food webs have very few predators, but are typical of phreatic cave systems due to scarce food supplies (Pohlman et al. 1997; Gibert and Deharveng 2002).

Allochthonous POC and associated microbes represent the primary source of energy in the typical detritus-based heterotrophic groundwater system (terrestrial caves and cave streams; Hobbs and Lodge 2010). The detrital material serves as a substrate for populations of bacteria, fungi, and actinomycetes (Gibert et al. 1994b) and constitutes the major food source for many troglobitic invertebrates (Dickson 1975). It is dissolved organic carbon (DOC); however, that represents the major heterotrophic pathway in phreatic cave systems because input of POC is often deficient (Brown et al. 1994). Dissolved organic carbon makes up approximately 90% of the total organic carbon in natural waters (Batiot 2003), and has been documented to represent a larger flux than POC in karst systems (Gibert 1986). Assimilation of DOC into microbial biomass and the consumption of this heterotrophic production by higher trophic levels allows microbes to be key players in mediating energy transfer between DOC and karst micro- and macroinvertebrates. This explains why microbial communities are capable of supporting the entire food web in many groundwater systems (Stanford and Gibert 1994).

Crayfish typically serve as both grazers and predators, feeding on groundwater microbes in the sediments, invertebrates, as well as freshwater fishes through consumption of juveniles and eggs (Jurcack et al. 2016). Because of the disproportionally large effect of their consumption across multiple trophic levels, they are usually characterized as playing keystone roles in energy transformation among trophic levels and producing important changes in ecosystem structure and function (Holdich 1987, Whitledge and Rabeni 1997, Hobbs 2001). Additionally, cannibalism among crayfish is common (Hobbs et al. 1977) as they often predate on newly molted conspecifics (Cruz et al. 2006). Low food availability has been shown to be the most important ecological factor favoring their switch to this feeding strategy (Elgar and Crespi 1992).

15.4.6 Proposed Foodweb Model for Roubidoux C. hubrichti Morphotypes

Stygobitic crayfish, like those found in Roubidoux, tend to aggregate in regions of the cave where food is spatially or temporally abundant (Culver et al. 2000, Hobbs 2001). Each aforementioned feeding pathway is associated with benthic substrates and areas of organic carbon deposition. This explains why crayfish sightings by the OCDA were consistently associated with sediment mounds located within large cavernous rooms of Roubidoux Spring. Hence, the distributional pattern of this new morphotype of C. hubrichti, as other authors have reported, is likely correlated with hydrogeomorphology because hydrogeomorphology is strongly correlated with the spatial and temporal abundance of food sources in karstic phreatic systems (Dole-Oliver et al. 2009).

Collectively, these patterns indicate that several food sources could be supporting these crayfish and serving as the base of the Roubidoux food web (Fig. 15.6). First, these crayfish could be foraging in areas of accumulated allochthonous debris as velocity of flow is greatly reduced in these large cavernous rooms, allowing for settling of POC. However, this is unlikely given that preliminary microscopic examination of multiple sediment samples (fine silt) from Roubidoux Spring revealed, as found in previous studies (Brown et al. 1994), little to no particulate organic material (vegetative or animal). Second, the consistent association with sediment mounds

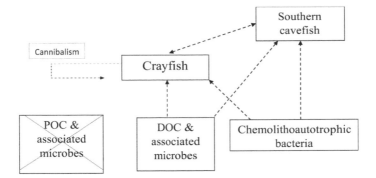

Figure 15.6 Proposed food web model for *Cambarus hubrichti* morphotypes from Roubidoux Spring. The least likely feeding source (given preliminary observations) is identified by X across the food source box. POC: particulate organic carbon; DOC: dissolved organic carbon.

suggests, as others have cited (Hobbs 2001, Cumberlidge et al. 2015), that our specimens could be foraging over the substrate in this resource-depressed system, straining cave sediments and feeding on microbial biofilms (fueled by DOC) in the silt. It is well documented that subterranean sediment microorganisms are an important food for stygobitic crustaceans, as both juvenile and adult crayfish are known to scrape them from hard substrates (Dickson et al. 1979, Mathieu et al. 1991, Hobbs 2001, Graening and Brown 2003). This is advantageous because this grazing practice promotes further biofilm production (Danielopol 1989). It is reasonable to postulate that the reliance on microbial films and DOC is due to the relative reliability of those foods in karst settings compared to allochthonous POC, which is patchily distributed (Simon 2008). Finally, the Roubidoux food web could be fueled by internal chemolithoautotrophy. Chemoautotrophic bacteria use inorganic energy sources such as hydrogen, ammonia, nitrite, sulfide, and iron. While most caves lack significant inputs of chemicals needed by chemoautotrophs (Simon et al. 2003), active gravel mines and inactive old iron mines still exist in the Roubidoux watershed and could, through groundwater input, be supporting this form of autotrophic production. Interestingly, little information is known about the microbiology of karstic groundwater systems even though their water is used as a source of drinking water (Griebler and Lueders 2009). More research is needed to address this hypothesis. Finally, collected crayfish are likely supported in their predatory role by feeding on the eggs and young of cave fish (Hobbs and Lodge 2010). Schools of the Southern Cavefish (*Typhlichthys subterraneus* Girard, 1859) have been cited by the OCDA swimming in and out of the same cavernous rooms where the crayfish were captured. It is not known if they were avoiding predation or adverse environmental conditions, mating, laying eggs, or feeding. Additional observations would be needed to determine if Roubidoux stygobitic crayfish were in fact feeding on the eggs and young of this cavefish. This is often prohibitory given the risks involved with these diving expeditions (see acknowledgments). Nonetheless, this is somewhat concerning given that *T. subterraneus* is currently ranked by the state of Missouri as Imperiled and

Vulnerable (S2, S3, MDC 2016). Additional research on the geochemistry and food web dynamics of this saturated karstic system is needed and is currently in the initial planning stages.

15.5 CONCLUSIONS

15.5.1 Conservation Concerns

Second to glaciers, groundwater represents the most extensive freshwater body in the world (Dole-Oliver et al. 2009). Yet the availability of groundwater is critical in many parts of the world (Danielopol et al. 2003). Subterranean habitats harbor 50% of North America's imperiled species, and 95% of them are considered vulnerable (Culver et al. 2000). These problems will become more exaggerated as karst water resources face increased water supply demands due to climate change, increases in human population size, urban development, and land use alterations (Banner et al. 2008).

These and other factors represent relevant potential threats to the survival of *C. hubrichti* in this phreatic system. First, several inactive iron mines are known to exist in Pulaski County. Research shows that effects of mining a century or more ago are still present today (Rösner 1998, Nimick et al. 2004, Church et al. 2007). The impacts of abandoned mines consist of distinctly increased concentrations of heavy metals (iron, lead, zinc, and manganese) and elevated levels of chemical substances used in ore processing such as arsenic and cadmium. All have been found in sediment, water, and aquatic organism samples in surface water and groundwater systems near areas where mining has occurred (Rösner 1998, Nimick et al. 2004). Concentrations have been found to exceed acute aquatic-life standards by as much as 200-fold in water, and to be more than 100 times greater in sediments than prior to historical mining (Nimick and Cleasby 2004, Besser et al. 2007). Interestingly, stream water has been reported to be less toxic than sediment and is attributed to the particulate adsorption of dissolved metals (Besser and Leib 2007). Additionally, sand and gravel mining in the Roubidoux watershed is still active and has the potential to impact the resident biota. Reports show that aquatic systems in sand and gravel mining areas typically have high specific conductivity, lower dissolved oxygen concentrations, and high concentrations of sulfate, iron, and toxic heavy metals (Saviour 2012). Moreover, stygobitic species are at a much greater risk of suffering from metal toxicity due to their relatively longer lifespans (or greater longevity) (Dickson et al. 1979). Research has shown that toxic metals accumulate in the tissues of *Cambarus* resulting in greater concentrations of these metals in their tissues in comparison to surface-water species from streams with similar levels of contamination (Taylor et al. 1995). Effects of long-term exposure include damage to crayfish gills and subsequent decrease in uptake of oxygen, reduced growth and reproduction, and damage to general metabolic processes (Taylor et al. 1995, Hobbs 2001).

Second, Pulaski County is one of the most populated areas in the upper Gasconade Watershed. Water from the Ozark aquifer is a principal source of water for public and domestic supply. Most of the riparian areas and watershed are privately owned. Increased cattle numbers per pastured acre has continued to the present (Nature Conservancy 2003). Threats to the Salem Plateau region include reduced or eliminated native riparian vegetation, nutrients and sedimentation from inappropriate cattle grazing, and groundwater pollution from urban and rural residential development (Nature Conservancy 2003). Hence, because karst topography allows rapid infiltration of surface pollutants (Graening and Brown 2003), these residential and agricultural land use alterations result in activities related to surface runoff or groundwater flow that result in substantial inputs of sewage or fecal material, pesticides and herbicides, sediment, and newly emerging contaminants such as steroids, antibiotics, and nonprescription drugs, all of which may lead to extinctions of stygobitic animals (Elliott 2000, Lewis 2002). This is concerning for the Roubidoux crayfish as livestock and poultry waste are a major source of nutrient loading, and application of animal waste for fertilizer is a widespread practice in the Ozark Plateaus (Davis and Bell 1998). These same authors also report that there are several municipal sewage treatment plants in the upper Gasconade Watershed that have effluents of 0.5 Mgal/day or more.

The physical fragmentation, characteristic of karstic groundwater systems, limits species range sizes (Lefébure et al. 2007). It is not surprising that limited natural range continues to be the primary factor responsible for the noted imperilment of crayfishes (Taylor et al. 2007). Of course, these crayfish have evolved for millions of years with restricted distributions. It is the combination of a limited range coupled with ecological impacts that create conditions of endangerment for the freshwater crayfish (Richman et al. 2015). Thus, a deeper understanding of the ecological requirements for the Roubidoux crayfish described here and the potential impacts of human activities will help inform management decisions for this crayfish. In general, our knowledge on the ecological requirements of stygobitic species is sparse (Buhay and Crandall 2005, Dole-Oliver et al. 2009) and, according to the USDA Forest Service, no monitoring, management, or conservation activities are currently being conducted on *C. hubrichti* (see Lewis 2002). Given the threats to groundwater systems and the small geographic ranges of their biotic inhabitants, it is essential that further studies focus on ecological needs to ensure the survival of these evolving species.

ACKNOWLEDGMENTS

Credit for collection of our crayfish specimens is given to members of the certified technical diving group known as the Ozark Cave Diving Alliance (OCDA). This not-for-profit team of cave divers has devoted their lives to studying phreatic cave systems in Missouri, and is dedicated to the exploration, preservation, and public education of Missouri Ozarks groundwater systems. The dives are extremely dangerous and require 6–8 h of dive time and nearly that in decompression time. We would also like to thank J.W. Fetzner Jr. (Carnegie Museum of Natural History) for his advice and consult regarding crayfish capture, preservation, and DNA analysis;

J. Koppelman (Missouri Department of Conservation) for his consultation during initial crayfish phylogenetic analysis; the Smithsonian Lab facility for sequencing services; and NSF DEB-1601631 for partial support of this work.

REFERENCES

Abascal, F., Zardoya, R. and Telford, M.J. 2010. Translator X: Multiple alignment of nucleotide sequences guided by amino acid translations. *Nucleic Acids Research* 38 (Web Server issue):W7–W13.

Adamski, J.C., Petersen, J.C., Freiwals, D.A. and Davis, J.D. 1995. Environmental and hydrologic setting of the Ozark Plateaus study unit, Arkansas, Kansas, Missouri, and Oklahoma. Water-Resources Investigations Report 94-4022:1–62. U.S. Geological Survey.

Banner, J.L., Musgrove, M., Rasmussen, J., Partin, J., Long, A., Katz, B., Mahler, B., Edwards, L., Cobb, K., James, E., Harmon, R.S., Herman, E. and Wicks, C.M. 2008. Geochemistry and climate change. In *Frontiers of Karst Research*, Special Publication 13, eds. J.B. Martin, and W.B. White. Karst Waters Institute: Leesburg, VA.

Bärlocher, F. and Murdoch, J.H. 1989. Hyporheic biofilms: A potential food source for interstitial animals. *Hydrobiologia* 184: 61–7.

Barr, T.C. and Holsinger, J.R. 1985. Speciation in cave faunas. *Annual Review of Ecology and Systematics* 16:313–37.

Besser, J.M. and Leib, K.J. 2007. Toxicity of metals in water and sediment to aquatic Biota. In *Integrated Investigations of Environmental Effects of Historical Mining in the Animas River Watershed, San Juan County*, Colorado, eds. S.E. Church, P. von Guerard, and S.E. Finger, U.S. Professional Paper 1651: 839–48. Reston, VA: Department of the Interior, U.S. Geological Survey.

Besser, J.M., Finger, S.E. and Church, S.E. 2007. Impacts of historical mining on aquatic ecosystems: An ecological risk assessment. In *Integrated Investigations of Environmental Effects of Historical Mining in the Animas River Watershed, San Juan County*, Colorado, eds. S.E. Church, P. von Guerard, and S.E. Finger, Professional Paper 1651: 89–104. Reston, VA: U.S. Department of the Interior, U.S. Geological Survey.

Breinholt, J., Pérez-Losada, M. and Crandall, K.A. 2009. The timing of the diversification of freshwater crayfishes. In *Decapod Crustacean Phylogenetics*, eds. J.W. Martin, K.A. Crandall, and D.L. Felder, 343–55. Boca Raton, FL: CRC Press/Taylor & Francis.

Breinholt, J.W., Porter, M.L. and Crandall, K.A. 2012. Testing phylogenetic hypotheses of the subgenus of the freshwater crayfish genus *Cambarus* (Decapoda: Cambaridae). *PLoS One* 7:e46105.

Brown, A., Pierson, W. and Brown, K. 1994. Organic carbon resources and the payoff-risk relationship in cave ecosystems. In 2nd International Conference on Ground Water Ecology, 67–76. Atlanta, GA: U.S. EPA and the American Water Resources Association.

Buhay, J.W. and Crandall, K.A. 2005. Subterranean phylogeography of freshwater crayfishes shows extensive gene flow and surprisingly large population sizes. *Molecular Ecology* 14:4259–73.

Church, S.E., von Guerard, P. and Finger, S.E. 2007. Abstract. In *Integrated Investigations of Environmental Effects of Historical Mining in the Animas River Watershed, San Juan County*, Colorado, eds. S.E. Church, P. von Guerard, and S.E. Finger, Professional Paper 1651, 1079. Reston, VA: U.S. Department of the Interior, U.S. Geological Survey.

Coineau, N. 2000. Adaptions to interstitial groundwater life. In *Ecosystems of the World*, Volume 30: Subterranean Ecosystems, eds. H. Wilkens, D.C. Culver, and W.F. Humphreys, 189–210. Amsterdam, The Netherlands: Elsevier.

Cooper, J.E. 1975. Ecological and behavioral studies in Shelta Cavern, Alabama, with emphasis on the decapod crustaceans. PhD thesis, University of Kentucky, USA.

Cooper, R.L., Li, H., Long, L.Y., Cole, J. and Hopper, H.L. 2001. Anatomical comparisons of neural systems in sighted epigean and troglobitic crayfish species. *Journal of Crustacean Biology* 21:360–74.

Cope, E.D. and Packard, A.S. 1881. The fauna of the Nickajack Cave. *The American Naturalist* 15:877–82.

Cordeiro, J., Jones, T. and Thoma, R.F. 2010. *Cambarus hubrichti*. In *The IUCN Red List of Threatened Species 2010*:e. T153903A4561335.

Crandall, K.A. 2016. Collecting and processing freshwater crayfishes. *Journal of Crustacean Biology* 36:761–66.

Crandall, K.A. and De Grave, S. 2017. An updated classificaiton of the freshwater crayfishes (Decapoda: Astacidea) of the world with a complete species list. *Journal of Crustacean Biology* 37:615–53.

Crandall, K.A. and Fitzpatrick, J.F. Jr. 1996. Crayfish molecular systematics: Using a combination of procedures to estimate phylogeny. *Systematic Biology* 45:1–26.

Crandall, K.A., Bininda-Emonds, O.R.P., Mace, G.M. and Wayne, R.K. 2000. Considering evolutionary processes in conservation biology. *Trends in Ecology and Evolution* 15:290–95.

Cruz, M.J., Pascoal, S., Tejedo, M. and Rebelo, R. 2006. Predation by an exotic crayfish, *Procambarus clarkii*, on Natterjack Toad, *Bufo calamita*, embryos: Its role on the exclusion of this amphibian from its breeding ponds. *Copeia* 2:274–80.

Culver, D.C. 1986. Cave faunas. In *Conservation Biology: The Science of Scarcity and Diversity*, ed. M.E. Soulé, 427–33. Sunderland, MA: Sinauer Associates.

Culver, D.C., Christman, M.C., Elliott, E.R., Hobbs, H.H., III. and Reddell, J.R. 2003. The North American obligate cave fauna: Regional patterns. *Biodiversity and Conservation* 12:441–68.

Culver, D.C., Master, L.L., Christman, M.C. and Hobbs, H.H., III. 2000. The obligate cave fauna of the 48 contiguous United States. *Conservation Biology* 14:386–401.

Culver, D.C. and Sket, B. 2000. Hotspots of subterranean biodiversity in caves and wells. *Journal of Cave and Karst Studies* 62:11–17.

Cumberlidge, N., Hobbs, H.H. and Lodge, D.M. 2015. Class Malacostraca, Order Decapoda. In *Thorp and Covich's Freshwater Invertebrates*.4th edition, Volume I: Ecology and General Biology, eds. J.H. Thorp, and D.C. Rogers, 797–847. Boston, MA: Academic Press.

Danielopol, D.L. 1989. Groundwater fauna associated with riverine aquifers. *Journal of the North American Benthological Society* 8:18–35.

Danielopol, D.L., Griebler, C. Gunatilaka, A. and Notenboom, J. 2003. Present state and future prospects for groundwater ecosystems. *Environmental Conservation* 30:1–27.

Datry, T., Malard, F. and Gibert, J. 2005. Response of invertebrate assemblages to increased groundwater recharge rates in a phreatic aquifer. *Journal of the North American Benthological Society* 24:461–77.

Davis, J.V. and Bell, R.W. 1998. U.S. Department of the Interior U.S. Geological Survey National Water-Quality Assessment Program water – Quality assessment of the Ozark Plateaus study unit, Arkansas, Kansas, Missouri, and Oklahoma – Nutrients, bacteria, organic carbon, and suspended sediment in surface water, 1993–95. Water-Resources Investigations Report 98–4164.

Death, R.G., Barquín J. and Scarsbrook M.R. 2004. Coldwater and geothermal springs. In *Freshwaters of New Zealand*, eds. J.S. Harding, M.P. Mosley, C. Pearson, and B. Sorrell 30.31–30.14. Christchurch, NZ: New Zealand Hydrological Society and New Zealand Limnological Society.

Dickson, G.W. 1975. A preliminary study of heterotrophic microorganisms as factors in substrate selection of troglobitic invertebrates. *The National Speleological Society Bulletin* 37:89–93.

Dickson, G.W., Briese, L.A. and Giesy, J.P. Jr. 1979. Tissue metal concentrations in two crayfish species cohabiting a Tennessee cave stream. *Oecologia* 44:8–12.

Dole-Oliver, M.J., Castellarini, F., Coineau, N., Galassi, D.M.P., Martin, P., Mori, N., Valdecasas, A. and Gibert, J. 2009. Towards an optimal sampling strategy to assess groundwater biodiversity: comparison across six European regions. *Freshwater Biology* 54:777–96.

Elgar, M.A. and Crespi, B.J. 1992. Ecology and evolution of cannibalism. In *Cannibalism Ecology and Evolution among Diverse Taxa*, eds. M.A. Elgar, and B.J. Crespi, 1–12. Oxford: Oxford University Press.

Elliott, W.R. 2000. Conservation of the North American cave and karst biota. In *Subterranean Ecosystems, Ecosystems of the World 30*, eds. H. Wilkens, D.C. Culver, and W.F. Humphreys, 664–689. Amsterdam, The Netherlands: Elsevier.

Engel, A.S., Porter, M.L., Stern, L.A., Quinlan, S. and Bennett, P.C. 2004. Bacterial diversity and ecosystem function of filamentous microbial mats from aphotic (cave) sulfidic springs dominated by chemolithoautotrophic "Epsilon proteobacteria". *FEMS Microbiology Ecology* 51:31–53.

Fetzner, J. W., Jr. and Crandall, K.A. 2003. Linear habitats and the nested clade analysis: An empirical evaluation of geographic vs. river distances using an Ozark crayfish (Decapoda: Cambaridae. *Evolution* 57:2101–18.

Folmer, O., Balck, M., Hoeh, W., Lutz, R. and Vrijenhoek, R. 1994. DNA primers for amplification of mitochondrial cytochrome c oxidase subunit I from diverse metazoan invertebrates. *Molecular Marine Biology and Biotechnology* 3:294–99.

Gibert, J. 1986. Ecologie d'un systeme karstique jurassien. Hydrogéologie, dérive animale, transits de matières, dynamique de la population de *Niphargus* (Crustacé Amphipode): *Memoires de Biospeologie* 13:1–379.

Gibert, J. and Culver, D.C. 2009. Groundwater biodiversity: An introduction. *Freshwater Biology* 54:639–48.

Gibert, J. and Deharveng, L. 2002. Subterranean ecosystems: A truncated functional biodiversity. *BioScience* 52:473–81.

Gibert, J., Danielopol, D.L. and Stanford, J.A. eds. 1994b. *Groundwater Ecology*. San Diego, CA: Academic Press.

Gibert, J., Stanford, J.A., Dole-Oliver, M.J. and Ward, J.V. 1994a. Basic attributes of groundwater ecosystems and prospects for research. In *Groundwater Ecology*, eds. J. Gibert, D.L. Danielopol, and J.A. Stanford, 7–40. San Diego, CA: Academic Press.

Glazier, D.S. 2012. Temperature affects food-chain length and macroinvertebrate species richness in spring ecosystems. *Freshwater Science* 31:575–85.

Graening, G.O. and Brown, A.V. 2003. Ecosystem dynamics and pollution effects in an Ozark cave stream. *Journal of the American Water Resources Association* 39: 1497–1507.

Griebler, C. and Lueders, T. 2009. Microbial biodiversity in groundwater ecosystems. *Freshwater Biology* 54:649–77.

Griebler, C., Mindl, B., Slezak, D. and Geiger-Kaiser, M. 2002. Distribution patterns of attached and suspended bacteria in pristine and contaminated shallow aquifers studied with an in-situ sediment exposure microcosm. *Aquatic Microbial Ecology* 28:117–29.

Haddaway, N.R., Mortimer, R.J.G., Christmas, M. and Dunn, A.M. 2015. Water chemistry and endangered white-clawed crayfish: A literature review and field study of water chemistry association in *Austropotamobius pallipes*. *Knowledge and Management of Aquatic Ecosystems* 416:01.

Hahn, H.J. 2002. Distribution of the aquatic meiofauna of the Marbling Brook catchment (Western Australia) with reference to landuse and hydrogeological features. *Archiv für Hydrobiologie. Supplementband. Monographische Beiträge* 139:237–63.

Hobbs, H.H. III. 2001. Decapoda. In *Ecology and Classification of North American Freshwater Invertebrates*. 2nd edition, eds. J.H. Thorp, and A.P. Covich, 955–1001. San Diego, CA: Academic Press.

Hobbs, H.H. III. and Lodge, D.M. 2010. Decapoda. In *Ecology and Classification of North American Freshwater Invertebrates*. 3rd edition, eds. J.H. Thorp, and A.P. Covich, 901–967. San Diego, CA: Academic Press.

Hobbs, H.H. Jr. 1952. A new albinistic crayfish of the genus *Cambarus* from southern Missouri with a key to the albinistic species of the genus (Decapoda, Astacidae). *The American Midland Naturalist* 48:689–93.

Hobbs, H.H. Jr. 1972. Crayfishes (Astacidae) of North and Middle America. In *Biota of Freshwater Ecosystems, Identification Manual, United States Environmental Protection Agency 9*. Washington, DC: United States Environmental Protection Agency.

Hobbs, H.H. Jr. 1989. An illustrated checklist of the American Crayfishes (Decapoda: Astacidae, Cambaridae, and Parastacidae). *Smithsonian Contributions to Zoology* 480:1–236.

Hobbs, H.H. Jr. and Barr, T. 1960. The origins and affinities of the troglobitic crayfishes of North America (Decapoda, Astacidae) I. The genus *Cambarus*. *American Midland Naturalist* 64:2–33.

Hobbs, H.H. Jr., Horton, H., Hobbs, H.H. III and Daniel, M.A. 1977. A review of the troglobitic decapod crustaceans of the Americas. *Smithsonian Contributions to Zoology* 244:1–183.

Hogger, J.B. 1988. Ecology, population biology and behavior. In *Freshwater Crayfish: Biology, Management and Exploitation*, eds. D.M. Holdich, and R.S. Lowery, 114–144. London: Croom Helm.

Holdich, D.M. 1987. The dangers of introducing alien animals with particular reference to crayfish. *Freshwater Crayfish* 7:25–30.

Holsinger, J.R. 1993. Biodiversity of subterranean amphipod crustaceans: Global patterns and zoogeographic implication. *Journal of Natural History* 27:821–35.

Jurcak, A.M., Lahman, S.E., Wofford, S.J. and Moore, P.A. 2016. Behavior of crayfish. In *Biology and Ecology of Crayfish*, eds. M. Longshaw, and P. Stebbing, 117–131. Boca Raton, FL: CRC Press/Taylor & Francis.

Katoh, K. and Standley, D.M. 2013. MAFFT multiple sequence alignment software version 7: Improvements in performance and usability. *Molecular Biology and Evolution* 30:772–80.

Koppleman, J.B. and Figg, D.E. 1995. Genetic estimates of variablity and relatedness for conservation of an Ozark cave crayfish species complex. *Conservation Biology* 9:1288–94.

Lanfear, R., Calcott, B., Ho, S.Y. and Guindon, S. 2012. PartitionFinder: Combined selection of partitioning schemes and substitution models for phylogenetic analyses. *Molecular Biology and Evolution* 29:1695–701.

Lefébure, T.C., Douady, J., Malard, F. and Gibert, J. 2007. Testing dispersal and cryptic diversity in a widely distributed groundwater amphipod (*Niphargus rhenorhodanensis*). *Molecular Phylogenetics and Evolution* 42:676–86.

Lewis, J.J. 2002. *Conservation Assessment for Salem Cave Crayfish (Cambarus hubrichti).* Eastern Region: USDA Forest Service.

Li, H. and Cooper, R.L. 2002. The effect of ambient light on blind cave crayfish: Social interactions. *Journal of Crustacean Biology* 22:449–58.

Malard, F., Boutin, C., Camacho, A.I, Ferreira, D., Michel, G., Sket, B. and Stoch, F. 2009. Diversity patterns of stygobiotic crustaceans across multiple spatial scales in Europe. *Freshwater Biology* 54:756–76.

Mathieu, J., Essafi, K. and Doledec, S. 1991. Dynamics of particulate organic matter in bed sediments of two karst streams. *Archiv für Hydrobiologie* 122:199–211.

MDC (Missouri Department of Conservation). 2016. *Missouri Species and Communities of Conservation Concern Checklist.* Missouri: Missouri Department of Conservation.

MDNR (Missouri Department of Natural Resources). 2014. *The Code of State Regulations (CSR); Division 20, Clean Water Commission. Chapter 7 - Water quality.* Missouri: Missouri Department of Conservation.

Mokady, M., Loya, Y., Achituv, Y. Geffen, E., Graur, D., Rozenblatt, S. and Brickner, I. 1999. Speciation versus phenotypic plasticity in coral inhabiting barnacles: Darwin's obervations in an ecological context. *Journal of Molecular Evolution* 49:367–75.

NatureServe. 2015. NatureServe Explorer: An online encyclopedia of life [web application]. Version 7.1. NatureServe, Arlington, VA. Available http://explorer.natureserve.org (accessed: June 21, 2016).

Nimick, D.A. and Cleasby, T.E. 2004. Trace elements in water in streams affected by historical mining. In *Integrated Investigations of Environmental Effects of Historical Mining in the Basin and Boulder Mining Districts, Boulder River Watershed, Jefferson County, Montana, U.S.*, eds. D.A. Nimick, S.E. Church, and S.E. Finger, Professional Paper 1652, 159–189. Montana: Geological, USDA Forest Service and U.S. Environmental Protection Agency.

Nimick, D.A., Church, S.E. and Finger, S.E. 2004. *Integrated Investigations of Environmental Effects of Historical Mining in the Basin and Boulder Mining Districts, Boulder River Watershed, Jefferson County,* Montana. U.S. Geological Survey Professional Paper 1652, 524 p. Washington, DC: US Department of the Interior.

Owen, C.L., Bracken-Grissom, H., Stern, D. and Crandall, K.A. 2015. A synthetic phylogeny of freshwater crayfish: Insights for conservation. *Philosophical Transactions of the Royal Society B Biological Sciences* 370:20140009.

Paran, F., Malard, F., Mathieu, J., Lafont, M., Glassi, D.M.P. and Marmonier, P. 2005. Distribution of groundwater invertebrates along an environmental gradient in shallow water-table aquifer. In Proceedings of an International Symposium on World Subterranean Biodiversity, ed. J. Gibert, 99–105. France: University of Lyon.

Peake, D.R., Pond, G.J. and McMurray, S.E. 2004. Development of tolerance values for Kentucky crayfish. In *Kentucky Environmental and Public Protection Cabinet.* Frankfort, KY: USA Department for Environmental Protection, Division of Water.

Pederson, K. 2001. Diversity and activity of microorganisms in deep igneous rock aquifers of the Fennoscandian Shield. In *Subsurface Microbiology and Biochemistry*, eds. J.K. Frederickson, and N. Fletcher, 97–139. New York: Wiley.

Pflieger, W.L. 1996. *The Crayfishes of Missouri.* Jefferson City, MO: Missouri Department of Conservation.

Pohlman, J.W., Iliffe, T.M. and Cifuentes, L.A. 1997. A stable isotope study of organic cycling and the ecology of an Anchialine cave ecosystem. *Marine Ecology Progress Series* 155:17–27.

Porter, M.L., Perez-Losada, M. and Crandall, K.A. 2005. Model based multi-locus estimation of Decapod phylogeny and divergence times. *Molecular Phylogenetics and Evolution* 37:355–69.

Posada, D. and Crandall, K.A. 2001. A comparison of different strategies for selecting models of DNA substitution. *Systematic Biology* 50:580–601.

Poulson, T.L. and White, W.B. 1969. The cave environment. *Science* 165:971–81.

Pusch, M. 1996. The metabolism of organic matter in the hyporheic zone of a mountain stream, and its spatial distribution. *Hydrobiologia* 323:107–18.

Reynolds J.D. and Souty-Grosset, C. 2012. *Management of Freshwater Biodiversity: Crayfish as Bioindicators.* Cambridge, UK: Cambridge University Press.

Reynolds, J.D., Souty-Grosset, C. and Richardson, A. 2013. Ecological roles of crayfish in freshwater and terrestrial habitats. *Freshwater Crayfish* 19:197–218.

Richman, N.I., Böhm, M., Adams, S., et al. 2015. Multiple drivers of decline in the global status of freshwater crayfish. *Philosophical Transactions of the Royal Society, Series B*, 370:20140060.

Romero, A. 2009. *Cave Biology; Life in the Darkness.* New York: Cambridge University Press.

Rösner, U. 1998. Effects of historical mining activities on surface water and groundwater: An example from northwest Arizona. *Environmental Geology* 33:224–30.

Sarbu, S.M., Kane, T.C. and Kinkel, B.F. 1996. A chemoautotrophically based cave ecosystem. *Science* 272:1953–55.

Saviour, M.N. 2012. Environmental impact of soil and sand mining: A review. *International Journal of Science, Environment and Technology* 1:125–34.

Schrimpf, A., Theissinger, K., Dahlem, J., Maguire, I., Pârvulescu, L., Schulz, J.K. and Schulz, R. 2014. Phylogeography of noble crayfish (*Astacus astacus*) reveals multiple refugia. *Freshwater Biology* 59:761–76.

Simon, K.S. 2008. Ecosystem science and karst systems. In: *Frontiers of Karst Research, Special Publication 13*, eds. J.B. Martin, and W.B. White, 49–53. Leesburg, VA: Karst Waters Institute, Inc.

Simon, K.S., Benfield, E.F. and Macko, S.A. 2003. Food web structure and the role of epilithic films in cave streams. *Ecology* 84:2395–406.

Skeet, B. 1999. The nature of biodiversity in hypogean waters and how it is endangered. *Biodiversity Conservation* 8:1319–38.

Song, H., Buhay, J.E., Whiting, M.F. and Crandall, K.A. 2008. Many species in one: DNA barcoding overestimates the number of species when nuclear mitochondrial pseudogenes are coamplified. *Proceedings of the National Academy of Sciences of the United States of America* 105:13486–91.

Stamatakis, A. 2014. RAxML version 8: A tool for phylogenetic analysis and post-analysis of large phylogenies. *Bioinformatics* 30:1312–13.

Stanford, D.J. and Gibert, J. 1994. Conclusions and perspective. In *Groundwater Ecology*, eds. J. Gibert, D. Danielopol, and J. Stanford, 543–47. San Diego, CA: Academic Press.

Strayer, D.L. 1994. Limits to biological distributions in groundwater. In *Groundwater Ecology*, eds. J. Gibert, D. Danielopol, and J. Stanford, J. 287–310. San Diego, CA: Academic Press.

Streever, W.J. 1996. Energy economy hypothesis and the troglobitic crayfish *Procambarus erythrops* in Sim's Sink Cave, Florida. *American Midland Naturalist* 135:357–66.

Taylor, C.A., Schuster, G.A., Cooper, J.E., DiStefano, R.J., Eversole, A.G., Hamr, P., Hobbs, H.H. III, Robison, H.W., Skelton, C.E. and Thom, R.F. 2007. A reassessment of the conservation status of crayfishes of the United Stated and Canada after 10+ years of increased awareness. *Fisheries* 32:371–89.

Taylor, R.M., Watson, G.D. and Alikhan, M.A. 1995. Comparative sub-lethal and lethal acute toxicity of copper to the freshwater crayfish, *Cambarus robustus* (Cambaridae, Decapoda, Crustacea) from an acidic metal-contaminated lake and a circumneutral uncontaminated stream. *Water Research* 29:401–8.

The Nature Conservancy, Ozarks Ecoregional Assessment Team. 2003. *Ozarks Ecoregional Conservation Assessment.* Minneapolis, MN: The Nature Conservancy Midwestern resource Office.

Thoma, R.F. 2016. Crustacea: Malacostraca: Decapoda: Astacidea: Cambaridae: *Cambarus*: Species. In *Thorp and Covich's Freshwater Invertebrates*, Volume II: Keys to Nearctic Fauna, eds. J.H. Thorp, and D.C. Rogers, 667–91. Amsterdam: Academic Press.

Toon, A., Finley, M., Staples, J. and Crandall, K.A. 2009. Decapod phylogenetics and molecular evolution. In *Decapod Crustacean Phylogenetics*, eds. J.W. Martin, K.A. Crandall, and D.L. Felder, 14–28. Boca Raton, FL: CRC Press/Taylor & Francis.

Venarsky, M.P., Huryn, A.D. and Benstead, J.P. 2012. Re-examining extreme longevity of the cave crayfish *Orconectes australis* using new mark-recapture data: A lesson on the limitations of iterative size-at-age models. *Freshwater Biology* 57:1471–81.

Veselý, L., Buřič, M. and Kouba, A. 2015. Hardy exotics species in temperate zone: can "warm water" crayfish invaders establish regardless of low temperatures? *Scientific Reports* 5:16340.

Vineyard, J.D. and Feder, J.L. 1982. Springs of Missouri. In *Missouri Department of Natural Resources, Division of Geology and Land Survey in Cooperation with U.S. Water Resources* Report No. 29, 212. Missouri: Geological Survey and Missouri Department of Conservation.

Weingartner, D.L. 1977. Production and trophic ecology of two crayfish species cohabiting an Indiana cave. PhD thesis. Michigan State University, East Lansing, MI, USA.

White, W.B. 1969. Conceptual models for carbonate aquifers. *Ground Water* 7:15–21.

White, W.B. 1988. *Geomorphology and Hydrology of Karst Terrains.* New York: Oxford Press.

Whitledge, G.W. and Rabeni, C.F. 1997. Energy sources and ecological role of crayfishes in an Ozark stream: insights from stable isotope and gut analyses. *Canadian Journal of Fisheries and Aquatic Sciences* 54:2555–63.

CHAPTER **16**

Historic Cultural Value of the Japanese Endangered Freshwater Crayfish, *Cambaroides japonicus* (De Haan, 1841) (Decapoda: Cambaroididae)

Tadashi Kawai and Jason Coughran

CONTENTS

16.1 INTRODUCTION

Freshwater crayfish represent an important component of lotic and lentic ecosystems in many parts of the world, with around 600 species occurring in Europe and the Middle East, North America, far eastern Asia, Oceania, Madagascar, and South America (Hobbs 1988, Kawai and Crandall 2016). Large and/or abundant species have often supported recreational or commercial fisheries in many areas (e.g., Henry and Lyle 2003, Jussila et al. 2013), and there is a notable aquarium trade for crayfish (Faulkes 2015, Kawai et al. 2020). *Cambaroides* includes five crayfish species distributed across the Korean Peninsula, Far East Russia, northern China, eastern Mongolia, and northern Japan (Fig. 16.1A).

The Japanese crayfish *Cambaroides japonicus* (De Haan 1841) is the sole indigenous species in Japan (Fig. 16.1A), mainly inhabiting small brooks or lakes in the northern terminal area of the Japanese Archipelago: Hokkaido island and Aomori Prefecture on Honshu Island. The species reaches approximately 5 cm in total length (Okada 1933, Kawai and Fitzpatrick 2004). It was designated as an endangered species by the Japanese government (Kawai et al. 2015, 2020) and several aspects of its biology have been studied (Nakata and Goshima 2006, Mrugala et al. 2016, Martin-Torrijos et al. 2018). Such information on the natural history of this endangered species will help toward its conservation, but these studies have not yet been sufficient.

To present another perspective on this endangered species, here we have reviewed information on its historic cultural value from the indigenous Ainu culture into the modern context. This effort is an initiative to consolidate information that the authors have previously documented in the grey literature and/or in Japanese literature. It is our hope that covering the connection between this crayfish species with Ainu and Japanese people in the peer-reviewed scientific literature will encourage similar reflections on the cultural values of other species around the world (particularly endangered species).

16.2 MATERIALS AND METHODS

Museum specimens of *C. japonicus-* or *C. japonicus*-related artifacts were observed and examined in museum collections between 2000 and 2012, as follows:

- RMNH: Naturalis, Nationaal Natuurhistorisch Museum (https://www.naturalis.nl/en/museum), Leiden, the Netherlands (Fig. 16.1B)
- USNM: Department of Invertebrate Zoology, National Museum of Natural History, Smithsonian Institution (https://naturalhistory.si.edu/research/invertebrate-zoology), Washington DC (Fig. 16.1B)
- Hakodate City Museum (http://hakohaku.com/top/guide/english/), Hakodate, Japan
- Matsuura Takeshiro Memorial Museum (www.city.matsusaka.mie.jp/site/takesiro/), Matsuzaka, Japan
- The Imperial Household Agency (宮内庁書陵部) (www.kunaicho.go.jp/culture/shoryobu/shoryobu.html), The Imperial Palace, Japan

In July 2015, one of the authors (TK) visited Vladivostok, in Far Eastern Russia, to observe a crayfish cultural object.

16.3 RESULTS AND DISCUSSION

16.3.1 Historical Context

The indigenous Ainu people lived in the northern areas of Japan, including Hokkaido, the Aomori Prefecture of Honshu Island, the Kuril Islands, Sakhalin Island, and

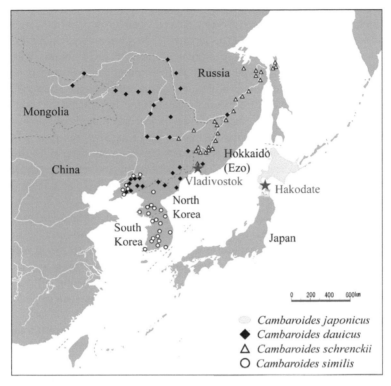

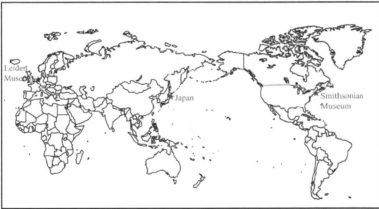

Figure 16.1   Museum specimens and distribution of species in the genus *Cambaroides*. (A) Distribution of Asian crayfish (*Cambaroides*). (B) The type series of *C. japonicus* is deposited in Leiden Museum, the Netherlands, and specimens from Hakodate are deposited in the Smithsonian Institution, National Museum of Natural History, USA.

Primorye (Poisson 2002, Chiba 2008). They primarily relied on fisheries. During the Edo Era (1603–1868), the northern indigenous Ainu were actively trading with the Japanese. Because Hokkaido held rich fisheries and forest resources, trading with the Ainu produced huge wealth for the Japanese Edo government. Under their policy of isolation, in which trading and communication with most nations was prohibited, the Edo government allowed trading and communication to continue with the Netherlands (as well as the United Kingdom, China, and Korea). Thus, information about Dutch medicine was introduced to Japan. One introduced method was the use of crayfish gastroliths as medicine, based on European medical knowledge, and these became remarkably expensive during the Edo Era. The importation of gastroliths from Hokkaido was important to the Edo government. The Edo government sought to obtain more exact knowledge of the geography and resources of Hokkaido, so as to continue trading with the Ainu and protect that northern area from invasions (e.g., by Russia).

During the Edo Era, Japanese adventurer Takeshiro Matsuura (1818–1888) recorded a typical scene of Ainu people collecting crayfish, with an inscription and illustration based on his six expeditions to Hokkaido (Fig. 16.2A, translation of the inscription given later). Matsuura had been employed by the Edo government, with three of his visits to Hokkaido (1844, 1846, and 1855) being official government visits. He documented that Ainu people gathered *C. japonicus* from brooks in Hokkaido using "miso" (a fermented soybean paste, a basic Japanese seasoning). The miso flavor lures *C. japonicus* from their burrows or from beneath rocky cover, making it easier to catch them. One of the authors (TK) has confirmed the effectiveness of miso to lure out *C. japonicus* from its natural habitat in Otaru, Hokkaido, Japan, on 15 September 1990 (Kawai and Nakata 2001) (Fig. 16.2B). Kawai et al. (2016) demonstrated that the natural habitats of *C. japonicus* are oligotrophic and of low productivity; broad leaf trees surround the natural habitat of *C. japonicus*, and hence their main food consists of fallen leaves. Since miso is made from a vegetable product, this may explain why it is such an attractive food item for *C. japonicus*. The saltiness of miso may also be an attracting factor in such fresh, oligotrophic streams. Luring *C. japonicus* with miso makes it much easier to collect the crayfish, particularly for Ainu children who would have had limited experience and no equipment to catch them. As shown (Kawai and Nakata 2001), using miso is a reasonably effective and sophisticated method of collecting *C. japonicus*.

The following translation of the inscription (Fig. 16.2A, by TK) is offered:

"Tapishitonpekorupe" (corrupted as "Tekunpekuru") in Ainu language, or "sarikani"
    in dialect of Matsumae region (the most southern part of Hokkaido Island, Ezo).

Their body shape is such as in the figure, resembling a shrimp shape but without longer antenna like shrimp. They have large chela like a crab. Their body color is like the Kuruma prawn (*Marsupenaeus japonicus*); if they were boiled or baked, their body color would be red. "Sharikani" seems to be a corruption of "Sawakani". Because they claw backward, so they were called Sharikani (in old Japanese language, "Shisaru" or "Sari" or "Saru" means "to move backward"; "kani" means

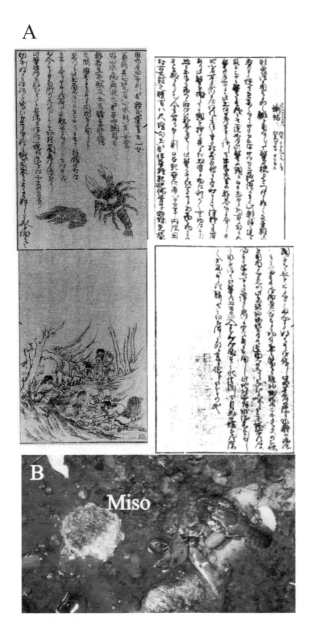

Figure 16.2 Crayfish collecting methods around 1800. (A) Antique archive literature in Memorial Museum of Takeshiro Matsuura, Matsuzaka, Japan, describing how Ainu children in Hokkaido, northern Japan, gathered crayfish by using Japanese "Miso" soup paste. (B) An experimental reconstruction confirming the viability of collecting crayfish using Miso soup paste in Hokkaido in 2000. A crayfish (right) is attracted by the Miso paste flavor and lured out from its burrow to the riverbank.

"crab"). Another theory states that the crayfish often have a milky white ball (the gastrolith), so they have been called Sharikani (in Buddhism, it is believed that small, white relics of Buddha's bones remain in the world, traditionally referred to as "shari" [Ruppert 2000, Trenson 2018], and otherwise known in Japanese Buddhism as "Busshari" [a construct word from "Buddha" and "Shari"]). Hence, within that cultural context, since crayfish gastroliths were thought to represent the small, white relics of Buddha's bones, the crayfish that produced them came to be called "Sharikani" ["kani" meaning 'crab']), and the cow, horse, dog, and sheep have the same calculus in their body, which Japanese people call "Satou".

In Hokkaido (*note:* the old place name given, Ezo, did not include the Hakodate area, see Fig. 16.1A), the freshwater crayfish always occur in brooks that have a gravel bottom. The freshwater crayfish also occur in Hakodate, and they often appear in the Tsugaru region (the western side of Aomori Prefecture). If we caught the freshwater crayfish, we should press the dorsal side of the carapace to check the hardening of body; if they have a gastrolith, their body is soft, whereas they have a hard body when not in the molting season. This is the same for that of cows and horses; when they have a disease, they have calculus in their body. So it is thought that crayfish that have a gastrolith is a diseased individual. This is same for humans, because diseased persons often develop calculus in their body. (This inscription inserts a citation of the classical Japanese text "Nihonshoki", as to the Japanese "Busshari" and the forming of calculus in animals and their status; citation untranslated, since it is not an original inscription.) I think that forming of calculus in animals is due to their disease. The gastrolith of freshwater crayfish is called "Okuriankiri" in the Dutch language. (As discussed in the results, trading and communication were maintained with the Netherlands during the Japanese policy of isolation.) The crayfish were caught to dedicate to the Shogun, but now we (people who have emigrated into Ezo) instruct the Ainu people to collect the freshwater crayfish to make the (gastrolith) product from Ezo; this medical efficiency is several times higher than (importing) medicine from the Netherlands.

The freshwater crayfish like "Miso". If we flow the Miso from upper streams of a brook to downstream areas, crayfish individuals will appear from crevices between stones to eat the Miso and aggregate for Miso. This state looks like moths aggregating to a torch fire at night. Ainu people collected the gathered crayfish and dissected them to collect their white ball (presumably, the gastrolith). The Ainu people like to eat the collected freshwater crayfish. In Ezo region, skin diseases are often prevalent from spring to summer. During this period, Ainu people must catch the freshwater crayfish, then boil the crayfish and eat them, so that they can obtain medicinal benefits. There is one medicinal mix of three components: amber, Okurikankiri, and Chikitaris (Chikitaris is a Japanese expression and means a medicinal plant from Europe, *Digitalis purpurea*, or Foxglove). In the Ezo area, the amber and Okurikankiri are produced, which can treat some local diseases. According to legend, if disease occurred, this place would always produce medicine for the disease; this theory came from our god. This is interesting. On the Okhotsk side of this place, we often have polluted water. We take the dirty water and put five to six freshwater crayfish into it, and they always make the dirty water clear and cool down; they filter natural suspended substance of the water. I [= Takeshiro Matsuura] experimented with this several times and confirmed this phenomenon, so I note this here.

16.3.2 Medical Use of Crayfish Gastroliths

Lowery (1988) reviewed the freshwater crayfish gastrolith in relation to crustacean growth with molting. After molting, the exoskeleton is relatively soft and hardening of the epicuticle and mineralization of the exocuticle starts. Freshwater has very low calcium concentration, so much of the calcium needed for the initial mineralization of the new exoskeleton after molting is derived from the old (molted) exoskeleton and the gastroliths that the crayfish had produced prior to molt as a calcium reserve. The gastroliths are milky white, paired disks of 7–10 mm in diameter that form in the wall of the cardiac region of the foregut (Scudamore 1947). The gastroliths are formed between the cuticular wall of the foregut and its underlying epidermis; when the lining of the gut is shed at molting, the gastrolith is shed into the foregut lumen and is gradually broken down and resorbed by the gut epithelium and the hepatopancreas (Travis 1960). The presence of gastroliths in *C. japonicus* is seasonal, with a maximum weight in June (Kawai and Fitzpatrick 2004).

Cambaroides japonicus gastroliths were regarded as expensive medicine for all diseases in the Edo Era; the finely powdered gastrolith was notably used for lung and diuretic diseases and exported to the Netherlands (Yamaguchi and Holthuis 2001). Philip Franz Balthasar von Siebold (1796–1866), a scientist of natural history and a medical doctor, came to Japan, where he taught medicine and ran a trading house in Nagasaki. Von Siebold made remarkable contributions to science for Japan and Europe, and part of his contributions have been published and stored in the Naturalis, Nationaal Natuurhistorisch Museum and library in Leiden, the Netherlands (e.g., "Nippon", von Siebold 1989–1995, 1994–1996). von Siebold returned to Europe with many types of specimens of Japanese fauna and flora (including the type series of *C. japonicus* he received in 1826), which were deposited, along with crayfish gastroliths, in the RMNH (Fig. 16.3) (Kawai and Fitzpatrick 2004). Since von Siebold worked as a medical doctor, bringing back the type series of *C. japonicus* and gastroliths seems likely to have been associated with the medicinal trade of gastroliths.

The crayfish and gastrolith trade between the Ainu and Japanese, and then between Japan and other countries, may also explain crayfish in early 19th-century ukiyo-e artwork, captured by renowned artist Katsushika Hokusai in 1814 (Kawai and Coughran 2018). The woodcut prints of "sarikani" (サリカニ) are sufficiently detailed (Fig. 16.4) that Hokusai presumably had observed these crayfish firsthand, either through personal travel to the northern part of Japan, where the species occurs, or by examining specimens that had been transported down to Edo for culture or sale.

To our knowledge, among the Japanese freshwater decapods, gastroliths are only produced by *C. japonicus*. If this is true, this could explain the intense interest in *C. japonicus* as the gastrolith source, their rarity, and thus high value, as a trade item between Japan and Europe.

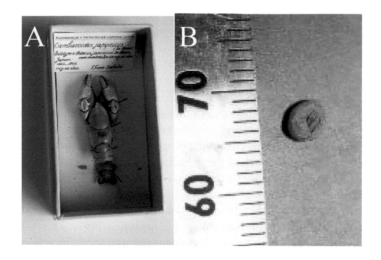

Figure 16.3   Type specimen of *Cambaroides japonicus* in RMNH. (A) The type series was preserved with gastroliths. (B) The gastrolith of *C. japonicus*.

A   B

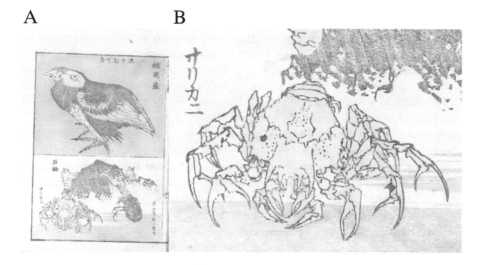

Figure 16.4   Woodcut print from the Ukiyo-e artwork of Katsushika Hokusai, 1814. (Permission from Sumida Ward, Tokyo). (A) Whole plate. (B) Close-up of the illustration of *Cambaroides japonicus*. Note the Japanese name in the script of that era, "Sarikani" (サリカニ).

16.3.3 Motifs on Ainu Ritual Objects

Ainu people do not have a written language, but they created various objects for their life and rituals. The objects were usually sculpted with animals on their surface, revealing which were popular in their culture during the Edo Era (Poisson 2002, Chiba 2008). Most natural history or folklore museums in Hokkaido house "Pasui"

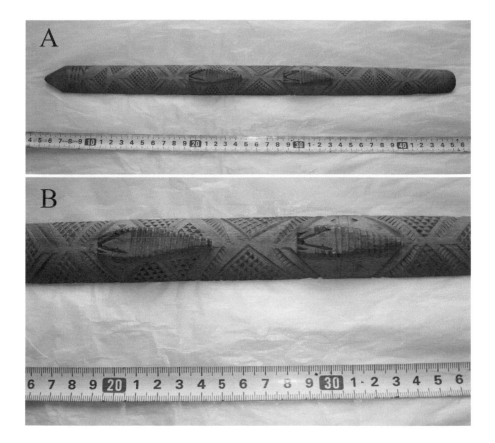

Figure 16.5   "Pasui" equipment used in rituals by the "Ainu" people – the aborigines of northern Japan. (A) Entire piece of Pasui. (B) Crayfish motif on the Pasui (deposited
and permission from Hakodate City Museum, Hokkaido, Japan).

objects used in Ainu rituals, including examples sculpted with freshwater crayfish
(Fig. 16.5); several Hokkaido museums have Pasui objects with freshwater crayfish
sculptures (Kawai and Onimaru 2006). It would seem from these artifacts that *C.
japonicus* was a very popular freshwater organism for the Ainu.

16.3.4 Native Crayfish for Emigrating Japanese People

Many Japanese immigrated to Hakodate, southern Hokkaido, to develop forestry and
fisheries resources during the latter half of the Edo Era. At that time, various manufac-
turing industries were established in Hakodate by the Japanese pioneers. Manufacturers
created materials such as dishware, needed for pioneer life in Hakodate. Various kinds
of popular organisms appeared as motifs on these items (Kawai and Onimaru 2006),
including a set of dishes in Hakodate City Museum with crayfish painted on them (Fig.
16.6). This suggests that *C. japonicus* was a familiar and popular aquatic organism for
Japanese immigrants in Hokkaido as well as the Ainu.

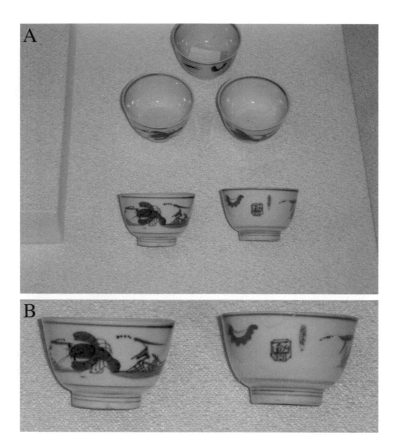

Figure 16.6 Japanese dish set made in Hokkaido, Japan, around 1859. (A) Complete set of dishes. (B) Crayfish motif painted on the dish (deposited and permission from Hakodate City Museum, Hokkaido, Japan).

The USNM houses *C. japonicus* specimens (lot number USNM 444239) that are longitudinally skewered on bamboo (Fig. 16.7). The label collection details are: "Streamer Albatross, Northwest Pacific, 1906, July 10, Hakodate Market, Hakodate Japan, *Cambaroides japonicus*". This was during the Meiji Era (1868–1912), with a reopened international trade and communication policy (Kawai and Onimaru 2006). These *C. japonicus* specimens in the USNM were simply sold in the public markets at Hakodate, Hokkaido, indicating that local citizens were using the crayfish. Moreover, the whole crayfish animal was sold rather than just the gastroliths. Thus, it seems that they were desired not just for medicine but also perhaps as food by local citizens (Fig. 16.7).

16.3.5 Native Crayfish for the Past Japanese Emperor

In Japan, a grand national festival is held when there is a change from one emperor to the next. The enthronement ceremony is the premier royal festival in Japan, called

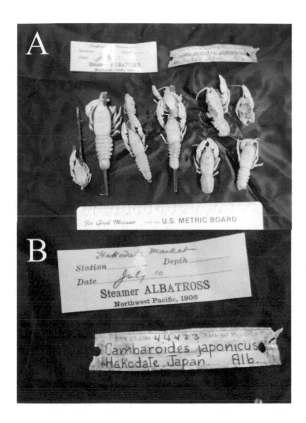

Figure 16.7 Specimen of *Cambaroides japonicus* in USNM. (A) Crayfishes bunched on a bamboo skewer and sold as medicine/food at the fish market in Hakodate City. (B) An American exploration ship, the Steamer Albatross, collected these specimens from Hakodate in 1906.

Gotaiten or Tairei. When Emperor Taisho became sovereign ruler of Japan (from 1912 to 1926), a Tairei was held in 1915, and the official banquet menu has been preserved in the Imperial Household Agency library (Fig. 16.8A). One menu item was a French cream pottage soup using *C. japonicus* from Hokkaido (Fig. 16.8B). Kawai and Takahata (2010) presented the cooking procedure for the royal banquet in detail: approximately 187 g of crayfish and 214 g of chicken meat with cream and butter were used for the soup. The crayfish were caught from Lake Shikotsu, Hokkaido.

Since the official banquet of that Tairei, Emperor Taisho may have taken a liking to the crayfish soup, because it was served at least four more times at royal banquets hosted by the Japanese Emperor, as recorded in the official documents of the Imperial Household Agency. Emperor Taisho often held banquets in the imperial household, after which a lunchbox was always presented to every invited guest. Two boiled crayfish were often included in the lunchbox. The document also notes that in June 1917, the imperial villa (the second house of the emperor) in Nikko City was equipped with a special aquarium to stock *C. japonicus* next to the kitchen.

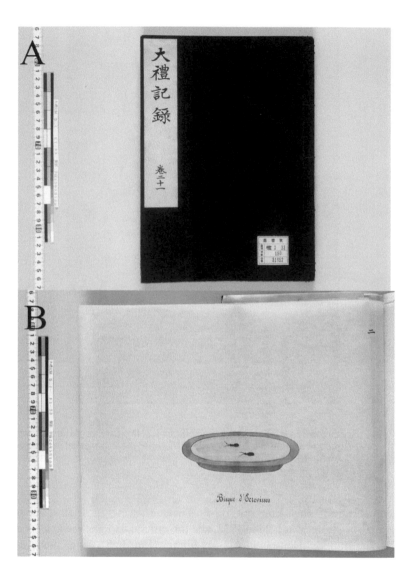

Figure 16.8   Menu of banquet in an enthronement ceremony (Tairei) for the Japanese
              Emperor Yoshihito in 1915. (Volume 21-84576). (A) Cover of menu of "Tairei
              Kiroku" (meaning official record of an enthronement ceremony) for Emperor
              Yoshihito (1879–1926). (B) Two red-colored Japanese freshwater crayfish,
              Cambaroides japonicas, in the French pottage soup (宮内庁書陵部所蔵).

Hokkaido National Fisheries Research Institute houses an antique official document, stating that *C. japonicus* were sent to Tokyo at least one to four times per year from 1917 to 1922, with 300–400 individuals transported at a time: a 1.5 day journey by train. Transportation of live *C. japonicus* was arranged throughout the year. Because *C. japonicus* likes cold water, a cooler box was used for transporting crayfish during the summer. The cooler box was made from double wood plates to insulate against temperature changes, and rock ice with peat moss were placed in the box to maintain cool, moist conditions during the summer shipments. Hokkaido National Fisheries Research Institute provided the authority for *C. japonicus* collections from Lake Shikotsu for shipping to Tokyo.

Since that time, two (American) crayfishes, the red swamp crayfish, *Procambarus clarkii* (Girard 1852), and the signal crayfish, *Pacifastacus leniusculus* (Dana 1852), were released into Hokkaido waters in 1927 and 1930, respectively. Unfortunately, *C. japonicus* became extinct in Lake Shikotsu, and the regular gift of the crayfish to the Japanese Imperial Household ceased.

16.3.6 Crayfish in Ainu Music and Folklore

Several hundred years ago, Ainu people had trade relations with Kamchatka, the Lower Amur River region, and Primorskii, in Far East Russia (Poisson 2002, Chiba 2008). This resulted in Ainu cultural influences spreading to these regions with the exchange of fisheries products. As the Ainu traditional lands overlap the distribution of *C. japonicus*, so too the lands of their mainland trading partners overlap the ranges of two Far East Russian *Cambaroides* species: *C. koshewnikow* and *C. schrenckii* (Poisson 2002, Chiba 2008, Kawai et al. 2016). Ainu cultural relics have influenced or have been influenced by their trading partners in the form of traditional music and folklore.

Fashioned like the chela of crayfish, an instrument called "kani-mukkuri" (translated as "Crayfish Flute"; Fig. 16.9) is a traditional musical instrument originating on the Far East Russia mainland and conveyed to Japan (Kawai and Onimaru 2006). Similar instruments are known from other traditional circumboreal cultures (Malm 2000), although it is unclear if they have a cultural connection with crayfish, like the Ainu kani-mukkuri. The instrument is essentially a lamellophone, or "jew's harp", and was one of the most important instruments used by the Ainu (Chiba 2008). Kani-mukkuri are traditionally played by Ainu women, by first holding the chela of the instrument in the mouth, and then using a combination of breathing and flicking or pulling a string from the instrument's central "tongue".

The kani-mukkuri is associated with two Ainu folktales. Rough summaries are offered below.

The first story: A long time ago in Hokkaido, an Ainu hunter went hunting bear and stayed in a small hunting hut. While there, he heard a strange sound coming from near the brook, so he left the hut to walk down to the brook. At the side of brook, a beautiful young woman was sitting on the edge of a well; she had in her hand some kind of instrument and was performing music. The hunter came toward her, and at that moment she suddenly fell to the bottom of the deep well, never to return. At the edge of the well, the hunter discovered that she had dropped the instrument, and it was a kani-mukkuri. That night, back in the hunter's hut, the hunter had a curious dream. The young woman at the well appeared in his dream and said: I am a crayfish. My parent always reproves me, because I play kanu-mukkuri too much. But I cannot stop playing, so my parents renounced me. I sadly played the kani-mukkuri

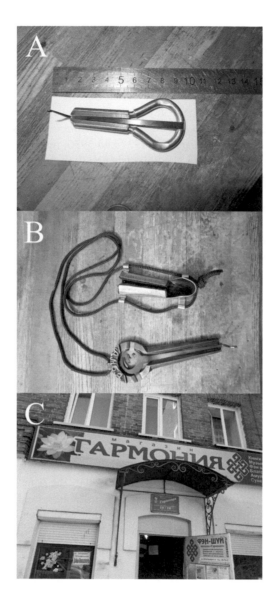

Figure 16.9   Crayfish flute – an instrument shared between northern Japan and Far East
Russia. (A) Crayfish flute in Japan as used by Ainu people. (B) Crayfish flute in
Vladivostok, Far East Russia. C. Folkcraft article shop in Vladivostok.

while sitting down on the edge of the well, just before daytime. Then, you came and
I could pass on to you the instrument. It was because of your action that I could give
up playing kani-mukkuri, so I am really pleased because I will now tell my better
situation to my parents.

    The second story: Two brothers went to the river to catch fish. In the river, the
older brother caught a crayfish that had the kani-mukkuri instrument. The older

brother took the instrument from the crayfish and passed it to his younger sibling. Then, he released the animal back to the river. After that, kani-mukkuri was played by the Ainu.

16.3.7 Conservation Situation

Cambaroides japonicus was designated as an endangered species by the Japanese Fisheries Agency in 1998 and by the Ministry of the Environment in 2000 (Table 16.1). A key threat to the species is the expanding distribution of the two alien crayfish species introduced from North America: the red swamp crayfish *Procambarus clarkii*, and the signal crayfish *Pacifastacus leniusculus*. Propagating and releasing alien crayfishes is now forbidden under the "Invasive Alien Species Act" (www.env .go.jp/nature/intro/1law/index.html) established by the Ministry of the Environment in 2005 (Kawai et al. 2015). However, the threat from *P. leniusculus* and *P. clarkii* is still increasing rapidly (Kawai et al. 2015, 2020), so additional measures are desperately needed to ensure the conservation of *C. japonicus*.

16.3.8 Conservation Impetus

The current economic value of this small and inconspicuous crayfish is virtually nonexistent. It has no value as a fisheries species nor for aquarium purposes (ironically, any potential value is dwarfed by the introduced species that endanger it).

Table 16.1 **Chronological Table of Historical Events of the Japanese Freshwater Crayfish, *Cambaroides japonicus*, as Outlined in This Study**

Year	Event
1800	Gastroliths of *C. japonicus* in Hokkaido, Japan, were taken from crayfish caught by Ainu aborigines using "Miso" and exported to foreign countries as expensive medicine
1841	Specimens of *C. japonicus* were brought to Europe and described
1859	Japanese people migrating to Hokkaido made traditional dishes with *C. japonicus* drawn as motif
1906	American exploration ship, the Steamer Albatross, collected samples of *C. japonicus* from the Hakodate market, Hokkaido, in July
1915	Menu of banquet in an enthronement ceremony (Tairei) for the Japanese Emperor Yoshihito included French Cream Pottage with *C. japonicus*
1927	The alien crayfish *Procambarus clarkii* was introduced into Japan from New Orleans, USA
1930	The alien crayfish *Pacifastacus leniusculus* was imported from Portland, Oregon, and released in Hokkaido
1980	In Hokkaido, the two alien crayfish species were increasing and the native *C. japonicus* decreasing
1998	The species was designated as an endangered species by Fisheries Agency
2000	The species was designated as an endangered species by Ministry of the Environment
2005	"Invasive Alien Species Act" was established by Ministry of the Environment to prohibit producing and releasing alien crayfishes

However, there is a strong cultural affinity between people and this crayfish, stretching across the animistic Ainu and Buddhist Japanese cultures, across the medicinal and culinary spheres of life, across music and folklore, and across the children playing in the creek, and the Japanese Emperor in the imperial palace.

The long-standing cultural value of this species is a reminder that economic considerations are but one factor in determining a species' value, in anthropocentric terms. While we refrain from attempting to assign a dollar value to that cultural history, the affinity Ainu and Japanese people have with the Japanese crayfish is certainly of great relevance to the species conservation. Aspects of its history such as the Emperor's Lake Shikotsu fishery loss of *C. japonicus* localities have occurred, or have been displaced. To that end, the species' cultural value should motivate further research initiatives. We suggest that the species' long cultural heritage would lend itself very well to public education, community awareness, and restoration activities.

The robust physiology of the crayfish also lends itself well to having citizens engaged with the species through touch communication, which is an ideal means of connecting with the important demographic of young children. Thus, there are opportunities to take children to natural *C. japonicus* habitat and teach them "on location" the meaning and importance of general conservation, and for this endangered crayfish species in particular. Indeed, engaging high school students in Odate City, Akita Prefecture has proven such opportunities are valuable and productive (Kudo et al. 2016). Public education events are a vital part of ongoing conservation for this species, instilling positive behavior in citizens around issues such as propagation, release, and dispersal of alien species. The affinity people always had for *C. japonicus* should readily translate to the modern era and could be the very thing that helps to secure this species into the future.

Moreover, the strong historical affinity between people and crayfish is not unique to *C. japonicus*. In presenting this case, we hope to encourage further consideration of historic cultural value for other crayfish species, and indeed for other threatened species generally. Tapping into such bonds, and where possible exploring and documenting the history of such bonds, ought to be beneficial for conservation.

ACKNOWLEDGMENTS

The authors would like to thank the following for allowing access to museum specimens: C. Fransen, RMNH; R. Lemaitre, USNM; M. Sato, Hakodate City Museum; K. Sakamoto, Imperial Institute of Biology and Imperial Household Agency, The Imperial Palace, Japan. We also thank the Memorial Museum of Takeshiro Matsuura, Matsuzaka, Japan, for permission to publish the picture of Fig. 16.2A.

REFERENCES

Chiba, N. 2008. The music of the Ainu. In *The Ashgate Research Companion to Japanese Music*, eds. A.M. Tokita, and D.W. Hughes, 323–344. England: Ashgate.
Faulkes, Z. 2015. The global trade in crayfish as pets. *Crustacean Research* 44:75–92.

Henry, G.W. and Lyle, J.M. 2003. The national recreational and indigenous fishing survey. In *FRDC Project No. 99*, 158. Canberra, Australia: Australian Government Department of Agriculture, Fisheries and Forestry.

Hobbs, H.H. Jr. 1988. Crayfish distribution, adaptation, and evolution. In *Freshwater Crayfish: Biology, Management and Exploitation*, eds. D.M. Holdich, and R.S. Lowery, 52–82. London: Croom Helm.

Jussila, J., Tiitinen, V., Fotedar, R. and Kokko, H. 2013. A simple and efficient cooling method for post-harvest transport of the commercial crayfish catch. *Freshwater Crayfish* 19:15–19.

Kawai, T. and Coughran, J. 2018. Knowledge of the natural history of the Japanese freshwater crayfish, *Cambaroides japonicus*, in Japan: Information from Ukiyo-e, "picture of the floating world" during the late Edo period. *Crayfish News* 40:7–8.

Kawai, T. and Crandall, K. 2016. Global diversity and conservation of freshwater crayfish (Crustacea: Decapoda: Astacoidea). In *A global Overview of the Conservation of Freshwater Decapod Crustaceans*, eds. T. Kawai, and N. Cumberlidge, 65–114. Switzerland: Springer.

Kawai, T. and Fitzpatrick, Jr. J.F. 2004. Redescription of *Cambaroides japonicus* (De Haan, 1841) (Crustacea: Decapoda: Cambaridae) with allocation of type locality and month of collection of types. *Proceedings of the Biological Society of Washington* 117:23–34.

Kawai, T. and Nakata, K. 2001. On a collection method using miso (soybean paste) and burrow utilization of the Japanese crayfish *Cambaroides japonicus* (De Haan, 1841). *Journal of the Natural History of Aomori* 6:49–52.

Kawai, T. and Onimaru, K. 2006. Natural history of the Japanese crayfish, *Cambaroides japonicus*. *Bulletin of Bihoro Museum* 14:63–86.

Kawai, T. and Takahata, M. (eds.). 2010. *Biology of crayfish*. Sapporo, Japan: Hokkaido University Press.

Kawai, T., Barabanshichikov, E. and Coughran, J. 2016. Crayfish music in Far-East Asia. *Crayfish News* 38:4–5.

Kawai, T., Grubb, B. and Grandjean F. 2020. Conservation issues for the Japanese Endangered Crayfish, *Cambaroides japonicus*. In *Recent Advances in Freshwater Crustacean Biodiversity and Conservation*. Roca Boca, FL): CRC Press/Taylor & Francis.

Kawai, T., Min, G.S., Barabanshchikov, E., Labay, V.S. and Ko, H.S. 2015. Asia. In *Global Overview of Freshwater Crayfish*, eds. T. Kawai, Z. Faulkes, and G. Scholtz, 313–368. Roca Boca, FL: Taylor & Francis.

Kudo, H., Hida, M., Torigata, Y. and Kawai, T. 2016. Natural history, present status of locality of southernmost distribution, and conservation situation of endangered freshwater crayfish *Cambaroides japonicus* (De Haan, 1841) in Odate City, Akita Prefecture, Japan. *Hinai (Scientific Reports of Odate City Museum)* 13:1–9.

Lowery, R.S. 1988. Growth, moulting and reproduction. In *Freshwater Crayfish: Biology, Management and Exploitation*, eds. D.M. Holdich, and R.S. Lowery, 83–113. London: Croom Helm.

Malm, W.P. 2000. *Traditional Japanese Music and Musical Instruments*. Tokyo: Kodansha International.

Martin-Torrijos, L., Kawai, T., Makkonen, J., Jussila, J., Kokko, H. and Diéguez-Uribeondo, J. 2018. Crayfish plague in Japan: A real threat to the endemic *Cambaroides japonicus*. *PLoS One* 13: e0195353.

Mrugała, A., Kawai, T., Kozubiková-Balcarová, E. and Petrusek, A. 2016. *Aphanomyces astaci* presence in Japan: A threat to the endemic and endangered crayfish species *Cambaroides japonicus*? *Aquatic Conservation: Marine and Freshwater Ecosystem* 27:103–14.

Nakata, K. and Goshima, S. 2006. Asymmetry in mutual predation between the endangered Japanese native crayfish *Cambaroides japonicus* and the North American invasive crayfish *Pacifastacus leniusculus*: A possible reason for species replacement. *Journal of Crustacean Biology* 26:134–40.

Okada, Y. 1933. Some observations of Japanese crayfishes. *Science Reports of the Tokyo Bunrika Daigaku*, Section B 1:155–58.

Poisson, B.A. 2002. *The Ainu of Japan*. Minneapolis: Lerner Publications Co.

Ruppert, B.D. 2000. *Jewel in the Ashes. Buddha Relics and Power in Early Medieval Japan*. Cambridge, MA: Harvard University Asia Centre.

Scudamore, H.H. 1947. The influence of the sinus glands upon molting and associated changes in the crayfish. *Physiological Zoology* 20:187–208.

Travis D.F. 1960. The deposition of skeleton structures in Crustacea. I. The histology of the gastrolith skeleton tissues complex and the gastrolith in the crayfish *Orconectes* (*Cambarus*) *virilis* Hagen–Decapoda. *Biological Bulletin* 118:137–49.

Trenson, S. 2018. Rice, relics, and jewels. The network and agency of rice grains in medieval Japanese esoteric Buddhism. *Japanese Journal of Religious Studies* 45:269–307.

von Siebold, P.F. 1989–1995. *Nippon*. Vol. 1–6. Philip Franz von Siebold Yumatsudo Press. Fukuoka Prefectural Library, Hazokai. http://www.lib.pref.fukuoka.jp/hp/gallery/nippon/nippon-top.html.

von Siebold, P.F. 1994–1996. *Nippon*. Pictorial Record Vol. 1–3. Yumatsudo Press. Fukuoka Prefectural Library, Hazokai. http://www.lib.pref.fukuoka.jp/hp/gallery/nippon/nippon-top.html.

Yamaguchi, T. and Holthuis, L.B. 2001. Kai-ka Rui Siyasin, a collection of pictures of crabs and shrimps, donated by Kurimoto Suiken to Ph. Von Siebold. *Calanus* III:1–156.

CHAPTER 17

Conservation of the Japanese Endangered Japanese Crayfish, *Cambaroides japonicus* (De Haan, 1841) (Decapoda; Cambaroididae)

Tadashi Kawai, Brooke Grubb, and Frederic Grandjean

CONTENTS

17.1 INTRODUCTION

Freshwater crayfish (Astacidea) are divided into two superfamilies – Astacoidea and Parastacoidea – and they comprise more than 600 species. The Parastacidae is the only member of the superfamily Parastacoidea and is found throughout Malagasy, Africa, Oceania, and southern South America. Astacoidea includes three extant families: Astacidea, Cambaridae, and Cambaroididae. The Astacidae are found throughout the majority of Europe and the Pacific drainages of North America. The

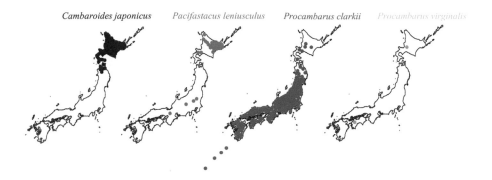

Figure 17.1 Current distribution of native *Cambaroides japonicus* and alien *Pacifastacus leniusculus*, *Procambarus clarkii*, and *Procambarus virginalis* in Japan (Kawai et al. 2004, 2016, Kawai and Koboyashi 2005, Usio 2007, Usio et al. 2016, Kawai 2017a, b, Kawai, unpublished data, Ohtaka et al. 2017).

Cambaridae are found throughout the eastern drainage systems of North America, and Cambaroididae are present throughout northeastern Asia. Cambaroididae is comprised of one genus, *Cambaroides*, with six recognized species. The Japanese crayfish, *Cambaroides japonicus* (De Haan 1841), is the only crayfish endemic to Japan with a geographic range extending over parts of Hokkaido, Aomori, Akita, and Iwate Prefectures in northern Japan (Fig. 17.1) (Kawai and Fitzpatrick 2004, Koizumi et al. 2012, Kawai et al. 2016).

Cambaroides japonicus is an ecological specialist found in depositional micro-habitats of high-gradient headwater streams (Usio 2007). Populations of *C. japonicus* have been in sharp decline due to increasing urbanization, canalization, and water contamination by agriculture and factory practices (Kawai et al. 2016). This sharp decline has led to *C. japonicus* being listed for conservation by the Japanese Fisheries Agency in 1988 and the Environmental Agency in 2000 (Kawai et al. 2002, Nakata et al. 2003).

Invasive crayfish species have been in Japan since the late 1930s when *Procambarus clarkii* (Girard 1852) (Cambaridae) and *Pacifastacus leniusculus* (Dana 1852) (Astacidea) were brought to Japan for aquaculture purposes and accidentally released into native streams (Kawai and Kobayashi 2006, Usio et al. 2016, Yi et al. 2018). *Procambarus clarkii* is a highly prolific invader now found on several continents through human-mediated introductions (Carreira et al. 2017, Nunes et al. 2017, Almerão et al. 2018, Yi et al. 2018); it has disrupted or displaced many native crayfish populations around the world (Gherardi and Daniels 2004, Bubb et al. 2006, Larson et al. 2009). *Pacifastacus leniusculus* is another prolific invader that has devastated native European crayfish populations (Nakata et al. 2005, Holdich et al. 2009, Lodge et al. 2012, Filipova et al. 2013, Grandjean et al. 2014, Vrålstad et al. 2014, Collas et al. 2016, Maguire et al. 2016, Grandjean et al. 2017, Kaldre et al. 2017). *Procambarus virginalis* (Lyko 2017), the marbled crayfish, was recently documented within the Hokkaido Prefecture (Usio et al. 2017, Vogt 2018). Due to parthenogenesis in *P. virginalis*, it may rival *P. clarkii* in its invasion potential as shown by its successful, rapid invasion in Madagascar (Gutekunst et al. 2018). Current known distributions of all three species in Japan are represented in Fig. 17.1.

North American crayfish are vectors for the crayfish plague, *Aphanomycres astaci* (Schikora 1906), an infectious disease that has devastated native European, Asian, and South American crayfish populations (Unestam 1969, Scholtz et al. 2003, Jones et al. 2009, Almerão et al. 2015, Martin et al. 2018). *Aphanomycres astaci* is considered to be among the top 100 worst invasive species worldwide (Lowe et al. 2004, DASIE 2009) and is one of the most notorious invertebrate pathogens (Diéguez-Uribeondo and Söderhäll 1993, Oidtmann et al. 2002, 2004, Bohman et al. 2006, Diéguez-Uribeondo 2006, Diéguez-Uribeondo et al. 2009, Kozubíkobá et al. 2010, Grandjean et al. 2014).

Although European countries are developing management procedures for *A. astaci* to prevent further outbreaks and mass mortalities (Taugbol et al. 1993, Furst 1995, Spink and Frayling 2000, Dieguez-Uribeondo 2006, Jussila et al. 2011), recovery plans for *C. japonicus* have not been fully assessed (Edsman 2004, Jussila et al. 2016, Rezinciuc et al. 2016). In this study, we report on the current sale of exotic (nonnative) crayfish in Japan to determine the risk of new invasive introductions. We also report on the recent extirpation of native *C. japonicus* populations due to *A. astaci* outbreaks. Finally, we propose a monitoring protocol for *C. japonicus* and invasive species populations and provide a plan to recolonize habitats where *C. japonicus* populations have been extirpated.

17.2 MATERIALS AND METHODS

17.2.1 Evaluation of the Current Status of the Exotic Crayfish Trade in Japan

Interviews were conducted with pet shop owners selling crayfish in Tokyo, Japan, only due to the density of pet shops compared to other prefectures. Interviewees remained anonymous to protect their privacy. Online searches of popular pet markets selling exotic animals were conducted using keywords "crayfish", "pet trade", and "ornamental". Results from the interviews and online searches (Table 17.1) were compared with previously published findings (Japan Crayfish Club 2003, Sasaki 2014) (Fig. 17.2).

17.2.2 *Cambaroides japonicus* Population Monitoring and Mass Mortality Occurrences in Hokkaido, Japan

Fifteen populations of *C. japonicus* were randomly selected for monitoring in Hokkaido Prefecture, Japan, in 1993 (Fig. 17.3). Sampling was conducted from June to September in 1993, 1999, 2005, 2011, and 2017. A single 30-m reach was sampled for a minimum of 30 min by a single individual. Each sampling event within reach was categorically marked as "present", "absent", or "mass mortality". Categories were defined as "finding one or more *C. japonicus* individuals", "none found in

Table 17.1   **Exotic Crayfish for Sale in Tokyo or Online in Japan in 2017**

Species		Pet shop	Online
Cambarellus patzcuarensis	Villalobos (1943)	1	8
Cambarellus shufeldtii	Faxon (1884)	0	3
Cambarellus texanus	Albaugh and Black (1973)	0	3
Procambarus acutus	Girard (1852)	0	3
Procambarus alleni	Faxon (1884)	0	2
Procambarus clarkii	Girard (1852)	8	8
Procambarus paenisulanus	Faxon (1914)	0	2
Procambarus vazquezae	Villalobos (1954)	0	3
Procambarus virginalis	Lyko (2017)	8	8

sampling period", and "dead specimens present and site positive for *A. astaci*", respectively. Dead crayfish were stored in 96% ethanol for further analysis. The detection of *A. astaci* was conducted upstream and downstream of the sampled reach following guidelines established by Vrålstad et al. (2009), Kozubíková et al. (2010), Murgala et al. (2016), and Martin-Torrijos et al. (2018).

Tissue samples were collected from the abdominal soft cuticle and uropod of each crayfish and stored in a single 1.5-ml tube at -80°C. Tissues were initially immersed in 360-µl Buffer ATL and crushed by stainless steel beads (1.6-mm diameter) in a BBX24B Bullet Blender (Next Advance) for 10 min. Afterward, DNA was extracted following the DNeasy Blood and Tissue Kit (Qiagen) manufacturer's instruction with reagent volumes doubled.

The presence of *A. astaci* in the extracted DNA was detected by the TaqMan MGB real-time PCR assay established by Vrålstad et al. (2009) in 25-µl reaction volumes using a LightCycler 480 Instrument (Roche). The level of *A. astaci* infection was determined by the strength of the qPCR signal and recalculated to correspond to

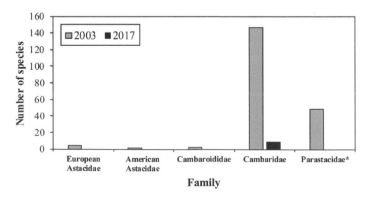

Figure 17.2   The number of exotic crayfishes sold in the Japanese exotic pet trade by continent and family. *Parastacidae have two species from South America, four species from Malagasy African, and 43 species from Australia and New Guinea.

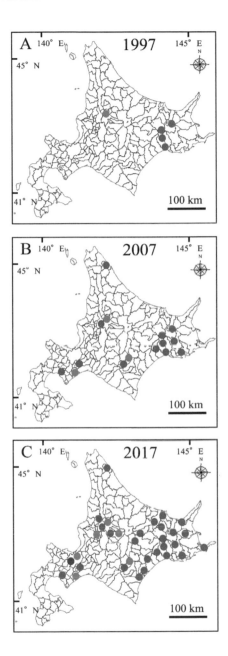

Figure 17.3 The current distribution of *Procambarus clarkii* (blue circles), *Pacifastacus leni-usculus* (red circles), and *Procambarus virginalis* (purple circle) on the island of Hokkaido, Japan (based on the present study, Kawai 2017a, b).

the amount of PCR forming units (PFUs) in the reaction and expressed as semiquantitative agent levels (following Vrålstad et al. 2009, Kozubíková et al. 2010, Filipova et al. 2013, Grandjean et al. 2017). Samples with semiquantitative agent levels A2 or higher were considered infected with *A. astaci* (Vrålstad et al. 2009). Habitat conditions, disturbance, and presence of invasive species were also recorded and presented in previous work (Kawai et al. 2002, Nakata et al. 2002, 2003).

Four additional sites within Otaru City (Fig. 17.5) were sampled in 2005 and 2017 to monitor *C. japonicus* habitat loss in greater detail. Otaru City's waterways are characterized by small spring-fed brooks that directly flow into the sea, providing a simpler system to study *C. japonicus* population fluctuations. Data collection followed the above-mentioned methods.

17.2.3 *Cambaroides japonicus* Recovery Potential

The recovery potential of *C. japonicus* populations that were lost was assessed. *Cambaroides japonicus* individuals were reintroduced into a previously inhabited site in the Tokachi region of central eastern Hokkaido from 22 August 2017 to 20 December 2017. This site had experienced a "mass mortality" event in 2016 (Fig. 17.4A). The study site is part of a complex riverine system with at least 10 branches (approximately less than 30 cm wide) originating from springs flowing directly into the main stem river (1.5 m width). Individuals of *C. japonicus* were translocated from the main stem river upstream of the study site on 22 August 2017. Ten cages, each housing an individual crayfish (four males, six females), were placed on the river bottom in low-flow areas. Cages were 15 cm × 10 cm × 8 cm. Crayfish were fed every two weeks using available leaf litter packs. Water temperature was recorded every 20 min for the duration of the experiment via a data logger (HOBO Onset, www .onsetcomp.com/applications) attached to the cages. All crayfish were preserved in 100% ethanol following the conclusion of the experiment to test for the presence of *A. astaci* following methods described above.

A laboratory experiment was carried out following similar methods to the field experiment to minimize environmental variables. Five cages, each housing an individual crayfish (two males, three females), were kept in one aquarium, and water temperature was maintained at similar levels to the field site using a chiller-pump system (Zensui Co. Zc-200a: www.zensui.co.jp/). The water was changed every two weeks and individual crayfish were given leaf packs similar to those at the field site. The laboratory experiment was conducted simultaneously with the field experiment: 22 August to 20 December 2017.

17.3 RESULTS

17.3.1 Current Status of the Exotic Crayfish Trade in Japan

Eight pet shops and eight online shops were recorded selling nine exotic crayfish species. This was ~4.4% of the exotic crayfish species reported being imported in 2003 (Fig.

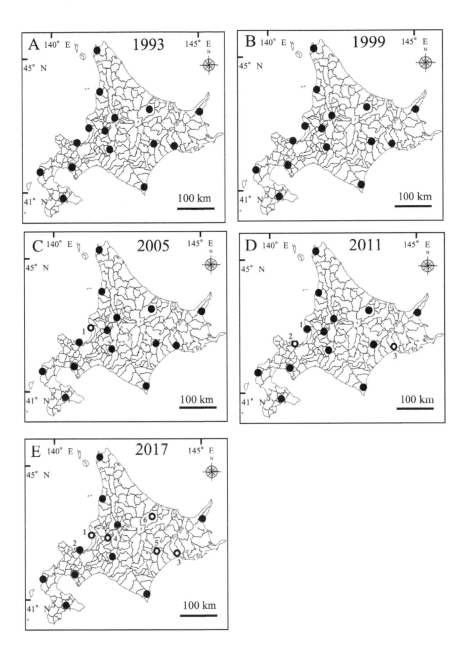

Figure 17.4 Sampled localities of *Cambaroides japonicus* from 1993 to 2017. Black circles represent *C. japonicus* presence. White circles represent *C. japonicus* mass mortality/absence. Numbers represent major cities. 1. Hamamasu City, 2. Otaru City, 3. Kushiro City, 4. Sunagawa City, 5. Tokachi, 6. Kitami City. Mass mortality occurred in 2 (Otaru City), 3 (Kushiro City), 5 (Tokachi), and 6 (Kitami City), from McCormack and Kawai (2016).

17.2). *Procambarus clarkii* and *P. virginalis* were sold by all observed shops. In addition, *Cambarellus patzcuarensis* (Villalobos 1943) was available from all online shops and one pet shop. Six additional species were available for sale in two to three shops (Table 17.1). The current distribution of native and alien species in Japan is illustrated in Fig. 17.1.

17.3.2 *Cambaroides japonicus* Population Monitoring and Mass Mortality Occurrences in Hokkaido

Procambarus clarkii populations were found at six localities throughout Hokkaido, Japan (Fig. 17.3). *Pacifastacus leniusculus* has spread through eastern and central Hokkaido, and one locality in northern Hokkaido. *Procambarus virginalis* was documented in Sapporo City, Hokkaido (Usio et al. 2017) (Fig. 17.3).

Cambaroides japonicus was observed at all 15 localities in 1993 and 1999. In 2005, *C. japonicus* was absent at Hamamasu City (Fig. 17.4D [1]). Two additional populations were recorded absent in 2011 in Otaru City and Kushiro City (Fig. 17.4E [2, 3]). In 2017, two additional populations were recorded absent at Sunagawa City and Kitami City (Fig. 4F [4, 6]). One population was undergoing an active reduction in Tokachi region (Fig. 17.4F [5]). One population of *C. japonicus* in Otaru City reported absent in 2011 (Fig. 17.4F [2]) had individuals present in 2017. Six sites experienced some form of disappearance since 1993 with one showing recovery. That is, 33% of observed populations have experienced declines or loss since 1993. There were no noticeable habitat disturbances observed at any of the six sites.

Within Otaru City, *C. japonicus* was found at all four sites in 2005: Akaiwa, Inaho, Otaru Midori, and Shioya. When resampled in 2017, *C. japonicus* was found at only two sites (Akaiwa and Otaru Midori), a 50% decrease in populations since 2005 (Fig. 17.5). There were no observable increases in water pollution, deforestation, or other environmental changes, nor were any alien crayfish populations recorded within the watersheds of Otaru City.

17.3.3 Mass Mortality Recovery Methods

All individual crayfish survived until project termination at the field site and in the laboratory (Fig. 17.6). No individuals exhibited any signs of melanization, and the infection level was stabilized at A0 (entirely negative) for all specimens in the field and laboratory. No infections of *A. astaci* were found. Additionally, there was no sign of melanization or other infections, nor any indications of injury or other stress or harm. Monitoring at the field site has continued and no mass mortality has been noted as of July 2019; ovigerous females were collected in spring 2019, and this is a sign of favorable environmental conditions for reproduction and translocation success.

17.4 DISCUSSION

In 2006, the Ministry of the Environment and the Ministry of Agriculture, Forest, and Fisheries of Japan designated members of the Cambaridae, *Cherax* species, and

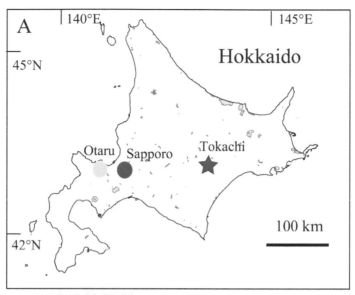

Figure 17.5   (A) Hokkaido, Japan, with Otaru City, and Sapporo City, Tokachi region. Red circle represents the locality of the alien *Procambarus clarkii* in Sapporo, red star represents the invasive *Pacifastacus leniusculus* in Tokachi region. (B) Otaru City. All circles represent known populations of *C. japonicus* in 2005, open circles denote extirpated populations in 2017. AK: Akaiwa, In: Inaho, Ot: Otaru Midori, Sh: SHioya. The map in part (B) was modified from www.gsi.go.jp/ downloaded on 1st April 2019.

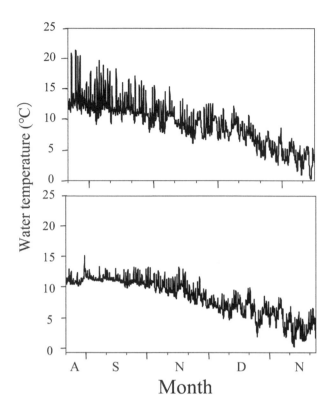

Figure 17.6   Water temperature at the experimental field station (upper, Tokachi region, Hokkaido, Japan, see Figure 17.5) and the aquarium maintained in the laboratory (lower) in 2017.

Pacifastacus leniusculus as "Invasive Alien Species" (IAS). Under current Japanese law, IAS are strictly prohibited from being imported, reared, transferred, or released throughout Japan. Despite this, many populations of invasive species (*P. clarkii, P. virginalis,* and *P. leniusculus*) are expanding throughout Hokkaido, and eight IAS-designated species are still being sold within the exotic crayfish pet trade in Japan. Invasive species can act as vectors of *A. astaci,* particularly *P. clarkii* and *P. leniusculus,* and continued importation of these species increases the risk of *A. astaci* outbreaks with resultant *C. japonicus* mortality. Dawkins and Furse (2012) suggest that all localities of native endangered *C. japonicus* have genetic uniqueness and different molecular genetic structure, such that each local population, or locality, with *C. japonicus* would be considered a minimum unit for conservation requirements. It indicates that any potential reintroductions of *C. japonicas* should be done with care to maintain or increase local genetic variability. The loss of genetic variability can inhibit a population's ability to persist and increase their vulnerability to future stochastic events such as anthropogenic disturbances, and climate change (Frankham 1995). This has important consequences for a native species already facing threats to their persistence from *A. astaci* and nonnative, invasive species.

Inaho and Otaru Midori are located less than 100 m apart. However, a loss of *C. japonicus* occurred only at Inaho. McCormack and Kawai (2016) remarked that mass mortality events of *C. japonicus* were spreading throughout Hokkaido and the present study confirms this. This study along with a previous study conducted in Sapporo (Martin-Torrijos et al. 2018) highlights the fact that loss of *C. japonicus* can occur in the absence of invasive crayfish (Fig. 17.3) and environmental changes. This suggests that other methods of *A. astaci* transmission have occurred.

Aphanomyces astaci thus far has been shown to infect only freshwater crayfish (Unestam 1969, 1972) and that cambarid crayfish act as vectors. However, recent studies have shown that invasive freshwater Chinese mitten crabs (*Eriocheir sinensis* Milne-Edwards 1853), the endangered freshwater crab *Potamon potamios* (Olivier 1804) and two Asian freshwater shrimps (*Macrobrachium dayanum* [Henderson 1893] and *Neocardina davidi* [Bouvier 1904]) can act as hosts of *A. astaci* under laboratory conditions (Schrimpf et al. 2014, Svoboda et al. 2014a, b). The short-lived zoospores of *A. astaci* can also be transmitted inadvertently by humans through contact with fishing gear, field sampling equipment, and pond cultures (Alderman and Polglase 1988, Matthews and Reynolds 1992, Alderman 1993, 1996, Oidtmann et al. 2002, 2004, 2006, Souty-Grosset et al. 2006).

Although native *Eriocheir japonicus* (De Haan 1835) is not known to occur throughout Hokkaido, they are widely distributed in rivers within Sapporo City and Otaru City (Kawai unpublished data). It is possible that *E. japonicus* has acquired *A. astaci* and moved into the sampling sites with it. The sites located in Otaru City are in highly urbanized areas with public access. It is possible that *A. astaci* zoospores were transmitted via bait buckets of anglers.

Results of the field experiment showed that *C. japonicus* reintroduction was possible after a mass mortality event. Similar procedures have been developed in Sweden and Finland to reintroduce native *Astacus astacus* (Linnaeus 1758) after *A. astaci* outbreaks (e.g., Rezinciuk et al. 2016). If caged individuals survived and showed no signs of infection, reintroduction plans can continue. Conservationists in the Tokachi region reintroduced *C. japonicus* to the field site in June 2018, and the species has persisted as of the last report in October 2018. To further improve the recovery program, sham tests will be performed as part of the field site procedures. Sham tests control for mortality due to handling, natural environmental fluctuations, etc. by placing individuals from a collection site in cages at the collection site. In the present study, no mortalities were observed, so a sham control was not needed, but these tests are important for future reintroductions to account for potential mortality caused by factors other than *A. astaci*.

Sniezko (1974) described a basic concept of infectious disease occurrence and prevention for aquatic organisms (Fig. 17.7). Based on the results of this study, a modified version of Sniezko's model is presented to prevent future crayfish plague outbreaks and restoration methods. Crayfish plague outbreaks in Japan have been attributed to four overlapping factors: (1) habitat deterioration due to human impacts, (2) decreasing tolerance to infectious diseases because of environmental stressors, (3) invasion of nonnative crayfish as competitors, and (4) introduction of *A. astaci* through nonnative species. By implementing appropriate conservation practices and

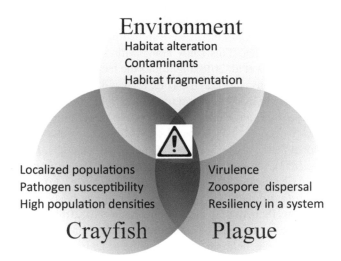

Figure 17.7 Venn diagram depicting the interaction between crayfish, *Aphanomyces astaci,* and the environment. Three-color gradients correspond to increasing bias in environmental conditions of the natural habitat, crayfish individuals in habitat, and crayfish plague in habitat. Modified from Sniezko (1974) and Shields (2013).

management actions, it will be possible to prevent further *A. astaci* outbreaks in Japan. For example, installing gabion walls along degraded riverbanks can provide additional shelters for native crayfish and other fauna (Kawai, personal information) prevents further habitat degradation (bank erosion, sedimentation, and loss of interstitial space) that impacts their use of shelters.

Conservationists in Japan range from biologists to public organization staff, to politicians, to the younger generation. The authors suggest a conservation strategy using citizen-oriented restoration methods to prevent crayfish plague outbreaks and recovery actions after loss of *C. japonicus.* The strategy builds upon previous plans (Unestam 1972, Edsman 2004, Diéguez-Uribeondo 2006, Jussila et al. 2016). Using conservationists, three actions are suggested (Fig. 17.8): Action 1 – monitor *C. japonicus* populations to determine if mass mortality has occurred and to what extent. This can be achieved by capturing individuals or through environmental DNA methods for both *C. japonicus* and *A. astaci.* Action 2 – determine if it is safe to relocate uninfected individuals from another location to a locality being considered for restoration question. This can be done by methods similar to the cage experiment presented in this chapter. It is best to obtain individuals from upstream of a site that has experienced the loss of *C. japonicus* to preserve the genetic character within the stream. Action 3: monitor released populations to prevent further mass mortalities. Throughout the process, invasive alien species populations should continue to be monitored by the appropriate official organizations.

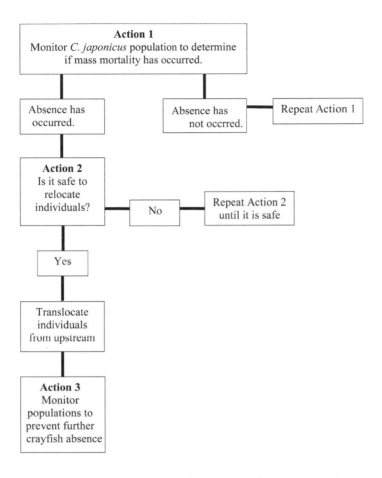

Figure 17.8 Flow diagram representing crayfish absent and recovery plan for *Cambaroides japonicus*.

Additional steps from a conservation management standpoint would need to be implemented for successful reintroduction of *C. japonicus* populations from mass mortality (Fig. 17.9): (1) Maintain natural environmental conditions for *C. japonicus* habitats. This can be achieved by preventing construction through critical habitats used by *C. japonicus* or if construction is conducted by installing gabions or other measures to provide habitat. (2) Monitor IAS populations, and if loss of *C. japonicus* occurs, implement the above conservation recovery plan. (3) Increase public awareness through education and stewardship. Pet stores should inform people about the harm of releasing nonnative species into local waters, be responsible about informing customers about proper pet care. This will create an appreciation for native Japanese fauna and hopefully instill a desire to preserve native fauna. Ultimately, this would lead to future stewards who can maintain the conservation recovery plan for *C. japonicus*.

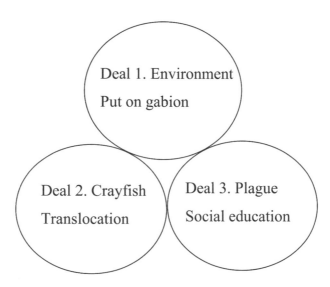

Figure 17.9   Schematic diagram of recovery from mass mortality due to crayfish plague.

ACKNOWLEDGMENTS

The authors sincerely thank Jörn Panteleit of University Koblenz-Landau, Institut für Umweltwissenschaften, and M. Sunagawa for critical comments. The authors also would like to thank M. Yamada for providing Fig. 17.5 and assisting with the field and laboratory experiments.

REFERENCES

Alderman, D.J. 1993. Crayfish plague in Britain, the first twelve years. *Freshwater Crayfish* 9:266–72.

Alderman, D.J. 1996. Geographical spread of bacterial and fungal diseases of crustaceans. *Revue Scientifique et Technique (OIE)* 15:603–32.

Alderman, D.J. and Polglase, J.L. 1988. Pathogens, parasites and commensals. In *Freshwater Crayfish: Biology, Management and Exploitation*, eds. D.M. Holdich, and R.S. Lowery, 167–212. London: Croom Helm.

Almerão, M.P., Delaunay. C., Coignet, A., Peiró, D.F., Pinet, F. and Souty-Grosset, C. 2018. Genetic diversity of the invasive crayfish *Procambarus clarkii* in France. *Limnologica* 69:135–41.

Almerão, M.P., Rudolph, E., Southy-Grosset, C., Crandall, K., Buckup, L., Amouret, J., Verdie, A., Santos, S. and De Araujo, P.B. 2015. The native South American crayfishes (Crustacea, Parastacidae): Start of knowledge and conservation status. *Aquatic Conservation: Marine and Freshwater Ecosystems* 25:288–301.

Bohman, P., Nordwall, F. and Edsman, L. 2006. The effect of the large-scale introduction of signal crayfish on the spread of crayfish plague in Sweden. *Bulletin Français de la Pêche et de la Pisciculture* 380–381:1291–302.

Bubb, D.H., Thom, T.J. and Lucas, M.C. 2006. Movement, dispersal and refuge use of co-occurring introduced and native crayfish. *Freshwater Biology* 51:1359–68.

Carreira, B.M., Segurado, P., Laurila, A. and Rebelo, R. 2017. Can heat waves change the trophic role of the world's most invasive crayfish? Diet shifts in *Procambarus clarkii*. *PLoS One* 12:e0183108.

Collas, M., Becking, T., Delpy, M., Pflieger, M., Bohn, P., Reynolds, J. and Grandjean, F. 2016. Monitoring of white-clawed crayfish (*Austropotamobius pallipes*) population during a crayfish plague outbreak followed by rescue. *Knowledge and Management of Aquatic Ecosystems* 417:1.

DAISE. 2009. *Handbook of Alien Species in* Europe. Dordrecht: Springer.

Dawkins, K.L. and Furse, J.M. 2012. Conservation genetics as a tool for conservation and management of the native Japanese freshwater crayfish *Cambaroides japonicus* (De Haan). *Crustacean Research, Special* Number 7:35–43.

Diéguez-Uribeondo, J. 2006.The dispersion of the *Aphanomyces astaci*-carrier *Pacifastacus leniusculus* by humans represents the main cause of disappearance of the indigenous crayfish *Austropotamobius pallipes* in Navarra. *Bulletin Français de la Pêche et de la Pisciculture* 380–381:1303–12.

Diéguez-Uribeondo, J. and Söderhäll, K. 1993. *Procambarus clarkii* Girard as a vector for the crayfish plague fungus, *Aphanomyces astaci* Schikora. *Aquaculture and Fisheries Management* 24:761–65.

Diéguez-Uribeondo, J., García, M.A., Cerenius, L., Kozubíková, E., Ballesteros, I., Windels, C., Welland, J., Kator, H., Söderhäll, K. and Martín, M.P. 2009. Phylogenetic relationships among plant and animal parasites, and saprotrophs in *Aphanomyces* (Oomycetes). *Fungal Genetics and Biology* 46:365–76.

Edsman, J. 2004. The Swedish history about import of live crayfish. *Bulletin Français de la Pêche et de la Pisciculture* 372–373:281–88.

Filipova, L., Petrusek, A., Matasova, K., Delaunay, C. and Grandjean, F. 2013. Prevalence of the crayfish plague pathogen *Aphanomyces astaci* in populations of the signal crayfish *Pacifastacus leniusculus* in France: Evaluating the threat to native crayfish. *PloS One* 8:e70157.

Frankham, R. 1995. Conservation genetics. *Annual Review Genetics* 29:305–27.

Fürst, M. 1995. On the recovery of *Astacus astacus* L. populations after an epizootic of the crayfish plague (*Aphanomyces astaci* Shikora). *Freshwater Crayfish* 8:65–576.

Gherardi, F. and Daniels, W.H. 2004. Agonism and shelter competition between invasive and indigenous crayfish species. *Canadian Journal of Zoology* 82:1923–32.

Grandjean, F., Roques, J., Delaunay, C., Petrusek, A., Becking, T. and Collas, M. 2017. Status of *Pacifastacus leniusculus* and its role in recent crayfish plague outbreaks in France: Improving distribution and crayfish plague infection patterns. *Aquatic Invasion* 12:541–49.

Grandjean, F., Vrålstad, T., Dieguez-Uribeondo, J., Jelic, M., Mangombi, J., Delaunay, C., Filipova, L., Rezinciuc, S., Kozubikova-Balcarova, E., Guyonnet, D., Viljamaa-Dirks, S. and Petrusek, A. 2014. Microsatellite markers for direct genotyping of the crayfish plague pathogen *Aphanomyces astaci* (Oomycetes) from infected host tissues. *Veterinary Microbiology* 170:317–24.

Gutekunst, J., Andriantsoa, R., Falckenhayn, C., Hanna, K., Stein, W., Rasamy, J. and Lyko, F. 2018. Clonal genome evolution and rapid invasive spread of the marbled crayfish. *Nature Ecology and Evolution* 2:567–73.

Holdich, D.M., Reynolds, J.D., Souty-Grosset, C. and Sibley, P.J. 2009. A review of the ever increasing threat to European crayfish from non-indigenous crayfish species. *Knowledge and Management of Aquatic Ecosystems* 11:394–95.

Japan Crayfish Club (ed.). 2003. *Crayfish of the World*. Tokyo: Marine Planning.

Jones, J.P.G., Rasamy, J.R., Harvey, A., Toon, A., Oidtmann, B., Randrianarison, M.H., Raminosoa.N. and Ravoahangimalala, O.R. 2009. The perfect invader: A parthenogenic crayfish poses a new threat to Madagascar's freshwater biodiversity. *Biological Invasions* 11:1475–82.

Jussila, J., Maguire, I., Kokko, H. and Makkonen, J. 2016. Chaos and adaptation in the pathogen-host relationship in relation to the conservation, the case of the crayfish plague and the noble crayfish. Section 2 Crayfish: New Development. In *Freshwater Crayfish: A Global Overview*, eds. T. Kawai, Z. Faulkes, and G. Scholtz, 246–74. Boca Raton, FL: CRC Press/Taylor & Francis.

Jussila, J., Makkonen, J., Vainikka, A., Kortet, R. and Kokko, H. 2011. Latent crayfish plague (*Aphanomyces astaci*) infection in a robust wild noble crayfish (*Astacus astacus*) population. *Aquaculture* 321:17–20.

Kaldre, K., Paaver, T., Hurt, M. and Grandjean, F. 2017. First records of the non-indigenous signal crayfish (*Pacifastacus leniusculus*) and its threat to noble crayfish (*Astacus astacus*) populations in Estonia. *Biological Invasions* 19:2771–76.

Kawai, T. 2017a. Observation on mandible and gill morphology in *Pacifastacus leniusculus* and *Cherax quadricarinatus* with a review of the introduction of alien crayfish into Japan. *Freshwater Crayfish* 23:29–39.

Kawai, T. 2017b. A review of the spread of *Procambarus clarkii* across Japan and its morphological observations. *Freshwater Crayfish* 23:41–53.

Kawai, T. and Fitzpatrick, J.F. Jr. 2004. Redescription of *Cambaroides japonicus* (De Haan, 1841) (Crustacea: Decapoda: Cambaridae) with allocation of a type locality and month of collection of types. *Procceedings of Biological Society of Washington* 117:23–34.

Kawai, T. and Kobayashi Y. 2006. Origin and current distribution of the alien crayfish, *Procambarus clarkii* (Girard, 1852) in Japan. *Crustaceana* 78:1143–49.

Kawai, T., Min, G.S., Barabanshchikov, B., Labay, V. and Ko, H.S. 2016. Asia, Section. In *Freshwater Crayfish: A Global Overview*, eds. T. Kawai, Z. Faulkes, and G. Scholtz, 313–68. Boca Raton, FL: CRC Press/Taylor & Francis.

Kawai, T., Mitamura, T. and Ohtaka, A. 2004. The taxonomic status of the introduced North American signal crayfish, *Pacifastacus leniusculus* (Dana, 1852) in Japan, and the source of specimens in the newly reported population in Fukushima Prefecture. *Crustaceana* 77:861–70.

Kawai, T., Nakata, K. and Hamano, T. 2002. Temporal changes of the density in two crayfish species, the native *Cambaroides japonicus* (De Haan) and the alien *Pacifastacus leniusculus* (Dana), in natural habitats of Hokkaido, Japan. *Freshwater Crayfish* 13:198–206.

Koizumi, I., Usio, N., Kawai, T., Azuma, N. and Masuda, R. 2012. Loss of genetic diversity means loss of geological information: the endangered Japanese crayfish exhibits remarkable historical footprints. *PLoS One* 7: e33986.

Kozubíková, E., Puky, M., Kiszely, P. and Petrusek, A. 2010. Crayfish plague pathogen in invasive North American crayfish species in Hungary. *Journal of Fish Diseases* 33:925–29.

Larson, E., Magoulick, D., Turner, C. and Laycock, K. 2009. Disturbance and species displacement: Different tolerances to stream drying and desiccation in a native and an invasive crayfish. *Freshwater Biology* 54:1899–908.

Lodge, D.M., Deines, A., Gherardi, F., Yeo, D.C.J., Arcella, T., Baldridge, A.K., et al. 2012. Global introductions of crayfishes: Evaluating the impact of species invasions on ecosystem services. *The Annual Review of Ecology, Evolution, and Systematics* 43:449–72.

Lowe, S., Browne, M., Boudjelas, S. and De Poorter, M. 2004. *100 of the World's Worst Invasive Alien Species. A Selection from the Global Invasive Species Database.* The invasive Species Specialist Group (ISSSG) a specialist group of the Species Survival Commission (SSC) of the World Conservation Union (IUCN). http://www.issg.org/ 'accessed on October 1, 2020)

Maguire, I., Jelic, M., Klobucar, G., Delpy, M., Delaunay, C. and Grandjean, F. 2016. Prevalence of the pathogen Aphanomyces astaci in freshwater crayfish populations in Croatia. *Diseases of Aquatic Organisms* 118:45–53.

Martin-Torrijos, L., Kawai, T., Makkonen, J., Jussila, J., Kokko, H. and Dieguez-Uribeondo, J. 2018. Crayfish plague in Japan: A real threat to the endemic *Cambaroides japonicus. PLoS One* 13:e0195353.

Matthews, M. and Reynolds, J.D. 1992. Ecological impact of crayfish plague in Ireland. *Hydrobiologia* 234:1–6.

McCormack, R. and Kawai, T. 2016. Crayfish plague *Aphanomyces astaci* in Japan and the growing threat to Australia. *Crayfish News* 38:5–9.

Mrugala, A. Kawai, T., Kozubikoba-Balcarova E. and Petrusek, A. 2016. *Aphanomyces astaci* in Japan: a threat to the endemic and endangered crayfish species. *Cambaroides japonicus? Aquatic Conservation: Marine and Freshwater Ecosystem* 27:103–114.

Nakata, K., Hamano, T., Hayashi, K. and Kawai, T. 2002. Lethal limits of high temperature for two crayfishes, the native species *Cambaroides japonicus* and the alien species *Pacifastacus leniusculus* in Japan. *Fisheries Science* 68:763–67.

Nakata, K., Hamano, T., Hayashi, K. and Kawai, T. 2003. Water velocity in artificial habitats of the Japanese crayfish *Cambaroides japonicus. Fisheries Science* 69:343–47.

Nakata, K., Tsutsumi, K., Kawai, T. and Goshima, S. 2005. Coexistence of two North American invasive crayfish species, *Pacifastacus leniusculus* (Dana, 1852) and *Procambarus clarkii* (Girard, 1852) in Japan. *Crustaceana* 78:1389–94.

Nunes, A., Hoffman, A., Zengeya, T., Measey, G. and Weyl, O. 2017. Red swamp crayfish, *Procambarus clarkii*, found in South Africa 22 years after attempted eradication. *Aquatic Conservation: Marine and Freshwater Ecosystems* 27:1–17.

Ohtaka, A., Gelder, S.R. and Smith, R.J. 2017. Long-anticipated new records of an ectosymbiotic branchiobdellidan and an ostracod on the North American red swamp crayfish, *Procambarus clarkii* (Girard, 1852) from an urban stream in Tokyo, Japan. *Plankton and Benthos Research* 12:123–28.

Oidtmann, B., Gelger, S., Steinbauer, P., Culas, A. and Hoffmann, R.W. 2006. Detection of *Aphanomyces astaci* in North American crayfish by polymerase chain reaction. *Diseases of Aquatic Organisms* 72:53–64.

Oidtmann, B., Heitz, E., Roger, D. and Hoffmann, R.W. 2002. Transmission of crayfish plague. *Diseases of Aquatic Organisms* 52:159–67.

Oidtmann, B., Schaefers, N., Cerenius, L., Söderhäll, K. and Hoffmann, R.W. 2004. Detection of genomic DNA of crayfish plague fungus *Aphanomyces astaci* (oomycete) in clinical samples by PCR. *Veterinary Microbiology* 100:269–82.

Rezinciuc, S., Sandoval-Sierra, J.V., Oidtmann, B. and Diéguez-Uribeondo, J. 2016. The biology of crayfish plague pathogen *Aphanomyces astaci*: current answers to most frequent questions, Section 2: Crayfish: New developments. In *Freshwater Crayfish: A Global Overview.* eds. T. Kawai, Z, Faulkes, and G. Scholtz, 182–204. Boca Raton, FL: CRC Press/Taylor & Francis.

Sasaki, J. 2014. Present state of ornamental crustaceans in Japan: symposium objectives and contents. *Cancer* 23:63–88.

Scholtz, G., Braband, A., Tolley, L., Reinman, A., Mittmann, B., Luckaup, C., Steuerwald, F. and Vogt, G. 2003. Parthenogenesis in an outsider crayfish. *Nature* 421:806.

Schrimpf, A., Schmidt, T. and Schulz, R. 2014. Invasive Chinese mitten crab (*Eriocheir sinensis*) transmit crayfish plague pathogen (*Aphanomyces astaci*). *Aquatic Invasions* 9:203–09.

Shields, J.D. 2013 Complex etiologies of emerging diseases in lobsters (*Homarus americanus*) from Long Island Sound 1. *Canadian Journal of Fisheries and Aquatic Sciences* 70:1576–87.

Sniezko, S.F. 1974. The effects of environmental stress on outbreaks of infectious diseases of fishes. *Journal of Fish Biology* 6:197–208.

Souty-Grosset, C., Holdich, D.M., Noel, P.Y., Reynolds, J.D. and Haffner, P. 2006. *Atlas of Crayfish in Europe*. Paris: Museum national d'Histoire naturelle.

Spink, J. and Frayling, M. 2000. An assessment of post-plague reintroduced native white-clawed crayfish, *Austropotamobius pallipes*, in the Sherston Avon and Tetbury Avon, Wiltshire. *Freshwater Forum* 14:59–69.

Svoboda, J., Mrugała, A., Kozubíková-Balcarová, E., Kouba, A., Diéguez-Uribeondo, J. and Petrusek, A. 2014a. Resistance to the crayfish plague pathogen, *Aphanomyces astaci*, in two freshwater shrimps. *Journal of Invertebrate Pathology* 121:91–104.

Svoboda, J., Strand, D.A., Vrålstad, T., Grandjean, F., Edsman, L., Kozak, P., Kouba, A., Fristad, R.F., Koca, S.B. and Petrusek, A. 2014b. The crayfish plague pathogen can infect freshwater-inhabiting crabs. *Freshwater Biology* 59:918–29.

Taugbol, T. and Skurdal, J. 1993. Noble crayfish in Norway: legislation and yield. *Freshwater Crayfish* 9:134–43.

Unestam, T. 1969. Resistance to the crayfish plague in some American, Japanese and European crayfishes. *Reports of the Institute of Freshwater Research, Drottingholm* 49:202–09.

Unestam, T. 1972. On the host range and origin of the crayfish plague fungus. *Reports of the Institute of Freshwater Research, Drottingholm* 52:192–98.

Usio, N. 2007. Endangered crayfish in northern Japan: Distribution, abundance and microhabitat specificity in relation to stream and riparian environment. *Biological Conservation* 134:517–26.

Usio, N., Azuma, N., Larson, E.R., Abbott, C.L., Olden, J.D., Akanuma, H., Takamura, K. and Takamura, N. 2016. Phylogeographic insights into the invasion history and secondary spread of the signal crayfish in Japan. *Ecology and Evolution* 6:5366–82.

Usio, N., Azuma, N., Sasaki, S., Oka, T. and Inoue, M. 2017. New record of Marmorkrebs from western Japan and its potential threats to freshwater ecosystem. *Cancer* 26:5–11. (in Japanese).

Vogt, G. 2018. Annotated bibliography of the parthenogenetic marbled crayfish *Procambarus virginalis*, a new research model, potent invader and popular pet. *Zootaxa* 4418:301–52.

Vrålstad, T., Knutsen, A.K., Tengs, T. and Holst-Jensen, A. 2009. A quantitative TaqMan MGB real-time polymerase chain reaction based assay for detection of the causative agent of crayfish plague *Aphanomyces astaci*. *Veterinary Microbiology* 137:146–55.

Vrålstad, T., Strand, D.A., Grandjean, F., Kvellestad, A., Hastein, T., Knutsen, A.K., Taugbol, T. and Skaar, I. 2014. Molecular detection and genotyping of *Aphanomyces astaci* directly from preserved crayfish samples uncovers the Norwegian crayfish plague disease history. *Veterinary Microbiology* 173:66–75.

Yi, S., Li, Y., Shi, L., Zhang, L., Li, Q. and Chen, J. 2018. Characterization of population genetic structure of red swamp crayfish, *Procambarus clarkii*, in China. *Scientific Reports* 8:5586.

CHAPTER 18

General Discussion

Tadashi Kawai and D. Christopher Rogers

CONTENTS

18.1 INTRODUCTION

The "Conservation Biology of Freshwater Crustaceans" symposium held at the Ninth International Crustacean Congress covered the conservation and biodiversity of many taxa worldwide. We present here a general distillation of the excellent research presented there by many researchers, several of whom provided magnificent chapters for this book. This final chapter summarizes the previous chapters and compares this book with the previous monograph (Kawai and Cumberlidge 2016) on freshwater decapod conservation. From this, we provide a general review of the status of freshwater crustacean biodiversity and conservation at a global level. Based on this, future directions of study are discussed.

18.2 REVIEW OF THE CHAPTERS

In Chapter 2, we learned that the almost exclusively marine barnacles (Cirripedia) have four species in the suborder Rhizocephala (parasitic barnacles), occurring in subterrestrial, brackish, or freshwaters. Molecular data indicate a single freshwater evolutionary invasion, and the morphology of the cypris larval stages support this scenario.

The Russian Sakhalin Island site in Far East Asia has numerous potentially endangered species, but biological surveys are few and published in Russian.

Chapter 3 provides a zoogeographical analysis of freshwater crustacean biodiversity of Sakhalin Island. Chapter 9 focuses on the distribution and species diversity of freshwater and brackish water Copepoda from Sakhalin Island. These two chapters provide the first comprehensive information on freshwater crustacean biodiversity in Far Eastern Russia.

Chapters 4 and 5 provide morphological and biodiversity information on Australian marsh-hoppers from north and eastern Australia and *Bellorchestia* from Tasmania, Australia (Amphipoda: Talitridae). This provides an important contribution for basic taxonomy and biodiversity in terrestrial and freshwater amphipods.

In Chapter 6, the malacostracan (Amphipoda, Isopoda, Thermosbaenacea, Bathynellacea, and Decapoda) biodiversity of the Mediterranean Region islands are reviewed, and the region is demonstrated to be a biodiversity hotspot for the Palaearctic. The origins, radiations, evolution, and extinctions were examined based on geographical history, and were used to construct conservation strategies.

The large branchiopod crustaceans (Branchiopoda: Anostraca, Notostraca, Laevicaudata, Spinicaudata, Cyclestherida) are flagship animals for the conservation of seasonal wetlands and salt lakes. However, the biodiversity of several groups is limited due to morphological plasticity, numerous cryptic taxa, and confounding genetic systems. Chapter 7 reviews the conservation status of the large branchiopods at local, regional, and national scales.

The zoogeography of Cladocera (Branchiopoda: Cladocera) on the Indian subcontinent with an examination of their body size distributions is reviewed in Chapter 8, revealing intriguing distribution and biodiversity patterns for the region.

Chapter 10 provides an excellent review on the conservation of continental Mysida and Stygiomysia, with a discussion on their current status by region, distribution type, native ranges, habitat status, and proposed conservation status for each species.

A decade long, monthly field survey was conducted for endangered freshwater *Aegla* species (Decapoda: Anomura) and is presented in Chapter 11, providing data on reproductive biology, environmental changes and effects, and population dynamics. These data allow for a reassessment of the conservation status of these taxa, which help to construct new conservation strategies for these endangered species.

In Chapter 12, previous studies of Indo-Pacific island amphidromous *Caridina* shrimp (Decapoda: Atyidae) are reviewed, and the taxonomy, phylogeny, habitat, and biogeography of these animals are discussed to provide new insights on biodiversity and conservation. Various threats to *Caridina*, including reduced habitat connectivity, introduced species, overharvesting, and climate change, are discussed.

The extinction risks to Colombian freshwater crabs (Brachyura: Pseudothelphusidae, Trichodactylidae) were originally assessed using the IUCN Red List Categories and Criteria in 2008 and 2015. Due to new conditions in Colombia, Chapter 13 reassesses these taxa using the same criteria. This reassessment revealed that the number of threatened species has increased. The major threats to Colombian freshwater crabs are deforestation and human population growth.

In Chapter 14, threats to endemic Columbian freshwater crabs (Brachyura: Pseudothelphusidae, Trichodactylidae) were analyzed using Maximum Entropy

Modeling methods, revealing that climate change and human-mediated activities stresses may be increasing. From this base, the authors provide recommendations for future conservation directions in Colombia.

In North America, blind crayfish dwell in deep caves and springs, and a study on them is presented in Chapter 15. Recent SCUBA surveys have demonstrated new morphological diversity that does not reflect genetic diversity in these isolated deep spring systems. This study raises some questions about the reliability of some morphological characters and provides new distribution, life history, food web interactions, and taxonomy of these cave crayfish. This study also suggests new directions for conservation of cave crayfish.

The historic culture of the Japanese endangered freshwater crayfish is reviewed, based on observations of museum specimens in the United States and Russia, coupled with historic documents in Japanese government archive, in Chapter 16. *Cambaroides japonicus* gastroliths were utilized as valuable medicine for Asian and European peoples, exported from Japan to Europe.

In Chapter 17, conservation strategies for the Japanese endangered freshwater crayfish are suggested from Japanese, American, and European specialists. Alien crayfish introduction in Japan from the pet trade is a new threat to conservation. Basic concepts and recovery programs from mass mortality due to infectious disease are suggested for protecting this endangered crayfish.

18.3 DISCUSSION

The most recent monograph on freshwater crustacean conservation gives a global overview on species diversity hotspots and conservation worldwide (Kawai and Cumberlidge 2016). However, this book focuses only on the Decapoda. Decapods are important taxa for conservation, due to their great alpha and beta diversity, the large number of narrow range endemics, the numerous species that have narrow habitat requirements, plus the many endangered species and the several invasive species, and their great economic importance. The freshwater decapod body often grows more than 10 cm in length, thus they often act as key species for freshwater ecosystems. However, smaller freshwater crustaceans (less than 1 cm body length) are often primary components of freshwater biodiversity in both species numbers and density, and they are important for ecosystem functions. Our book covers mostly these smaller sized freshwater crustaceans, which is a great advance over the previous book (Kawai and Cumberlidge 2016).

Advanced techniques, such as molecular analyses (Chapters 1, 12, 15, and 16), analytical modeling methods (Chapter 14), and ultrafine structure analyses using scanning electron microscopy (Chapters 3 and 4), were employed by the authors for further conservation and our understanding of biodiversity in these animals. Biodiversity hotspots for these smaller crustacean groups, along with analyses of their zoogeography (Chapters 2, 5, 6, 7, 8, 9, and 10), are more fundamental tools for conservation analysis. For the decapods, comparisons of listed endangered species and longer-term monitoring demonstrate an increase in extinction threat (Chapters

11, 13, and 14). Interdisciplinary research with remarkably different scientific disciplines was employed for decapod conservation (Chapters 5, 12, 15, and 16).

The previous monograph mainly dealt with basic conservation biology and provided lists of freshwater decapod biodiversity hotspots (Kawai and Cumberlidge 2016). However, this book presents research that makes realistic suggestions concerning conservation strategies and programs, and demonstrates how modern techniques, analysis, and interdisciplinary approaches can have a positive impact on biodiversity.

The conservation of biodiversity is particularly important for the next generation of crustacean biologists as well as the economies of many nations. The editors hope that this book will help the studies of early career scientists of freshwater crustacean biology.

ACKNOWLEDGMENT

Professor Ingo Wehrtmann of Universidad de Costa Rica provided excellent advice; without his help, this book would never have been published.

REFERENCE

Kawai, T. and Cumberlidge, N. 2016. *A Global Overview of the Conservation of Freshwater Decapod Crustaceans*. Switzerland: Springer.

T - #0319 - 071024 - C532 - 234/156/23 - PB - 9780367689049 - Gloss Lamination